Pro Spring Boot 3

An Authoritative Guide with Best Practices

Third Edition

Felipe Gutierrez

Pro Spring Boot 3: An Authoritative Guide with Best Practices, Third Edition

Felipe Gutierrez
4109 Rillcrest Grove Way Fuquay Varina, NC 27526-3562
Albuquerque, NM, USA

ISBN-13 (pbk): 978-1-4842-9293-8   ISBN-13 (electronic): 978-1-4842-9294-5
https://doi.org/10.1007/978-1-4842-9294-5

Managing Director, Apress Media LLC: Welmoed Spahr
Acquisitions Editor: Melissa Duffy
Development Editor: Jim Markham
Coordinating Editor: Gryffin Winkler

Cover image designed by kentaro-tachikawa on Unsplash (https://unsplash.com/)

Distributed to the book trade worldwide by Springer Science+Business Media LLC, 1 New York Plaza, Suite 4600, New York, NY 10004. Phone 1-800-SPRINGER, fax (201) 348-4505, e-mail orders-ny@springer-sbm. com, or visit www.springeronline.com. Apress Media, LLC is a California LLC and the sole member (owner) is Springer Science + Business Media Finance Inc (SSBM Finance Inc). SSBM Finance Inc is a **Delaware** corporation.

For information on translations, please e-mail booktranslations@springernature.com; for reprint, paperback, or audio rights, please e-mail bookpermissions@springernature.com.

Apress titles may be purchased in bulk for academic, corporate, or promotional use. eBook versions and licenses are also available for most titles. For more information, reference our Print and eBook Bulk Sales web page at https://www.apress.com/bulk-sales.

Any source code or other supplementary material referenced by the author in this book is available to readers on the GitHub repository. For more detailed information, please visit https://www.apress.com/gp/services/source-code.

If disposing of this product, please recycle the paper

*Dedicated in loving memory of my aunt, Fabiola Cerón,
and Simón Cruz, my uncle; I miss you so much!
Thanks uncle, for everything you taught me!*

Table of Contents

About the Author

Felipe Gutierrez is a solutions software architect with bachelor's and master's degrees in computer science from Instituto Tecnologico y de Estudios Superiores de Monterrey Campus Ciudad de Mexico. Felipe has over 25 years of IT experience and has developed programs for companies in multiple vertical industries, such as government, retail, healthcare, education, and banking. He is currently working as Staff Engineer for VMware, specializing in content development for Tanzu Learning and the new Spring Academy learning site, Spring Framework, Spring Cloud Native Applications, Groovy, and RabbitMQ, among other technologies. He has also worked as a solutions architect for big companies like Nokia, Apple, Redbox, IBM, and Qualcomm. He is the author of *Spring Boot Messaging* (Apress, 2017) and *Spring Cloud Data Flow* (Apress, 2020).

About the Technical Reviewer

Manuel Jordan Elera is an autodidactic developer and researcher who enjoys learning new technologies for his own experiments and creating new integrations. Manuel won the Springy Award – Community Champion and Spring Champion 2013. In his little free time, he reads the Bible and composes music on his guitar. Manuel is known as dr_pompeii. He has tech-reviewed numerous books, including *Pro Spring MVC with WebFlux* (Apress, 2020), *Pro Spring Boot 2* (Apress, 2019), *Rapid Java Persistence and Microservices* (Apress, 2019), *Java Language Features* (Apress, 2018), *Spring Boot 2 Recipes* (Apress, 2018), and *Java APIs, Extensions and Libraries* (Apress, 2018). You can read his detailed tutorials on Spring technologies and contact him through his blog at `www.manueljordanelera.blogspot.com`. You can follow Manuel on his Twitter account, @dr_pompeii.

Acknowledgments

Thanks to the Spring team for creating such an amazing Java Framework!

Thanks to my technical reviewer, Manuel Jordan, for always excelling in his reviews! I also want to thank the Apress editorial team for their patience and excellent work.

Finally, I want to thank all of my family for their support and a special dedication to my loving aunt Fabiola and my super awesome uncle Simon (wife and husband), who passed away too soon! We miss you.

PART I

Introductions

CHAPTER 1

Spring Boot Quick Start

Welcome to the first chapter of the book, which will quickly immerse you in Spring Boot and demonstrate how easy it is to use by walking you through a simple project that exposes an API over the Web. If you are new to Spring Boot, this chapter will help you to rapidly familiarize yourself with the framework. If you are an experienced developer, feel free to quickly review the setup of the project (which will be referenced throughout this book) and move on to the next chapter.

Project: Users App

The project that we are going to build, named Users App, will expose a simple CRUD (create, read, update, and delete) API over the Web. These are the requirements for the Users App project:

- A user must have a name and an email address.

- A map is used to hold the information, using the email address as the key.

- It exposes an API that uses CRUD over the Web.

Initial Setup

To start with Spring Boot, you need to have the following installed:

- *Java*: You can install, for example, OpenJDK (https://jdk.java.net/archive/) or Eclipse Temurin (https://adoptium.net/temurin/releases/).

  - If you are Unix user, you can use SDKMAN! (https://sdkman.io/), which works for Linux and macOS.

© Felipe Gutierrez 2024
F. Gutierrez, *Pro Spring Boot 3*, https://doi.org/10.1007/978-1-4842-9294-5_1

- If you are Windows user, you can use Chocolatey (`https://chocolatey.org/`).

- *An integrated development environment (IDE)*: As a suggestion, you can use Microsoft Visual Studio Code (`https://code.visualstudio.com/download`), the Community edition of IntelliJ IDEA from JetBrains (`https://www.jetbrains.com/idea/download/`), or Spring Tools (`https://spring.io/tools`).

- *The `curl` or `http` command*: For `http`, you can install HTTPie (`https://httpie.io/`). Both commands are demonstrated later in this chapter.

- *The `jq` command*: You can install it using the instructions at `https://stedolan.github.io/jq/`.

Start @ start.spring.io

start.spring.io is the official web-based tool for generating Spring Boot projects. It provides a user-friendly interface to quickly set up a new Spring Boot project with your desired dependencies and configurations. Key features and benefits:

- **Streamlined project creation**: Eliminates the need to manually configure a project structure and dependencies.

- **Curated dependencies**: Offers a selection of common libraries and frameworks to easily add to your project.

- **Customization**: Allows you to choose the build tool (Maven or Gradle), language (Java, Kotlin, Groovy), and Spring Boot version.

- **Downloadable project**: Generates a ZIP file containing the configured project ready to be imported into your IDE.

- **Spring Boot integration**: Leverages Spring Boot's auto-configuration and convention-over-configuration principles for rapid development.

Open a browser and go to `https://start.spring.io`. You should see the home page of the Spring Initializr, as shown in Figure 1-1.

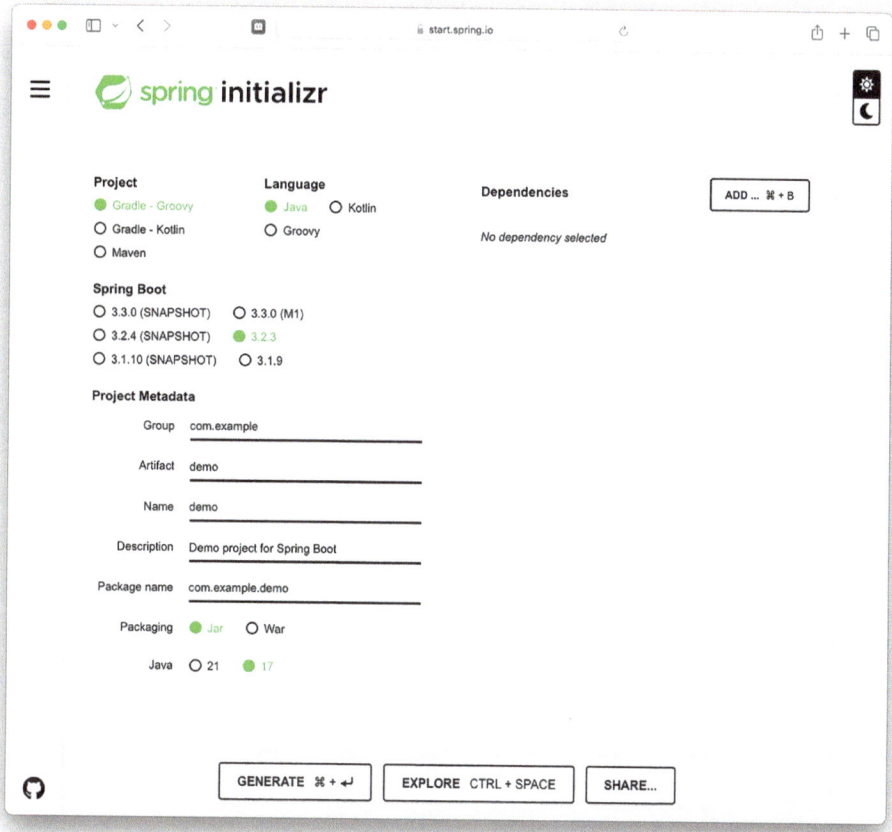

Figure 1-1. *Spring Initializr home page*

Notice that by default the Spring Initializr uses Gradle – Groovy as the project builder, Java as the programming language, JAR for packaging, Java 17, and Spring Boot 3 (at the time of this writing, I'm using Spring Boot 3.2.3).

Modify the Project Metadata section with the following values, as shown in Figure 1-2 (The value of the Package name field will change automatically based on the values of the Group and Artifact fields).

- Group: `com.apress`

- Artifact: `users`

- Name: `users`

- Dependencies: Spring Web (click Add to find it)

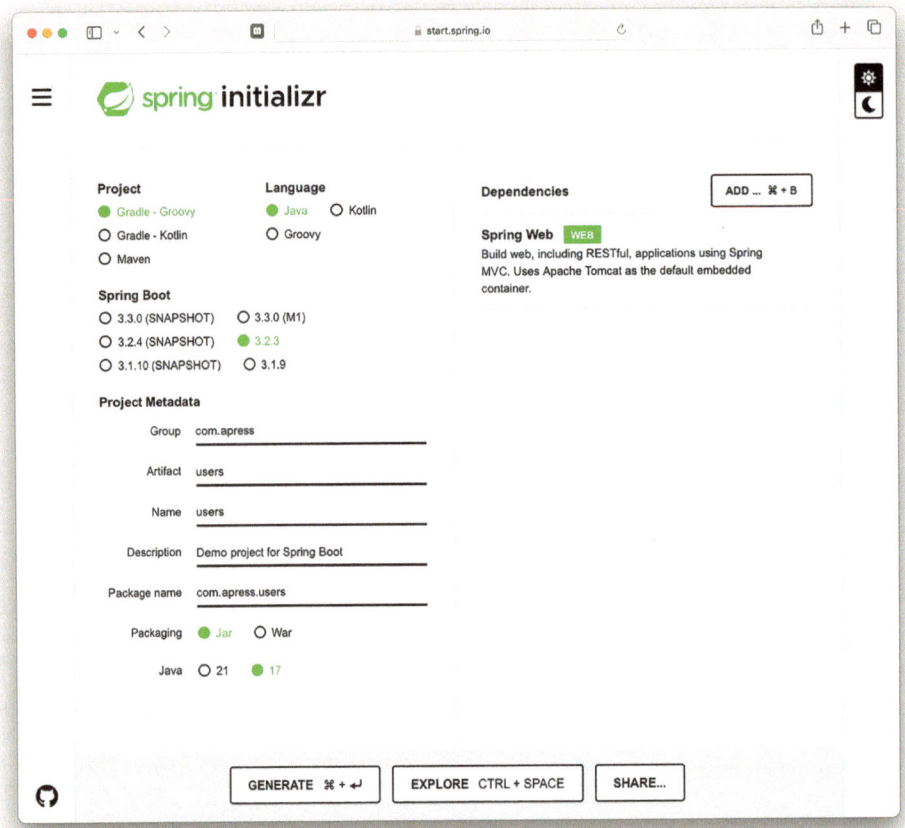

Figure 1-2. *Spring Initializr: Users App project*

Figure 1-2 shows all the necessary information to create the Users App project. Click the Generate button to zip the project and save it to your computer. Then, unzip it and import it to your favorite IDE. (I am using IntelliJ IDEA Community Edition, so that is what you will see in the screenshots in this book.)

Note   You can download or fork the source code from the Apress GitHub site.

When you open the project in your IDE, you should see the project structure shown in Figure 1-3.

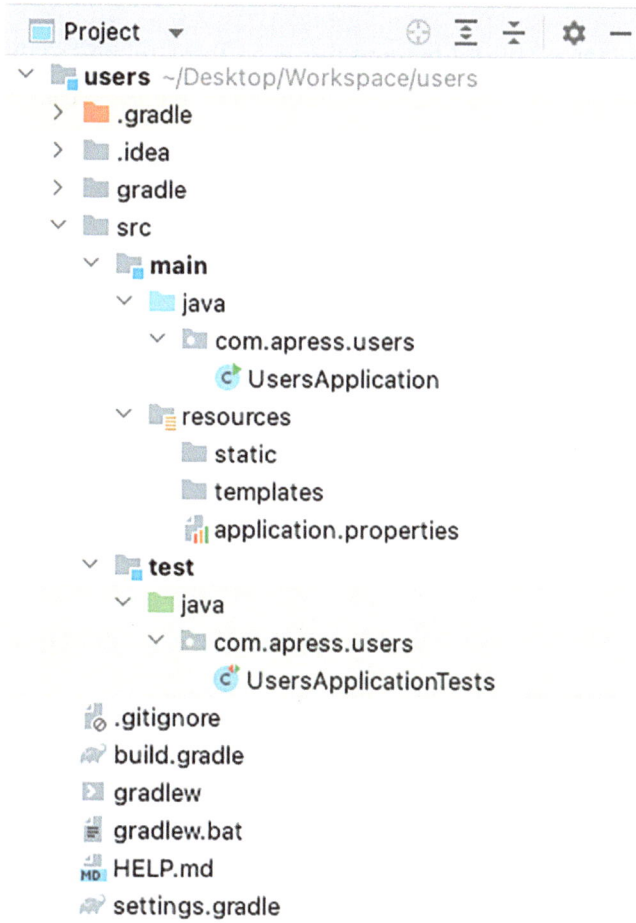

Figure 1-3. *Users App project structure*

Figure 1-3 shows the following three folders in the project structure:

- `src/main/java`: This folder contains all the source code. By default, the Spring Initializr creates the `UsersApplication.java` file. This has the main entry point where the application will start.

- `src/main/resources`: This folder contains one of the most important files, `application.properties`, which is used to modify configuration. We will use this file repeatedly throughout the book; for now, just know that it's located in this folder. This folder also contains subfolders that normally hold assets such as HTML pages, images, JavaScript, etc. (more details are provided in upcoming chapters).

- src/test/java: This folder contains everything related to the unit and integration testing that you can perform to ensure that your project does what you expect it to do. By default, the Spring Initializr creates the UsersApplicationTests.java file.

Let's take a look at some of the files generated by the Spring Initializr, starting with the build.gradle file, shown in Listing 1-1.

Listing 1-1. build.gradle - File Generated by Spring Initializr

```
plugins {
    id 'java'
    id 'org.springframework.boot' version '3.2.3'
    id 'io.spring.dependency-management' version '1.1.4'
}

group = 'com.apress'
version = '0.0.1-SNAPSHOT'
sourceCompatibility = '17'

repositories {
    mavenCentral()
    maven { url 'https://repo.spring.io/milestone' }
    maven { url 'https://repo.spring.io/snapshot' }
}

dependencies {
    implementation 'org.springframework.boot:spring-boot-starter-web'
    testImplementation 'org.springframework.boot:spring-boot-starter-test'
}

tasks.named('test') {
    useJUnitPlatform()
}
```

The `build.gradle` file is important because it contains all the details about our project, some attributes, and the dependencies that will be used to generate everything we need, from compiling our code to creating an executable JAR for the Java virtual machine (JVM). This file first declares the plugins we are going to use and the dependencies repository (in this case, Maven Central, the milestone and snapshot). It then declares the dependencies (in this case, `spring-boot-starter-web` and `spring-boot-starter-test`; one for all related to a web application and the unit/integration tests, respectively). Don't worry about these dependencies for now; we are going to discuss them in detail in the following chapters.

The `build.gradle` file also comes with a few wrappers (`gradlew` for Unix users and `gradlew.bat` for Windows users). These wrappers will bring the Gradle engine without any prior installation of such builder. We will be using these files when running our application.

Before we continue to review the files generated by the Spring Initializr, modify the `build.gradle` file as shown in Listing 1-2 and described next.

Listing 1-2. build.gradle Modified

```
//...
dependencies {
    // ... previous dependencies
    implementation 'org.webjars:bootstrap:5.2.3'
}
//...


test {
    testLogging {
        events "passed", "skipped", "failed"
        showExceptions true
        exceptionFormat "full"
        showCauses true
        showStackTraces true
        showStandardStreams = false
    }
}
```

First, add the bootstrap dependency that will help us to create a nice style for a home page; we are going to use bootstrap (https://getbootstrap.com/). At the time of writing, the version is 5.2.3, but you should choose the latest version in the maven repository. Then, add the test section, which tells Gradle to show the keywords passed, skipped, or failed when running a test. If you set showStandardStreams to true, you can use System.out.println statements in the tests.

Next, open the UsersApplication.java file generated by the Spring Initializr and view the contents, shown in Listing 1-3.

Listing 1-3. src/main/java/com/apress/users/UsersApplication.java

```java
package com.apress.users;

import org.springframework.boot.SpringApplication;
import org.springframework.boot.autoconfigure.SpringBootApplication;

@SpringBootApplication
public class UsersApplication {

    public static void main(String[] args) {
        SpringApplication.run(UsersApplication.class, args);
    }

}
```

UsersApplication.java is the main file for this project because it contains the main(String[] args) method that is necessary to run any Java application. As shown in Listing 1-3, it uses the @SpringBootApplication annotation and uses the static SpringApplication class that invokes the run method that accepts two parameters, one if the configuration class, and the arguments that we can pass at time of executing as parameters.

Next, add a new class that will hold the information of our user. Create the User. java file in the src/main/java/com/apress/users folder with the content shown in Listing 1-4.

Listing 1-4. src/main/java/com/apress/users/User.java

```java
package com.apress.users;

public class User {
    private String email;
    private String name;

    public User() {
    }

    public User(String email, String name) {
        this.email = email;
        this.name = name;
    }

    public String getEmail() {
        return email;
    }

    public void setEmail(String email) {
        this.email = email;
    }

    public String getName() {
        return name;
    }

    public void setName(String name) {
        this.name = name;
    }
}
```

As Listing 1-4 shows, the User.java class has two fields/properties, Email and Name. Notice that we have some constructors and the getters and setters for these fields/properties.

Next, let's add another class that will expose our API and manage the CRUD for our users. In the same folder, src/main/java/com/apress/users, create the UsersController.java file with the content shown in Listing 1-5.

Listing 1-5. src/main/java/com/apress/users/UsersController.java

```java
package com.apress.users;

import org.springframework.web.bind.annotation.*;

import java.util.Collection;
import java.util.HashMap;
import java.util.Map;

@RestController
@RequestMapping("/users")
public class UsersController {
    private Map<String,User> users = new HashMap() {{
        put("ximena@email.com",new User("ximena@email.com","Ximena"));
        put("norma@email.com",new User("norma@email.com","Norma"));

    }};

    @GetMapping
    public Collection<User> getAll(){
        return this.users.values();
    }

    @GetMapping("/{email}")
    public User findUserByEmail(@PathVariable String email){
        return this.users.get(email);
    }

    @PostMapping
    public User save(@RequestBody User user){
        this.users.put(user.getEmail(),user);
        return user;
    }

    @DeleteMapping("/{email}")
    public void save(@PathVariable String email){
        this.users.remove(email);
    }
}
```

The UsersController.java class will expose the API. This class is marked with the annotation @RestController, telling Spring Boot that this class is responsible for accepting any incoming requests. It's also marked with @RequestMapping("/users"), which tells Spring Boot that it will have a based /users endpoint for every request made with any of the HTTP methods (GET, POST, PUT, DELETE, PATCH, etc.). This class also has some methods that are marked with special annotation such as @GetMapping, @PostMapping, and @DeleteMapping, which are used to tell Spring Boot that those methods will be executed once a request is made.

This class also has a java.util.Map interface that is being initialized with some users. That will be covering one of the requirements.

Now, let's take a look at the following endpoints:

- /users: This will handle the read part of CRUD. The method that will be executed is getAll(). This method is executed when an HTTP GET request is made. This endpoint is also used when an HTTP POST request is made (can be taken as the create and update parts of CRUD); the save(@RequestBody User user) method will be executed and it will add or update a user based on the user's email address. Note that the argument of this method has an annotation, @RequestBody, meaning that for every POST request, Spring Boot will convert the data sent over into the class type, in this case the User class. The data sent is in JSON format by default, unless you modify the HTTP Header Content-Type and inform Spring Boot of that change.

- /users/{email}: This will also be our read, but it will look for an email in the Map instance. The method executed will be findUserByEmail(@PathVariable String email), which will return the user found. This method also is called when an HTTP GET request is made. This endpoint is also used by the deleteByEmail (@PathVariable String email) method when an HTTP DELETE request is made. This will remove the user from the map. This method contains the @PathVariable annotation that will translate the path that matches the name, in this case {email} match, with the String email parameter. This means that when there is a request, either the GET or DELETE such as /users/ximena@email.com for example, it will execute any of this methods (depending on the HTTP method request) and it will assign ximena@email.com to the email variable.

Next, let's create a landing page that will open when we run this project. For this purpose, add an index.html file in the src/main/resources/static folder with the content shown in Listing 1-6.

Listing 1-6. src/main/resources/static/index.html

```html
<!DOCTYPE html>
<html lang="en">
<head>
    <meta charset="UTF-8">
    <link rel="stylesheet" type="text/css"
        href="webjars/bootstrap/5.2.3/css/bootstrap.min.css">
    <title>Welcome - Users App</title>
</head>
<body class="d-flex h-100 text-center">

<div class="cover-container d-flex w-100 h-100 p-3 mx-auto flex-column">
    <header class="mb-auto">
        <div>
            <h3 class="float-md-start mb-0">Users</h3>
        </div>
    </header>

    <main class="px-3">
        <h1>Simple Users Rest Application</h1>
        <p class="lead">This is a simple Users app where you can access any
        information from a user</p>
        <p class="lead">
            <a href="/users">Get All Users</a>
        </p>
    </main>
```

```
<footer class="mt-auto text-black-50">
    <p>Powered by Spring Boot 3</p>
</footer>
</div>
</body>
</html>
```

As you can see, we are using bootstrap (https://getbootstrap.com/) to style our home page; we are passing the path for the CSS that begins with webjars. There is a convention to this, but we are going to talk about it later. For now, you can take this as a recipe. If you want to use any webjars tech, such as jQuery or any other, you must provide the path starting with webjars.

So, you now have in place for the Users App project everything that you need to run a web API using Spring Boot. Next, we'll look at how to test the application to make sure that it performs as expected.

Testing the Users App Project

Let's test our code using the Testing framework that Spring Boot provides. In the src/test/java folder structure, open the UserApplicationTests.java file. Listing 1-7 shows its content.

Listing 1-7. src/test/java/com/apress/users/UsersApplicationTests.java

```
package com.apress.users;

import org.junit.jupiter.api.Test;
import org.springframework.boot.test.context.SpringBootTest;

@SpringBootTest
class UsersApplicationTests {

    @Test
    void contextLoads() {
    }

}
```

Note in Listing 1-7 that UserApplicationTests.java uses the @SpringBootTest annotation, which prepares your environment for executing any unit or integration testing; for every test that you want to conduct, you only need to use the @Test annotation for the method that you want to use to execute that test.

Next, replace all the code in UserApplicationTests.java with the content shown in Listing 1-8.

Listing 1-8. Modified src/test/java/com/apress/users/ UsersApplicationTests.java

```
package com.apress.users;

import org.junit.jupiter.api.Test;
import org.springframework.beans.factory.annotation.Autowired;
import org.springframework.boot.test.context.SpringBootTest;
import org.springframework.boot.test.web.client.TestRestTemplate;
import org.springframework.beans.factory.annotation.Value;
import org.springframework.http.ResponseEntity;

import java.util.Collection;

import static org.assertj.core.api.Assertions.assertThat;

@SpringBootTest(webEnvironment = SpringBootTest.WebEnvironment.RANDOM_PORT)
public class UsersApplicationTests {

    @Value("${local.server.port}")
    private int port;

    private final String BASE_URL = "http://localhost:";
    private final String USERS_PATH = "/users";

    @Autowired
    private TestRestTemplate restTemplate;

    @Test
    public void indexPageShouldReturnHeaderOneContent() throws Exception {
        assertThat(this.restTemplate.getForObject(BASE_URL + port,
                String.class)).contains("Simple Users Rest Application");
    }
```

```java
@Test
public void usersEndPointShouldReturnCollectionWithTwoUsers() throws
Exception {
    Collection<User> response = this.restTemplate.
            getForObject(BASE_URL + port + USERS_PATH,
            Collection.class);

    assertThat(response.size()).isEqualTo(2);
}

@Test
public void userEndPointPostNewUserShouldReturnUser() throws
Exception {
    User user =  new User("dummy@email.com","Dummy");
    User response =  this.restTemplate.postForObject(BASE_URL + port +
    USERS_PATH,user,User.class);

    assertThat(response).isNotNull();
    assertThat(response.getEmail()).isEqualTo(user.getEmail());

    Collection<User> users = this.restTemplate.
            getForObject(BASE_URL + port + USERS_PATH,
            Collection.class);

    assertThat(users.size()).isGreaterThanOrEqualTo(2);

}

@Test
public void userEndPointDeleteUserShouldReturnVoid() throws Exception {
    this.restTemplate.delete(BASE_URL + port + USERS_PATH + "/norma@
    email.com");

    Collection<User> users = this.restTemplate.
            getForObject(BASE_URL + port + USERS_PATH,
            Collection.class);

    assertThat(users.size()).isLessThanOrEqualTo(2);
}
```

```
@Test
public void userEndPointFindUserShouldReturnUser() throws Exception{
    User user = this.restTemplate.getForObject(BASE_URL + port + USERS_
    PATH + "/ximena@email.com",User.class);

    assertThat(user).isNotNull();
    assertThat(user.getEmail()).isEqualTo("ximena@email.com");
  }
}
```

The following list explains the code in the revised UsersApplicationTest.java file:

- @SpringBootTest: To create a test for Spring Boot, you need to annotate your class with this annotation. This annotation accepts multiple parameters, one of which is the webEnvironment, which randomly assigns a port to Tomcat when the test runs. Clearly you can see that this is an integration test.

- @Value: This annotation injects the port number into the variable port by using the local.server.port property that is set by the webEnvironment parameter declaration from the @SpringBootTest annotation. This enables you to avoid any port collision.

- @Autowired: This annotation is also used to inject a new instance of a class, in this case the TestRestTemplate type class, which allows you to execute any remote request to an external web API. In this case, we are using it in every test method.

- @Test: This annotation marks a method to be a Test case where we can do any assertions; from doing a request for getting a list of users, to post, or delete some user data. Spring Boot comes with some libraries to conduct unit and integration tests and to execute assertions. In this case, we are using the library AssertJ for every test method.

- TestRestTemplate: This class provides several methods that are convenient for executing any HTTP method requests. These methods include .getForObject or .postForObject where behind the scenes wires everything up and do the right conversion type to get the objects we need, in this case the User instance.

Chapter 8 is dedicated to coverage of unit and integration testing. It discusses every detail and describes what else you can do with the Spring Boot Testing Framework.

Running the UserApplicationTests Class

Let's run our tests. If you are using an IDE, you should be able to right-click the `UsersApplicationTests.java` file and click a Run option. If you are running IntelliJ IDEA, you should see a window similar to the one shown in Figure 1-4.

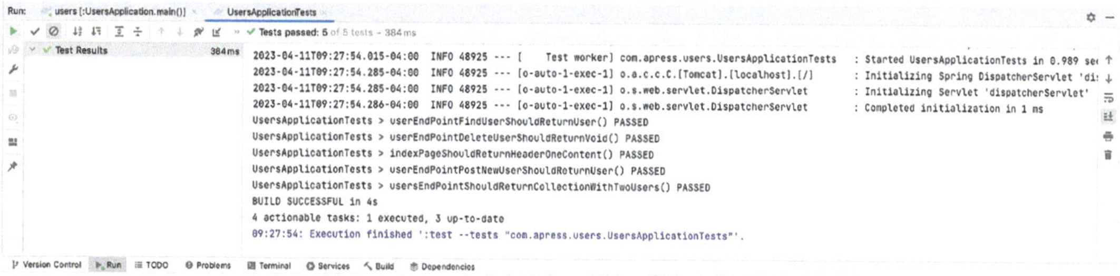

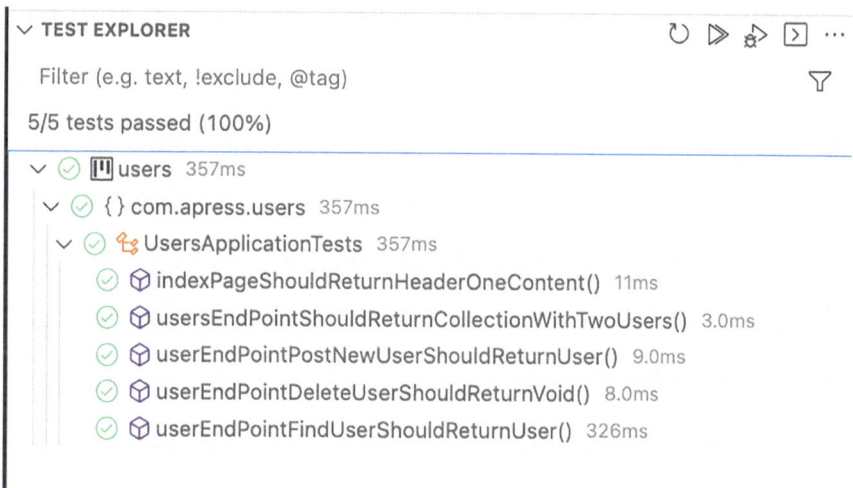

Figure 1-4. *IntelliJ IDEA tests*

If you have imported the project into Visual Studio Code (a.k.a. VS Code), you can run the test in the Test Explorer, which will produce a result similar to that shown in Figure 1-5.

Figure 1-5. *Microsoft VS Code Test Explorer*

Also, you can execute the tests from the command line using the gradlew wrapper and expect the same results:

```
./gradlew test
UsersApplicationTests > userEndPointFindUserShouldReturnUser() PASSED
UsersApplicationTests > userEndPointDeleteUserShouldReturnVoid() PASSED
UsersApplicationTests > indexPageShouldReturnHeaderOneContent() PASSED
UsersApplicationTests > userEndPointPostNewUserShouldReturnUser() PASSED
UsersApplicationTests >
usersEndPointShouldReturnCollectionWithTwoUsers() PASSED

BUILD SUCCESSFUL in 3s
4 actionable tasks: 1 executed, 3 up-to-date
```

Now that your tests have passed, it's time to run the Users App project.

Running the Users App Project

To run the Users App project, you can use the IDE to which you imported the project by right-clicking the UsersApplication.java file where is our main function. Alternatively, you can run the project from the command line by using the following command:

```
./gradlew bootRun
```

```
> Task :bootRun


  .   ____          _            __ _ _
 /\\ / ___'_ __ _ _(_)_ __  __ _ \ \ \ \
( ( )\___ | '_ | '_| | '_ \/ _` | \ \ \ \
 \\/  ___)| |_)| | | | | || (_| |  ) ) ) )
  '  |____| .__|_| |_|_| |_\__, | / / / /
 =========|_|==============|___/=/_/_/_/
 :: Spring Boot ::        (v3.1.0)

INFO 66966 - [main] com.apress.users.UsersApplication          : Starting
UsersApplication using Java 17.0.5 ....
INFO 66966 - [main] com.apress.users.UsersApplication          : No active
profile set, falling back to 1 default profile: "default"
```

```
INFO 66966 - [main] o.s.b.w.embedded.tomcat.TomcatWebServer  : Tomcat
initialized with port(s): 8080 (http)
INFO 66966 - [main] o.a.c.core.StandardService               : Starting
service [Tomcat]
INFO 66966 - [main] o.a.c.core.StandardEngine                : Starting
Servlet engine: [Apache Tomcat/10.1.7]
INFO 66966 - [main] o.a.c.c.C.[Tomcat].[localhost].          : Initializing
Spring embedded WebApplicationContext
INFO 66966 - [main] w.s.c.ServletWebServerApplicationContext : Root
WebApplicationContext: initialization completed in 390 ms
INFO 66966 - [main] o.s.b.a.w.s.WelcomePageHandlerMapping    : Adding
welcome page: class path resource [static/index.html]
INFO 66966 - [main] o.s.b.w.embedded.tomcat.TomcatWebServer  : Tomcat
started on port(s): 8080 (http) with context path ''
INFO 66966 - [main] com.apress.users.UsersApplication        : Started
UsersApplication in 0.706 seconds (process running for 0.856)
```

Now you can open a browser and go to localhost:8080 to see the home page, shown in Figure 1-6.

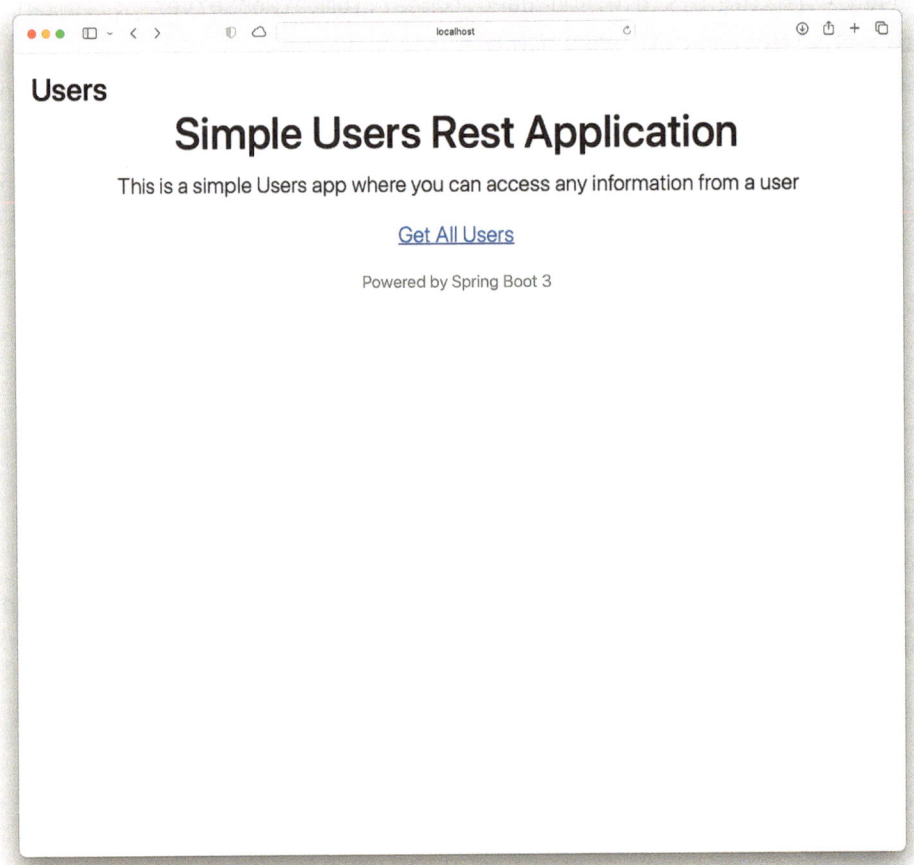

Figure 1-6. *http://localhost:8080: home page*

Click the Get All Users link, and you should see the response of the /users endpoint, shown in Figure 1-7.

```
[
  {
    "email": "ximena@email.com",
    "name": "Ximena"
  },
  {
    "email": "norma@email.com",
    "name": "Norma"
  }
]
```

Figure 1-7. *http://localhost:8080/users: /users endpoint*

Spring Boot responds using JSON by default (as described further in the following chapters).

You can interact with the Users App (list, add, update, or remove users) by using the command line and executing the instructions in the following list; to do so, you need to have the curl or http command and the jq command (as indicated earlier in this chapter).

- Listing users using the curl command:

```
curl -XGET -s http://localhost:8080/users | jq .
[
  {
```

```
      "email": "ximena@email.com",
      "name": "Ximena"
  },
  {
      "email": "norma@email.com",
      "name": "Norma"
  }
]
```

- Listing users using the http command:

```
http :8080/users
```

```
HTTP/1.1 200
Connection: keep-alive
Content-Type: application/json
Date: Tue, 11 Apr 2023 18:44:11 GMT
Keep-Alive: timeout=60
Transfer-Encoding: chunked
```

```
[
    {
        "email": "ximena@email.com",
        "name": "Ximena"
    },
    {
        "email": "norma@email.com",
        "name": "Norma"
    }
]
```

- Adding a new user using the curl command:

```
curl -XPOST -s -H "Content-Type: application/json" -d
'{"email":"nayely@email.com","name":"Nayely"}' http://
localhost:8080/users | jq .
```

```
{
    "email": "nayely@email.com",
    "name": "Nayely"
}
```

- Adding a new user using the http command:

```
http :8080/users email=laura@email.com name=Laura
```

```
HTTP/1.1 200
Connection: keep-alive
Content-Type: application/json
Date: Tue, 11 Apr 2023 18:48:20 GMT
Keep-Alive: timeout=60
Transfer-Encoding: chunked

{
    "email": "laura@email.com",
    "name": "Laura"
}
```

- Finding a user using the curl command:

```
curl -XGET -s http://localhost:8080/users/ximena@email.com | jq .

{
    "email": "ximena@email.com",
    "name": "Ximena"
}
```

- Finding a user using the http command:

```
http :8080/users/ximena@email.com
```

```
HTTP/1.1 200
Connection: keep-alive
Content-Disposition: inline;filename=f.txt
Content-Type: application/json
Date: Tue, 11 Apr 2023 18:50:52 GMT
```

```
Keep-Alive: timeout=60
Transfer-Encoding: chunked

{
    "email": "ximena@email.com",
    "name": "Ximena"
}
```

- Deleting a user using the `curl` command:

```
curl -XDELETE http://localhost:8080/users/ximena@email.com
```

- Deleting a user using the `http` command:

```
http DELETE :8080/users/laura@email.com

HTTP/1.1 200
Connection: keep-alive
Content-Length: 0
Date: Tue, 11 Apr 2023 18:53:02 GMT
Keep-Alive: timeout=60
```

Congratulations! This was a quick start for creating a well-defined application in Spring Boot that just runs.

Creating the Users App Project Using Kotlin

If Kotlin is your preferred programming language (and assuming you already have Kotlin installed), this section demonstrates how to create the Users App project using Kotlin instead of Java. If you open the start.spring.io and use now the Kotlin programming language, you can generated and imported into your favorite IDE. See Figure 1-8.

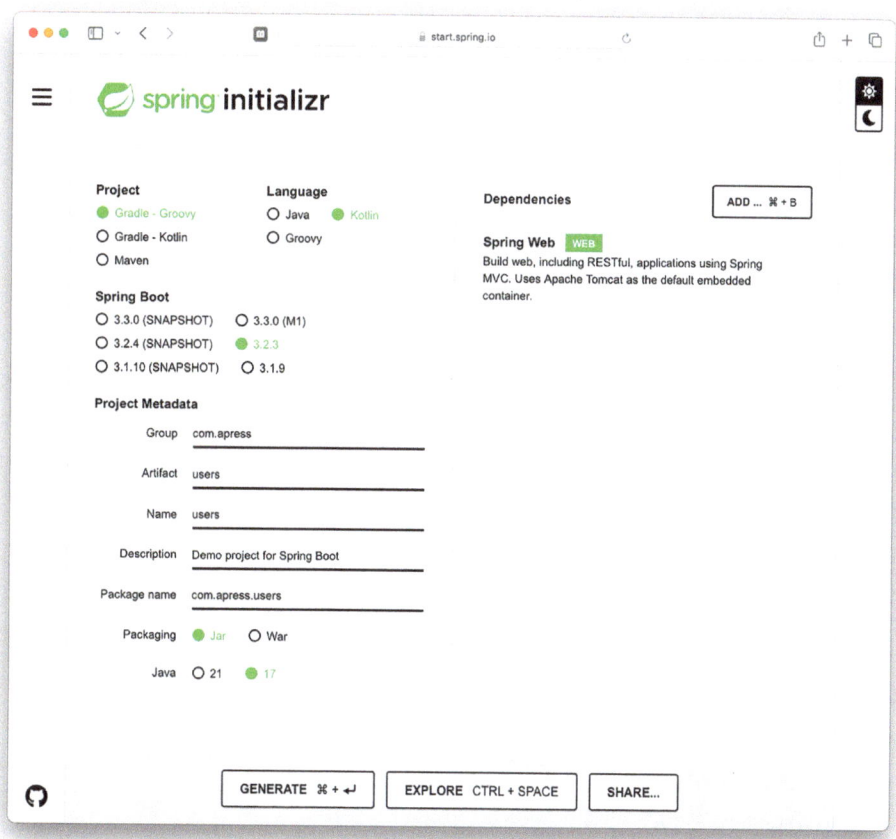

Figure 1-8. *Users App project using Kotlin*

After you have imported the project into your favorite IDE, open the `build.gradle` file and you should have the following content. Look that we are adding the extra information to it, such as the `bootstrap` dependency and the `test` declarations. See Listing 1-9.

Listing 1-9. build.gradle for Kotlin

```
import org.jetbrains.kotlin.gradle.tasks.KotlinCompile

plugins {
    id 'org.springframework.boot' version '3.2.4'
    id 'io.spring.dependency-management' version '1.1.4'
    id 'org.jetbrains.kotlin.jvm' version '1.9.22'
    id 'org.jetbrains.kotlin.plugin.spring' version '1.9.22'
}
```

```
group = 'com.apress.users'
version = '0.0.1-SNAPSHOT'
sourceCompatibility = '17'

repositories {
    mavenCentral()
    maven { url 'https://repo.spring.io/milestone' }
    maven { url 'https://repo.spring.io/snapshot' }
}

dependencies {
    implementation 'org.springframework.boot:spring-boot-starter-web'
    implementation 'com.fasterxml.jackson.module:jackson-module-kotlin'
    implementation 'org.jetbrains.kotlin:kotlin-reflect'
    implementation 'org.webjars:bootstrap:5.2.3'
    testImplementation 'org.springframework.boot:spring-boot-starter-test'
}

tasks.withType(KotlinCompile) {
    kotlinOptions {
        freeCompilerArgs = ['-Xjsr305=strict']
        jvmTarget = '17'
    }
}

tasks.named('test') {
    useJUnitPlatform()
}

test {
    testLogging {
        events "passed", "skipped", "failed"
        showExceptions true
        exceptionFormat "full"
        showCauses true
        showStackTraces true
        showStandardStreams = false
    }
}
```

Next, open the `UsersApplication.kt` file (note the new path, `src/main/kotlin`). See Listing 1-10.

Listing 1-10. src/main/kotlin/com/apress/users/UsersApplication.kt

```
package com.apress.users

import org.springframework.boot.autoconfigure.SpringBootApplication
import org.springframework.boot.runApplication

@SpringBootApplication
class UsersApplication

fun main(args: Array<String>) {
    runApplication<UsersApplication>(*args)
}
```

Listing 1-10 shows the main app. Comparing it to `UsersApplication.java` in Listing 1-3, note that there is still a `@SpringBootApplication` annotation and the `runApplication` call passing the arguments (any argument from the command line) and the configuration class, in this case the `UserApplication` class itself.

Next, add the `User.kt` file with the content shown in Listing 1-11.

Listing 1-11. src/main/kotlin/com/apress/users/User.kt

```
package com.apress.users

data class User(var email:String? = null, var name:String? = null)
```

As you can see, Kotlin removes all the boilerplate that Java has (as shown in Listing 1-4). Of course, there are pros and cons to both Kotlin and Java, but we are not going to discuss them in this book. I recommend selecting the programming language that you find easier to use and/or that is widely used in your organization.

Next, add the `UsersController.kt` file with the content shown in Listing 1-12.

Listing 1-12. src/main/kotlin/com/apress/users/UsersController.kt

```kotlin
package com.apress.users

import org.springframework.web.bind.annotation.*

@RestController
@RequestMapping("/users")
class UsersController {
    private val users: HashMap<String?, User?> = object : HashMap<String?,
User?>() {
        init {
            put("ximena@email.com", User("ximena@email.com", "Ximena"))
            put("norma@email.com", User("norma@email.com", "Norma"))
        }
    }

    @GetMapping
    fun getAll(): MutableCollection<User?> = users.values

    @GetMapping("/{email}")
    fun findUserByEmail(@PathVariable email: String?): User? {
        return users[email]
    }

    @PostMapping
    fun save(@RequestBody user: User): User {
        users[user.email] = user
        return user
    }

    @DeleteMapping("/{email}")
    fun save(@PathVariable email: String?) {
        users.remove(email)
    }
}
```

The UsersController.kt class will expose the API to the world. As you can see, it is pretty much the same as the version in Java (see Listing 1-5).

Testing the Users App Project with Kotlin

To test the Users App project, add the code shown in Listing 1-13 to the
UserApplicationTest.kt file.

Listing 1-13. src/test/kotlin/com/apress/users/UsersApplicationTests.kt

```
package com.apress.users

import org.assertj.core.api.Assertions
import org.junit.jupiter.api.Test
import org.springframework.beans.factory.annotation.Autowired
import org.springframework.beans.factory.annotation.Value
import org.springframework.boot.test.context.SpringBootTest
import org.springframework.boot.test.web.client.TestRestTemplate

@SpringBootTest(webEnvironment = SpringBootTest.WebEnvironment.RANDOM_PORT)
class UsersApplicationTests {
    @Value("${local.server.port}")
    private val port = 0
    private val BASE_URL = "http://localhost:"
    private val USERS_PATH = "/users"

    @Autowired
    private val restTemplate: TestRestTemplate? = null

    @Test
    @Throws(Exception::class)
    fun indexPageShouldReturnHeaderOneContent() {
        Assertions.assertThat(
            restTemplate!!.getForObject(
                BASE_URL + port,
                String::class.java
            )
        ).contains("Simple Users Rest Application")
    }

    @Test
    @Throws(Exception::class)
```

```kotlin
    fun usersEndPointShouldReturnCollectionWithTwoUsers() {
        val response: MutableCollection<User> = restTemplate!!.
getForObject(
            BASE_URL + port + USERS_PATH,
            MutableCollection::class.java
        ) as MutableCollection<User>
        Assertions.assertThat(response.size).isEqualTo(2)
    }

    @Test
    @Throws(Exception::class)
    fun userEndPointPostNewUserShouldReturnUser() {
        val user = User("dummy@email.com", "Dummy")
        val response = restTemplate!!.postForObject(
            BASE_URL + port + USERS_PATH, user,
            User::class.java
        )
        Assertions.assertThat(response).isNotNull
        Assertions.assertThat(response.email).isEqualTo(user.email)
        val users: MutableCollection<User> = restTemplate!!.getForObject(
            BASE_URL + port + USERS_PATH,
            MutableCollection::class.java
        ) as MutableCollection<User>
        Assertions.assertThat(users.size).isGreaterThanOrEqualTo(2)
    }

    @Test
    @Throws(Exception::class)
    fun userEndPointDeleteUserShouldReturnVoid() {
        restTemplate!!.delete("$BASE_URL$port$USERS_PATH/norma@email.com")
        val users: MutableCollection<User> = restTemplate!!.getForObject(
            BASE_URL + port + USERS_PATH,
            MutableCollection::class.java
        ) as MutableCollection<User>
        Assertions.assertThat(users.size).isLessThanOrEqualTo(2)
    }
```

```kotlin
@Test
@Throws(Exception::class)
fun userEndPointFindUserShouldReturnUser() {
    val user = restTemplate!!.getForObject(
        "$BASE_URL$port$USERS_PATH/ximena@email.com",
        User::class.java
    )
    Assertions.assertThat(user).isNotNull
    Assertions.assertThat(user.email).isEqualTo("ximena@email.com")
}
}
```

The UsersApplicationTests.kt class is the Kotlin version of UserApplicationTests.java (see Listing 1-8). The only new annotation is @Throws, which is self-explanatory for this test.

Now, you can execute the tests by right-clicking this file and clicking the Run option, which should produce results similar to shown earlier in Figure 1-4. The same for the command line. That doesn't change.

```
./gradlew test
```

Running the Users App Project with Kotlin

Running the Users App project with Kotlin is also the same as with Java: either run it within your IDE or run it with the following command:

```
./graldew bootRun
```

Nice! Now you have the same Users App project with a different programming language using Spring Boot.

Why Spring Boot?

Spring Boot is an opinionated runtime that enables you to create amazing, enterprise-grade applications faster and with ease. Behind the scenes I can say that gets all the best practices of the Spring Framework and uses some default to configure everything in your behalf.

Even though we are talking about Spring Boot, it is important to understand what the Spring Framework is and why it is so important for Spring Boot. The Spring Framework has the following guiding principles (quoted verbatim from `https://docs.spring.io/spring-framework/reference/overview.html`):

- Provide choice at every level. Spring lets you defer design decisions as late as possible. For example, you can switch persistence providers through configuration without changing your code. The same is true for many other infrastructure concerns and integration with third-party APIs.

- Accommodate diverse perspectives. Spring embraces flexibility and is not opinionated about how things should be done. It supports a wide range of application needs with different perspectives.

- Maintain strong backward compatibility. Spring's evolution has been carefully managed to force few breaking changes between versions. Spring supports a carefully chosen range of JDK versions and third-party libraries to facilitate maintenance of applications and libraries that depend on Spring.

- Care about API design. The Spring team puts a lot of thought and time into making APIs that are intuitive and that hold up across many versions and many years.

- Set high standards for code quality. The Spring Framework puts a strong emphasis on meaningful, current, and accurate javadoc. It is one of very few projects that can claim clean code structure with no circular dependencies between packages.

The Spring Framework implements several design patterns, such as Dependency of Injection and Inversion of Control, Factories, Abstract Factories, Strategies, Singletons, Templates, MVC, and many more.

One of the main features of the Spring Framework is that it allows you to work with plain old Java objects (POJOs), making your apps easy to extend without any dependency; in other words, the Spring Framework is not invasive.

As depicted in Figure 1-9, to create an application using only the Spring Framework, you need your classes (Java, Groovy, Kotlin), some configuration that tells the Spring Framework how to wire every class (which are called Spring beans) to set the Spring Context, and have your application ready.

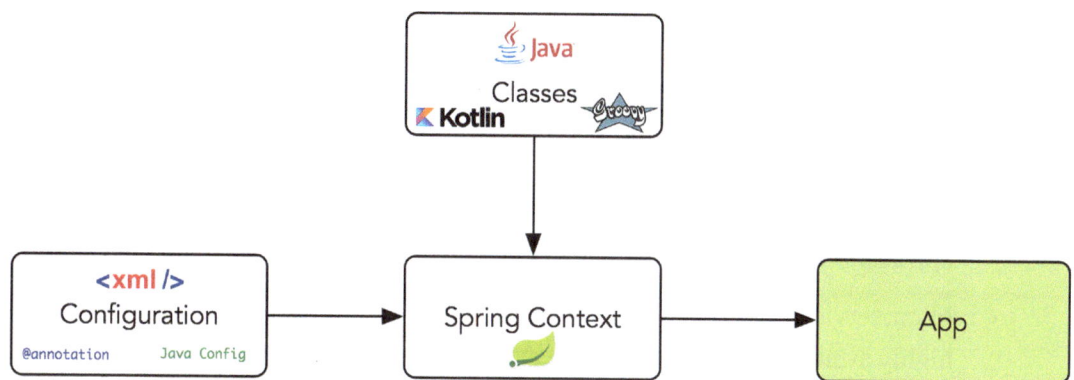

Figure 1-9. *Spring Framework: Spring Context*

This process of configuration sometimes can be challenging if you have many classes and you want to use features such as Web Mvc, Security, or persistence.

So, suppose that your application requires some web and persistence features; you must tell the Spring Framework (with configuration: XML, Annotations, or JavaConfig classes) about your `TransactionManager`, an `EntityManagerFactory`, a `DataSource`, a `ResourceViewResolver` (for your web views), a `MessageConverter` (for your web responses, based on the `Content-Type`), a `ResourceHandlerRegistry`, and some other extra configuration...yes, that's a lot of configuration, but it keeps your code clean and easy to maintain.

Then, why Spring Boot? The biggest advantage of using Spring Boot is that it is based on the Spring Framework, making it enterprise-ready with all the best practices applied. And because Spring Boot is opinionated, it will configure everything for you, resulting in (sometimes) zero configuration. How does it do it? You are going to see how in the next chapter.

Still questioning why to use Spring Boot or what can you do with it? Here's a sampling of what Spring Boot is suitable for:

- Cloud-native applications that follow the twelve-factor app developed by the Heroku engineering team (see `https://12factor.net`)

- Native images with the new Ahead Of Time (AOT) and GraalVM support.

- Better productivity by reducing time of development and deployment.

- Enterprise, production-ready Spring applications.

- Nonfunctional requirements, such as the Spring Boot Actuator (a module that provides metrics with the new platform-agnostic Micrometer [`https://micrometer.io`], health checks, and management) and embedded containers for running web applications (Tomcat, Netty, Undertow, Jetty, etc.).

- Microservices, which are getting attention for creating scalable, highly available, and robust applications. Spring Boot allows developers to focus only on business logic, leaving the heavy lifting to the Spring Framework.

Spring Boot Features

The following list provides a brief introduction to several of the many Spring Boot features that I will describe in more depth in the following chapters:

- The `SpringApplication` class provides a convenient way to initiate a Spring application. As you saw earlier in this chapter, in a Java Spring Boot application, the `main` method executes this singleton class.

- Spring Boot allows you to create applications without requiring any XML configuration. Spring Boot doesn't generate any code.

- Spring Boot provides a fluent builder API through the `SpringApplicationBuilder` singleton class, which allows you to create hierarchies with multiple application contexts. This feature is more related to the Spring Framework and how it works internally. I will explain this feature in more detail in the following chapters, but if you are new to Spring and Spring Boot, at this point you only need to know that you can extend Spring Boot to get more control over your applications.

- Spring Boot offers more ways to configure the Spring application events and listeners.

- As an "opinionated" technology (as previously mentioned), Spring Boot attempts to create the right type of application, either as a web application (by embedded a Tomcat, Netty, Undertow, or Jetty container) or as a single application.

- The Spring Boot `org.springframework.boot.ApplicationArguments` interface allows you to access any application argument. This is a useful feature when you try to run your application with parameters.

- Spring Boot allows you to execute code after the application has started. You only need to implement the `CommandLineRunner` interface and provide the implementation of the `run(String ...args)` method. For example, this feature enables you to initialize records in a database during the start, or check if particular services are running before your application executes.

- Spring Boot enables you to externalize configurations by using `application.properties` or `application.yml` files.

- Spring Boot allows you to add administration-related features, normally through JMX, by enabling the `spring.application.admin.enabled` property in the `application.properties` or `application.yml` files.

- Spring Boot offers *profiles*, which help your application to run in different environments.

- Spring Boot allows you to configure and use logging in a very simple way.

- Spring Boot provides a simple way to configure and manage your dependencies by using starter poms. In other words, if you are going to create a web application, you only need to include the `spring-boot-start-web` dependency in your Maven `pom.xml` or `build.gradle` file.

- Spring Boot provides out-of-the-box nonfunctional requirements by using the Spring Boot Actuator together with the new Micrometer platform-agnostic framework that allows you to instrument your apps.

- Spring Boot provides @Enable<feature> annotations that help you to include, configure, and use technologies such as databases (SQL and NoSQL), Caching, Scheduling, Messaging, Spring Integration, Batch, Cloud, and more.

The new Spring Framework 6 brings a lot of improvements and new features, such as support for Java 17, Jakarta EE 9+, Servlet 6, JPA 3.1, Tomcat 10.x, and AOT transformations; the latter enables first-class support for GraalVM native images with Spring Boot 3 and an HTTP interface client, among other improvements.

Summary

In this chapter you learned how easy it is to set up a Spring Boot project by creating a simple web API project that shows the user's email address and name, using a Map as an in-memory solution.

You learned that you can include HTML pages in the src/main/resources/static folder and they can be rendered by Spring Boot. You also learned that, to expose a web API, you need to use some annotation that are a marker for the class that will have methods that will be executed when there is a request for a particular endpoint. You learned that using the @RestController, @RequestMapping, @GetMapping, @PostMapping and @DeleteMapping mark the class and methods for every HTTP method request, that in this case correlate with the CRUD (Create, Read, Update and Delete).

You learned how testing works with Spring Boot by creating an integration test that uses some classes to execute requests to your application. You saw some of the assertations that are declared in each method to tests the CRUD.

As an alternative to using Java, this chapter showed you how to use Kotlin for the Users App project. You saw that the behavior is practically the same in the two programming language. The Kotlin syntax can be strange at first, but if you keep working with it, it will make a lot of sense. Whatever programming language you choose, it must make sense to you and your team.

The most important part of this chapter was to show you how easy is to create a Spring Boot application with ease and understand the basics.

In Chapter 2 we are going to explore all the main features that make Spring Boot an awesome technology to create enterprise-ready applications with ease.

CHAPTER 2

Spring Boot Internals

Chapter 1 provided a quick Spring Boot introduction by showing you a fast way to create an enterprise-grade application with the Spring Initializr. In this chapter, we are going to explore the internals of what Spring Boot is doing behind the scenes with a project named My Retro App. But first, let's see what you need to create a Spring Boot app (if you are an experienced Spring Boot developer, you can skip this section).

Requirements to Create a Spring Boot App

Creating a Spring Boot application requires the following:

- A dependency management tool that allows you to use any dependency and verify, compile, test, and build your application with ease. Popular dependency management tools that support Spring Boot (through plugins) include Apache Maven (`https://maven.apache.org/`), Gradle (`https://gradle.org/`), and Apache Ant (`https://ant.apache.org/`). In this book we will be using Gradle as the default dependency management tool.

- A *starter*, which is a convenient set of dependency descriptors that comes with what you need depending on the technology used. A starter brings a curated set of libraries to your application, and the good part is that you don't need to worry about the version (no more dependency madness). The Spring Boot team established a naming convention that is very useful: `spring-boot-starter` as the prefix, followed by the name of the technology to use. For example, if you need a web application, then the starter to use is `spring-bootstarter-web`. If you need data JPA, then it will be `spring-boot-start-data-jpa`. The minimal for a Spring Boot app is the

© Felipe Gutierrez 2024
F. Gutierrez, *Pro Spring Boot 3*, https://doi.org/10.1007/978-1-4842-9294-5_2

spring-boot-starter dependency, which brings spring-core, spring-beans, spring-context, spring-aop, spring-test, and many other libraries.

- A structure in which you can define your classes, your tests, and any other assets or properties for your application. The following is the default structure for any Java application:

```
demo
  src
   +- main
       +- java
           +- com
               +- example
                   +- demo
                       +- <other-structure>
                       +- DemoApplication.java
       +- resources
           +- application.properties
   +- test
       +- java
           +- com
               +- example
                   +- demo
                       +- DemoApplicationTests.java
```

You can create any package structure, but just make sure that your main app (DemoApplication.java in this example) is at the top level of your main structure.

- Usage of the @SpringBootApplication annotation. This tells Spring Boot to auto-configure everything on your behalf using some defaults. More details on this annotation are provided in the following sections.

- Execution of the SpringApplication.run(<config>, <args>)
  method. This will boot your application and, together with the
  @SpringBootApplication annotation, will wire everything up for
  your application. The first parameter is the Configuration class that
  provides any other configuration, and the subsequent parameter
  or parameters (if any) are the arguments (if any) passed in the
  command line when running the application. The following is an
  example:

```
@SpringBootApplication
public class DemoApplication {

  public static void main(String[] args) {
    SpringApplication.run(DemoApplication.class, args);
  }

}
```

Of course, you normally don't need do this manually, because you have the Spring
Initializr to do it for you. As you saw in Chapter 1, this web tool enables you to create
the project quickly by choosing what you need, including the dependency management
tool, the dependencies (spring-boot-starter-<tech>), the structure, and the main
class with the annotation and the execution of the application; you can create and run
the project in a matter of minutes, and it works!

Project: My Retro App

Before going into the internals of Spring Boot, let me introduce our main project: My
Retro App. This project and the Users App project are referenced throughout the book.
In every chapter, we are going to add to both projects features that correspond to the
chapter topic, so that by the end of the book, we will have a complete solution.

The My Retro App project is all about retrospection: review past events, share
thoughts, synthesize group ideas, discuss what is important, commit and take any action
to a better process in the future. This simple app has three primary components: Retro
Board, Card, and Card Type. The following are the main requirements and features of
this project:

- Retro Board, which contains the following:

  - UID, a unique identification

  - Name (the name of the Retro Board)

  - Multiple cards grouped by the card type (Happy, Meh, and Sad)

- Card, which includes a comment box in which users can express their ideas based on the Card Type

- Card Type:

  - *Happy card*: All the positive thoughts, actions, or events

  - *Meh card*: Ideas that are questionable or thoughts that probably don't mean anything or don't need much attention

  - *Sad card*: Ideas that the user does not agree with, or any event that was bad or should not be happening

- Expose a web API and a UI

- UI requirements:

  - Login/Logout option.

  - Admin and User roles.

  - Admin can manage Users and Multiple Retro boards.

  - A Retro Board that displays three columns (HAPPY, MEH, SAD) in which users can add cards.

  - A User is assigned to a Retro Board.

  - A User can add any Cards in the right column.

Figure 2-1 shows the basic classes we are going to use.

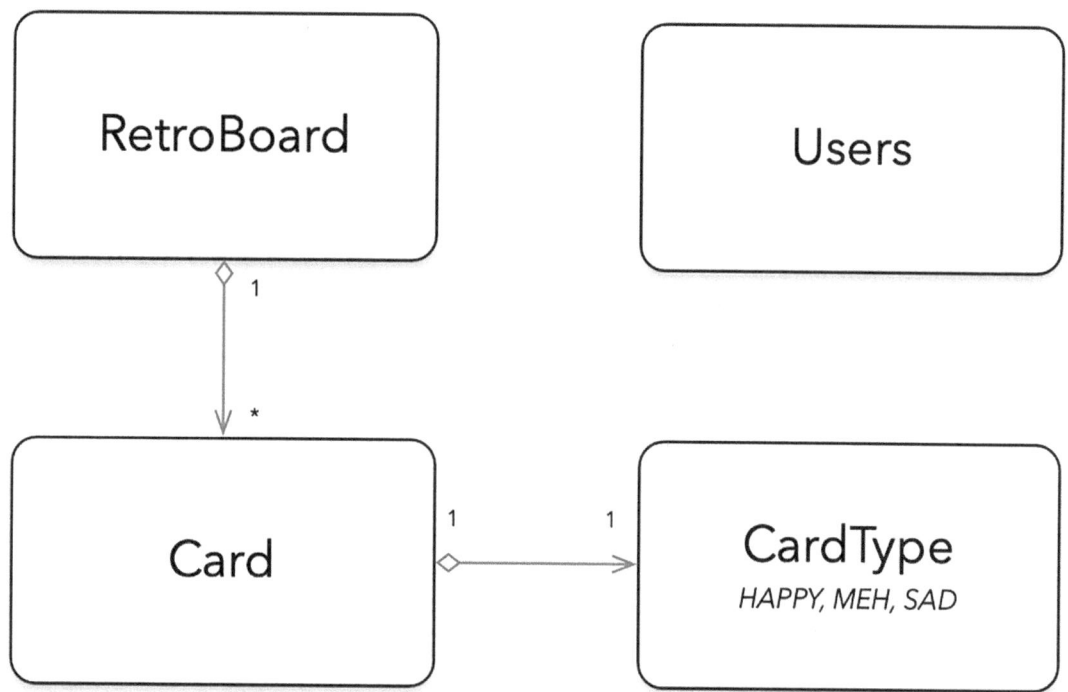

Figure 2-1. *Basic classes for My Retro App project*

Figure 2-2 shows the UI of the My Retro App that we'll be building throughout this book.

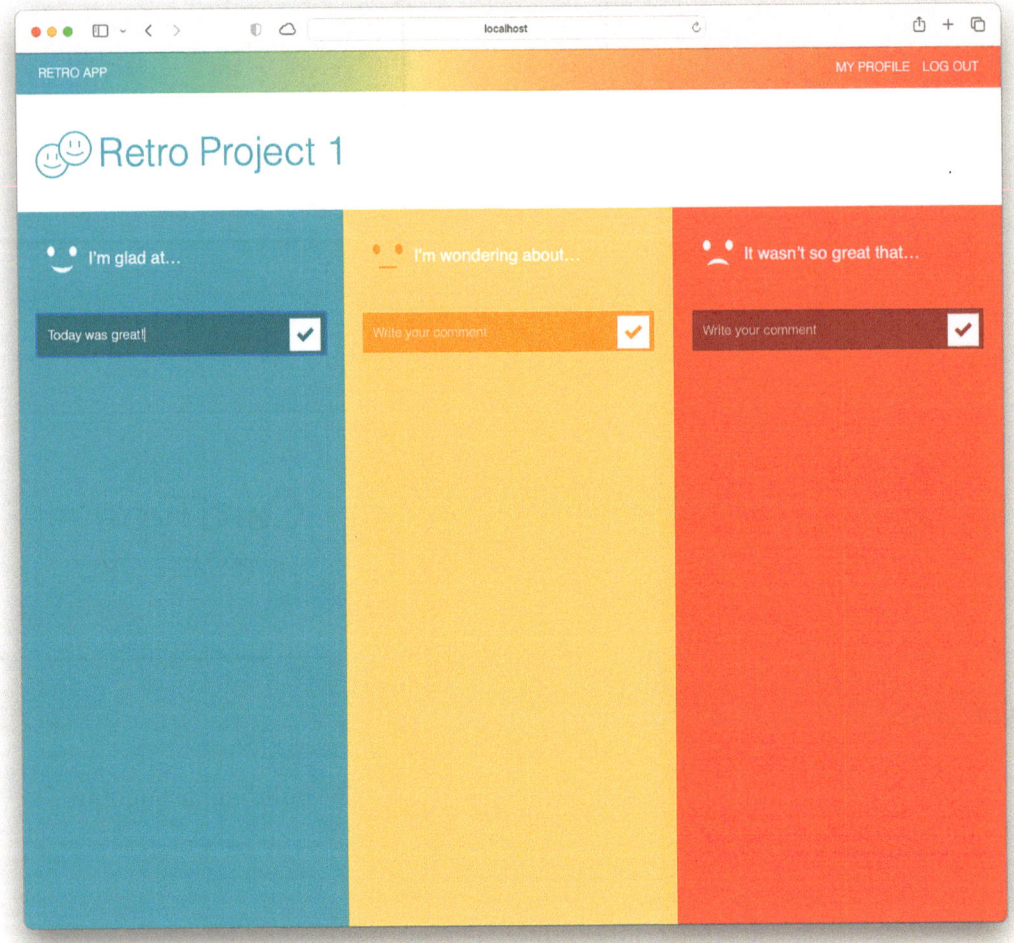

Figure 2-2. *The final My Retro App project UI*

To start this project with the Spring Initializr, open a browser and go to `https://start.spring.io`. Use the following configuration, as shown in Figure 2-3:

- Project: Gradle – Groovy

- Language: Java

- Spring Boot: 3.2.3

- Project Metadata:

  - Group: `com.apress`

  - Artifact: `myretro`

- Name: `myretro`

- Packaging: Jar

- Java: 17

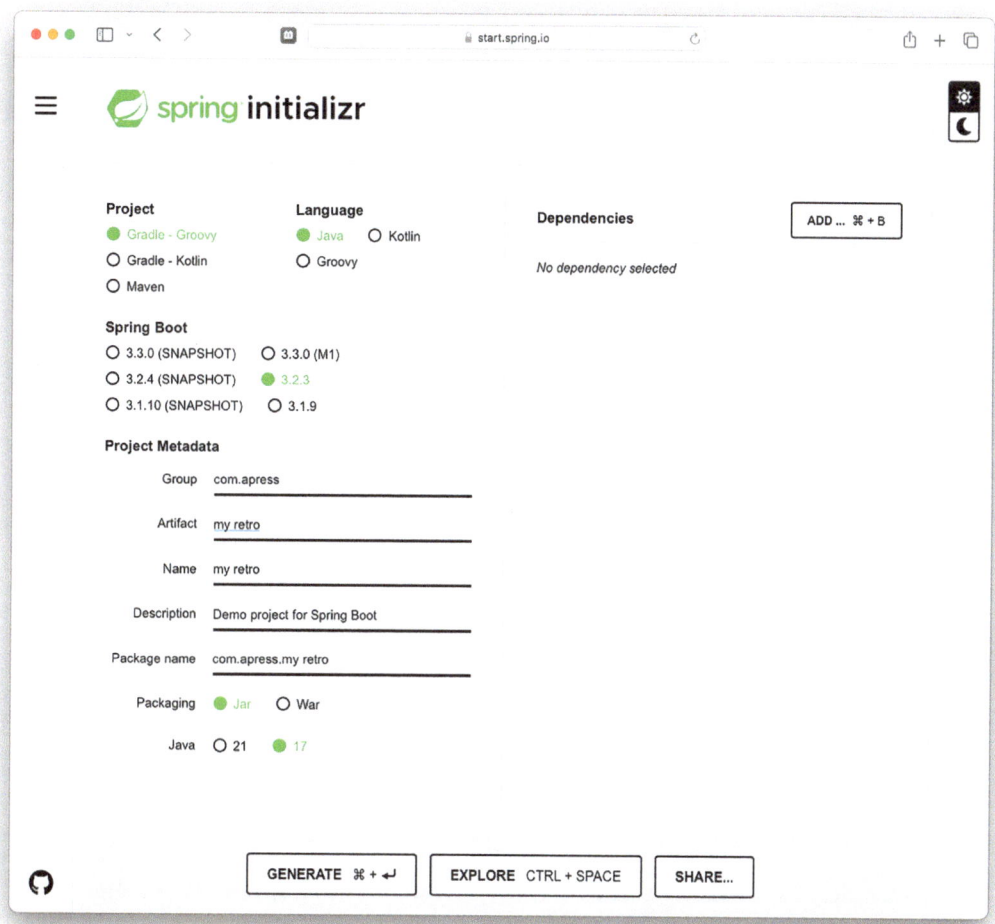

Figure 2-3. *Spring Initializr: My Retro App project configuration*

As you can see in Figure 2-3, we don't have any dependencies selected yet. We are going to be adding the features and technologies in the following sections and chapters. Click Generate to download the `myretro.zip` file, and then import it into your favorite IDE. Figure 2-4 shows the structure that was generated by the Spring Initializr, discussed next.

Figure 2-4. *My Retro App project structure*

The Gradle dependency management tool comes with the following:

- gradlew (Gradle Wrapper, a command-line tool that allows you to interact with your dependency management tool).

- The build.gradle file, in which we declare all the dependencies that we are going to need.

- settings.gradle, to add more configuration to it.

- The .gradle/ folder, which comes with some JARs already (meaning that we don't need to install Gradle; this is a light version that downloads all the necessary files).

Open the build.gradle file and analyze what's in it. See Listing 2-1.

Listing 2-1. build.gradle

```
plugins {
    id 'java'
    id 'org.springframework.boot' version '3.2.3'
    id 'io.spring.dependency-management' version '1.1.4'
}

group = 'com.apress'
version = '0.0.1-SNAPSHOT'
sourceCompatibility = '17'

repositories {
    mavenCentral()
    maven { url 'https://repo.spring.io/milestone' }
    maven { url 'https://repo.spring.io/snapshot' }
}

dependencies {
    implementation 'org.springframework.boot:spring-boot-starter'
    testImplementation 'org.springframework.boot:spring-boot-starter-test'
}

tasks.named('test') {
    useJUnitPlatform()
}
```

As described in Chapter 1, in build.gradle we declare the plugins, app information (group, version, and compatibility), where to search (repository), and the dependencies to download. Looking at the dependencies section in Listing 2-1, it declares the starters spring-boot-starter and spring-boot-starter-test that will bring all related dependencies for Spring Framework, Spring Boot, Testing, and other useful libraries. Practically we are all set, we are ready for the next part. Let's analyze the main application.

Auto-Configuration

One of the main features of Spring Boot is auto-configuration. Recall from Chapter 1 that if you want to create a Spring app using just the Spring Framework, you need to supply not only your classes but also how you want them to interact, the dependencies, and some extra configuration depending on the type of application.

Spring Boot helps with all the configuration via auto-configuration. Spring Boot favors Java-based configuration, which means that it looks for any class that is marked with the @Configuration annotation and any dependencies that are in your application (in your classpath). The @Configuration annotation triggers (during the life cycle of the Spring Boot app) logic to find any @Bean (creating the famous Spring beans) declaration that can help the Spring Boot app to know what you need to use your application.

Within the My Retro App, open the MyretroApplication.java class, shown in Listing 2-2.

Listing 2-2. src/main/com/apress/myretro/MyretroApplication.java

```
package com.apress.myretro;

import org.springframework.boot.SpringApplication;
import org.springframework.boot.autoconfigure.SpringBootApplication;

@SpringBootApplication
public class MyretroApplication {

    public static void main(String[] args) {
        SpringApplication.run(MyretroApplication.class, args);
    }

}
```

When the MyretroApplication.java class is executed by the SpringApplication. run call, it triggers the auto-configuration feature. This call needs the configuration class that has any @Configuration annotation and the arguments (which are passed in the command line when executing the program). Now, you may wonder why we are passing the MyretroApplication.class if this class doesn't have such @Configuration annotation, right?

Behind the scenes, the @SpringBootApplication interface annotation is declared as shown in Listing 2-3.

Listing 2-3.   org.springframework.boot.autoconfigure.
SpringBootApplication.java

```
@Target(ElementType.TYPE)
@Retention(RetentionPolicy.RUNTIME)
@Documented
@Inherited
@SpringBootConfiguration
@EnableAutoConfiguration
@ComponentScan(excludeFilters = { @Filter(type = FilterType.CUSTOM,
classes = TypeExcludeFilter.class),
        @Filter(type = FilterType.CUSTOM, classes =
AutoConfigurationExcludeFilter.class) })
public @interface SpringBootApplication {
  //... some other declarations
}
```

The important parts of Listing 2-3 are as follows:

- @SpringBootConfiguration: This annotation is just an alias
  for the @Configuration annotation that will search for any
  @Bean (Spring bean) declarations that you have in your class. In
  this case, we don't have any declarations yet. Listing 2-4 shows
  what the SpringBootConfiguration annotation look like.

Listing 2-4.   org.springframework.boot.SpringBootConfiguration.java

```
@Target(ElementType.TYPE)
@Retention(RetentionPolicy.RUNTIME)
@Documented
@Configuration
@Indexed
public @interface SpringBootConfiguration {
    // ... some other declarations
}
```

Practically, you can create your own custom annotation and
extend its functionality.

- @EnableAutoConfiguration: This annotation triggers even more functionality because it reads the file META-INF/spring/org. springframework.boot.autoconfigure.AutoConfiguration. imports and executes every class that has the AutoConfiguration, name at the end of the class name (this is just a naming convention) to identify any defaults and set everything up so that you don't have to. This file is part of the spring-boot-starter and belongs to the spring-boot-autoconfigure.jar dependency. See Listing 2-5.

Listing 2-5. META-INF/spring/org.springframework.boot.autoconfigure. AutoConfiguration.imports

```
//...
org.springframework.boot.autoconfigure.data.jdbc.
JdbcRepositoriesAutoConfiguration
org.springframework.boot.autoconfigure.data.jpa.JpaRepositoriesAutoConfiguration
org.springframework.boot.autoconfigure.data.ldap.
LdapRepositoriesAutoConfiguration
org.springframework.boot.autoconfigure.data.mongo.MongoDataAutoConfiguration
org.springframework.boot.autoconfigure.data.mongo.
MongoReactiveDataAutoConfiguration
org.springframework.boot.autoconfigure.data.mongo.
MongoReactiveRepositoriesAutoConfiguration
org.springframework.boot.autoconfigure.data.mongo.
MongoRepositoriesAutoConfiguration
org.springframework.boot.autoconfigure.data.neo4j.Neo4jDataAutoConfiguration
org.springframework.boot.autoconfigure.data.neo4j.
Neo4jReactiveDataAutoConfiguration
org.springframework.boot.autoconfigure.data.neo4j.
Neo4jReactiveRepositoriesAutoConfiguration
org.springframework.boot.autoconfigure.data.neo4j.
Neo4jRepositoriesAutoConfiguration
org.springframework.boot.autoconfigure.data.r2dbc.R2dbcDataAutoConfiguration
org
//...
```

Around 146 `AutoConfiguration` classes that inspect what dependencies you have declared either in your classes or as dependencies, and then it will auto-configure your app so that you don't have to. For example, if you add the Data JPA and two drivers, H2 and PostgreSQL, the Spring Boot auto-configuration mechanism will go to the JDBC and JPA `AutoConfiguration` classes and auto-configure your transaction manager, manager factory, and the `DataSource`. Now, the `DataSource` requires some extra settings, including the URL of the database, the database name, the password, and the dialect, so how will it be configured if you have two drivers declared? That's when the defaults (opinions) kick in. If you don't have any properties that set all the values previously mentioned, it will default to use the H2 that it can be set as in-memory database, with some default values such as username: sa and password (empty) and the url: jdbc:h2:mem:<guid-database-name>. This means that your application will work out-of-the-box without doing any configuration at all. Amazing!

- `@ComponentScan`: This annotation will help to search for specific package base, (if you have multiple package base name) or any other external library that require to be recognized as Spring bean (looking for @Bean declarations). For example, if you have a package with the path `com.mycompany` and also `com.other.company`, you can use the parameter `base Packages` and set `basePackages={ "com.mycompany", "com.other.company"}`. There are some best practices for this, but we are going about this later.

@EnableAutoConfiguration, @Enable<Technology>, and @Conditional* Annotations

Spring Boot is highly customizable, meaning that you can disable some of the `AutoConfiguration` classes. This allows you to have even more control over how you configure your application. Although the process of `AutoConfiguration` is fast (it doesn't affect performance at all because it occurs when your application is booting up), sometimes you don't need some of the defaults.

@EnableAutoConfiguration

As you already know, the @SpringBootApplication annotation inherits from the @EnableAutoConfiguration annotation, which has two parameters that can be used to disable some of the defaults, as shown in Listing 2-6.

Listing 2-6. org.springframework.boot.autoconfigure. EnableAutoConfiguration.java

```
@Target(ElementType.TYPE)
@Retention(RetentionPolicy.RUNTIME)
@Documented
@Inherited
@AutoConfigurationPackage
@Import(AutoConfigurationImportSelector.class)
public @interface EnableAutoConfiguration {

    String ENABLED_OVERRIDE_PROPERTY = "spring.boot.
enableautoconfiguration";

    Class<?>[] exclude() default {};

    String[] excludeName() default {};

}
```

As described in the previous section, the @EnableAutoConfiguration annotation triggers AutoConfiguration classes to configure your application based on the dependencies that your application uses. This annotation also brings the exclude and the excludeName parameters that can be used to avoid any other defaults (AutoConfiguration).

For example, if you have a Spring Boot Web MVC application and you want to do the configuration manually, you can add the following to the @SpringBootApplication annotation:

```
@SpringBootApplication(exclude = {WebMvcAutoConfiguration.class})
```

@Enable<Technology>

If you are working only with the Spring Framework, you can use the auto-configuration feature for your apps. Every Spring technology has an @Enable<Technology> annotation that auto-configures some class so that you don't have to. For example, if you have the spring-rabbit dependency, this library brings the @EnableRabbit annotation that will trigger auto-configuration such as adding the default message converters, adding a simple listener, and so forth. And, of course, if you are using Spring Boot with the spring-boot-starter-amqp, you don't need to use the @EnableRabbit annotation; the AutoConfiguration will do that for you by configuring all the beans necessary to connect to Rabbit using the defaults (host, port, username, and password).

@Conditional*

In almost every AutoConfiguration class, you will find the @Conditional* annotation that helps to configure the necessary beans for the application. For example, the WebMvcAutoConfiguration class has @ConditionalOnClass({ Servlet.class, DispatcherServlet.class, WebMvcConfigurer.class }), which means that the Web MVC auto-configuration logic will execute only if these classes are being used (in the classpath) in your application (My Retro App project). In the My Retro App project, we are not using any web dependency, so the WebMvcAutoConfiguration will fail the @ConditionalOnWebApplication and the @ConditionalOnClass because there is no Servlet.class, or DispatcherServlet.class, or any other dependency declared in the build. gradle file. Is there any way to see how this? Yes, we can run our application just like it is with the following parameter: --debug.

If you are using the IntelliJ, you can run the application and add the --debug parameter as "Program arguments". See Figure 2-5.

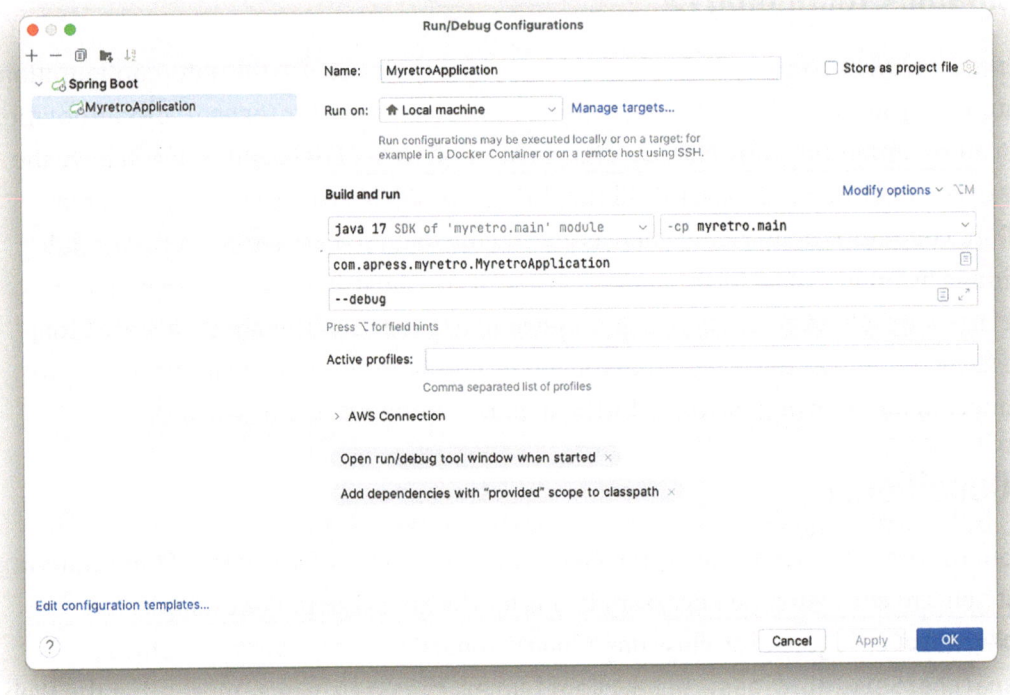

Figure 2-5. *IntelliJ IDEA configuration to pass the --debug argument to the launch process*

If you are using VS Code, you need to add the key `"args"`: "--debug" to the `launch.json` file. See Figure 2-6.

```
{} launch.json  ×

.vscode > {} launch.json > ...
   1    {
   2        "version": "0.2.0",
   3        "configurations": [
   4            {
   5                "type": "java",
   6                "name": "MyretroApplication",
   7                "request": "launch",
   8                "mainClass": "com.apress.myretro.MyretroApplication",
   9                "projectName": "myretro",
  10                "args": "--debug"
  11            }
  12        ]
  13    }
```

Figure 2-6. *VS Code configuration to pass the --debug argument to the launch process*

Alternatively, you can run the application using the command line. In the root of the project, execute the following command:

```
./gradlew bootRun -args="--debug"
```

Whichever way you choose to run the application, you will see the same output:

```
...
...
WebMvcAutoConfiguration:
    Did not match:
        - @ConditionalOnClass did not find required class 'jakarta.
        servlet.Servlet' (OnClassCondition)
...
...
```

In this case, WebMvcAutoConfiguration did not match the criteria; it the AutoConfiguration class did not find the Servlet.class by using the @ConditionalOnClass annotation. Additional @Conditional* annotations will be covered in other chapters, but for now you simply need to know that this annotation is used to set or not set some of the opinionated defaults for your Spring Boot apps.

Now, if you add @SpringBootApplication(exclude = {WebMvcAutoConfiguration.class}) to the application and rerun it, you will see something different:

```
...
Exclusions:
-----------
    org.springframework.boot.autoconfigure.web.servlet.
WebMvcAutoConfiguration
...
```

The Exclusions section shows that the WebMvcAutoConfiguration class was not executed at all.

So, auto-configuration involves no magic after all! Spring Boot uses the power of auto-configuration to create everything you need for your application, so you don't have to, making this a faster way to create amazing apps with ease and with little effort.

Spring Boot Features

Now, let's explore some of the Spring Boot features by using our My Retro App project. If you look at the main class, you will find the SpringApplication.run method. This can be instantiated in a different way to take advantage of other features; take a look at Listing 2-7.

Listing 2-7. src/main/java/com/apress/myretro/MyretroApplication.java

```java
package com.apress.myretro;

import org.springframework.boot.SpringApplication;
import org.springframework.boot.autoconfigure.SpringBootApplication;

@SpringBootApplication
public class MyretroApplication {

    public static void main(String[] args) {
        SpringApplication sa = new SpringApplication(MyretroApplication.class);
                // Spring Application features ...
                sa.run(args);
    }
}
```

Listing 2-7 shows a small modification from Listing 2-2. Here, we are not using the static `run(...args)` method call. If you view the options the `SpringApplication` class offers (using the code-completion feature of your IDE by typing the instance variable `sa.`), you will see a list similar to the (partial) list shown in Figure 2-7.

Figure 2-7. *SpringApplication class*

Figure 2-7 shows you the code completion for the `SpringApplication` instance variable. The following sections introduce some of the most common features you can add to your Spring Boot application.

Custom Banner

When you start your application, you will see a banner displayed. The default is the ASCII art of Spring Boot. See Figure 2-8.

```
  .   ____          _            __ _ _
 /\\ / ___'_ __ _ _(_)_ __  __ _ \ \ \ \
( ( )\___ | '_ | '_| | '_ \/ _` | \ \ \ \
 \\/  ___)| |_)| | | | | || (_| |  ) ) ) )
  '  |____| .__|_| |_|_| |_\__, | / / / /
 =========|_|==============|___/=/_/_/_/
 :: Spring Boot ::        (v3.1.0         )

2023-04-26T07:29:33.556-04:00  INFO 27679 --- [          main] com.apress.myretro.MyretroApplication       :
2023-04-26T07:29:33.558-04:00  INFO 27679 --- [          main] com.apress.myretro.MyretroApplication       :
2023-04-26T07:29:33.880-04:00  WARN 27679 --- [          main] ocalVariableTableParameterNameDiscoverer    :
2023-04-26T07:29:33.892-04:00  INFO 27679 --- [          main] com.apress.myretro.MyretroApplication       :

BUILD SUCCESSFUL in 1s
3 actionable tasks: 1 executed, 2 up-to-date
```

Figure 2-8. *Spring Boot banner*

Within the `SpringApplication` class, you can add a custom banner in different ways. You can do it programmatically, as shown in Listing 2-8.

Listing 2-8. src/main/java/com/apress/myretro/MyretroApplication.java

```java
package com.apress.myretro;

import org.springframework.boot.Banner;
import org.springframework.boot.SpringApplication;
import org.springframework.boot.autoconfigure.SpringBootApplication;
import org.springframework.core.env.Environment;

import java.io.PrintStream;

@SpringBootApplication
public class MyretroApplication {

    public static void main(String[] args) {
```

```
SpringApplication sa = new SpringApplication(MyretroApplicati
on.class);
sa.setBanner(new Banner() {
            @Override
            public void printBanner(Environment environment,
            Class<?> sourceClass, PrintStream out) {
                out.println("\n\n\tThis is my custom Banner!\n\n".
                toUpperCase());
        }
    });
    sa.run(args);
    }
}
```

Listing 2-8 shows the implementation of a custom banner by using the setBanner(Banner) method call. If you run this program, you will see something like Figure 2-9.

```
THIS IS MY CUSTOM BANNER!


2023-04-26T07:40:14.949-04:00  INFO 30030 --- [        main] com.apress.myretro.MyretroApplication          :
2023-04-26T07:40:14.950-04:00  INFO 30030 --- [        main] com.apress.myretro.MyretroApplication          :
2023-04-26T07:40:15.283-04:00  WARN 30030 --- [        main] ocalVariableTableParameterNameDiscoverer       :
2023-04-26T07:40:15.295-04:00  INFO 30030 --- [        main] com.apress.myretro.MyretroApplication          :

BUILD SUCCESSFUL in 1s
```

Figure 2-9. *Custom banner*

Another way that you can create a custom banner is by using ASCII art. There are websites that can generate this for you. For example, go to https://www.patorjk.com, scroll down the page, click the Text to ASCII Art Generator link, and type My Retro. Figure 2-10 shows the results of selecting different fonts for the text.

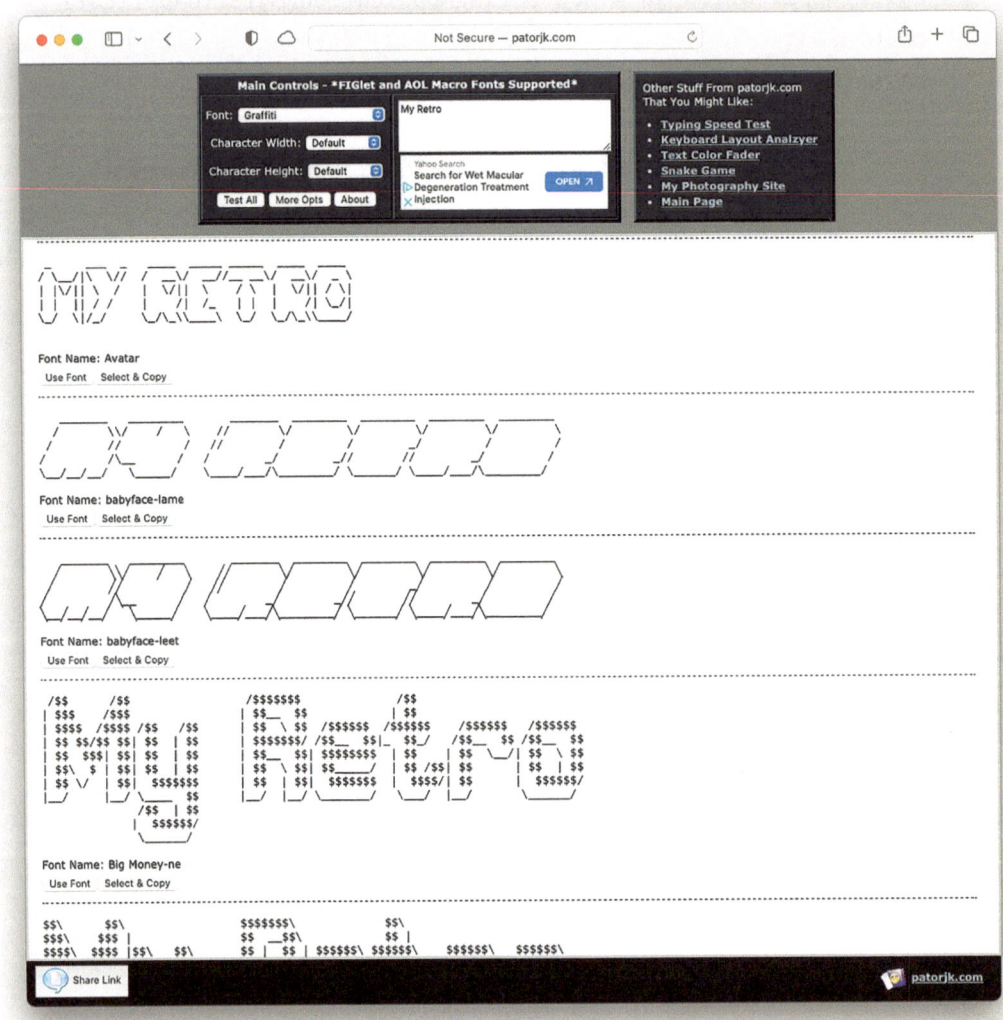

Figure 2-10. *Example Text to ASCII Art Generator output (`https://www.patorjk.com`)*

You can browse, tryout various fonts, and select the one you like. Click the Select & Copy button. Then you need to create in the `src/main/resources` directory a `banner.txt` file and paste the ASCII art. Figure 2-11 shows the `banner.txt` file and the ASCII art.

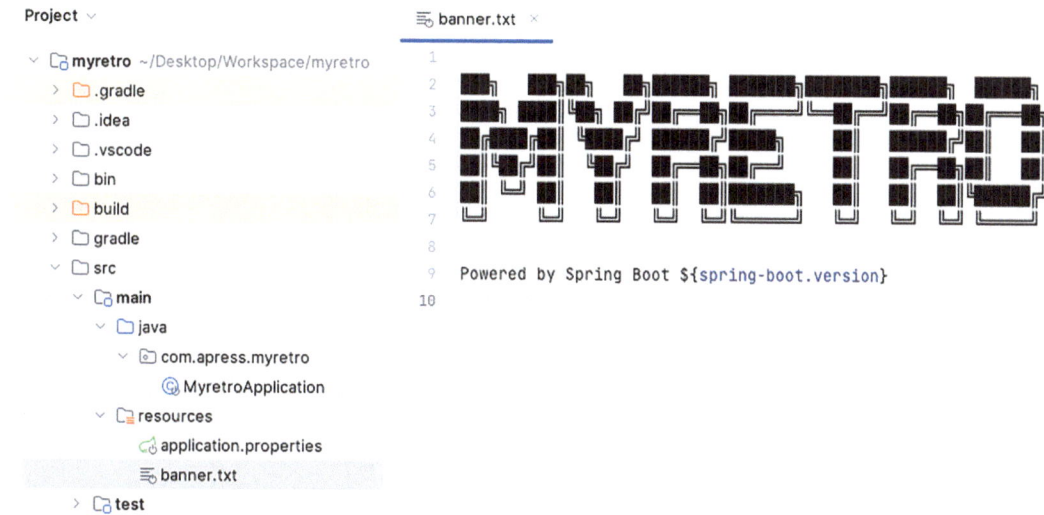

Figure 2-11. *src/main/resources/banner.txt*

Also note in Figure 2-11 that we added a caption ${spring-boot.version} in the My Retro App. This is one of the global properties that are set at runtime, and it will display the Spring Boot version used. If you run the My Retro App, you should see the same result as shown in Figure 2-12.

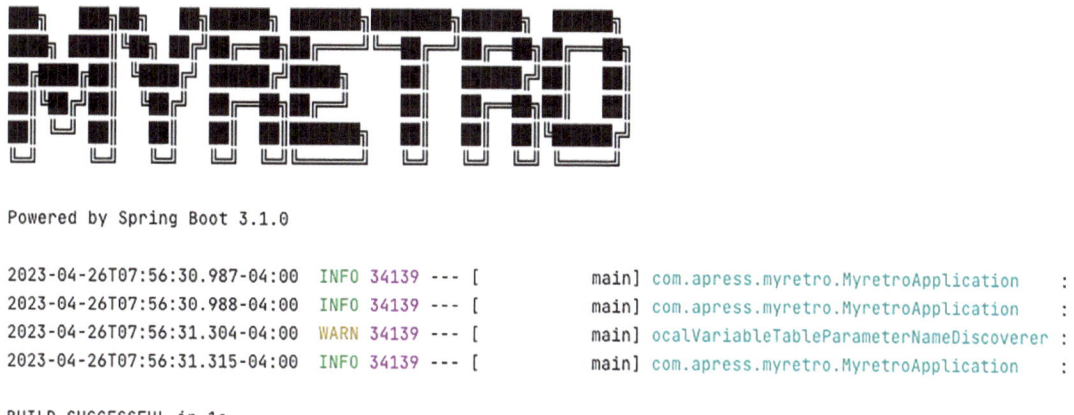

Figure 2-12. *Custom banner*

By default, Spring Boot locates the banner.txt file in the resources folder, but you can specify another location or extension. For example, you can create the src/main/resources/META-INF folder and add the banner.txt file. See Figure 2-13.

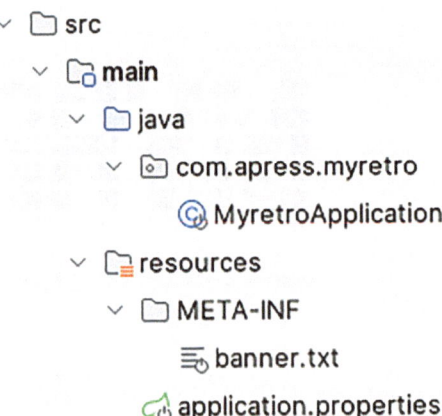

Figure 2-13. *src/main/resources/META-INF/banner.txt*

To make this work, you need to use the `spring.banner.location` property. If you want to use your IDE, you need to figure out how to override Spring Boot configuration properties. If you are using IntelliJ IDEA, choose Edit Configurations ➤ Modify Options, locate the Override Configuration Properties section, and add `spring.banner.location` and `classpath:/META-INF/banner.txt` as shown in Figure 2-14.

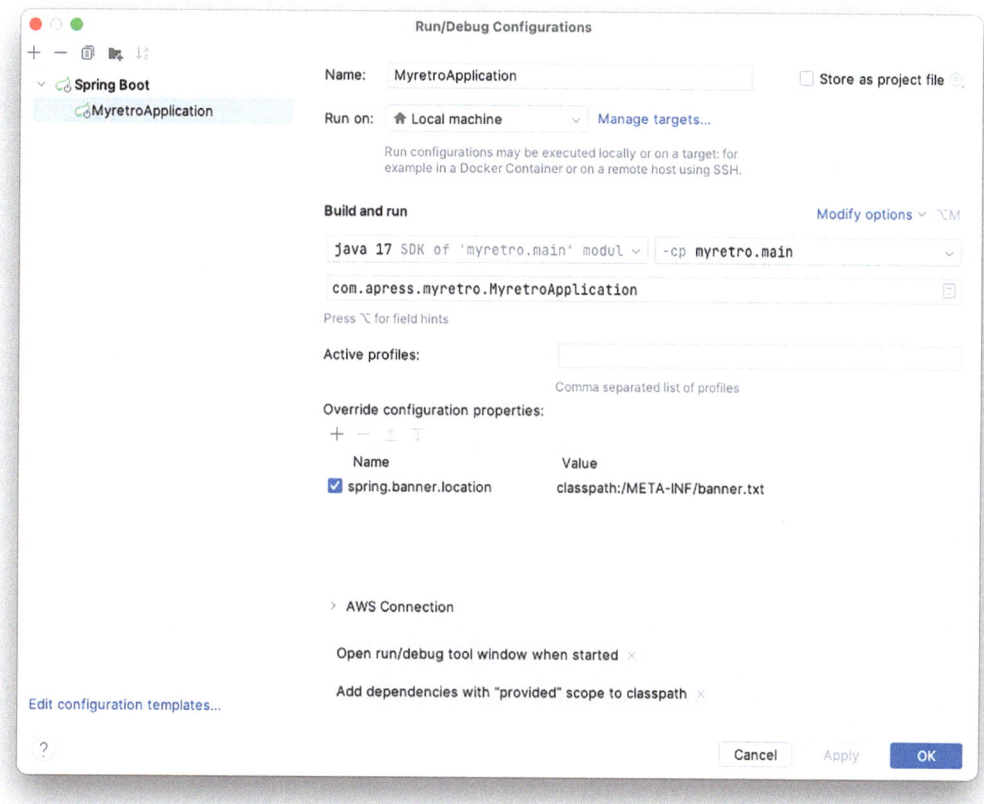

Figure 2-14. *Overriding configuration properties in IntelliJ IDEA*

If you are using VS Code, you need to modify `launch.json` and add the env key with the values `spring.banner.location` and `classpath:/META-INF/banner.txt`. See Figure 2-15.

```
{} launch.json  ×

.vscode > {} launch.json > ...
   1  ∨ {
   2          "version": "0.2.0",
   3  ∨       "configurations": [
   4  ∨          {
   5                  "type": "java",
   6                  "name": "MyretroApplication",
   7                  "request": "launch",
   8                  "mainClass": "com.apress.myretro.MyretroApplication",
   9                  "projectName": "myretro",
  10  ∨               "env": {
  11                      "spring.banner.location":"classpath:/META-INF/banner.txt"
  12                  }
  13              }
  14          ]
  15  }
```

Figure 2-15. *Modifying launch.json in VS Code*

And if you like to run your application via the command line, you need to add at the end of the build.gradle file the following snippet:

```
bootRun {
    systemProperties = System.properties
}
```

Then you can execute the following command and expect the same result as shown earlier in Figure 2-12:

```
./gradlew bootRun -Dspring.banner.location=classpath:/META-INF/banner.txt
```

As you can see, there are many ways to configure these features. And if you don't need the banner at all, you can use the setBanner(Mode) method call as shown in Listing 2-9.

Listing 2-9. src/main/java/com/apress/myretro/MyretroApplication.java

```
package com.apress.myretro;

import org.springframework.boot.Banner;
import org.springframework.boot.SpringApplication;
```

```
import org.springframework.boot.autoconfigure.SpringBootApplication;
import org.springframework.core.env.Environment;

import java.io.PrintStream;

@SpringBootApplication
public class MyretroApplication {

    public static void main(String[] args) {
        SpringApplication sa = new SpringApplication(MyretroApplicati
        on.class);
        sa.setBannerMode(Banner.Mode.OFF);
        sa.run(args);
    }

}
```

SpringApplicationBuilder

Spring Boot offers a Fluent Builder API that allows you to configure your application. The API brings the ApplicationContext (from the Spring Framework), making your app even more customizable. Listing 2-10 shows the SpringApplicationBuilder class, in which you can customize the startup.

Listing 2-10. src/main/java/com/apress/myretro/MyretroApplication.java

```
package com.apress.myretro;

import org.slf4j.Logger;
import org.slf4j.LoggerFactory;
import org.springframework.boot.Banner;
import org.springframework.boot.SpringApplication;
import org.springframework.boot.WebApplicationType;
import org.springframework.boot.autoconfigure.SpringBootApplication;
import org.springframework.boot.builder.SpringApplicationBuilder;
import org.springframework.context.ApplicationEvent;
import org.springframework.context.ApplicationListener;
import org.springframework.core.env.ConfigurableEnvironment;
import org.springframework.core.env.Environment;
```

```
import java.io.PrintStream;

@SpringBootApplication
public class MyretroApplication {

    static Logger  log = LoggerFactory.getLogger(MyretroApplication.class);

    public static void main(String[] args) {
        new SpringApplicationBuilder()
                    .sources(MyretroApplication.class)
                    .logStartupInfo(false)
                    .bannerMode(Banner.Mode.OFF)
                    .lazyInitialization(true)
                    .web(WebApplicationType.NONE)
                    .profiles("cloud")
                    .listeners(event -> log.info("Event: {}",event.
                    getClass().getCanonicalName()))
                    .run(args);
    }
}
```

Let's analyze the code shown in Listing 2-10:

- `.sources(Class<?>...)`: This method is where you add all the necessary configuration classes and components of your application. Remember, these classes need to be marked as `@Configuration` or as one of the other Spring-related markers such as `@Component`, `@Service`, `@Repository`, etc.

- `.logStartup(boolean)`: This method prints out to console all the logging information about the startup. It accepts a `boolean`, so if you set it to `false`, you won't see anything (or only your own logging, if any).

- `.bannerMode(Mode)`: This method accepts a `Mode` for showing or not showing the banner. The possible values are `Banner.Mode.OFF` (disable printing of the banner), `Banner.Mode.CONSOLE` (print the banner to System.out), and `Banner.Mode.LOG` (print the banner to the log file).

- `.lazyInitialization(boolean)`: This method accepts a `boolean`, which by default is set to `false`, but if set to `true`, then the bean creation will not happen until the bean is needed.

- `.web(WebApplicationType)`: This method defines the type of web application. The possible values are `WebApplicationType.NONE` (instructs that the application should not run as a web application and should not start an embedded web server), `WebApplicationType.SERVLET` (means that the application should run as a servlet-based web app and start an embedded servlet web server; this will be true if you have the `spring-boot-starter-web` starter as a dependency/classpath), and `WebApplicationType.REACTIVE` (means that the application should run as a reactive web application and should start an embedded reactive web server, and you need the `spring-boot-starter-webflux` starter as a dependency/classpath).

- `.profiles(String...)`: This method lists Spring profiles that you can use. When running the application, you should see the output `The following 1 profile is active: "cloud"`. (The final section of this chapter discusses application profiles in detail.)

- `.listeners(ApplicationListener<?>...)`: This method enables you to define Spring Events. In this case (in My Retro App) we are listening to all implementations of the `ApplicationListener` interface, so if you run the application, you should have output like the following:

```
Event: o.s.boot.context.event.ApplicationPreparedEvent
Event: o.s.context.event.ContextRefreshedEvent
Event: o.s.boot.context.event.ApplicationStartedEvent
Event: o.s.boot.availability.AvailabilityChangeEvent
Event: o.s.boot.context.event.ApplicationReadyEvent
Event: o.s.boot.availability.AvailabilityChangeEvent
Event: o.s.context.event.ContextClosedEvent
```

- `run(String...)`: This method enables you to pass any argument, as you saw earlier in this chapter. In fact, you can add the following to the `args` if you are still using the `META-INF/banner.txt` and the banner mode is set to `Banner.Mode.Console`:

```
.run("--spring.banner.location=classpath:/META-INF/banner.txt");
```

Application Arguments

In any Java application, you can pass useful arguments to your main class, and Spring Boot will help you access those arguments by providing the `ApplicationArguments` interface. Let's see how to use it.

Create a new class named `MyRetroConfiguration` under the `myretro` package. See Listing 2-11.

Listing 2-11. src/main/java/com/apress/myretro/MyRetroConfiguration.java

```
package com.apress.myretro;

import org.slf4j.Logger;
import org.slf4j.LoggerFactory;
import org.springframework.boot.ApplicationArguments;
import org.springframework.context.annotation.Configuration;

@Configuration
public class MyRetroConfiguration {
    Logger log = LoggerFactory.getLogger(MyretroApplication.class);

    MyRetroConfiguration(ApplicationArguments arguments){
log.info("Option Args: {}", arguments.getOptionNames());
            log.info("Option Arg Values: {}", arguments.
            getOptionValues("option"));
            log.info("Non Option: {}",arguments.getNonOptionArgs());
    }
}
```

In the MyRetroConfiguration class, we are going to use and pass the following arguments:

```
--enable --remote --option=value1 --option=value2 update upgrade
```

You can use your IDE (IntelliJ IDEA or VS Code) to add the preceding arguments (as you already learned). If you are using the command line, you can use

```
./gradlew bootRun --args="--enable --remote --option=value1 --option=value2 update upgrade"
```

When you execute the application, the logs of the MyRetroConfiguration class should be printed out:

```
Option Args: [enable, remote, option]
Option Arg Values: [value1, value2]
Option: [update, upgrade]
```

Let's analyze the code in Listing 2-11 and the result:

- @Configuration: Auto-configuration will recognize this annotation, evaluate it, and identify (in this case) that it has a constructor that uses the ApplicationArguments interface as a parameter. This will be injected automatically by the Spring app life cycle; and in this case, it will pass the args declared by the .run(args) method call in the main class. The ApplicationArguments interface has different methods that can help to identify all the arguments if there are options with values or not (denoted by -- and the = with their values) and just parameters known as non-option arguments.

- getOptionNames(). This method will get all the options denoted by --<argument name>. In this example, they are enable, remote, and option.

- getOptionValues(String). This method accepts the argument name that has one or more values, denoted by --<argument-name>=<value>. In this example we are looking for the argument named option and its values, and the result is value1 and value2 (because it repeats more than once).

- getNonOptionArgs(). This method gets all the arguments that are not denoted by -- (that is, only regular arguments). In this example, they are update and upgrade.

Executable JAR

This feature is more related to the Spring Boot plugin of whatever dependency management tool you chose (e.g., Maven or Gradle). Again, this book assumes the use of Gradle. To create an executable JAR in the build/libs directory, run the following command:

```
./gradlew build
```

Then, you can run it with

```
java -jar build/libs/myretro-0.0.1-SNAPSHOT.jar
```

If you want to pass some parameters/arguments to your JAR, you can execute the following command:

```
java -jar build/libs/myretro-0.0.1-SNAPSHOT.jar --enable --remote
--option=value1 --option=value2 update upgrade
```

and you should have the same result as before.

ApplicationRunner, CommandLineRunner, and ApplicationReadyEvent

As developers, we are required to start a process or run certain logic before our application is ready to accept any request or any other interaction. Spring Boot has several interfaces and events that enable us to execute code before our application is ready. You can add the code shown in Listing 2-12 to the MyRetroConfiguration class.

Listing 2-12. src/main/java/com/apress/myretro/MyRetroConfiguration.java

```
package com.apress.myretro;

import org.slf4j.Logger;
import org.slf4j.LoggerFactory;
```

```
import org.springframework.boot.ApplicationRunner;
import org.springframework.boot.CommandLineRunner;
import org.springframework.boot.context.event.ApplicationReadyEvent;
import org.springframework.context.ApplicationListener;
import org.springframework.context.annotation.Bean;
import org.springframework.context.annotation.Configuration;

import java.util.Arrays;

@Configuration
public class MyRetroConfiguration {
    Logger log = LoggerFactory.getLogger(MyretroApplication.class);

@Bean
        CommandLineRunner commandLineRunner(){
                return args -> {
                        log.info("[CLR] Args: {}",Arrays.toString(args));
                };
        }

        @Bean
        ApplicationRunner applicationRunner(){
            return args -> {
                    log.info("[AR] Option Args: {}", args.getOptionNames());
                    log.info("[AR] Option Arg Values: {}", args.
                    getOptionValues("option"));
                    log.info("[AR] Non Option: {}",args.getNonOptionArgs());
            };
    }

    @Bean
    ApplicationListener<ApplicationReadyEvent>
    applicationReadyEventApplicationListener(){
            return event -> {
                    log.info("[AL] Im ready to interact...");
            };
    }

}
```

As you can see, we have eliminated the constructor, which is not needed because the ApplicationRunner interface brings the application arguments. Note that there are now methods that are marked with the @Bean annotation that Spring Boot auto-configuration will use to create the necessary Spring beans. Let's review the code

- CommandLineRunner: This interface implementation will be called after Spring and Spring Boot wires everything up. This is a functional interface that has a callback (run method) that accepts the arguments passed to the application. In this case the result is every single argument.

- ApplicationRunner: This interface implementation will be called before the CommandLineRunner interface implementation. This is a functional interface that has a callback (run method) that has ApplicationArguments as a parameter, making it a good candidate if we want to use the arguments.

- ApplicationListener<ApplicationReadyEvent>: This event will be called when the Spring Boot app has finished wiring everything up and is ready to interact. So, this event will be the last to be called.

If you run your application as before (with the same arguments), you should have the following output:

```
...
[AR] Option Args: [enable, remote, option]
[AR] Option Arg Values: [value1, value2]
[AR] Non Option: [update, upgrade]
[CLR] Args: [--enable, --remote, --option=value1, --option=value2, update,
upgrade]
[AL] Im ready to interact...
...
```

Note that the ApplicationRunner is called first, then the CommandLineRunner, and lastly the ApplicationReadyEvent. Now, you can have multiple CommandLineRunner classes that implement the run method, and they must be marked as @Component and, if needed, listed in order is required, the you can use the @Order annotation (with the Ordered enum as a parameter depending of the precedence).

Application Configuration

Sometimes it is necessary to have very specific access to a remote server, or have credentials to connect to a remote server, or even sensitive data that we need to store in a database. We can hard-code all this information (credentials, remote ips, sesnsitive data), but this is not a best practice because this info can change in a snap and probably, we're ending doing something bad trying to redeploy our app with bad consequences, such as connecting to the wrong server for example.

Spring Boot uses `application.properties` or `application.yaml` to externalize the configuration we need, and Spring Boot also uses these files to override some of the defaults (due to the auto-configuration and `conditionals`) in our application. You will see how to change several of these defaults throughout this book, but for now let's talk about how we can create our own properties and how to use them.

My Retro App and Users App Project Integration

To see how we can use external configuration, let's resume our journey with our two projects, My Retro App and Users App. Although we haven't yet discussed how the projects will be integrated later in this book, at some point the My Retro App will need to reach out to the Users App to authenticate and authorize users who want to use the My Retro App. So, in this case, it's necessary to have the server, port, and other useful information.

In the My Retro App, open the `application.properties` file and add the content shown in Listing 2-13.

Listing 2-13. src/main/java/resources/application.properties

```
# Users Properties
users.server=127.0.0.1
users.port=8081
users.username=admin
users.password=aW3sOm3
```

Listing 2-13 shows some of the properties we are going to need for the My Retro App project. Next, open the `MyRetroConfiguration` class and replace the content with the code shown in Listing 2-14.

Listing 2-14. src/main/java/com/apress/myretro/MyRetroConfiguration.java

```java
package com.apress.myretro;

import org.slf4j.Logger;
import org.slf4j.LoggerFactory;
import org.springframework.beans.factory.annotation.Value;
import org.springframework.boot.context.event.ApplicationReadyEvent;
import org.springframework.context.ApplicationListener;
import org.springframework.context.annotation.Bean;
import org.springframework.context.annotation.Configuration;

@Configuration
public class MyRetroConfiguration {

    Logger log = LoggerFactory.getLogger(MyRetroConfiguration.class);

    @Value("${users.server}")
    String server;

    @Value("${users.port}")
    Integer port;

    @Value("${users.username}")
    String username;

    @Value("${users.password}")
    String password;

    @Bean
    ApplicationListener<ApplicationReadyEvent> init(){
        return event -> {
            log.info("\nThe users service properties are:\n- Server: {}\n-
            Port: {}\n- Username: {}\n- Password: {}",server,port,username,
            password);
        };
    }

}
```

Listing 2-14 shows the updated MyRetroConfiguration class. Let's analyze it:

- @Value("${property-name}"): This is a new annotation that collects the value of the property specified between " and using the SpEL (Spring Expression Language) that start with $ and the property name between {}. During the Spring app life cycle (gathering information and creating the beans), it will search for any @Value annotation and attempt to look for it in the default file: application.properties or application.yaml. For example, we have declared an Integer port variable and annotated it with @Value("${users.port}"), so it will get the value 8081 and be assigned to the port variable when this class get instantiated.

- ApplicationListener: this interface will execute the method init when the ApplicationReadyEvent is fired (meaning that the app is ready to interact with other logic).

If you run the application, you should see something similar to the following output:

```
The users service properties are:
- Server: 127.0.0.1
- Port: 8081
- Username: admin
- Password: aW3s0m3
```

Now, what will happen if we have more than four properties? We could add more variables and mark them with @Value, but that can get messy quickly. Spring Boot offers a better solution in such cases, described next.

Configuration Properties

Spring Boot offers a simple solution for multiple and more complex properties. It defines a class marker @ConfigurationProperties annotation, which binds the fields of the class to the properties defined externally (in our case, to the application.properties).

Let's create a new class, MyRetroProperties, in the src/main/java/com/apress/myretro folder with the content shown in Listing 2-15.

Listing 2-15. src/main/java/com/apress/myretro/MyRetroProperties.java

```java
package com.apress.myretro;

import org.springframework.boot.context.properties.ConfigurationProperties;

@ConfigurationProperties
public class MyRetroProperties {
    Users users;

    public Users getUsers() {
        return users;
    }

    public void setUsers(Users users) {
        this.users = users;
    }
}

class Users {
    String server;
    Integer port;
    String username;
    String password;

    public String getServer() {
        return server;
    }

    public void setServer(String server) {
        this.server = server;
    }

    public Integer getPort() {
        return port;
    }

    public void setPort(Integer port) {
        this.port = port;
    }
```

```
    public String getUsername() {
        return username;
    }

    public void setUsername(String username) {
        this.username = username;
    }

    public String getPassword() {
        return password;
    }

    public void setPassword(String password) {
        this.password = password;
    }
}
```

Listing 2-15 shows the MyRetroProperties class that thanks to the @ConfigurationProperties it will be binding the properties found in application. properties (or application.yaml) to every field in this class. Notice that this class is just a POJO (plain old Java object) with setters and getters. The idea is to have the same name for every field so that they match. To tell Spring Boot that this class is a ConfigurationProperties class, we need to give a hint by either also marking the class as @Component or using the @EnableConfigurationProperties annotation in a Configuration class. Listing 2-16 shows the modified MyRetroConfiguration class.

Listing 2-16. src/main/java/com/apress/myretro/MyRetroConfiguration.java

```
package com.apress.myretro;

import org.slf4j.Logger;
import org.slf4j.LoggerFactory;
import org.springframework.boot.context.event.ApplicationReadyEvent;
import org.springframework.boot.context.properties.
EnableConfigurationProperties;
import org.springframework.context.ApplicationListener;
import org.springframework.context.annotation.Bean;
import org.springframework.context.annotation.Configuration;
```

```
@EnableConfigurationProperties({MyRetroProperties.class})
@Configuration
public class MyRetroConfiguration {

    Logger log = LoggerFactory.getLogger(MyRetroConfiguration.class);

    @Bean
    ApplicationListener<ApplicationReadyEvent> init(MyRetroProperties
    myRetroProperties){
        return event -> {
            log.info("\nThe users service properties are:\n- Server: {}\n-
            Port: {}\n- Username: {}\n- Password: {}",
                    myRetroProperties.getUsers().getServer(),
                    myRetroProperties.getUsers().getPort(),
                    myRetroProperties.getUsers().getUsername(),
                    myRetroProperties.getUsers().getPassword());
        };
    }

}
```

The modified MyRetroConfiguration class, we are using the
@EnableConfigurationProperties annotation that accepts an array of
ConfigurationProperties marked classes. Also, note that the init method now has
a parameter MyRetroProperties, this class (MyRetroConfiguration.class) will be
automatically injected by Spring (because it's marked using the @Bean annotation), In
this way you can access these properties like any other regular class with its getters.

What happens if we have multiple services (not only Users). It will be nice to identify
them as service, right? Meaning that we can add a prefix to our configuration properties
class. See the next snippet.

```
@ConfigurationProperties(prefix="service")
public class MyRetroProperties {
    Users users;

    public Users getUsers() {
        return users;
    }
```

```
    public void setUsers(Users users) {
        this.users = users;
    }
}
```

The `prefix` parameter will use the `service.*` properties binding match on what you have in your properties files. This means that our `application.properties` should be like the next snippet:

```
service.users.server=127.0.0.1
service.users.port=8081
service.users.username=admin
service.users.password=aW3sOm3
```

If you run the application, you should have the same result.

Relaxed Binding

I previously mentioned that the properties must match the name of the field in the class, right? Well, Spring Boot offers a relaxed binding approach; it provides some relaxed rules for binding `Environment` properties to classes marked with `@ConfigurationProperties`. If you have a field name that is camel case too long, you can use camel case, underscore, kebab case, or uppercase format. Consider the following example:

```
private String hostNameServer;
```

You can use the following notation:

- Camel case: `hostNameServer`

- Kebab case: `host-name-server`

- Underscore: `host_name_server`

- Uppercase: `HOST_NAME_SERVER`

Configuration Precedence

Another benefit of using externalized configuration is that you can use different variants to define your properties. You can use environment variables, Java system properties, JDNI, servlet content, config parameters, command-line parameters, and much more.

But what happens if I have defined the same properties in all these mechanisms? Well, the good part is that Spring Boot has some precedences that can be applied when running your application.

The following list shows the precedence that will take place when running your application and the binding process for your properties:

- Default properties (specified by setting `SpringApplication.setDefaultProperties`).

- `@PropertySource` annotations on your `@Configuration` classes. Note that such property sources are not added to the environment until the application context is being refreshed. This is too late to configure certain properties such as `logging.*` and `spring.main.*`, which are read before the refresh begins.

- Config data (such as `application.properties` files).

- A `RandomValuePropertySource` that has properties only in `random.*`.

- OS environment variables.

- Java System properties (`System.getProperties()`).

- JNDI attributes from `java:comp/env`.

- `ServletContext` init parameters.

- `ServletConfig` init parameters.

- Properties from `SPRING_APPLICATION_JSON` (inline `JSON` embedded in an environment variable or system property).

- Command-line arguments.

- properties attribute on your tests (available on `@SpringBootTest` and the test annotations for testing a particular slice of your application).

- `@TestPropertySource` annotations on your tests.

- Devtools global settings properties in the `$HOME/.config/spring-boot` directory when devtools is active.

Using the `application.properties`, you have the following precedence:

- Application properties packaged inside your JAR (`application.properties` and YAML variants)

- Profile-specific application properties packaged inside your JAR (`application-{profile}.properties` and YAML variants)

- Application properties outside of your packaged JAR (`application.properties` and YAML variants)

- Profile-specific application properties outside of your packaged JAR (`application-{profile}.properties` and YAML variants)

Of course, there are some many options for using external configuration, and even you can establish where to read such file with a location, using the following property/ argument: `--spring.config.location=<location>`. I think the most important point here it to look at the precedence. To sum up this section, if you have a JAR that contains the `application.properties`, and where you have run your app has another file (with the same name) it will override what you have, or any environment variable as well will be overriding the properties from the JAR app.

Changing Defaults

As you already know, Spring Boot uses defaults to configure your application, but sometimes these defaults are not necessarily what your application needs. One of the many Spring Boot features is the ability to change these defaults. You already saw that you can exclude some of the auto-configuration by using the `exclude={}` parameter in the `@SpringBootApplication` annotation. But there also is a way to specify how you want those defaults to change.

There is a significant list of application properties that you can access here: `https:// docs.spring.io/spring-boot/docs/current/reference/html/application-properties.html`. For example, if you want to override the default port, 8080, you can do so by using the property `server.port=8082`.

The following command achieves the same result:

```
java -jar build/libs/myretro-0.0.1-SNAPSHOT.jar --server.port=8082

SERVER_POST=8082  java -jar build/libs/myretro-0.0.1-SNAPSHOT.jar

java -Dspring.application.json='{"server.port":8082}' -jar build/libs/
myretro-0.0.1-SNAPSHOT.jar
```

Application Profiles

Spring Boot can handle profiles as well, meaning that you not only can use your custom properties or override defaults, but also use profiles for creating beans. And during the following chapters we will talk about Spring Profile as well.

As you saw earlier, you can create application-<profile>.properties (or YAML files). This is handy when you have multiple environments and want to keep track of each of them with different values. To use these features, it is necessary to activate the profile either by using SpringApplicationBuilder.profiles (as you saw earlier) or by using the property spring.profiles.active=<profiles>.

For example, you can create an additional application-cloud.properties file in the resources folder with other values:

```
service.users.server=cloud.server.com
service.users.port=1089
service.users.username=root
service.users.password=Sup3RaW3sOm3
```

And if you run the application by passing the property (either at the command line or in your favorite IDE)

```
--spring.profiles.active="cloud"
```

you should see only the application-cloud.properties file values.

Using a YAML format can simplify the properties by having only one file. For example, an application.yaml file can look like this:

```
service:
  users:
    server: 127.0.0.1
    port: 8081
    username: admin
    password: aW3sOm3
---
spring:
  config:
    activate:
      on-profile: cloud
```

```
service:
  users:
    server: cloud.server.com
    port: 1089
    username: root
    password: Sup3RaW3sOm3
```

One of the important declarations here is the `spring.config.activate.on-profile` property, that specify the section divided by three dash characters, this section will be only activated when the `cloud` profile is called. If there is no `on-profile`, it will be the `default` profile set.

Summary

This chapter covered a lot of features, focusing on the most important, or what most of us developers will use. You learned about the internals of Spring Boot and discovered that Spring Boot uses the auto-configuration feature to set defaults that incorporate the best practices for your application. You also learned that Spring Boot, through the auto-configuration feature, will check out the dependencies or classpath your application is using and decide which defaults to apply. All of this is thanks not only to the auto-configuration classes but also the `@Conditional*` annotations that filter what your app needs and how to wire everything up, removing a lot of configuration that normally you would need to do if you were creating just Spring apps.

This chapter also showed you that you can customize your Spring Boot app by using either the `SpringApplication` class or the `SpringApplicationBuilder` fluent API. You also discovered how to use arguments in your application and how to set up and use the `application.properties` and even filter them by using `profiles`.

In Chapter 3 we are going to talk about web apps with Spring Boot and how to create web end points using our My Retro App and Users App projects.

CHAPTER 3

Spring Boot Web Development

In this chapter we'll review features for creating web applications with Spring Boot using our two projects, My Retro App and Users App. The Spring ecosystem has two ways to do web development: using the Servlet stack or using Reactive stack. In this chapter we'll review web development using the Servlet stack and in chapter 7 will cover the Reactive Stack.

Spring MVC

The Spring MVC technology has been around since the Spring Framework was created, bringing the MVC (Model View Controller) pattern for web applications, not only for the back end but for the front end as well, using HTML template engines to create views (Views/JSP pages) that can access objects (Models) from the back end (Controllers) and make use of them.

Creating a Spring Boot web application means including the `spring-boot-starter-web` dependency, which brings all the Spring MVC dependencies, such as `spring-web`, `spring-web-mvc`, an embedded application server (which can run immediately, so there's no need to deploy it externally), by default will use a web server (Apache `Tomcat` by default), and much more.

One of the main classes for the Spring MVC is the `DispatcherServlet` servlet class. This servlet implements a `front-controller` pattern that provides all the request processing and delegation of responsibilities to components (classes marked with @`Controller` or @`RestController` annotations).

In a regular Spring web app, it is necessary to declare the `DispatcherServlet` class (in a `web.xml` file) and add configuration (by declaring a context XML file) that allows the discovery of components that serve for the request mappings (for HTTP methods such as `GET`, `POST`, `PUT`, `DELETE`, `PATCH`, etc.), view resolution, exception handling, and much more.

© Felipe Gutierrez 2024
F. Gutierrez, *Pro Spring Boot 3*, https://doi.org/10.1007/978-1-4842-9294-5_3

The Spring MVC provides several ways to register components that serve the HTTP request mappings (through XML, JavaConfig, or annotations). It brings the @Controller and @RestController annotations.

Normally with the @Controller annotation (a class marker annotation), your class methods must return a Model object, or the name View interface to be rendered. If you want to respond with another object (such as JSON, XML, etc.), you must add the @ResponseBody annotation to your methods as well; this will set up everything you need for your content negotiation during the request/response scenario. If you use the @RestController annotation, you no longer need the @ResponseBody annotation, because Spring Boot will default to JSON as the content resolver; in other words, if you need to create a RESTful API, you need to use @RestController in your class.

As previously mentioned, the @Controller and @RestController annotations are markers for classes that register components to serve all the HTTP requests, and all the methods that will serve the request should be marked with @GetMapping (for an HTTP GET request), @PostMapping (for an HTTP POST request), @PutMapping, @DeleteMapping, or one of many other annotation. There is one annotation, @RequestMapping, that can be configured for multiple requests scenarios at once (this annotation can be used as a class marker as well).

Spring Boot MVC Auto-Configuration

With Spring Boot, you don't need anything of the previous secrtion, because when your application runs, the auto-configuration feature will set all the defaults. Let's see what auto-configuration is doing behind the scenes:

- *Static content support*: This means that you can add static content, such as HTML, JavaScript, CSS, media, and so forth, in a directory named /static (by default) or /public, /resources, or /META-INF/ resources, which should be in your classpath or in your current directory. Spring Boot picks it up the static content and serves it upon request. You can change this easily by modifying the spring. mvc.static-path-pattern property or the spring.web.resources. static-locations property. One of the cool features with Spring Boot and web applications is that if you create an index.html file, Spring Boot serves it automatically without registering any other bean or the need for extra configuration.

- `HttpMessageConverters`: If you are using a regular Spring MVC application and you want to get a JSON response, you need to create the necessary configuration (XML or JavaConfig) for the `HttpMessageConverters` bean. Spring Boot adds this support by default, so you don't have to; this means you get the JSON format by default (due to the Jackson libraries that the `spring-boot-starter-web` starter provides as dependencies). And if Spring Boot auto-configuration finds that you have the Jackson XML extension in your classpath, it aggregates an XML `HttpMessageConverter` to the converters, meaning that your application can serve based on your `content-type` request, either `application/JSON` or `application/XML`.

- *JSON serializers and deserializers*: If you want to have more control over the serialization/deserialization to/from JSON, Spring Boot provides an easy way to create your own by extending from `JsonSerializer<T>` and/or `JsonDeserializer<T>` and annotating your class with the `@JsonComponent` so that it can be registered for usage. Another feature of Spring Boot is the Jackson support; by default, Spring Boot serializes the date fields as `2024-05-01T23:31:38.141+0000`, but you can change this default behavior by changing the `spring.jackson.date-format=yyyy-MM-dd` property, for example (you can apply any date format pattern); the previous value generates the output `2024-05-01`.

- *Path matching and content negotiation*: One of the Spring MVC application features is the ability to respond to any suffix to represent the `content-type` response and its content negotiation. If you have something like `/api/retros.json` or `/api/retros.pdf`, the `content-type` is set to `application/json` and `application/pdf`; the response is JSON format or a PDF file, respectively. In other words, Spring MVC performs `.*` suffix pattern matching, such as `/ api/ retros.*`. Spring Boot disables this by default. If you prefer, you can either enable it or manually add a parameter by using the `spring. mvc.contentnegotiation.favor-parameter=true` property (false by default); you can do something like /api/retros?format=xml (format is the default parameter name, but you can change it with spring.mvc.contentnegotiation.parameter-name=myparam). This triggers the content-type to application/xml.

- *Error handling*: Spring Boot uses /error mapping to create a white labeled page to show all the global errors. You can change the behavior by creating your own custom pages. You need to create your custom HTML page in the src/main/resources/public/error/ location, so you can create 500.html or 404.html pages, for example. If you are creating a RESTful application, Spring Boot responds as JSON format. Spring Boot also supports Spring MVC to handle errors when you are using @ControllerAdvice or @ExceptionHandler annotations. You can register custom ErrorPages by implementing ErrorPageRegistrar and declaring it as a Spring bean.

- *Template engine support*: Spring Boot supports Apache FreeMarker, Groovy Templates, Thymeleaf, and Mustache. When you include the spring- boot-starter-<template engine> dependency, Spring Boot will auto-configure all the beans to enable all the view resolvers and file handlers. By default, Spring Boot looks at the src/main/resources/templates/ path. This can be overridden by using the spring.<template-engine>.prefix property.

Let's review some of these features using our main projects.

My Retro App Project

In Chapter 2, we created the basic My Retro App configuration, with no behavior or any other class, just some properties to demonstrate a few Spring Boot features. In this section, we will complete the project by adding dependencies, more classes, and, of course, everything about the Web, a Rest API.

Let's start by adding some dependencies to our project. Open the build.gradle file and replace all the content with the content in Listing 3-1.

Listing 3-1. build.gradle

```
plugins {
    id 'java'
    id 'org.springframework.boot' version '3.2.3'
    id 'io.spring.dependency-management' version '1.1.4'
}

group = 'com.apress'
version = '0.0.1-SNAPSHOT'
sourceCompatibility = '17'

repositories {
    mavenCentral()
    maven { url 'https://repo.spring.io/milestone' }
    maven { url 'https://repo.spring.io/snapshot' }
}

dependencies {
    implementation 'org.springframework.boot:spring-boot-starter-web'
        implementation 'org.springframework.boot:spring-boot-starter-
        validation'
        implementation 'org.springframework.boot:spring-boot-starter-aop'

    // Lombok
        compileOnly 'org.projectlombok:lombok'
        annotationProcessor 'org.projectlombok:lombok'

        // Properties
        annotationProcessor "org.springframework.boot:spring-boot-
        configuration-processor"

        // Web
        implementation 'org.webjars:bootstrap:5.2.3'

    testImplementation 'org.springframework.boot:spring-boot-starter-test'
}

tasks.named('test') {
    useJUnitPlatform()
}
```

```
test {
    testLogging {
        events "passed", "skipped", "failed"
        showExceptions true
        exceptionFormat "full"
        showCauses true
        showStackTraces true
        showStandardStreams = false
    }
}
```

Let's review the new build.gradle file:

- spring-boot-starter-web: This dependency brings all the necessary JARs to My Retro App. It brings the spring-web, spring-webmvc, tomcat, Jackson (for JSON manipulation) modules, and much more. This starter enables us to use the web annotations (@RestController, @GetMapping, and many more) that are necessary to create a web application. And, of course, everything will be auto-configured for us.

- spring-boot-starter-validation: This dependency brings the jakarta-validation JARs that will help us to validate that we are sending and creating the objects we need for our domain, the RetroBoard and Card classes (discussed later in this section).

- spring-boot-starter-aop: This dependency brings JARs to use *aspect-oriented programming (AOP)*, which is what Spring uses to perform some of the amazing logic, behavior, and auto-configuration to make your applications run smoothly. AOP will help us to remove some of the code scattering and code tangling, so that our code looks readable and understandable.

- lombok: This library will help us to remove all the Java verbosity (like having all the getters and setters, toString(), constructors, and more), making our code more readable. It won't affect the performance of our app (and is easy to learn).

- `spring-boot-starter-configuration-processor`: This dependency adds features such as preprocessing our custom properties (used in the previous chapter) and adding hints for IDEs that have *IntelliSense* and thus know the meaning of the property we are setting up.

- `bootstrap`: This dependency brings the css, javascript files from the Bootstrap Project (`https://getbootstrap.com/`) and adds them as resources in the webjars/resource folder. Although we are not going to build a front end with Spring (because we are going to use JavaScript), we want to add some nice looking into our front page.

- `test > testLogging`: These declarations help with all the logging when we are executing the `test` task from the command line. It will give us more information about each individual test. We added this in previous chapters.

Adding Dependencies

When Spring Boot starts, auto-configuration will review all the dependencies in `build.gradle` and configure and wire up everything for us so that our web app is ready to use.

Note   Lombok (`https://projectlombok.org/`) will be used throughout the entire book. This library doesn't affect the performance of the application in any way. Of course, you are free to remove it and put up with the Java verbosity.

So far, the My Retro App project has the directory and package structure shown in Figure 3-1. This structure was sufficient to demonstrate the features we discussed earlier.

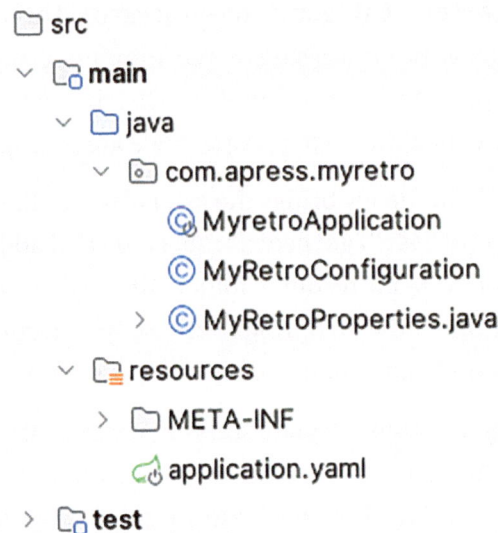

Figure 3-1. *Original directory/package structure*

Now, we are going to transform our project by using the directory/package structure shown in Figure 3-2.

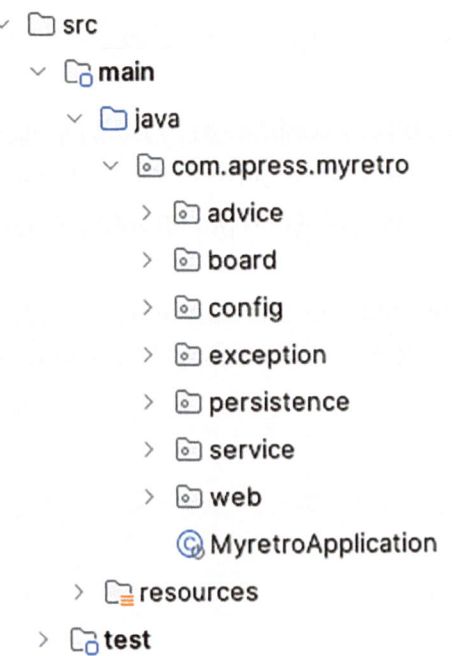

Figure 3-2. *Updated directory/package structure*

Next, we are going to examine every package in detail and all the classes that we are going to create. Let's start with our domain/model classes. Recall from Chapter 2 that we have two main classes, RetroBoard and Card, and one enum, CardType.

So, let's create these classes and enum in the board package, beginning with RetroBoard. Listing 3-2 shows the RetroBoard class.

Listing 3-2. src/main/java/com/apress/myretro/board/RetroBoard.java

```java
package com.apress.myretro.board;

import jakarta.validation.constraints.NotBlank;
import jakarta.validation.constraints.NotNull;
import lombok.Builder;
import lombok.Data;
import lombok.Singular;

import java.util.ArrayList;
import java.util.List;
import java.util.UUID;

@Builder
@Data
public class RetroBoard {

    private UUID id;

    @NotNull
        @NotBlank(message = "A name must be provided")
    private String name;

    @Singular
    private List<Card> cards;
}
```

The RetroBoard class has just three fields: a UUID (for the user ID), a String (for the name), and a List (for the Cards); remember that the RetroBoard can have from 0 to N Cards. RetroBoard also has the following annotations:

- @Builder: This annotation will create a Fluent API to create a RetroBoard instance. This annotation belongs to the Lombok library.

- @Data: This annotation generates all the getters and setters and the toString(), hashCode(), and equals(Object o) methods. This makes our code cleaner, and it won't impact the performance of the application. This annotation also belongs to the Lombok library.

- @NotNull: This annotation, part of the Jakarta project, allows you to mark a field as requiring a value (i.e., indicate that it can't have a null value).

- @NotBlank: This annotation, also part of the Jakarta project, allows you to mark a field for further processing to check that it is not empty (i.e., it needs to have a value). You can add a message explaining why that field cannot be blank or null.

- @Singular: This annotation from the Lombok library helps to build single elements by adding them to the list. It's worth mentioning that once the RetroBoard instance is built, the List<Card> is an unmodifiable list; keep this in mind.

Listing 3-3 shows the Card class.

Listing 3-3. src/main/java/com/apress/myretro/board/Card.java

```java
package com.apress.myretro.board;

import jakarta.validation.constraints.NotBlank;
import jakarta.validation.constraints.NotNull;
import lombok.Builder;
import lombok.Data;

import java.util.UUID;

@Builder
@Data
public class Card {
```

```
    private UUID id;

    @NotBlank(message = "A comment must be provided always")
    @NotNull
    private String comment;

    @NotNull(message = "A CardType HAPPY|MEH|SAD must be provided")
    private CardType cardType;
}
```

Again, Card is a very simple class with three fields. It has a relationship of type CardType that is marked to indicate that this field must not be null.

Listing 3-4 shows CardType, an enum that provides only three elements: HAPPY, MEH, and SAD.

Listing 3-4. src/main/java/com/apress/myretro/board/CardType.java

```
package com.apress.myretro.board;

public enum CardType {
    HAPPY,MEH,SAD
}
```

We need to hold our information about the RetroBoard and the Cards objects. So, in this chapter, we are going to do everything in memory; we are going to use a java.util. Map that allows us to use keys for faster lookup. Let's look at our persistence package. We will create the Repository interface (see Listing 3-5) and the RetroBoardRepository that implements the Repository interface (see Listing 3-6).

Listing 3-5. src/main/com/apress/myretro/persistence/Repository.java

```
package com.apress.myretro.persistence;

import java.util.Optional;

public interface Repository<D,ID> {
    D save(D domain);
```

```
    Optional<D> findById(ID id);

    Iterable<D> findAll();

    void delete(ID id);
}
```

As Listing 3-5 shows, the Repository interface has placeholders so that you can use the actual domain/model and the type as primary key.

Listing 3-6 shows the RetroBoardRepository class that implements the Repository interface.

Listing 3-6. src/main/com/apress/myretro/persistence/ RetroBoardRepository.java

```
package com.apress.myretro.persistence;

import com.apress.myretro.board.Card;
import com.apress.myretro.board.CardType;
import com.apress.myretro.board.RetroBoard;
import org.springframework.stereotype.Repository;

import java.util.HashMap;
import java.util.Map;
import java.util.Optional;
import java.util.UUID;

@Respository
public class RetroBoardRepository implements Repository<RetroBoard,UUID> {

    private Map<UUID,RetroBoard> retroBoardMap = new HashMap<>(){{
        put(UUID.fromString("9DC9B71B-A07E-418B-B972-40225449AFF2"),

                RetroBoard.builder()
                        .id(UUID.fromString("9DC9B71B-A07E-418B-
                        B972-40225449AFF2"))
                        .name("Spring Boot 3.0 Meeting")
                        .card(Card.builder()
                                .id(UUID.fromString("BB2A80A5-A0F5-4180-
                                A6DC-80C84BC014C9"))
```

```java
                            .comment("Happy to meet the team")
                            .cardType(CardType.HAPPY)
                            .build())
                    .card(Card.builder()
                            .id(UUID.fromString("011EF086-7645
                            -4534-9512-B9BC4CCFB688"))
                            .comment("New projects")
                            .cardType(CardType.HAPPY)
                            .build())
                    .card(Card.builder()
                            .id(UUID.fromString("775A3905-D6BE-49AB-
                            A3C4-EBE287B51539"))
                            .comment("When to meet again??")
                            .cardType(CardType.MEH)
                            .build())
                    .card(Card.builder()
                            .id(UUID.fromString("896C093D-1C50-49A3-
                            A58A-6F1008789632"))
                            .comment("We need more time to finish")
                            .cardType(CardType.SAD)
                            .build())
                    .build()
            );
}};

@Override
public RetroBoard save(RetroBoard domain) {
    if (domain.getId()==null)
        domain.setId(UUID.randomUUID());

    this.retroBoardMap.put(domain.getId(),domain);

    return domain;
}
```

```
@Override
public Optional<RetroBoard> findById(UUID uuid) {
    return Optional.ofNullable(this.retroBoardMap.get(uuid));
}

@Override
public Iterable<RetroBoard> findAll() {
    return this.retroBoardMap.values();
}

@Override
public void delete(UUID uuid) {
    this.retroBoardMap.remove(uuid);
}
}
```

The RetroBoardRepository class has a marker, the @Repository annotation, that will be essential when the application starts, because it will be injected in the service that we will create later in this section. Note in Listing 3-6 the usage of the builder() method, which was generated thanks to the @Builder annotation in the RetroBoard domain/model class. This method allows us to create the instance using a Fluent API. The retroBoardMap will act as in-memory persistence.

In the My Retro App, we are going to be able to have a nice way to show any error that occurs when a RetroBoard or a Card is not found or when the validation we set already (using @NotNull and @NotBlank) is triggered if conditions are not met.

Now create the exception package and add the RetroBoardNotFoundException, CardNotFoundException, and RetroBoardResponseEntityExceptionHandler classes. First, Listing 3-7 shows the RetroBoardNotFoundException class.

Listing 3-7. src/main/com/apress/myretro/exception/
RetroBoardNotFoundException.java

```
package com.apress.myretro.exception;

public class RetroBoardNotFoundException extends RuntimeException{

    public RetroBoardNotFoundException(){
        super("RetroBoard Not Found");
    }
```

```
    public RetroBoardNotFoundException(String message) {
        super(String.format("RetroBoard Not Found: {}", message));
    }

    public RetroBoardNotFoundException(String message, Throwable cause) {
        super(String.format("RetroBoard Not Found: {}", message), cause);
    }
}
```

The RetroBoardNotFoundException class extends the RunTimeException and has some fixed messages for easy handling. An instance of this class will be thrown when we try to search for a RetroBoard with a UUID that doesn't exist.

Listing 3-8 shows the CardNotFoundException class, which will be thrown when a Card within the RetroBoard is not found.

Listing 3-8. src/main/com/apress/myretro/exception/ CardNotFoundException.java

```
package com.apress.myretro.exception;

public class CardNotFoundException extends RuntimeException{
    public CardNotFoundException() {
        super("Card Not Found");
    }

    public CardNotFoundException(String message) {
        super(String.format("Card Not Found: {}", message));
    }

    public CardNotFoundException(String message, Throwable cause) {
        super(String.format("Card Not Found: {}", message), cause);
    }
}
```

Listing 3-9 shows the RetroBoardResponseEntityExceptionHandler class, which extends from the ResponseEntityExceptionHandler class, meaning that when it finds an error, it will know how to handle it and how to create the *response* for the requester.

Listing 3-9. src/main/com/apress/myretro/exception/
RetroBoardResponseEntityException.java

```
package com.apress.myretro.exception;

import org.springframework.http.HttpHeaders;
import org.springframework.http.HttpStatus;
import org.springframework.http.ResponseEntity;
import org.springframework.web.bind.annotation.ControllerAdvice;
import org.springframework.web.bind.annotation.ExceptionHandler;
import org.springframework.web.context.request.WebRequest;
import org.springframework.web.servlet.mvc.method.annotation.
ResponseEntityExceptionHandler;

import java.time.LocalDateTime;
import java.time.format.DateTimeFormatter;
import java.util.HashMap;
import java.util.Map;

@ControllerAdvice
public class RetroBoardResponseEntityExceptionHandler extends
ResponseEntityExceptionHandler {

    @ExceptionHandler(value
            = { CardNotFoundException.class,RetroBoardNotFoundException.
            class })
    protected ResponseEntity<Object> handleNotFound(
            RuntimeException ex, WebRequest request) {

        Map<String, Object> response = new HashMap<>();

        response.put("msg","There is an error");
        response.put("code",HttpStatus.NOT_FOUND.value());
        response.put("time", LocalDateTime.now().format(DateTimeFormatter.
        ofPattern("yyyy-mm-dd HH:mm:ss")));

        Map<String, String> errors = new HashMap<>();
        errors.put("msg",ex.getMessage());
        response.put("errors",errors);
```

```
    return handleExceptionInternal(ex, response,
        new HttpHeaders(), HttpStatus.NOT_FOUND, request);
  }
}
```

The following list explains the RetroBoardResponseEntityExceptionHandler class:

- @ControllerAdvice: This annotation uses AOP to implement the *Around Advice* (this means that it will intercept all the method calls defined in the controller and execute them inside a try and catch, and if there is any exception, the method handleNotFound will be called; here is where you add you own logic to handle the error), and is registered to catch any error thrown during runtime. It declares a method that is marked with the @ExceptionHandler annotation.

- @ExceptionHandler: This annotation catches any exception declared as part of the parameter value. In this case, is telling to execute the handleNotFound method when a RetroBoardNotFoundException or CardNotFoundException is thrown.

  ResponseEntity<T>: This annotation extends from an HttpEntity class that represents an HTTP request or response entity, consisting of headers and body. As you can see in Listing 3-9, we are creating a Map, which will be translated by default as a JSON content-type.

- handleExceptionInternal: This method belongs to the extended class, and it will prepare all the common handling and will create the ResponseEntity.

As you can see, the RetroBoardResponseEntityExceptionHandler class provides a special way to handle web errors that occur during the process of looking for a RetroBoard or Card objects, or even at the beginning of the request or response.

Next, create the service package and add the RetroBoardService class as shown in Listing 3-10.

Listing 3-10. src/main/com/apress/myretro/service/RetroBoardService.java

```java
package com.apress.myretro.service;

import com.apress.myretro.board.Card;
import com.apress.myretro.board.RetroBoard;
import com.apress.myretro.exception.CardNotFoundException;
import com.apress.myretro.persistence.Repository;
import lombok.AllArgsConstructor;
import org.springframework.stereotype.Service;

import java.util.*;

@AllArgsConstructor
@Service
public class RetroBoardService {

    Repository<RetroBoard,UUID> repository;

    public RetroBoard save(RetroBoard domain) {
        if (domain.getCards() == null)
            domain.setCards(new ArrayList<>());
        return this.repository.save(domain);
    }

    public RetroBoard findById(UUID uuid) {
        return this.repository.findById(uuid).get();
    }

    public Iterable<RetroBoard> findAll() {
        return this.repository.findAll();
    }

    public void delete(UUID uuid) {
        this.repository.delete(uuid);
    }

    public Iterable<Card> findAllCardsFromRetroBoard(UUID uuid) {
        return this.findById(uuid).getCards();
    }
```

```
public Card addCardToRetroBoard(UUID uuid, Card card){
    if (card.getId() == null)
        card.setId(UUID.randomUUID());

    RetroBoard retroBoard = this.findById(uuid);
    List<Card> cardList = new ArrayList<>(retroBoard.getCards());
    cardList.add(card);

    retroBoard.setCards(cardList);
    return card;
}

public Card findCardByUUIDFromRetroBoard(UUID uuid, UUID uuidCard){
    RetroBoard retroBoard = this.findById(uuid);
    Optional<Card> card = retroBoard.getCards().stream().filter(c ->
    c.getId().equals(uuidCard)).findFirst();
    if (card.isPresent())
        return card.get();
    throw new CardNotFoundException();
}

public void removeCardFromRetroBoard(UUID uuid, UUID cardUUID){
    RetroBoard retroBoard = this.findById(uuid);
    List<Card> cardList = new ArrayList<>(retroBoard.getCards());
    cardList.removeIf(card -> card.getId().equals(cardUUID));
    retroBoard.setCards(cardList);
}
}
}
```

The RetroBoardService class will help us to drive all the business logic we need for the application. Let's analyze it:

- @AllArgsConstructor: This annotation from the Lombok library creates the constructor that uses the fields as parameters. In other words, it creates a constructor with a Repository<RetroBoard,UUID> as a parameter. It will look something like the following code snippet:

```
public RetroBoardService(Repository<RetroBoard,UUID> repository) {
    this.repository = repository;
}
```

We are using the `Repository<D,ID>` interface, so the Spring Framework will inject the implementation of this interface by looking at all the Spring beans registered; we registered this bean in Listing 3-6 using the `@Respository` annotation, meaning that the `RetroBoardRepository` implementation will be injected and have access to the in-memory persistence. This capability to use or swap a different implementations and still use the same code is one of the features that make the Spring Framework awesome.

- `@Service`: This annotation is another stereotype that marks the class as a Spring bean so that it can be injected or used somewhere in your code. In this case, we will be using this service in our web controller.

Now, before continuing with the other classes in My Retro App, review the `RetroBoardService` code (see Listing 3-10) again and look for the method `findById(UUID)`; first, notice that we call the `repository.findById(UUID)` method. And if you take a look at Listing 3-6, the function returns an `Optional.ofNullable`, meaning that the value can be null, and we are calling a `.get()` right away, which will result in a `NullPointerException`. So, how can we handle this and translate it into a Not Found exception? We are going to use AOP for this. Also notice in the `RetroBoardService` code that `findById` is called in several other methods, and we can implement a simple logic to catch the error, handle it, or throw the exception. With this idea, we will be scattering and tangling the code all over the place, and we don't want to do this. The next section provides the solution.

AOP to the Rescue

Aspect-oriented programming can help us to treat this concern in a separate class and make this logic more readable and avoid any code scattering and tangling.

Next, create the `advice` package and the `RetroBoardAdvice` class. See Listing 3-11.

Listing 3-11. src/main/com/apress/myretro/advice/RetroBoardAdvice.java

```
package com.apress.myretro.advice;

import com.apress.myretro.board.RetroBoard;
import com.apress.myretro.exception.RetroBoardNotFoundException;
import com.apress.myretro.persistence.RetroBoardRepository;
import com.apress.myretro.service.RetroBoardService;
```

```java
import lombok.extern.slf4j.Slf4j;
import org.aspectj.lang.JoinPoint;
import org.aspectj.lang.ProceedingJoinPoint;
import org.aspectj.lang.annotation.Around;
import org.aspectj.lang.annotation.Aspect;
import org.aspectj.lang.annotation.Before;
import org.aspectj.lang.annotation.Pointcut;
import org.springframework.stereotype.Component;

import java.util.Optional;
import java.util.UUID;

@Slf4j
@Component
@Aspect
public class RetroBoardAdvice {

    @Around("execution(* com.apress.myretro.persistence.
    RetroBoardRepository.findById(java.util.UUID))")
    public Object checkFindRetroBoard(ProceedingJoinPoint
    proceedingJoinPoint) throws Throwable {
        log.info("[ADVICE] findRetroBoardById");
Optional<RetroBoard> retroBoard = (Optional<RetroBoard>)
proceedingJoinPoint.proceed(new Object[]{
                    UUID.fromString(proceedingJoinPoint.getArgs()[0].
                    toString())
            });
            if (retroBoard.isEmpty())
                    throw new RetroBoardNotFoundException();

        return retroBoard;
    }
}
```

The following list breaks down the RetroBoardAdvice class:

- @Slf4j: This annotation from the Lombok library creates an instance
  of a logger and you will be able to use it. In this case is just logging
  the name of the method (by calling the proceedingJoinPoint.
  getSignature().getName() method) being intercepted.

105

- @Component: This annotation is a marker that tell the Spring Framework to register it as a Spring bean and use it when needed.

- @Aspect: This annotation is a marker for a class and tells the Spring Framework that the class has an *Aspect* that will contain an *Advice* (Before, After, Around, AfterReturn, AfterThrowing) that will be intercepting a particular method depending on the matcher declaration. Behind the scenes, Spring creates proxies for these classes and applies everything related to AOP.

- @Around. This annotation is one of the many annotations to create an *Advice*. In this case, it will intercept the call (based on the execution declaration that matches the method) before it gets executed, create an instance of the ProceedingJoinPoint, and execute the method you marked with this annotation and execute your logic. Then, you can execute the actual call, do some more logic, and return the result. Related annotations are the @Before advice, which will execute your method logic before the actual call happens, the @After advice, which will execute your method logic after the call, and the @AfterReturning advice, which will execute the logic when

- execution: This is the key to the Advice. This keyword needs a pattern matching that identifies the method to be advised. In this case, we are looking for every method with any return type (*) and looking for that specific method in the com.apress.myretro. repository.RetroBoardRepository.findById that has as a parameter the UUID. In this case, this is very straightforward, but we can have an expression like this: * com.apress.*..*.find*(*). This means finding any class that is between the package apress and up and any class that has the find prefix for the method that accepts any number of parameters, no matter the type.

- ProceedingJoinPoint: This is an interface, and its implementation knows how to get the object that is being advised (RetroBoardRepository.findById); it has the actual object, and we can call the proceed that will execute it, and we can get the result and return it. See that we can manipulate the result or even the parameters that we are sending. Only the *Around Advice* must have this ProceedingJoinPoint.

So, when we tell the service to look for a RetroBoard instance (saved in-memory), it will intercept the call (*advice/around*) to the RetroBoardRepository and it will execute the method checkFindRetroBoard, it will gather all the information from the ProcedingJoinPoint, we are proceeding with the actual call and get back the possible nullable from the Optional class, and we check if this is empty, if it is then we throw the RetroBoardNotFoundException, and we return the result otherwise.

With this *Advice* we are isolating our concern about a check, avoiding any repetition of code (tangling and scattering), and making our code more understandable and cleaner.

This is just a small example of the power of the AOP paradigm. You can use your own custom annotation and advice on methods. For example, you can create a @Cache annotation and use the Around Advice for every method that uses that annotation.

Tip   If you want to know more about AOP, I recommend *Pro Spring 6* (Apress, 2023) or the Spring documentation, where you can find a very good explanation of what advice types are and what else you can do with them (https://docs. spring.io/spring-framework/docs/current/reference/html/core. html#aop).

Now, it's time to do some web logic!

Spring Web Annotated Controllers

Spring MVC provides annotation-based programming and includes two useful annotations, @Controller and @RestController. These annotations are used for request mappings, request input, exception handling, and much more.

The @Controller annotation is used with the Model and ModelAndView classes and the View interface. When using together, you have access to Session attributes and to objects that you can use in your HTML pages. With this functionality, you need to return the name of the view, and the Spring Web will handle the resolution and the rendering of the HTML page. See the following snippet:

```
// more ...
@Controller
public class MyRetroBoardController {
```

```
    private RetroBoardService retroBoardService;

    @GetMapping("/retros")
    public String handle(Model model) {
        model.addAttribute("retros", retroBoardService.findAll());
        return "listRetros";
    }

    // more ...
}
```

The preceding snippet marks the `MyRetroBoardController` as a *web controller* (using the `@Controller` annotation). When we access the endpoint `"/retros"`, it will create the necessary `Model` class to which we can add an attribute (in this case, the `retros` attribute with the value of all the retros), then return the name of the HTML page (`listRetros`, which will live in `src/main/resources/templates/listRetros.html`) that will do the rendering and use the retros data within the page. The Spring Web will know how to resolve the location of the page and how to render it by using a template engine. If you need to respond with a particular value (different from the View), then you need to add the `@ResponseBody` annotation to the return type (declared in the method) so that the Spring Web will use the HTTP message converter and respond properly.

The `@RestController` annotation is another class marker (this is a compose annotation from `@Controller`) and will write directly to the response body (so there is no need for the `@ResponseBody` annotation). We are going to use this annotation throughout the book, because we will use Spring Boot much more for the back end.

The Spring Web technology also includes the `@RequestMapping` annotation, which can be a marker for a class or a method because it is useful for mapping requests to controllers. One of the benefits of `@RequestMapping` is that it brings access to request parameters, headers, and media types. `@RequestMapping` can be used for every method, but sometimes you can use shortcuts: `@GetMapping`, `@PostMapping`, `@PutMapping`, `@DeleteMapping`, and many others. These shortcuts are described following their appearance in Listing 3-12.

Now, it's time to code our web controller. Create the package `web` and the class `RetroBoardController` as shown in Listing 3-12.

Listing 3-12. src/main/java/com/apress/myretro/web/
RetroBoardController.java

```
package com.apress.myretro.web;

import com.apress.myretro.board.Card;
import com.apress.myretro.board.RetroBoard;
import com.apress.myretro.service.RetroBoardService;
import jakarta.validation.Valid;
import lombok.AllArgsConstructor;
import org.springframework.http.HttpStatus;
import org.springframework.http.ResponseEntity;
import org.springframework.validation.FieldError;
import org.springframework.web.bind.MethodArgumentNotValidException;
import org.springframework.web.bind.annotation.*;
import org.springframework.web.servlet.ModelAndView;
import org.springframework.web.servlet.View;
import org.springframework.web.servlet.support.ServletUriComponentsBuilder;

import java.net.URI;
import java.time.LocalDateTime;
import java.time.format.DateTimeFormatter;
import java.util.HashMap;
import java.util.Map;
import java.util.UUID;

@AllArgsConstructor
@RestController
@RequestMapping("/retros")
public class RetroBoardController {

    private RetroBoardService retroBoardService;

    @GetMapping
    public ResponseEntity<Iterable<RetroBoard>> getAllRetroBoards(){
        return ResponseEntity.ok(retroBoardService.findAll());
    }
```

```java
@PostMapping
public ResponseEntity<RetroBoard> saveRetroBoard(@Valid @RequestBody
RetroBoard retroBoard){
    RetroBoard result = retroBoardService.save(retroBoard);
    URI location = ServletUriComponentsBuilder
            .fromCurrentRequest()
            .path("/{uuid}")
            .buildAndExpand(result.getId().toString())
            .toUri();
    return ResponseEntity.created(location).body(result);
}

@GetMapping("/{uuid}")
public ResponseEntity<RetroBoard> findRetroBoardById(@PathVariable
UUID uuid){
    return ResponseEntity.ok(retroBoardService.findById(uuid));
}

@GetMapping("/{uuid}/cards")
public ResponseEntity<Iterable<Card>> getAllCardsFromBoard(@
PathVariable UUID uuid){
    return ResponseEntity.ok(retroBoardService.findAllCardsFromRetroBoa
    rd(uuid));
}

@PutMapping("/{uuid}/cards")
public ResponseEntity<Card> addCardToRetroBoard(@PathVariable UUID
uuid,@Valid @RequestBody Card card){
    Card result = retroBoardService.addCardToRetroBoard(uuid,card);
    URI location = ServletUriComponentsBuilder
            .fromCurrentRequest()
            .path("/{uuid}/cards/{uuidCard}")
            .buildAndExpand(uuid.toString(),result.getId().toString())
            .toUri();
    return ResponseEntity.created(location).body(result);
}
```

```java
@GetMapping("/{uuid}/cards/{uuidCard}")
public ResponseEntity<Card> getCardFromRetroBoard(@PathVariable UUID
uuid,@PathVariable UUID uuidCard){
    return ResponseEntity.ok(retroBoardService.findCardByUUIDFromRetroB
    oard(uuid,uuidCard));
}

@ResponseStatus(HttpStatus.NO_CONTENT)
@DeleteMapping("/{uuid}/cards/{uuidCard}")
public void deleteCardFromRetroBoard(@PathVariable UUID uuid,@
PathVariable UUID uuidCard){
    retroBoardService.removeCardFromRetroBoard(uuid,uuidCard);
}

@ExceptionHandler(MethodArgumentNotValidException.class)
@ResponseStatus(HttpStatus.BAD_REQUEST)
public Map<String, Object> handleValidationExceptions(MethodArgumentNot
ValidException ex) {
    Map<String, Object> response = new HashMap<>();

    response.put("msg","There is an error");
    response.put("code",HttpStatus.BAD_REQUEST.value());
    response.put("time", LocalDateTime.now().format(DateTimeFormatter.
    ofPattern("yyyy-MM-dd HH:mm:ss")));

    Map<String, String> errors = new HashMap<>();
    ex.getBindingResult().getAllErrors().forEach((error) -> {
        String fieldName = ((FieldError) error).getField();
        String errorMessage = error.getDefaultMessage();
        errors.put(fieldName, errorMessage);
    });
    response.put("errors",errors);

    return response;
}
}
```

Let's analyze the `RetroBoardController` class:

- **@AllArgsConstructor**: This annotation from the Lombok library creates the constructor using the fields as parameters. In this case, it will use `RetroBoardService` as a parameter. Spring will *inject* the `RetroBoardService` bean using this constructor, so you have access in this controller class.

- **@RestController**: We are marking this class using the `@RestController` annotation. This annotation writes directly to the response body using all the methods declared.

- **@RequestMapping**: This annotation marks the class as the one that will respond to any request with the right HTTP method, and it will have the `/retros` endpoint as a base for any other path declared.

- **@GetMapping**: This annotation marks several methods that will respond with the HTTP GET method to the `/retros` endpoint. This is a shortcut of `@RequestMapping(method = RequestMethod.GET)`, which has more parameters that you can use. If you look at the `findRetroBoardById` method, you will see that `@GetMapping` is using the value `"/{uuid}"`; this is a path variable that is mapped to URL patterns. In this case, is using the `PathPattern` that allows to use matching patterns such as the following:

  - `"/retros/docu?ent.doc"`: Match only one character in a path.

  - `"/retros/*.jpg"`: Match zero or more characters in a path.

  - `"/retros/**"`: Match multiple path segments.

  - `"/retros/{project}/versions"`: Match a path segment and capture it as a variable.

  - `"/retros/{project:[a-z]+}/versions"`: Match and capture a variable with a regex.

  You can have something like this in your method: `@GetMapping ("/{product:[a-z-]+}-{version:\\d\\.\\d\\.\\d}{ext:\\.[a-z]+}")`. As you can see, you have options to declare how you want to access your endpoint, and this applies for every `@RequestMapping` and its shortcuts.

- ResponseEntity: This class extends from the HttpEntity (a generic) class that represents an HTTP request or response entity that brings the headers and the body. If you look around the code, you will see that the type varies. Including the ResponseEntity class is one of the common practices to use for a web app in Spring Boot or Spring Web applications. By default, Spring Boot will respond using an HTTP JSON message converter, so you can always expect a JSON response. Of course, you can override this default and respond in another format, such as XML.

- @PostMapping: This is another annotation that responds to the HTTP POST method request, sending a body (data) in the HTTP packet. This is the same as @RequestMapping(method = RequestMethod.POST). For this type of request, normally, you send data, so you will probably be required to use the @RequestBody annotation.

- @RequestBody: This annotation looks at the body of the web request and tries to bind it (using the HttpMessageConverter) to the instance that is marked with this annotation. And in this controller, we have two: one for the RetroBoard and another for the Card classes. You can use validation to validate the data using the @Valid annotation.

- @Valid: This annotation marks a parameter to do a cascading validation to see if the parameter passes validation based on the constraints used, such as @NotNull, @NotEmpty, @NotBlank, etc. All these annotations are present in the RetroBoard and Card classes and belong to the Jakarta library. Behind the scenes, will do the validation when the web request is happening, and it will throw an exception that we can catch.

- @PathVariable: This annotation does the binding with the path declared in the @RequestMapping or the shortcuts used, like in the findRetroBoardById method, where the value of the path /retros/ {uuid} is bound to the UUID instance. So, we can access it using something like this:

GET http://localhost:8080/retros/**9dc9b71b-a07e-418b-b972-40225449aff2**

9dc9b71b-a07e-418b-b972-40225449aff2 will be set to the UUID instance variable. The name of the path ({uuid}) must match with the name of the parameter declared in the function.

- ServletUriComponentsBuilder: This is a helper class that can be used to create a URI and expand to the base path with the keys and values. In this case, this class is used in the saveRetroBoard method, where it will create the location needed in the Header that will be set as part of the response with the ResponseEntity.create(<URI>) method call.

- @PutMapping: This annotation responds to the HTTP PUT method request, which normally brings an HTTP Body into the request. In our case, it is a combination of a particular path (with a URL pattern) and an HTTP Body (@RequestBody); see the addCardToRetroBoard method that also brings some validation with the @Valid annotation.

- @ResponseStatus: This annotation is used to customize the HTTP status code returned in a response for a given controller method or exception handler. Sometimes, regardless of the operation, we can return a particular *HTTP Status code*. In this case, in the deleteCardFromRetroBoard method, we are returning the HttpStatus.NO_CONTENT status (204 code).

- @DeleteMapping: This annotation responds to the HTTP DELETE request method. In our code, we also declare a URL path using the @PathVariable.

- @ExceptionHandler: This annotation is used to define methods that handle specific exceptions thrown during the execution of controller methods (or within the scope of a @ControllerAdvice class for global exception handling). When an exception occurs that matches the type declared in the @ExceptionHandler method's parameter, Spring MVC will invoke this method to handle the exception and generate an appropriate response. Sometimes, it is necessary to respond in a specific way to any error or any exception that has occurred in our application. By default, the Spring will answer with an exception that our Tomcat server (embedded) will throw as "Server Internal error" (or any other error) with not too much information about what happened. To avoid this, we can catch these types of errors, such as the error that occurs if we don't have valid data (from the

RetroBoard or Card), and explain what happened. We can annotate a method that can take care of that. In our code, we have the handleValidationExceptions method. Our @ExceptionHandler has the MethodArgumentNotValidException class declared as a parameter, meaning that it will be triggered only if the validation that is happening throws this error.

- MethodArgumentNotValidException: This class is thrown when the @Valid cascading validation is happening. The handleValidationExceptions method will create the necessary response with the errors that happened, and it will return a Map that will be converted as JSON format, which is a better way to understand what is happening.

We have just implemented a web application with Spring Boot. Before you continue to testing it, take another look at the RetroBoardController class. Check out every detail.

Testing My Retro App

To test our application, let's start by testing the RetroBoardService class, which is the core of the application. So, in the test folder, replace the content of the MyretroApplicationTests using the code shown in Listing 3-13.

Listing 3-13. src/test/java/com/apress/myretro/MyRetroApplicationTests.java

```
package com.apress.myretro;

import com.apress.myretro.board.Card;
import com.apress.myretro.board.CardType;
import com.apress.myretro.board.RetroBoard;
import com.apress.myretro.exception.CardNotFoundException;
import com.apress.myretro.exception.RetroBoardNotFoundException;
import com.apress.myretro.service.RetroBoardService;
import org.junit.jupiter.api.Test;
import org.springframework.beans.factory.annotation.Autowired;
import org.springframework.boot.test.context.SpringBootTest;

import java.util.Collection;
import java.util.UUID;
```

```java
import static org.assertj.core.api.Assertions.assertThat;
import static org.assertj.core.api.AssertionsForClassTypes.
assertThatThrownBy;

@SpringBootTest
class MyretroApplicationTests {

    @Autowired
    RetroBoardService service;

    UUID retroBoardUUID = UUID.fromString("9DC9B71B-A07E-418B-
    B972-40225449AFF2");
    UUID cardUUID = UUID.fromString("BB2A80A5-A0F5-4180-
    A6DC-80C84BC014C9");
    UUID mehCardUUID = UUID.fromString("775A3905-D6BE-49AB-A3C4-
    EBE287B51539");

    @Test
    void saveRetroBoardTest(){
        RetroBoard retroBoard = service.save(RetroBoard.builder().
        name("Gathering 2023").build());
        assertThat(retroBoard).isNotNull();
        assertThat(retroBoard.getId()).isNotNull();
    }

    @Test
    void findAllRetroBoardsTest(){
        Iterable<RetroBoard> retroBoards = service.findAll();
        assertThat(retroBoards).isNotNull();
        assertThat(retroBoards).isNotEmpty();
    }

    @Test
    void cardsRetroBoardNotFoundTest() {
        assertThatThrownBy(() -> {
            service.findAllCardsFromRetroBoard(UUID.randomUUID());
                }).isInstanceOf(RetroBoardNotFoundException.class);
    }
```

```java
@Test
void  findRetroBoardTest(){
    RetroBoard retroBoard = service.findById(retroBoardUUID);
    assertThat(retroBoard).isNotNull();
    assertThat(retroBoard.getName()).isEqualTo("Spring Boot 3.0 Meeting");
    assertThat(retroBoard.getId()).isEqualTo(retroBoardUUID);
}

@Test
void findCardsInRetroBoardTest(){
    RetroBoard retroBoard = service.findById(retroBoardUUID);
    assertThat(retroBoard).isNotNull();
    assertThat(retroBoard.getCards()).isNotEmpty();
}

@Test
void addCardToRetroBoardTest(){
    Card card = service.addCardToRetroBoard(retroBoardUUID, Card.
    builder()
                    .comment("Amazing session")
                    .cardType(CardType.HAPPY)
            .build());
    assertThat(card).isNotNull();
    assertThat(card.getId()).isNotNull();

    RetroBoard retroBoard = service.findById(retroBoardUUID);
    assertThat(retroBoard).isNotNull();
    assertThat(retroBoard.getCards()).isNotEmpty();
}

@Test
void findAllCardsFromRetroBoardTest() {
    Iterable<Card> cardList = service.findAllCardsFromRetroBoard(retroB
    oardUUID);
    assertThat(cardList).isNotNull();
    assertThat(((Collection) cardList).size()).isGreaterThan(3);
}
```

```java
@Test
void removeCardsFromRetroBoardTest(){
    service.removeCardFromRetroBoard(retroBoardUUID,cardUUID);
    RetroBoard retroBoard = service.findById(retroBoardUUID);
    assertThat(retroBoard).isNotNull();
    assertThat(retroBoard.getCards()).isNotEmpty();
    assertThat(retroBoard.getCards()).hasSizeLessThan(4);
}

@Test
void findCardByIdInRetroBoardTesT(){
    Card card = service.findCardByUUIDFromRetroBoard(retroBoardUUID,meh
    CardUUID);
    assertThat(card).isNotNull();
    assertThat(card.getId()).isEqualTo(mehCardUUID);
}

@Test
void notFoundCardInRetroBoardTest(){
    assertThatThrownBy(() -> {
        service.findCardByUUIDFromRetroBoard(retroBoardUUID,UUID.
        randomUUID());
    }).isInstanceOf(CardNotFoundException.class);
}

}
```

As you can see, the MyretroApplicationTests class only tests the service. In this test we are using very simple assertions from the *AssertJ library*. To test whether My Retro App returns the correct exception, we can use the assertThatThrownBy method. You can run these tests either by using your IDE or by running the following command:

```
./gradlew clean test
Starting a Gradle Daemon, 2 incompatible Daemons could not be reused,
use --status for details

> Task :compileJava
> Task :test
```

```
MyretroApplicationTests > saveRetroBoardTest() PASSED
MyretroApplicationTests > findAllRetroBoardsTest() PASSED
MyretroApplicationTests > findRetroBoardTest() PASSED
MyretroApplicationTests > removeCardsFromRetroBoardTest() PASSED
MyretroApplicationTests > cardsRetroBoardNotFoundTest() PASSED
MyretroApplicationTests > notFoundCardInRetroBoardTest() PASSED
MyretroApplicationTests > findCardsInRetroBoardTest() PASSED
MyretroApplicationTests > addCardToRetroBoardTest() PASSED
MyretroApplicationTests > findCardByIdInRetroBoardTesT() PASSED
MyretroApplicationTests > findAllCardsFromRetroBoardTest() PASSED

BUILD SUCCESSFUL in 10s
5 actionable tasks: 5 executed
```

Now, what we really want to do is test our web API, right? There are various tools that can help us to do this, such as PostMan (`https://www.postman.com`) and Insomnia (`https://insomnia.rest`), but I want to introduce you to a tool called REST Client that can help us to do HTTP requests directly. There is an open source version and a paid version, sharing the same style. The paid version is within the *IntelliJ IDEA Enterprise Edition*, the *Rest Client*. And if you are using VS Code, you must install the *REST Client (v0.25.x)* from the plugins tab, the author of this plugin is Huachao Mao (see Figure 3-3).

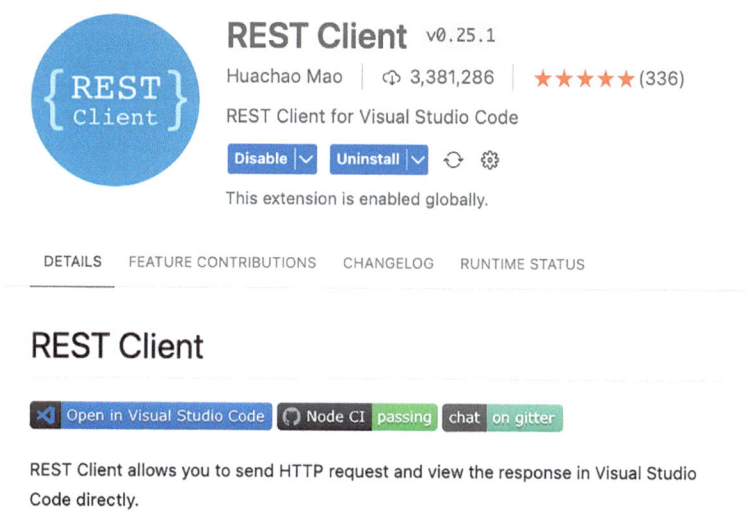

Figure 3-3. VS Code plugin REST Client from Huachao Mao

I'll demonstrate the use of the VS Code REST Client plugin to test the My Retro App. To follow along, download the plugin and then create the src/http folder and add the myretro.http file with the content shown in Listing 3-14.

Listing 3-14. src/http/myretro.http

```
### Get All Retro Boards
GET http://localhost:8080/retros
Content-Type: application/json

### Get Retro Board
GET http://localhost:8080/retros/9dc9b71b-a07e-418b-b972-40225449aff2
Content-Type: application/json

### Get All Cards from Retro Board
GET http://localhost:8080/retros/9dc9b71b-a07e-418b-b972-40225449aff2/cards
Content-Type: application/json

### Get Single Card from Retro Board
GET http://localhost:8080/retros/9dc9b71b-a07e-418b-b972-40225449aff2/
cards/bb2a80a5-a0f5-4180-a6dc-80c84bc014c9
Content-Type: application/json

### Create a Retro Board
POST http://localhost:8080/retros
Content-Type: application/json

{
  "name": "Spring Boot Conference"
}

### Add Card to Retro
PUT http://localhost:8080/retros/9dc9b71b-a07e-418b-b972-40225449aff2/cards
Content-Type: application/json

{
  "comment": "We are back in business",
  "cardType": "HAPPY"
}
```

```
### Delete Card from Retro
DELETE http://localhost:8080/retros/9dc9b71b-a07e-418b-b972-40225449aff2/
cards/bb2a80a5-a0f5-4180-a6dc-80c84bc014c9
Content-Type: application/json
```

Listing 3-14 shows all the necessary calls to cover the My Retro App. At the top of every call, you should have enabled a Send Request link; if you click it, you will see the response in another window. Experiment with each call.

Using this type of client is very easy. This plugin can do much more than described in this brief introduction. If you want to learn more about how to use it, visit `https://github.com/Huachao/vscode-restclient`.

Users App Project

Now, it's time to transform our Users App project that we started in Chapter 1. In this section, we are going to modify it to be more functional. Figure 3-4 shows the directory structure that you will end with after completing this section.

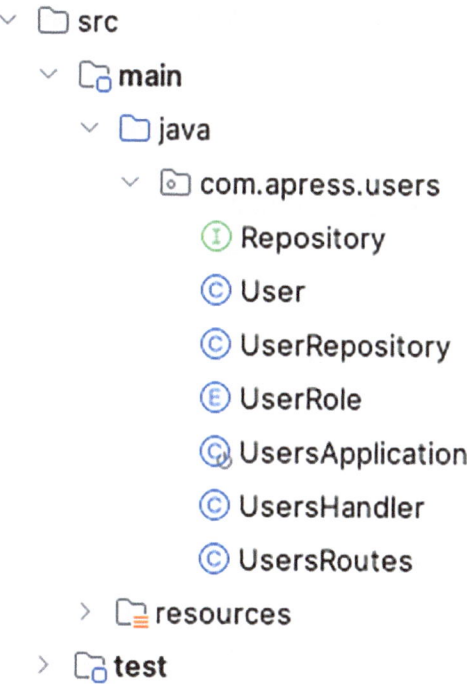

Figure 3-4. *Users App project*

Open the build.gradle file and modify the existing content with the content shown in Listing 3-15.

Listing 3-15. build.gradle

```
plugins {
    id 'java'
    id 'org.springframework.boot' version '3.2.4'
    id 'io.spring.dependency-management' version '1.1.4'
}

group = 'com.apress'
version = '0.0.1-SNAPSHOT'
sourceCompatibility = '17'

repositories {
    mavenCentral()
}

dependencies {
    implementation 'org.springframework.boot:spring-boot-starter-web'
    implementation 'org.springframework.boot:spring-boot-starter-
    validation'

    // Lombok
    compileOnly 'org.projectlombok:lombok'
    annotationProcessor 'org.projectlombok:lombok'

    // Web
    implementation 'org.webjars:bootstrap:5.2.3'

    testImplementation 'org.springframework.boot:spring-boot-starter-test'
}

tasks.named('test') {
    useJUnitPlatform()
}
```

```
test {
    testLogging {
        events "passed", "skipped", "failed"
        showExceptions true
        exceptionFormat "full"
        showCauses true
        showStackTraces true
        showStandardStreams = false
    }
}
```

This new build.gradle file incorporates validation and Lombok.

Next, open the User class and modify its content with the code shown in Listing 3-16.

Listing 3-16. src/main/java/com/apress/users/User.java

```
package com.apress.users;

import jakarta.validation.constraints.NotBlank;
import jakarta.validation.constraints.Pattern;
import lombok.Builder;
import lombok.Data;
import lombok.Singular;

import java.util.List;

@Builder
@Data
public class User {

    @NotBlank(message = "Email can not be empty")
    private String email;

    @NotBlank(message = "Name can not be empty")
    private String name;

    private String gravatarUrl;
```

```
@Pattern(message = "Password must be at least 8 characters long and
contain at least one number, one uppercase, one lowercase and one
special character",
        regexp = "^(?=.*[0-9])(?=.*[a-z])(?=.*[A-Z])(?=.*[@#$%^&+=!])
        (?=\\S+$).{8,}$")
private String password;

@Singular("role")
private List<UserRole> userRole;

private boolean active;

}
```

The following list describes the annotations in the new User class:

- @Builder/@Data: We are using the @Builder and @Data annotations from the Lombok library; recall that the @Builder annotation creates a Fluent API that can help us to build a User instance, and the @Data annotation generates all the setters and getters and the toString(), equals(Object), and hashCode() methods for the class. Using this annotation won't impact the performance of the application.

- @NotBlank/@Pattern: These annotations are used when the validation starts for a web request. @NotBlank was introduced earlier. The @Pattern annotation does a match pattern validation, and it fails, produces an error message.

- @Singular: This Lombok library annotation helps to add single roles to the list of user roles.

As you can see, we are adding a few more features to our app, such as validation. Next, create the UserRole enum, the Repository interface, and the UserRepository class. Listing 3-17 shows the enum for the User roles.

Listing 3-17. src/main/java/com/apress/users/UserRole.java

```
package com.apress.users;

public enum UserRole {
    USER, ADMIN, INFO
}
```

Listing 3-18 shows the `Repository` interface, which in fact is the same as the one in the My Retro App (see Listing 3-5).

Listing 3-18. src/main/java/com/apress/users/Repository.java

```java
package com.apress.users;

import java.util.Optional;

public interface Repository<D,ID>{

    D save(D domain);

    Optional<D> findById(ID id);

    Iterable<D> findAll();

    void deleteById(ID id);
}
```

Listing 3-19 shows the `UserRepository` implementation class.

Listing 3-19. src/main/java/com/apress/users/UserRepository.java

```java
package com.apress.users;

import org.springframework.stereotype.Repository;

import java.util.*;

@Repository
public class UserRepository implements Repository<User,String>{

    private Map<String,User> users = new HashMap<>() {{
        put("ximena@email.com",User.builder()
                .email("ximena@email.com")
                .name("Ximena")
                .gravatarUrl("https://www.gravatar.com/avatar/23bb62a7d0ca6
                3c9a804908e57bf6bd4?d=wavatar")
                .password("aw2s0meR!")
                .role(UserRole.USER)
                .active(true)
                .build());
```

```java
            put("norma@email.com",User.builder()
                    .name("Norma")
                    .email("norma@email.com")
                    .gravatarUrl("https://www.gravatar.com/avatar/f07f7e55326
                    4c9710105edebe6c465e7?d=wavatar")
                    .password("aw2sOmeR!")
                    .role(UserRole.USER)
                    .role(UserRole.ADMIN)
                    .active(true)
                    .build());
    }};

    @Override
    public User save(User user) {
        if (user.getGravatarUrl()==null)
            user.setGravatarUrl("https://www.gravatar.com/avatar/23bb62a7d0
            ca63c9a804908e57bf6bd4?d=wavatar");
        if (user.getUserRole() == null)
            user.setUserRole(Collections.emptyList());

        return this.users.put(user.getEmail(),user);
    }

    @Override
    public Optional<User> findById(String id) {
        return Optional.of(this.users.get(id));
    }

    @Override
    public Iterable<User> findAll() {
        return this.users.values();
    }

    @Override
    public void deleteById(String id) {
        this.users.remove(id);
    }
}
```

Review the listings closely and you'll see that the updated Users App is very similar to the My Retro App, and that we are using in-memory persistence.

Spring Web Functional Endpoints

Spring MVC provides functional programming as well to define web endpoints. Functions are used to define routes and handle requests. Every HTTP request is handled by a HandlerFunction (RouterFunction) that takes a ServerRequest and returns a ServerResponse.

These requests are routed to a RouterFunction that takes the ServerRequest and returns an optional HandlerFunction. You can consider this RouterFunction as being equivalent to the @RequestMapping annotation but with the advantage that it deals not only with data but also with behavior.

Next, add the UsersRoutes class. See Listing 3-20.

Listing 3-20. src/main/java/com/apress/users/UsersRoutes.java

```
package com.apress.users;

import org.springframework.context.annotation.Bean;
import org.springframework.context.annotation.Configuration;
import org.springframework.validation.Validator;
import org.springframework.validation.beanvalidation.
LocalValidatorFactoryBean;
import org.springframework.web.servlet.function.RequestPredicates;
import org.springframework.web.servlet.function.RouterFunction;
import org.springframework.web.servlet.function.ServerResponse;

import static org.springframework.http.MediaType.APPLICATION_JSON;
import static org.springframework.web.servlet.function.
RequestPredicates.accept;
import static org.springframework.web.servlet.function.
RouterFunctions.route;

@Configuration
public class UsersRoutes {
```

```
@Bean
public RouterFunction<ServerResponse> userRoutes(UsersHandler
usersHandler) {
    return route().nest(RequestPredicates.path("/users"), builder -> {
        builder.GET("",accept(APPLICATION_JSON), usersHandler::findAll);
        builder.GET("/{email}",accept(APPLICATION_JSON), usersHandler::f
        indUserByEmail);
        builder.POST("", usersHandler::save);
        builder.DELETE("/{email}", usersHandler::deleteByEmail);
    }).build();
}

@Bean
        public Validator validator() {
          return new LocalValidatorFactoryBean();
          }
}
```

The UsersRoutes class has the following components:

- @Configuration: Spring Boot looks for this annotation when the
  application starts; it helps to identify any possible Spring beans,
  marked with the @Bean annotation. In this class we are defining two
  beans, the userRoutes and the validator.

- RouterFunction<ServerResponse>: This is an interface that defines
  several methods that help to build a RouterFunction. The common
  way it to use the Fluent API that defines the RouterFunction
  interface. Note the userRoutes(UsersHandler) method, which is
  expecting the Spring bean UsersHandler; for Spring to know about
  this bean, it must be declared (using @Bean or marking the class using
  the @Component annotation).

- route(): This is a Fluent API that defines the necessary routing
  depending on the endpoint defined. In this case, we are creating a
  common path /users using the RequestPredicates abstract class.

- builder: We are using a builder (a java.util.function.Consumer) that allows us to define which HTTP methods will be routed to the handler, in this case to the UsersHandler class. Here we are declaring the GET, POST, and DELETE HTTP methods and defining the method that will be used from the handler.

- @Bean: This annotation is used to declare the Spring beans used in our application.

- Validator/LocalValidatorFactoryBean: This bean validator returns a LocalValidatorFactoryBean class that is used when using the constraints (@NotBlank and @Pattern) in the User class. This is the only way to do validation using the functional way to declare web API endpoints. This validator is going to be used in the UsersHandler class, covered next.

Next, add the UsersHandler class. See Listing 3-21.

Listing 3-21. src/main/java/com/apress/users/UsersHandler.java

```
package com.apress.users;

import jakarta.servlet.ServletException;
import lombok.RequiredArgsConstructor;
import org.springframework.http.HttpStatus;
import org.springframework.http.MediaType;
import org.springframework.stereotype.Component;
import org.springframework.validation.BindingResult;
import org.springframework.validation.DataBinder;
import org.springframework.validation.Validator;
import org.springframework.web.servlet.function.ServerRequest;
import org.springframework.web.servlet.function.ServerResponse;
import org.springframework.web.servlet.support.ServletUriComponentsBuilder;

import java.io.IOException;
import java.net.URI;
import java.time.LocalDateTime;
import java.time.format.DateTimeFormatter;
import java.util.HashMap;
import java.util.Map;
```

```java
@RequiredArgsConstructor
@Component
public class UsersHandler {

    private final Repository userRepository;
    private final Validator validator;

    public ServerResponse findAll(ServerRequest request) {
        return ServerResponse
                .ok()
                .contentType(MediaType.APPLICATION_JSON)
                .body(this.userRepository.findAll());
    }

    public ServerResponse findUserByEmail(ServerRequest request) {
        return ServerResponse
                .ok()
                .contentType(MediaType.APPLICATION_JSON)
                .body(this.userRepository.findById(request.
                pathVariable("email")));
    }

    public ServerResponse save(ServerRequest request) throws
    ServletException, IOException {
        User user = request.body(User.class);

        BindingResult bindingResult = validate(user);
                if (bindingResult.hasErrors()) {
                  return prepareErrorResponse(bindingResult);
                }

        this.userRepository.save(user);

        URI location = ServletUriComponentsBuilder
                .fromCurrentRequest()
                .path("/{email}")
                .buildAndExpand(user.getEmail())
                .toUri();
```

```java
        return ServerResponse.created(location).body(user);
    }

    public ServerResponse deleteByEmail(ServerRequest request) {
        this.userRepository.deleteById(request.pathVariable("email"));
        return ServerResponse.noContent().build();
    }

    private BindingResult validate(User user) {
            DataBinder binder = new DataBinder(user);
            binder.addValidators(validator);
            binder.validate();
            return binder.getBindingResult();
        }

    private ServerResponse prepareErrorResponse(BindingResult
    bindingResult) {
        Map<String, Object> response = new HashMap<>();

        response.put("msg","There is an error");
        response.put("code", HttpStatus.BAD_REQUEST.value());
        response.put("time", LocalDateTime.now().format(DateTimeFormatter.
        ofPattern("yyyy-MM-dd HH:mm:ss")));

        Map<String, String> errors = new HashMap<>();
        bindingResult.getFieldErrors().forEach(fieldError -> {
            errors.put(fieldError.getField(), fieldError.
            getDefaultMessage());
        });
        response.put("errors",errors);

        return ServerResponse.badRequest().body(response);
    }
}
```

The `UsersHandler` class has the following parts:

- `@RequiredArgsConstructor`: This annotation from the Lombok library creates a constructor using the fields `Repository` and `Validator`. With this constructor in place, Spring will inject the beans that correspond to each field. One will be our `UserRepository` and the other the bean declared in the `UsersRoutes` class, the validator (`LocalValidationFactoryBean`).

- `@Component`: This annotation is the marker that identifies this class as Spring bean to be used in the web app when needed. In this case, this is the handler that will attend every web request.

- `ServerRequest`: This is an interface that represents a server-side HTTP request and it's being handled by a `HandlerFunction`. It has access to the headers and body. The method `save(ServerRequest)` has access to the HTTP body when using `request.body(<class-type>)`. Behind the scenes its using all the necessary HTTP message converters to get the right class-type (in this case the User class).

- `ServerResponse`: This interface represents the server-side HTTP response, as returned by a `HandlerFunction`. Similar to the `ResponseEntity` class discussed earlier in the chapter, it can build the whole response with several helpful methods.

- `BindingResult`: In the `save` method, we are using the `validate(user);` this will return a `BindingResult` interface that invokes the validator for every constraint set in the class that is being validated. Look at the `prepareErrorResponse`; if the validation caught any errors, we can iterate over and prepare the message using the `ServerResponse`.

Testing the Users App

It's time to test our Users App. Add the `UsersHttpRequestTests` class in the `src/main/test/java/com/apress/users` folder. See Listing 3-22.

Listing 3-22. src/test/java/com/apress/users/UsersHttpRequestTests.java

```
package com.apress.users;

import org.junit.jupiter.api.Test;
import org.springframework.beans.factory.annotation.Autowired;
import org.springframework.beans.factory.annotation.Value;
import org.springframework.boot.test.context.SpringBootTest;
import org.springframework.boot.test.web.client.TestRestTemplate;

import java.util.Collection;
import java.util.Map;

import static org.assertj.core.api.Assertions.assertThat;

@SpringBootTest(webEnvironment = SpringBootTest.WebEnvironment.RANDOM_PORT)
public class UsersHttpRequestTests {

    @Value("${local.server.port}")
    private int port;

    private final String BASE_URL = "http://localhost:";
    private final String USERS_PATH = "/users";

    @Autowired
    private TestRestTemplate restTemplate;

    @Test
    public void indexPageShouldReturnHeaderOneContent() throws Exception {
        assertThat(this.restTemplate.getForObject(BASE_URL + port,
                String.class)).contains("Simple Users Rest Application");
    }

    @Test
    public void usersEndPointShouldReturnCollectionWithTwoUsers() throws
    Exception {
        Collection<User> response = this.restTemplate.
                getForObject(BASE_URL + port + USERS_PATH,
                Collection.class);
```

```
        assertThat(response).isNotNull();
        assertThat(response).isNotEmpty();
    }

    @Test
    public void userEndPointPostNewUserShouldReturnUser() throws
    Exception {
        User user =  User.builder().email("dummy@email.com").name("Dummy").
        password("aw2sOm3R!").build();
        User response =  this.restTemplate.postForObject(BASE_URL + port +
        USERS_PATH,user,User.class);

        assertThat(response).isNotNull();
        assertThat(response.getEmail()).isEqualTo(user.getEmail());

        Collection<User> users = this.restTemplate.
                getForObject(BASE_URL + port + USERS_PATH,
                Collection.class);

        assertThat(users.size()).isGreaterThanOrEqualTo(2);

    }

    @Test
    public void userEndPointDeleteUserShouldReturnVoid() throws Exception {
        this.restTemplate.delete(BASE_URL + port + USERS_PATH + "/norma@
        email.com");

        Collection<User> users = this.restTemplate.
                getForObject(BASE_URL + port + USERS_PATH,
                Collection.class);

        assertThat(users.size()).isLessThanOrEqualTo(2);
    }

    @Test
    public void userEndPointFindUserShouldReturnUser() throws Exception{
        User user = this.restTemplate.getForObject(BASE_URL + port + USERS_
        PATH + "/ximena@email.com",User.class);
```

```
        assertThat(user).isNotNull();
        assertThat(user.getEmail()).isEqualTo("ximena@email.com");
    }

    @Test
    public void userEndPointPostNewUserShouldReturnBadUserResponse() throws
    Exception {
        User user =  User.builder().email("dummy@email.com").name("Dummy").
        password("aw2s0m").build();
        Map response =  this.restTemplate.postForObject(BASE_URL + port +
        USERS_PATH,user, Map.class);

        assertThat(response).isNotNull();
        assertThat(response.get("errors")).isNotNull();
        Map errors = (Map) response.get("errors");
        assertThat(errors.get("password")).isNotNull();
        assertThat(errors.get("password")).isEqualTo("Password must be
        at least 8 characters long and contain at least one number, one
        uppercase, one lowercase and one special character");
    }
}
```

The UsersHttpRequestTests class includes the following:

- @SpringBootTest. This annotation sets up our integration tests that let you interact with a runnign web application. As you can see, is using a parameter webEnvironment = SpringBootTest. WebEnvironment.RANDOM_PORT, that instructs Spring Boot to start the application with a real web server listening on a randomly chosen available port.

- @Value: this annotation is used to inject values into fields, method parameters, or constructor arguments in Spring-managed beans. It provides a convenient way to externalize configuration values and inject them directly where they are needed.

- @Autowired: this annotation is used to enable automatic dependency injection. It instructs Spring to resolve and inject collaborating beans into other beans, automatically wiring them together.

- `TestRestTemplate`: This class is a convenient alternative to the standard RestTemplate specifically designed for integration testing of RESTful web services. It provides several features that make it easier to test your application's endpoints in a controlled environment. It automatically configures an HTTP client (Apache HttpClient or OkHttp) for testing purposes, eliminating the need for manual setup. Unlike RestTemplate, which throws exceptions for 4xx and 5xx status codes, TestRestTemplate handles these errors gracefully by returning a ResponseEntity object. This allows you to easily check the response status and handle errors in your test code. It provides convenient methods for handling basic authentication (withBasicAuth) and OAuth2 authentication (withOAuth2Client).

If you look at the last method (`userEndPointPostNewUserShouldReturnBad UserResponse()`), we are testing the actual response. In this case, there's no need for an exception, practically we are receiving a JSON (based on the `Map` interface. In other words, the `Map` interface can be converted as a JSON with no issues), and we can assert that the password does not comply with the constraints. If you want to see this in action using the REST Client (VS Code or IntelliJ plugin), you can find the `users.http` file in the `src/http` folder.

Congratulations! You have created two awesome web applications with Spring Boot!

Note Remember that you have access to all the source code for this book. You can go to the Apress site under the book's name and click Resources/Source Code.

Spring Boot Web: Overriding Defaults

Now that we have completed our two apps, we can explore some of the defaults that we can override. The main goal with the two apps is that the My Retro App will use the Users App for authentication and authorization, plus some other features. If we want to run the two apps on the same machine, it won't be possible with the default settings, because both apps use the same port to run. But no worries, Spring Boot allows us to override this default, as described next. Spring Boot also allows us to override the default JSON date format and the default application container, as described in the subsequent sections.

Overriding Default Server Settings

By default, the embedded Tomcat server starts on port 8080, but you can easily change that by using the following property:

```
server.port=8082
```

One of the cool features of Spring is that you can apply the Spring Expression Language (SpEL) to these properties. For example, when you create an executable JAR (./gradlew build), you can pass parameters when running your application. You can do the following:

```
java -jar users-0.0.1-SNAPSHOT.jar --port=8082
```

and in your application.properties file, you have something like this:

```
server.port=${port:8082}
```

This expression means that if you pass the --port argument, the application takes that value for the port; if not, it's set to 8182.

> **Tip**   This is just a small taste of what you can do with SpEL. If you want to know more, go to https://docs.spring.io/spring/docs/current/spring-framework-reference/core.html#expressions.

You can also change the server address, which is useful when you want to run your application using a particular IP address. For example:

```
server.address=10.0.0.7
```

You can also change the context of your application:

```
server.servlet.context-path=/contacts-app
```

And you can execute a cUrl command like this:

```
curl -I http://localhost:8080/contacts/users
```

You can have Tomcat with SSL by using the following properties:

```
server.port=8443
server.ssl.key-store=classpath:keystore.jks
server.ssl.key-store-password=secret
server.ssl.key-password=secret
```

We will revisit these properties and make our app work with SSL in Chapter. You can manage a session by using the following properties:

```
server.servlet.session.store-dir=/tmp
server.servlet.session.persistent=true
server.servlet.session.timeout=15
server.servlet.session.cookie.name=todo-cookie.dat
server.servlet.session.cookie.path=/tmp/cookies
```

You can enable HTTP/2 support as follows if your environment supports it:

```
server.http2.enabled=true
```

JSON Date Format

By default, the date types are exposed in the JSON response in a long format, but you can change that by providing your own pattern in the following properties:

```
spring.jackson.date-format=yyyy-MM-dd HH:mm:ss
spring.jackson.time-zone=MST7MDT
```

These properties format the date and also use the time zone that you specify.

Tip   If you want to know more about the available time zone IDs, execute `java.util.TimeZone#getAvailableIDs`. If you want to know more about which properties exist, check out `https://docs.spring.io/spring-boot/docs/current/reference/html/common-application-properties.html`.

Using a Different Application Container

By default, Spring Boot uses Tomcat (for web servlet apps) as an application container and sets up an embedded server. If you want to override this default, you can do it by modifying the Gradle build.gradle file, as shown in Listing 3-23.

Listing 3-23. build.gradle

```
plugins {
    id 'java'
    id 'org.springframework.boot' version '3.2.3'
    id 'io.spring.dependency-management' version '1.1.4'
}

group = 'com.apress'
version = '0.0.1-SNAPSHOT'
sourceCompatibility = '17'

repositories {
    mavenCentral()
}

ext['jakarta-servlet.version'] = '5.0.0'

dependencies {

    implementation('org.springframework.boot:spring-boot-starter-web')
    modules {
                module("org.springframework.boot:spring-boot-starter-
                tomcat") {
                        replacedBy("org.springframework.boot:spring-boot-
                        starter-jetty",

                                               "Use Jetty instead of
                                               Tomcat")

        }
    }
```

```
    implementation 'org.springframework.boot:spring-boot-starter-
validation'
```

// Jetty
**   implementation 'org.springframework.boot:spring-boot-**
**   starter-jetty'**

```
    // Lombok
    compileOnly 'org.projectlombok:lombok'
    annotationProcessor 'org.projectlombok:lombok'

    // Web
    implementation 'org.webjars:bootstrap:5.2.3'

    testImplementation 'org.springframework.boot:spring-boot-starter-test'
}

tasks.named('test') {
    useJUnitPlatform()
}

test {
    testLogging {
        events "passed", "skipped", "failed"
        showExceptions true
        exceptionFormat "full"
        showCauses true
        showStackTraces true
        showStandardStreams = false
    }
}
```

Spring Boot Web Clients

Now we are going to create a client that will connect to the Users App and obtain user information. We are going to use the RestTemplate class to connect externally to a service.

To create this client, we need to modify the current configuration of My Retro App. In the `config` package, we previously had only two classes: `MyRetroConfiguration` and `MyRetroProperties`. We need to add a third class, `UsersConfiguration`. In `MyRetroProperties`, we currently have some class declarations that allow us to bind the properties, but the problem is just the visibility, we need access to them. Figure 3-5 shows the new classes and packages for the client that the following listings will produce.

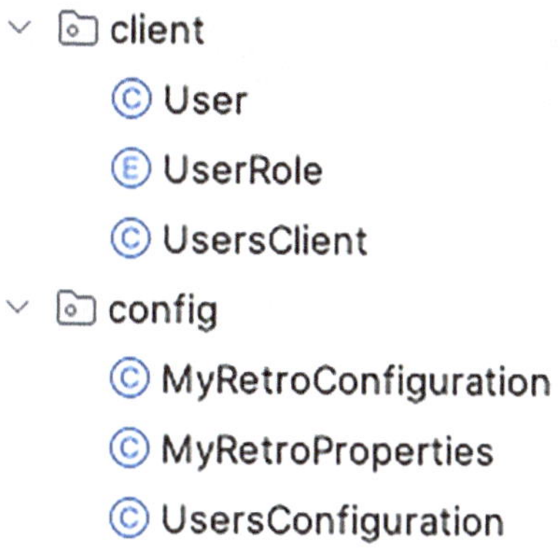

Figure 3-5. *The config and client packages*

The `MyRetroConfiguration.java`, `MyRetroProperties.java`, and `UsersConfiguration.java` files for the `config` package are shown in Listing 3-24, 3-25, 3-26, respectively.

Listing 3-24. src/main/java/com/apress/myretro/config/ MyRetroConfiguration.java

```
package com.apress.myretro.config;

import org.springframework.boot.context.properties.
EnableConfigurationProperties;
import org.springframework.context.annotation.Configuration;

@EnableConfigurationProperties({MyRetroProperties.class})
@Configuration
public class MyRetroConfiguration {
}
```

Listing 3-25. src/main/java/com/apress/myretro/config/
MyRetroProperties.java

```
package com.apress.myretro.config;

import lombok.Data;
import org.springframework.boot.context.properties.ConfigurationProperties;

@ConfigurationProperties(prefix="service")
@Data
public class MyRetroProperties {
    UsersConfiguration users;
}
```

Listing 3-26. src/main/java/com/apress/myretro/config/UsersConfiguration.java

```
package com.apress.myretro.config;

import lombok.Data;

@Data
public class UsersConfiguration {
    String server;
    Integer port;
    String username;
    String password;
}
```

As you can see, the only change is the visibility to add separated classes.

Next, let's look at the client package. Listing 3-27 and 3-28 show the User class and the UserRole enum. As you can see, are the same fields, except that we don't need the password.

Listing 3-27. src/main/java/com/apress/myretro/client/User.java

```
package com.apress.myretro.client;

import lombok.AllArgsConstructor;
import lombok.Data;
import lombok.NoArgsConstructor;
```

```java
import java.util.List;

@AllArgsConstructor
@NoArgsConstructor
@Data
public class User {
    private String email;

    private String name;

    private String gravatarUrl;

    private List<UserRole> userRole;

    private boolean active;
}
```

Listing 3-28. src/main/java/com/apress/myretro/client/UserRole.java

```java
package com.apress.myretro.client;

public enum UserRole {
    ADMIN, USER, INFO
}
```

The important class in the client package is UsersClient, shown in Listing 3-29, which will connect remotely to our other web app (Users App).

Listing 3-29. src/main/java/com/apress/myretro/client/UsersClient.java

```java
package com.apress.myretro.client;

import com.apress.myretro.config.MyRetroProperties;
import lombok.AllArgsConstructor;
import org.springframework.stereotype.Component;
import org.springframework.web.client.RestTemplate;

import java.text.MessageFormat;

@AllArgsConstructor
@Component
public class UsersClient {
```

```
private final String USERS_URL = "/users";
private final RestTemplate restTemplate = new RestTemplate();

private MyRetroProperties myRetroProperties;

public User findUserByEmail(String email) {
    String uri = MessageFormat.format("{0}:{1}{2}/{3}",
            myRetroProperties.getUsers().getServer(),
            myRetroProperties.getUsers().getPort().toString(),
            USERS_URL,email);
    return restTemplate.getForObject(uri, User.class);

}

}
```

Let's analyze the UsersClient class:

- MyRetroProperties: These are the properties that defined the external services to which we can add them using the prefix service and the class UsersConfiguration. So, our properties will be like this (from the src/main/resources/application.yaml file):

```
service:
  users:
    server: http://localhost
    port: 8082
    username: admin
    password: aW3sOm3
```

- RestTemplate: This class uses the Template pattern that hides all the boilerplate of creating a connection and dealing with exceptions. The RestTemplate provides several useful methods that allow us to get the objects. We are going to use more features of this class throughout the book.

As you can see this is very simple, right now we only required a method to look up for a user using the user's email and bring the data (except for the password, for now).

Testing the Client

To test the client, create the UsersClientTest class in the test structure. See Listing 3-30.

Listing 3-30. src/main/test/com/apress/myretro/UsersClientTest.java

```java
package com.apress.myretro;

import com.apress.myretro.client.User;
import com.apress.myretro.client.UsersClient;
import org.junit.jupiter.api.Test;
import org.springframework.beans.factory.annotation.Autowired;
import org.springframework.boot.test.context.SpringBootTest;

import static org.assertj.core.api.Assertions.assertThat;

@SpringBootTest
public class UsersClientTest {

    @Autowired
    UsersClient usersClient;

    @Test
    public void findUserTest() {
        User user = usersClient.findUserByEmail("norma@email.com");
        assertThat(user).isNotNull();
        assertThat(user.getName()).isEqualTo("Norma");
        assertThat(user.getEmail()).isEqualTo("norma@email.com");
    }

}
```

Listing 3-30 shows the test for the UsersClient. Of course, to run it, you need to have the Users app running on port 8082.

Summary

In this chapter you learned how to create a Spring Boot web application. You learned that Spring Web and Spring Web MVC are the foundation for Spring Boot. Spring Boot will use the default settings to configure your web application, but you learned how to change those defaults.

You learned the two different methods of using Web Servlet applications: using Annotation based and Functional Base. You completed the formal structure of the two projects, Users App and My Retro App, which means that adding new features in subsequent chapters will be relatively easy.

Currently, our applications hold the data using in-memory persistence. In Chapter 4, we are going to explore how to store the data on a SQL database and how Spring Boot can help us manage that data using a database engine.

CHAPTER 4

Spring Boot SQL Data Access

In the previous chapters, you learned how to create applications that hold data using in-memory persistence. In this chapter, you'll learn how to use Spring Boot to create applications that store data in a SQL database. Spring Boot relies on the *Spring Framework Data Access*, which provides access to SQL databases using the JdbcTemplate class. This class removes all the boilerplate of connecting to the database engine, session management, transaction management, and more.

Spring Boot can also use the power of the *Spring Data* project, which provides additional functionality such as using the Repository interface without worrying about the implementation because Spring Data takes care of it behind the scenes.

In this chapter, we will continue working with our two projects, Users App and My Retro App.

Spring Boot Features for SQL Databases

After you create a project in Spring Boot and add the Data Drivers dependencies (e.g., org.postgresql:postgresql), when the app starts, the Spring Boot *auto-configuration* will try to configure everything to create a DataSource implementation. If you are new to data development, normally you will require some information about your database: the URL (in the form jdbc:<engine>://<server>:<port>/<database>[/|?<additio nal-parameters>]), the *username* and *password* for the database, and sometimes the database engine *driver*. All these will get configured by Spring Boot unless you override those default values.

Another important feature is that Spring Boot uses *connection pools* that allow the application to have a better management, better concurrency, and performance when you application has persistence. By default, the *HikariCP* connection pool is selected if

© Felipe Gutierrez 2024
F. Gutierrez, *Pro Spring Boot 3*, https://doi.org/10.1007/978-1-4842-9294-5_4

present; normally this happens when you use the `spring-boot-starter-jdbc` starter or `spring-boot-starter-data-jpa` starter as a dependency. You have the option to change the connection pool from the default.

If you deploy your Spring Boot application to an application server, such as Tomcat, IBM WebSphere, Jetty, GlassFish, WildFly, or JBoss, among others, you can use the JNDI connection property in the `DataSource` (`spring.datasource.jndi-name`).

This chapter covers many new features for our main applications, so let's get started with the Spring Framework Data Access.

Spring Framework Data Access

As part of its core, the Spring Framework has Data Access, which offers the following features:

- *Transaction management*: Spring provides a complete abstraction for transaction management with features such as a consistent programming model across different APIs (such as JTA, JDBC, Hibernate, and JPA, supports a declarative transaction management or annotation based (using `@Transactional`), a way to configure transaction isolation levels, and excellent integration with Spring data access abstractions.

- *Data Access Object (DAO) support*: DAO support provides a consistent way to interact with different APIs (such as JDBC, Hibernate, and JPA), so any switch between them is easy to do and maintain. Spring also supplies an exception translator that is consistent between technologies, so you don't have to worry about specific API errors. To get all the benefits of DAO support, it's necessary to use the `@Repository` annotation.

- *Data Access with JDBC*: Using *Spring JDBC* you will only take care of specifying the SQL statements, declare parameters and provide the values, and provide the connection parameters. You can use the `JdbcTemplate` class to take care of the boilerplate of interacting with the database, and it can give you a better error interpretation when there is one. Spring JDBC provides features such as executing operations for batch processing, using the `SimpleJDBC` classes, and modeling JDBC operations as Java objects. It also provides embedded support and a way to initialize the `DataSource`. The easy way to start with Spring JDBC is to provide the `DataSource` (which requires the URL, username, password, and driver of your database engine), and then you can use the `JdbcTemplate` (this class requires the `DataSource`). Another important Spring JDBC feature is embedded database support (for database engines such as H2, HSQL, and Derby, among others), and you can initialize a database by providing SQL files with schema definitions and data.

- *Data Access with R2DBC*: Spring supports the implementation of reactive patterns for databases that use SQL for non-blocking scenarios. The Spring R2DBC brings the `DatabaseClient` class that is the core for control basic R2DBC processing and error handling, among other utility classes; also brings the `ConnectionFactory` implementation R2DBC connectivity among other utility classes.

- *Object Relational Mapping (ORM)*: Spring's ORM supports integration with the Java Persistence API (JPA) and Hibernate, Data Access Object (DAO) implementations, and transactions strategies. One of the most popular features of this technology is the use of reverse engineering, because it can create the table relationships based on the classes without any XML mapping.

- *Object-XML mappers*: Spring supports Object-XML mapping to convert an XML document to and from an object.

As you can see, the Spring Framework Data Access is the core of several data technologies that enable developers to create enterprise-ready data applications following the consistent Spring programming model. And, of course, Spring Boot takes advantage of all of this to help developers apply the common and default practices to avoid any errors and help with development.

JDBC with Spring Boot

As previously mentioned, Spring JDBC requires you to provide the connection parameters, specify the SQL statements, declare parameters and parameter values, and do some work when you have a result set, among other settings. With Spring Boot, you are covered with some of these features, where you don't need to specify any configuration at all.

By using the `spring-boot-starter-jdbc` starter dependency, Spring's `JdbcTemplate` and `NamedParameterJdbcTemplate` are auto-configured, enabling us to use Spring JDBC directly in our classes using the constructor. There are also some `spring.jdbc.template.*` properties that you can modify when needed.

Users App: Using Spring Boot JDBC

Our Users App currently is using in-memory persistence, so now it's time to switch and use the JDBC. I recommend creating an empty project from the Spring Initializr (`https://start.spring.io`) and starting from there. After you download the project and unzip it, you can import it into your favorite IDE. If you feel comfortable modifying your existing code, that will be fine as well. Figure 4-1 shows the structure and code that we'll develop in this section.

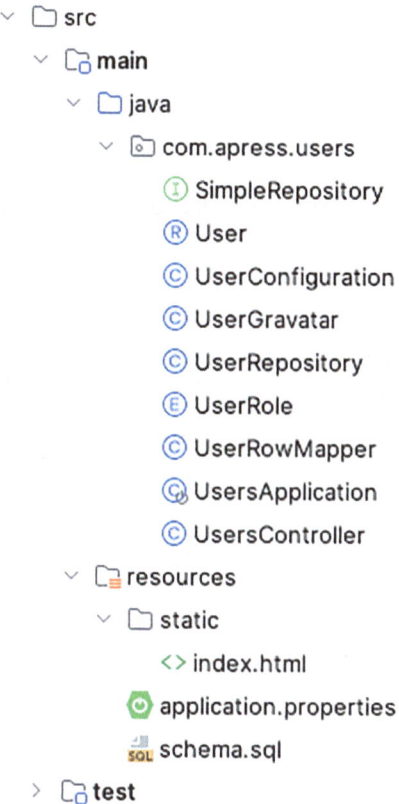

Figure 4-1. *Users App Project - directory structure*

As you can see, we are going to go back to using the controller programming.

Start by opening the build.gradle file and replacing its content with the content shown in Listing 4-1.

Listing 4-1. build.gradle

```
plugins {
    id 'java'
    id 'org.springframework.boot' version '3.2.3'
    id 'io.spring.dependency-management' version '1.1.4'
}

group = 'com.apress'
version = '0.0.1-SNAPSHOT'
sourceCompatibility = '17'
```

```
repositories {
    mavenCentral()
}

dependencies {
    implementation 'org.springframework.boot:spring-boot-starter-web'
    implementation 'org.springframework.boot:spring-boot-starter-
    validation'

    implementation 'org.springframework.boot:spring-boot-starter-jdbc'
    runtimeOnly 'com.h2database:h2'
    runtimeOnly 'org.postgresql:postgresql'

    compileOnly 'org.projectlombok:lombok'
    annotationProcessor 'org.projectlombok:lombok'

    // Web
    implementation 'org.webjars:bootstrap:5.2.3'

    testImplementation 'org.springframework.boot:spring-boot-starter-test'
}

tasks.named('test') {
    useJUnitPlatform()
}

test {
    testLogging {
        events "passed", "skipped", "failed"
        showExceptions true
        exceptionFormat "full"
        showCauses true
        showStackTraces true
        showStandardStreams = false
    }
}
```

The important change to the build.gradle file (from the previous version) is that we are adding three additional dependencies: the spring-boot-starter-jdbc dependency and the drivers h2 and postgresql (which are going to be used only at runtime). Keep in mind that we are using two drivers. If you are wondering which database driver Spring Boot will configure, h2 or postgresql, that answer is provided later in this chapter.

Next, create the SimpleRepository interface with the code shown in Listing 4-2.

Listing 4-2. src/main/java/com/apress/users/SimpleRepository.java

```java
package com.apress.users;

import java.util.Optional;

public interface SimpleRepository<D,ID>{
    Optional<D> findById(ID id);
    Iterable<D> findAll();
    D save(D d);
    void deleteById(ID id);
}
```

Note that the SimpleRepository interface is the same as in the previous version.

Gravatar - https://gravatar.com: Identifying the User

Create the UserGravatar class as shown in Listing 4-3.

Listing 4-3. src/main/java/com/apress/users/UserGravatar.java

```java
package com.apress.users;

import java.io.UnsupportedEncodingException;
import java.security.MessageDigest;
import java.security.NoSuchAlgorithmException;

public class UserGravatar {

    public static String getGravatarUrlFromEmail(String email){
        return String.format("https://www.gravatar.com/
        avatar/%s?d=wavatar", md5Hex(email));
    }
```

```java
    private static String hex(byte[] array) {
        StringBuffer sb = new StringBuffer();
        for (int i = 0; i < array.length; ++i) {
            sb.append(Integer.toHexString((array[i]
                    & 0xFF) | 0x100).substring(1, 3));
        }
        return sb.toString();
    }

    private static String md5Hex(String message) {
        try {
            MessageDigest md =
                    MessageDigest.getInstance("MD5");
            return hex(md.digest(message.getBytes("CP1252")));
        } catch (NoSuchAlgorithmException e) {
        } catch (UnsupportedEncodingException e) {
        }
        return "23bb62a7d0ca63c9a804908e57bf6bd4";
    }
}
```

The UserGravatar class is a simple utility class that will be useful to add a Gravatar icon to our web application based on the user's email address.

Model: enum and record Types

Create the UserRole enum as shown in Listing 4-4. This enum is very simple and is the same as in the previous versions.

Listing 4-4. src/main/java/com/apress/users/User.java

```java
package com.apress.users;

public enum UserRole {
    USER, ADMIN, INFO
}
```

Next, we are going to use the new Java record type from our User. Open the User record and replace its content with the content shown in Listing 4-5.

Listing 4-5. src/main/java/com/apress/users/User.java

```
package com.apress.users;

import lombok.Builder;
import lombok.Singular;

import java.util.ArrayList;
import java.util.List;
import java.util.Objects;
import java.util.regex.Matcher;
import java.util.regex.Pattern;

@Builder
public record User(Integer id, String email, String name, String password,
boolean active, String gravatarUrl, @Singular("role") List<UserRole>
userRole) {
    public User {
        Objects.requireNonNull(email);
        Objects.requireNonNull(name);
        Objects.requireNonNull(password);

        Pattern pattern = Pattern.compile("^(?=.*[0-9])(?=.*[a-z])
        (?=.*[A-Z])(?=.*[@#$%^&+=!])(?=\\S+$).{8,}$");
        Matcher matcher = pattern.matcher(password);
        if (!matcher.matches())
            throw new IllegalArgumentException("Password must be at least 8
            characters long and contain at least one number, one uppercase,
            one lowercase and one special character");

        pattern = Pattern.compile("^[a-zA-Z0-9_!#$%&'*+/=?`{|}~^.-]
        +@[a-zA-Z0-9.-]+$");
        matcher = pattern.matcher(email);
        if (!matcher.matches())
            throw new IllegalArgumentException("Email must be a valid email
            address");
```

```
        if (gravatarUrl == null) {
            gravatarUrl = UserGravatar.getGravatarUrlFromEmail(email);
        }

        if (userRole == null) {
            userRole = new ArrayList<>(){{ add(UserRole.INFO); }};
        }
    }

    public User withId(Integer id){
        return new User(id, this.email(), this.name(), this.password(),
        this.active(), this.gravatarUrl(), this.userRole());
    }
}
```

Listing 4-5 shows the User *record* that is based on the Java 14 features. Records are immutable data types, and you only need the name of the field and its type. In a previous version we were using *Lombok*, which generates the setters, getters, and toString(), equals(), and hashCode() methods when using the @Data annotation. Here, using record, we have the same result, and a little different way to get the value. Remember, because it's immutable, you cannot alter the object once it is created. Let's analyze this record:

- public record User(Integer id, String email, String name, String password, boolean active, String gravatarUrl, @Singular("role") List<UserRole> userRole){}: This is how you declare a record, providing the name of the fields and their types. Note that we can also use annotations, in this case the @Singular from Lombok.

- public User{}: For this new record type, you can have a canonical constructor, a compact constructor, or a custom constructor. In this case, we are using a compact constructor, enabling us to omit all the arguments and apply certain logic.

- Objects.requireNonNull: We are using this method call because it enables us ensure that the object is built correctly, which ensures controlled behavior; this method call also is easier to debug.

- public User withId(Integer id){}. With a record type, you are allowed to use methods that can return a copy of the object, like in this case, we are creating one copy based on its Integer id type.

Note that we are still using the Lombok annotations. I think they are still useful. You might consider this to be unnecessary code, but remember that we are dealing with immutability, and this is what we need to do. It's up to you to choose what is more convenient for you and your apps.

JdbcTemplate and RowMapper

Create the UserRowMapper class as shown in Listing 4-6.

Listing 4-6. src/main/java/com/apress/users/UserRowMapper.java

```
package com.apress.users;

import org.springframework.jdbc.core.RowMapper;

import java.sql.Array;
import java.sql.ResultSet;
import java.sql.SQLException;
import java.util.Arrays;
import java.util.List;
import java.util.stream.Collectors;

public class UserRowMapper implements RowMapper<User> {
    @Override
    public User mapRow(ResultSet rs, int rowNum) throws SQLException {
        Array array = rs.getArray("user_role");
        String[] rolesArray = Arrays.copyOf((Object[])array.getArray(),
        ((Object[])array.getArray()).length, String[].class);
        List<UserRole> roles = Arrays.stream(rolesArray).
        map(UserRole::valueOf).collect(Collectors.toList());

        User newUser = User.builder()
                .id(rs.getInt("id"))
                .name(rs.getString("name"))
                .email(rs.getString("email"))
```

```
                .password(rs.getString("password"))
                .userRole(roles)
                .build();

        return newUser;
    }
}
```

The UserRowMapper class implements the RowMapper functional interface (with the method mapRow(ResultSet,int)), which is being used along with the JdbcTemplate class (introduced shortly) to map rows of a java.sql.ResultSet result one row at a time. As you can see, we are building our User object and returning it.

Next, create the UserRepository class. See Listing 4-7.

Listing 4-7. src/main/java/apress/com/users/UserRepository.java

```
package com.apress.users;

import lombok.AllArgsConstructor;
import org.springframework.jdbc.core.JdbcTemplate;
import org.springframework.jdbc.support.GeneratedKeyHolder;
import org.springframework.jdbc.support.KeyHolder;
import org.springframework.stereotype.Repository;

import java.sql.PreparedStatement;
import java.sql.Statement;
import java.sql.Types;
import java.util.Optional;

@AllArgsConstructor
@Repository
public class UserRepository implements SimpleRepository<User, Integer> {

    private JdbcTemplate jdbcTemplate;

    @Override
    public Optional<User> findById(Integer id) {
        String sql = "SELECT * FROM users WHERE id = ?";
        Object[] params = new Object[] { id };
```

```java
    User user = jdbcTemplate.queryForObject(sql, params, new int[] {
    Types.INTEGER }, new UserRowMapper());
    return Optional.ofNullable(user);
}

@Override
public Iterable<User> findAll() {
    String sql = "SELECT * FROM users";
    return this.jdbcTemplate.query(sql, new UserRowMapper());
}

@Override
public User save(User user) {
    String sql = "INSERT INTO users (name, email, password, gravatar_
    url,user_role,active) VALUES (?, ?, ?, ?, ?, ?)";

    KeyHolder keyHolder = new GeneratedKeyHolder();

    jdbcTemplate.update(connection -> {
        String[] array = user.userRole().stream().map(Enum::name).
        toArray(String[]::new);
        PreparedStatement ps = connection.prepareStatement(sql,
        Statement.RETURN_GENERATED_KEYS);
        ps.setString(1, user.name());
        ps.setString(2, user.email());
        ps.setString(3, user.password());
        ps.setString(4, user.gravatarUrl());
        ps.setArray(5, connection.createArrayOf("varchar", array));
        ps.setBoolean(6, user.active());
        return ps;
    }, keyHolder);

    User userCreated = user.withId((Integer)keyHolder.getKeys().
    get("id"));
    return userCreated;
}
```

```
    @Override
    public void deleteById(Long id) {
        String sql = "DELETE FROM users WHERE id = ?";
        jdbcTemplate.update(sql, id);
    }
}
```

The UserRepository class implements the SimpleRepository interface. In this class we are using the JdbcTemplate class. This class will be *autowired* by Spring Boot auto-configuration. Remember, this class will do the heavy lifting of the interaction with the database through JDBC calls. With JdbcTemplate, we have several ways to interact with the database:

- query: This is one of several overload methods that executes a SQL query and maps each row to a result object via RowMapper, in this case our UserRowMapper.

- queryForObject: This is one of several overload methods that query a given SQL statement to create a prepared statement from a list of arguments to bind to the query, and it maps a single result row to a result object via a RowMapper.

- update: This is one of several overload methods that issues a single SQL update operation (you can use INSERT, UPDATE, or DELETE statements) via a prepared statement, binding the given arguments.

It's worth mentioning that in the save(User user) method we are using the record withId method call, so we add the Id from the keyHolder (this is how we get an *auto-increment* value back). The keyHolder will be populated with different keys, and the one we want will be generated.

If you look at the JdbcTemplate documentation (https://docs.spring.io/spring-framework/docs/current/javadoc-api/org/springframework/jdbc/core/JdbcTemplate.html), you'll see that brings implementation to methods such as queryForMap, queryForList, queryForRowSet, queryForStream, execute, batchUpdate, and many more.

Remember that the JdbcTemplate already knows how to interact with the database based on the DataSource information (URL, username, password, etc.).

Adding the Web Controller

Create the UsersController class as shown in Listing 4-8.

Listing 4-8. src/main/java/apress/com/users/UsersController.java

```java
package com.apress.users;

import jakarta.validation.Valid;
import lombok.AllArgsConstructor;
import org.springframework.http.HttpStatus;
import org.springframework.http.ResponseEntity;
import org.springframework.http.converter.HttpMessageNotReadableException;
import org.springframework.validation.FieldError;
import org.springframework.web.bind.MethodArgumentNotValidException;
import org.springframework.web.bind.annotation.*;
import org.springframework.web.servlet.support.ServletUriComponentsBuilder;

import java.net.URI;
import java.time.LocalDateTime;
import java.util.HashMap;
import java.util.Map;

@AllArgsConstructor
@RestController
@RequestMapping("/users")
public class UsersController {

    private SimpleRepository<User,Integer> userRepository;

    @GetMapping
    public ResponseEntity<Iterable<User>> getAll(){
        return ResponseEntity.ok(this.userRepository.findAll());
    }

    @GetMapping("/{id}")
    public ResponseEntity<User> findUserById(@PathVariable Integer id){
        return ResponseEntity.of(this.userRepository.findById(id));
    }
```

```java
@RequestMapping(method = {RequestMethod.POST,RequestMethod.PUT})
public ResponseEntity<User> save(@RequestBody @Valid User user){
    User result = this.userRepository.save(user);

    URI location = ServletUriComponentsBuilder
            .fromCurrentRequest()
            .path("/{id}")
            .buildAndExpand(user)
            .toUri();
    return ResponseEntity.created(location).body(this.userRepository.
    findById(result.id()).get());
}

@DeleteMapping("/{id}")
@ResponseStatus(HttpStatus.NO_CONTENT)
public void delete(@PathVariable Integer id){
    this.userRepository.deleteById(id);
}

@ExceptionHandler(MethodArgumentNotValidException.class)
@ResponseStatus(HttpStatus.BAD_REQUEST)
public Map<String, String> handleValidationExceptions(MethodArgumentNot
ValidException ex) {
    Map<String, String> errors = new HashMap<>();
    ex.getBindingResult().getAllErrors().forEach((error) -> {
        String fieldName = ((FieldError) error).getField();
        String errorMessage = error.getDefaultMessage();
        errors.put(fieldName, errorMessage);
    });
    errors.put("time", LocalDateTime.now().format(java.time.format.
    DateTimeFormatter.ISO_LOCAL_DATE_TIME));
    return errors;
}

@ExceptionHandler(HttpMessageNotReadableException.class)
@ResponseStatus(HttpStatus.BAD_REQUEST)
public Map<String,Object> handleHttpMessageNotReadableException
(HttpMessageNotReadableException ex){
```

```
        Map<String,Object> errors = new HashMap<>();
        errors.put("code",HttpStatus.BAD_REQUEST.value());
        errors.put("message",ex.getMessage());
        errors.put("time", LocalDateTime.now().format(java.time.format.
        DateTimeFormatter.ISO_LOCAL_DATE_TIME));
        return errors;
    }
}
```

The UsersController class should be familiar at this point. We are using annotation-based programming to create our web controller that will respond to any of the / users endpoint requests. Note that the save method has the @Valid annotation for the User class. Actually, the validation will never be trigger and this is because what is happening first is the construction of the object, then the validation, and if you remember we have the new User as record type, and we have some validation in the compact constructor; what this means is that the constructor will send the error if the fields fail in the Objects.requireNonNull call or the pattern matcher logic, if the logic fails then the constructor will throw the IllegalArgumentException, which will cascade to the HttpMessageNotReadableException, and the handleHttpMessageNotReadableException will be called.

Adding Users when the App Is Ready

Create the UserConfiguration class as shown in Listing 4-9.

Listing 4-9. src/main/java/apress/com/users/UserConfiguration.java

```
package com.apress.users;

import org.springframework.boot.context.event.ApplicationReadyEvent;
import org.springframework.context.ApplicationListener;
import org.springframework.context.annotation.Bean;
import org.springframework.context.annotation.Configuration;

@Configuration
public class UserConfiguration {
```

```
@Bean
ApplicationListener<ApplicationReadyEvent> init(SimpleRepository
userRepository) {
    return applicationReadyEvent -> {
        User ximena = User.builder()
                .email("ximena@email.com")
                .name("Ximena")
                .password("aw2sOmeR!")
                .active(true)
                .role(UserRole.USER)
                .build();

        userRepository.save(ximena);

        User norma = User.builder()
                .email("norma@email.com")
                .name("Norma")
                .password("aw2sOmeR!")
                .active(true)
                .role(UserRole.USER)
                .role(UserRole.ADMIN)
                .build();

        userRepository.save(norma);
    };
}
}
```

The UserConfiguration class is marked with the @Configuration annotation, which means that Spring Boot will execute any declaration of @Value, @Bean, etc. In this case we have the @Bean that will create an ApplicationListener<Application ReadyEvent>, which means that it will get executed when the application is ready. Also, look that it has the SimpleRepository interface as a parameter, and right now, we have only the UserRepository implementation, so it will be injected here. This method creates two users and saves them into our database, but which one? H2 or PostgreSQL? If there are no properties declared (driver, username, password) the auto-configuration will setup the H2 by default, but if we provide the driver, username and password, the auto-configuration will use those values and try to create the DataSource object to connect to the database and perform any SQL statement.

Next, open the `application.properties` file and add the content shown in Listing 4-10.

Listing 4-10. src/main/resources/application.properties

```
# H2
spring.h2.console.enabled=true

# DataSource
spring.datasource.generate-unique-name=false
spring.datasource.name=test-db
```

The following are the new properties in the `application.properties` file:

- `spring.h2.console.enable`: This property allows us to have in development the `/h2-console` endpoint that brings a small UI to manipulate in-memory databases for the H2 engine. The default is `false`, but we are enabling it.

- `spring.datasource.generate-unique-name`: By default, Spring Boot auto-generates the name of the database, so by setting this property to `false` we are stopping that logic.

- `spring.datasource.name`: This property names our database as `test-db`, regardless of the database engine.

We'll discuss a few more properties that are important for our project later in this chapter.

Database Initialization

One of the most important features of Spring Boot when you have the `spring-boot-starter-jdbc` starter is that you can add SQL files that will be executed automatically if found in the classpath. They must follow a specific naming convention. In this case, the `schema.sql` file (for `CREATE`, `DROP`, etc. statements to manipulate the database), and a `data.sql` file (where you can have all the `INSERT` or `UPDATE` statements). You can have multiple schema files and even can be recognized by database engine, for example you can have `schema-h2.sql` and `schema-postgresql.sql`; the same applies for the `data-{engine}.sql` files. To use this feature, it is important to use the `spring.sql.init.platform` property and either `h2`, `postgresql`, `mysql`, `oracle`, `hsqldb`, etc.

Next, create the schema.sql file as shown in Listing 4-11. We don't need the data. sql file because we already created some users in the UserConfiguration class.

Listing 4-11. src/main/resources/schema.sql

```
DROP TABLE IF EXISTS USERS CASCADE;
CREATE TABLE USERS
(
    ID            INTEGER        NOT NULL AUTO_INCREMENT,
    EMAIL         VARCHAR(255)   NOT NULL UNIQUE,
    NAME          VARCHAR(100)   NOT NULL,
    GRAVATAR_URL  VARCHAR(255)   NOT NULL,
    PASSWORD      VARCHAR(255)   NOT NULL,
    USER_ROLE     VARCHAR(5) ARRAY NOT NULL DEFAULT ARRAY ['INFO'],
    ACTIVE        BOOLEAN        NOT NULL,
    PRIMARY KEY (ID)
);
```

Listing 4-11 shows the SQL statement that will be executed when the application starts. As you can see, it is very simple.

Next, create the index.html file (this is just to render something when we go to the / endpoint). You can copy from the other projects and put/create it in the src/main/ resources/static folder. Remember that in a web app with Spring Boot, if it finds an index.html in the static (or public/) folder, it will be rendered. Now we are ready!

Testing the Users App

As this point we need to test our application again. Create the UsersHttpRequestTests class shown in Listing 4-12.

Listing 4-12. src/test/java/apress/com/users/UsersHttpRequestTests.java

```
package com.apress.users;

import org.junit.jupiter.api.Test;
import org.springframework.beans.factory.annotation.Autowired;
import org.springframework.boot.test.context.SpringBootTest;
import org.springframework.boot.test.web.client.TestRestTemplate;
```

```java
import org.springframework.beans.factory.annotation.Value;
import org.springframework.http.ResponseEntity;

import java.util.Arrays;
import java.util.Collection;

import static org.assertj.core.api.Assertions.assertThat;
import static org.assertj.core.api.Assertions.assertThatThrownBy;

@SpringBootTest(webEnvironment = SpringBootTest.WebEnvironment.RANDOM_PORT)
public class UsersHttpRequestTests {

    @Value("${local.server.port}")
    private int port;

    private final String BASE_URL = "http://localhost:";
    private final String USERS_PATH = "/users";

    @Autowired
    private TestRestTemplate restTemplate;

    @Test
    public void indexPageShouldReturnHeaderOneContent() throws Exception {
        assertThat(this.restTemplate.getForObject(BASE_URL + port,
                String.class)).contains("Simple Users Rest Application");
    }

    @Test
    public void usersEndPointShouldReturnCollectionWithTwoUsers() throws
    Exception {
        Collection<User> response = this.restTemplate.
                getForObject(BASE_URL + port + USERS_PATH,
                Collection.class);

        assertThat(response.size()).isGreaterThan(1);
    }

    @Test
    public void shouldRetrunErrorWhenPostBadUserForm() throws Exception {
        assertThatThrownBy(() -> {
```

```java
        User user =  User.builder()
                .email("bademail")
                .name("Dummy")
                .active(true)
                .password("aw2sO")
                .build();
    }).isInstanceOf(IllegalArgumentException.class)
            .hasMessageContaining("Password must be at least 8
            characters long and contain at least one number, one
            uppercase, one lowercase and one special character");
}

@Test
public void userEndPointPostNewUserShouldReturnUser() throws
Exception {

    User user =  User.builder()
            .email("dummy@email.com")
            .name("Dummy")
            .password("aw2sOmeR!")
            .active(true)
            .role(UserRole.USER)
            .build();

    User response =  this.restTemplate.postForObject(BASE_URL + port +
    USERS_PATH,user,User.class);

    assertThat(response).isNotNull();
    assertThat(response.email()).isEqualTo(user.email());

    Collection<User> users = this.restTemplate.
            getForObject(BASE_URL + port + USERS_PATH,
            Collection.class);

    assertThat(users.size()).isGreaterThanOrEqualTo(2);

}
```

```
@Test
public void userEndPointDeleteUserShouldReturnVoid() throws Exception {
    this.restTemplate.delete(BASE_URL + port + USERS_PATH + "/norma@
    email.com");

    Collection<User> users = this.restTemplate.
            getForObject(BASE_URL + port + USERS_PATH,
            Collection.class);

    assertThat(users.size()).isLessThanOrEqualTo(2);
}

@Test
public void userEndPointFindUserShouldReturnUser() throws Exception{
    User user = this.restTemplate.getForObject(BASE_URL + port + USERS_
    PATH + "/1",User.class);

    assertThat(user).isNotNull();
    assertThat(user.email()).isEqualTo("ximena@email.com");
}
}
```

Listing 4-12 shows the tests for our Users app. The only difference from previous versions is the User instance creation. Before you run the code, analyze the various tests.

When you are ready, you can run the tests either in your IDE or by executing the following command:

```
./gradlew clean test

..
UsersHttpRequestTests > userEndPointFindUserShouldReturnUser() PASSED
UsersHttpRequestTests > userEndPointDeleteUserShouldReturnVoid() PASSED
UsersHttpRequestTests > shouldRetrunErrorWhenPostBadUserForm() PASSED
UsersHttpRequestTests > indexPageShouldReturnHeaderOneContent() PASSED
UsersHttpRequestTests > userEndPointPostNewUserShouldReturnUser() PASSED
UsersHttpRequestTests >
usersEndPointShouldReturnCollectionWithTwoUsers() PASSED

..
```

But wait... what happen with our Tests?

Our tests passed, but how? Where does the data persist? Which database engine was used if we have two drivers declared? Do you know the answer? Did you notice that we are missing something?

We are missing the connection parameters! The URL, username, password, and database driver are missing! We need to set up the DataSource with these parameters (the JdbcTemplate has the DataSource dependency, so it's necessary to declare it).

Normally, we declare these parameters in the command line, as environment variables, or in the application.properties/yaml file, and it looks something like this:

```
spring.datasource.url=jdbc:h2:mem:test-db;DB_CLOSE_DELAY=-1;DB_CLOSE_ON_
EXIT=FALSE
spring.datasource.username=sa
spring.datasource.password=
spring.datasource.driver-class-name=org.h2.Driver
```

Thanks to the magic of Spring Boot auto-configuration, we no longer need to do this. It will detect that we have two drivers but no connection parameters set, so it will use the H2 embedded driver to set up the default values and create the DataSource for the JdbcTemplate class, execute the schema.sql script, and that's it! We have a fully functional application that is using an H2 engine in-memory database.

This database initialization (execution of the schema.sql and data.sql) happens *only* with embedded databases such as H2, HSQL, and Derby. If we want initialization regardless of the database engine we use, we need to set the spring.sql.init.mode property to always.

Note   Remember that you have access to all the source code for this book (https://github.com/felipeg48/pro-spring-boot-3rd). You can find it in the Apress site and look for the right chapter https://www.apress.com/gp/services/source-code.

Running the Users App

Let's run our application and see in action some of the endpoints. You can run it from your IDE or you can use the following command in the terminal:

```
./gradlew bootRun
```

You should see the following in the console output:

```
..
o.s.b.a.h2.H2ConsoleAutoConfiguration    : H2 console available at '/h2-
console'. Database available at 'jdbc:h2:mem:test-db'
..
```

It shows that the /h2-console endpoint is available, so open it in the browser by going to http://localhost:8080/h2-console. See Figure 4-2.

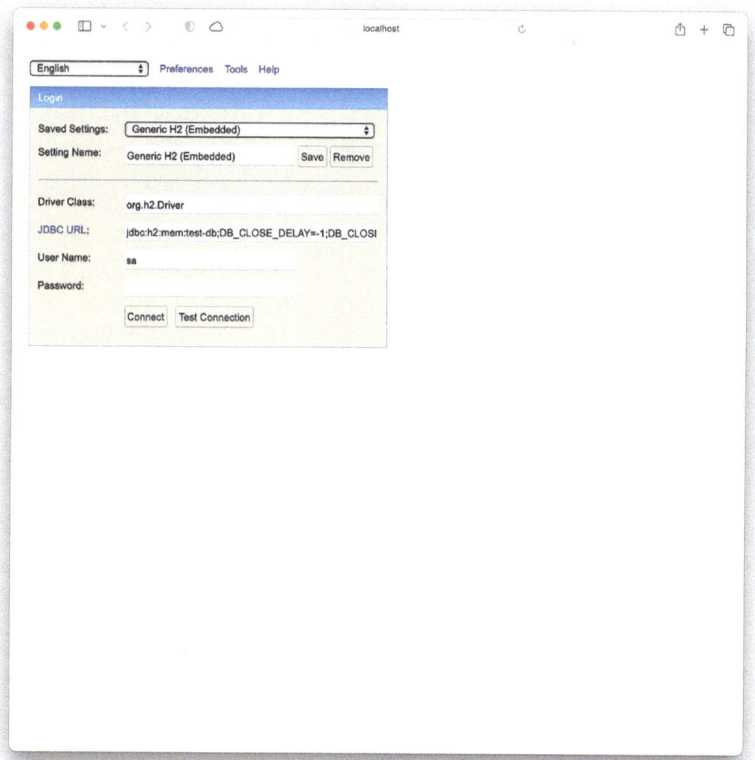

Figure 4-2. *http://localhost:8080/h2-console*

Make sure the JDBC URL field has the following value and then click the Connect button:

```
jdbc:h2:mem:test-db;DB_CLOSE_DELAY=-1;DB_CLOSE_ON_EXIT=FALSE
```

After you connect, you'll see the table USERS and the data, as shown in Figure 4-3.

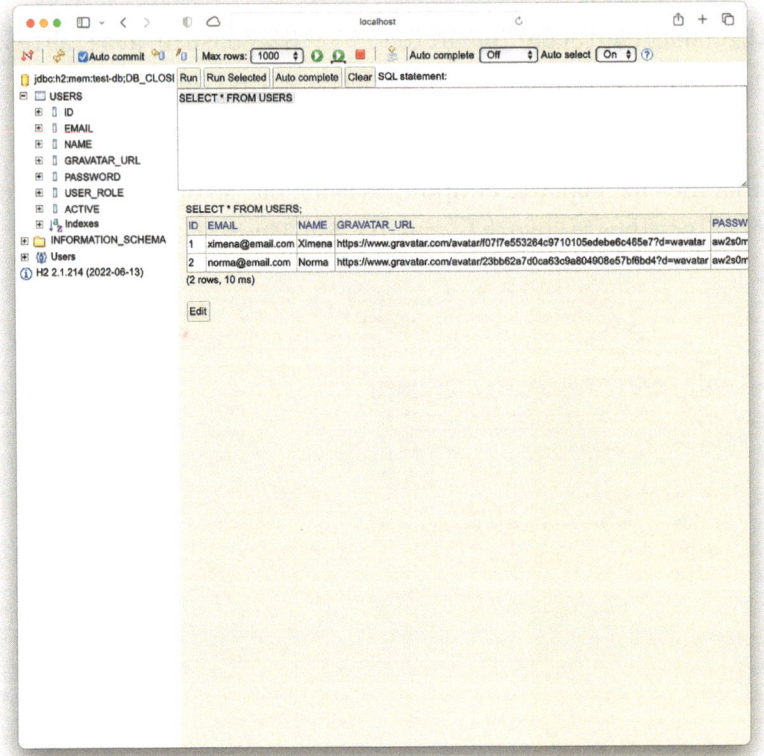

Figure 4-3. *USERS table*

And if you direct your browser to the http://localhost:8080/users endpoint, you should see the JSON response shown in Figure 4-4.

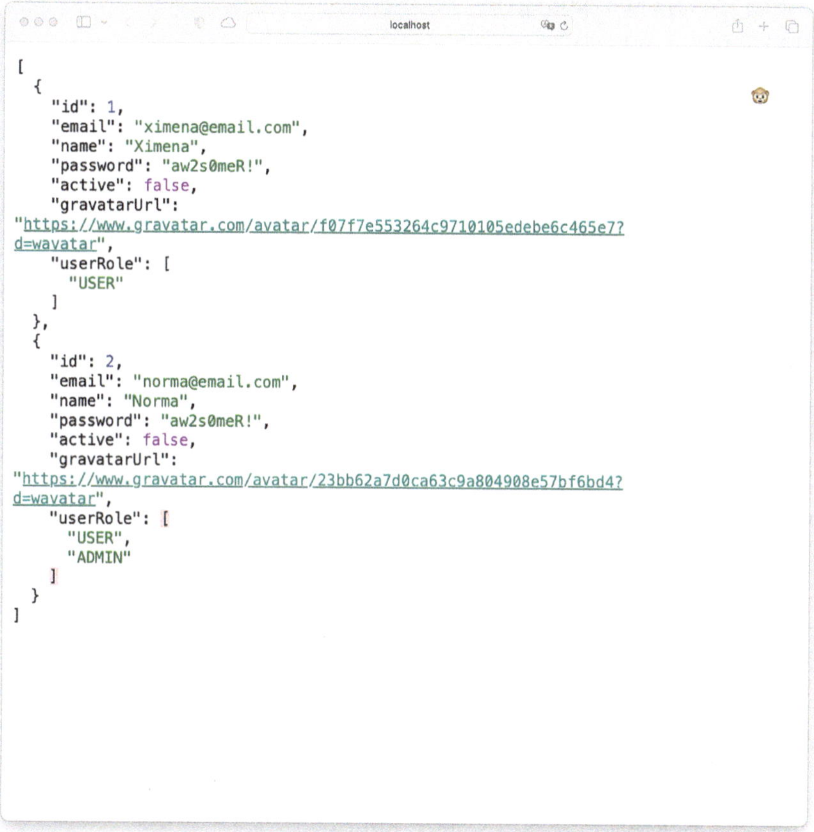

```
[
  {
    "id": 1,
    "email": "ximena@email.com",
    "name": "Ximena",
    "password": "aw2s0meR!",
    "active": false,
    "gravatarUrl":
"https://www.gravatar.com/avatar/f07f7e553264c9710105edebe6c465e7?
d=wavatar",
    "userRole": [
      "USER"
    ]
  },
  {
    "id": 2,
    "email": "norma@email.com",
    "name": "Norma",
    "password": "aw2s0meR!",
    "active": false,
    "gravatarUrl":
"https://www.gravatar.com/avatar/23bb62a7d0ca63c9a804908e57bf6bd4?
d=wavatar",
    "userRole": [
      "USER",
      "ADMIN"
    ]
  }
]
```

Figure 4-4. `http://localhost:8080/users`

Using PostgreSQL

To use PostgreSQL, we are going to use Docker Compose, as shown in Listing 4-13.

Listing 4-13. docker-compose.yaml

```yaml
version: "3"
services:
  postgres:
    image: postgres
    restart: always
    environment:
      POSTGRES_USER: postgres
```

173

```
    POSTGRES_PASSWORD: postgres
    POSTGRES_DB: test-db
  ports:
    - 5432:5432
```

Using Docker Compose is our best option and avoid to install multiple external services, is better this way, and actually can emulate a real-world scenario of using containers. Execute the following command in the terminal:

```
docker compose up
```

```
[+] Running 1/1
Container users-postgres-1   Started
```

Now, we need to add a new SQL script for PostgreSQL. Create the schema-postgresql.sql script shown in Listing 4-14.

Listing 4-14. src/main/resources/schema-postgresql.sql

```
DROP TABLE IF EXISTS USERS CASCADE;
CREATE TABLE USERS
(
    ID           SERIAL          NOT NULL,
    EMAIL        VARCHAR(255)    NOT NULL UNIQUE,
    NAME         VARCHAR(100)    NOT NULL,
    GRAVATAR_URL VARCHAR(255)    NOT NULL,
    PASSWORD     VARCHAR(255)    NOT NULL,
    USER_ROLE    VARCHAR[]       NOT NULL,
    ACTIVE       BOOLEAN         NOT NULL,
    PRIMARY KEY (ID)
);
```

Listing 4-14 shows the PostgreSQL schema version. Notice that the ID changed from AUTO_INCREMENT to SERIAL, and the USER_ROLE from ARRAY to VARCHAR[] declaration.

Next, rename the current schema.sql file to schema-h2.sql. This is very important! Next, open your application.properties file and replace its content with the content shown in Listing 4-15.

Listing 4-15. src/main/resources/application.properties

```
# H2
# spring.h2.console.enabled=true

# DataSource
spring.datasource.generate-unique-name=false
spring.datasource.name=test-db

# SQL init
spring.sql.init.mode=always
spring.sql.init.platform=postgresql

# Postgresql
spring.datasource.url=jdbc:postgresql://localhost:5432/test-db
spring.datasource.username=postgres
spring.datasource.password=postgres
spring.datasource.driver-class-name=org.postgresql.Driver
```

Let's analyze the new application.properties file:

- We added a comment for the spring.h2.console.enable property, we don't need it anymore.

- The new # SQL init section has the spring.sql.init.mode property set to always, meaning that it will do the initialization of the database and it will search for the schema.sql and data.sql files. However, we renamed the schema.sql file to schema-h2.sql and we added the schema-postgresql, so we need to specify which platform the app needs to use; in this case we are using the spring.sql.init.platform set to postgresql, which means that it will execute the schema-postgresql.sql script.

- The new # Postgresql section shows all the properties required for the Postgres engine to run (with Docker Compose). All these variables can be omitted and use environment variables as well.

If you execute the tests from Listing 4-14 again, they should pass without any issue, and if you run your application, you should have the same result.

My Retro App: Using Spring Boot JDBC

Now let's see how we can use JDBC in the My Retro App. In this case we won't be using the record type for our class model, but you are welcome to experiment on your own. Again, I recommend creating an empty project from the Spring Initializr (https://start.spring.io) and starting from there. After you download the project and unzip it, you can import it into your favorite IDE. If you feel comfortable modifying your existing code, that will be fine as well. Figure 4-5 shows the structure and code that we'll develop in this section.

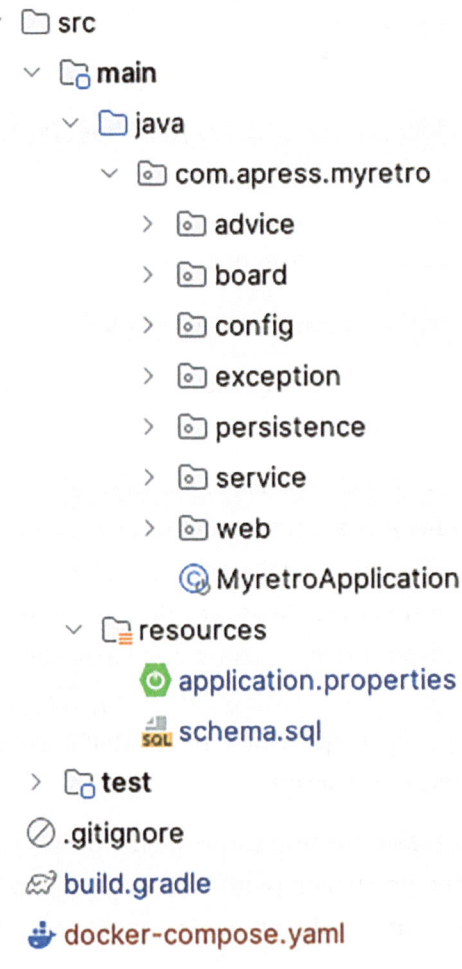

Figure 4-5. *Myretro App with JDBC project structure*

Make sure to enter com.apress in the Group field and myretro in the Artifact and Name fields. After you download the project, unzip it, and import it to your favorite IDE, open the build.gradle and replace its content with the content shown in Listing 4-16.

Listing 4-16. build.gradle

```
plugins {
    id 'java'
    id 'org.springframework.boot' version '3.2.3'
    id 'io.spring.dependency-management' version '1.1.4'
}

group = 'com.apress'
version = '0.0.1-SNAPSHOT'
sourceCompatibility = '17'

repositories {
    mavenCentral()
}

dependencies {
    implementation 'org.springframework.boot:spring-boot-starter-web'
    implementation 'org.springframework.boot:spring-boot-starter-
    validation'
    implementation 'org.springframework.boot:spring-boot-starter-aop'

    implementation 'org.springframework.boot:spring-boot-starter-jdbc'

    developmentOnly 'org.springframework.boot:spring-boot-docker-compose'

    annotationProcessor 'org.springframework.boot:spring-boot-
    configuration-processor'

    runtimeOnly 'org.postgresql:postgresql'

    compileOnly 'org.projectlombok:lombok'
    annotationProcessor 'org.projectlombok:lombok'

    testImplementation 'org.springframework.boot:spring-boot-starter-test'
}
```

```
tasks.named('test') {
    useJUnitPlatform()
}
```

As you can see in Listing 4-16, we are using the spring-boot-starter-jdbc starter dependency, the postgresql, and a new spring-boot-docker-compose starter dependency. Note that we are not using the H2 dependency. Also note that the spring-boot-docker-compose dependency will be used as developmentOnly. We'll discuss what this new dependency does when we run our application. This is a new feature in Spring Boot 3.1.

Next, create the board package with the following domain/model classes: CardType, Card, and RetroBoard (see Listings 4-17, 4-18, and 4-19, respectively).

Listing 4-17. src/main/java/apress/com/myretro/board/CardType.java

```
package com.apress.myretro.board;

public enum CardType {
    HAPPY,MEH,SAD
}
```

Listing 4-18. src/main/java/apress/com/myretro/board/Card.java

```
package com.apress.myretro.board;

import jakarta.validation.constraints.NotBlank;
import jakarta.validation.constraints.NotNull;
import lombok.AllArgsConstructor;
import lombok.Builder;
import lombok.Data;
import lombok.NoArgsConstructor;

import java.util.UUID;


@Builder
@AllArgsConstructor
@NoArgsConstructor
@Data
```

```java
public class Card {

    private UUID id;

    @NotBlank
    private String comment;

    @NotNull
    private CardType cardType;

    private UUID retroBoardId;
}
```

Listing 4-19. src/main/java/apress/com/myretro/board/RetroBoard.java

```java
package com.apress.myretro.board;

import jakarta.validation.constraints.NotBlank;
import lombok.*;

import java.util.HashMap;
import java.util.Map;
import java.util.UUID;

@Builder
@AllArgsConstructor
@NoArgsConstructor
@Data
public class RetroBoard {

    private UUID id;

    @NotBlank(message = "A name must be provided")
    private String name;

    @Singular
    private Map<UUID,Card> cards = new HashMap<>();
}
```

Note in Listing 4-19 that the cards field now is a Map type, which provides an easy way to find Card objects by UUID.

Next, create the persistence package with the SimpleRepository interface (see Listing 4-20) and RetroBoardRowMapper and RetroBoardRepository classes (see Listings 4-21 and 4-22 respectively).

Listing 4-20. src/main/java/apress/com/myretro/persistence/SimpleRepository.java

```java
package com.apress.myretro.persistence;

import java.util.Optional;

public interface SimpleRepository <D,ID>{

    Optional<D> findById(ID id);

    Iterable<D> findAll();

    D save(D d);

    void deleteById(ID id);
}
```

***Listi**ng **4-21**. src/main/java/apress/com/myretro/persistence/RetroBoardRowMapper.java

```java
package com.apress.myretro.persistence;

import com.apress.myretro.board.Card;
import com.apress.myretro.board.CardType;
import com.apress.myretro.board.RetroBoard;
import org.springframework.jdbc.core.RowMapper;

import java.sql.ResultSet;
import java.sql.SQLException;
import java.util.HashMap;
import java.util.Map;
import java.util.UUID;

public class RetroBoardRowMapper implements RowMapper<RetroBoard> {
    @Override
    public RetroBoard mapRow(ResultSet rs, int rowNum) throws
    SQLException {
```

```
        RetroBoard retroBoard = new RetroBoard();
        retroBoard.setId(UUID.fromString(rs.getString("id")));
        retroBoard.setName(rs.getString("name"));

        Map<UUID, Card> cards = new HashMap<>();
        do {
            Card card = new Card();
            card.setId(UUID.fromString(rs.getString("card_id")));
            card.setComment(rs.getString("comment"));
            card.setCardType(CardType.valueOf(rs.getString("card_type")));
            card.setRetroBoardId(retroBoard.getId());
            cards.put(card.getId(), card);
        } while (rs.next() && retroBoard.getId().equals(UUID.fromString(rs.
        getString("id"))));
        retroBoard.setCards(cards);

        return retroBoard;
    }
}
```

Listing 4-22. src/main/java/apress/com/myretro/persistence/
RetroBoardRepository.java

```
package com.apress.myretro.persistence;

import com.apress.myretro.board.Card;
import com.apress.myretro.board.RetroBoard;
import lombok.AllArgsConstructor;
import org.springframework.jdbc.core.JdbcTemplate;
import org.springframework.stereotype.Repository;
import org.springframework.transaction.annotation.Transactional;

import java.sql.Types;
import java.util.*;

@AllArgsConstructor
@Repository
```

```java
public class RetroBoardRepository implements
SimpleRepository<RetroBoard, UUID> {

    private JdbcTemplate jdbcTemplate;

    @Override
    public Optional<RetroBoard> findById(UUID uuid) {
        String sql = """
                SELECT r.ID AS id, r.NAME, c.ID AS card_id, c.CARD_TYPE AS
                card_type, c.COMMENT AS comment
                FROM RETRO_BOARD r
                LEFT JOIN CARD c ON r.ID = c.RETRO_BOARD_ID
                WHERE r.ID = ?
                """;
        List<RetroBoard> results = jdbcTemplate.query(sql, new Object[]
        {uuid}, new int[]{Types.OTHER}, new RetroBoardRowMapper());
        return results.isEmpty() ? Optional.empty() : Optional.of(results.
        get(0));
    }

    @Override
    public Iterable<RetroBoard> findAll() {
        String sql = """
                SELECT r.ID AS id, r.NAME, c.ID AS card_id, c.CARD_TYPE,
                c.COMMENT
                FROM RETRO_BOARD r
                LEFT JOIN CARD c ON r.ID = c.RETRO_BOARD_ID
                """;
        return jdbcTemplate.query(sql, new RetroBoardRowMapper());
    }

    @Override
    @Transactional
    public RetroBoard save(RetroBoard retroBoard) {

        if (retroBoard.getId() == null) {
            retroBoard.setId(UUID.randomUUID());
        }
```

```
    String sql = "INSERT INTO RETRO_BOARD (ID, NAME) VALUES (?, ?)";
    jdbcTemplate.update(sql, retroBoard.getId(), retroBoard.getName());

    Map<UUID, Card> mutableMap = new HashMap<>(retroBoard.getCards());

    for (Card card : retroBoard.getCards().values()) {
        card.setRetroBoardId(retroBoard.getId());
        card = saveCard(card);
        mutableMap.put(card.getId(), card);
    }
    retroBoard.setCards(mutableMap);
    return retroBoard;
}

@Override
@Transactional
public void deleteById(UUID uuid) {
    String sql = "DELETE FROM CARD WHERE RETRO_BOARD_ID = ?";
    jdbcTemplate.update(sql, uuid);

    sql = "DELETE FROM RETRO_BOARD WHERE ID = ?";
    jdbcTemplate.update(sql, uuid);
}

private Card saveCard(Card card) {
    if (card.getId() == null) {
        card.setId(UUID.randomUUID());
    }

    String sql = "INSERT INTO CARD (ID, CARD_TYPE, COMMENT, RETRO_
    BOARD_ID) VALUES (?, ?, ?, ?)";
    jdbcTemplate.update(sql, card.getId(), card.getCardType().name(),
    card.getComment(), card.getRetroBoardId());

    return card;
}

}
```

Listings 4-20, 4-21, and 4-22 show familiar classes from the previous version of the project. Notice that we are using the @Repository annotation that sets all the necessary DAO support, which includes a convenient translation from tech-specific exceptions such as SQLException, a more understandable way to know what's going on in case of error.

If you look closer at the save(RetroBoard) method in Listing 4-22, you will see that it is necessary to save in a different statement the Card, which means that we have two tables: one is the RetroBoard, and the other is the Card that must have an association with the RetroBoard. Also notice that we are using the @Transactional annotation, a declarative way to add transactions to your app. This annotation is based on aspect-oriented programming (AOP) and performs the transaction flow, the begin transaction, commit, and rollback if necessary.

Tip   If you want to know more about transactions, consult the Spring Data Access documentation at https://docs.spring.io/spring-framework/reference/data-access/transaction/declarative/tx-decl-explained.html.

After you have thoroughly analyzed the preceding classes, create the exception package with the following classes: CardNotFoundException, RetroBoardNotFoundException, and RetroBoardResponseEntityExceptionHandler. See Listings 4-23, 4-24, and 4-25, respectively.

Listing 4-23. src/main/java/apress/com/myretro/exception/CardNotFoundException.java

```
package com.apress.myretro.exception;

public class CardNotFoundException extends RuntimeException{
    public CardNotFoundException() {
        super("Card Not Found");
    }

    public CardNotFoundException(String message) {
        super(String.format("Card Not Found: {}", message));
    }
```

```java
    public CardNotFoundException(String message, Throwable cause) {
        super(String.format("Card Not Found: {}", message), cause);
    }
}
```

Listing 4-24. src/main/java/apress/com/myretro/exception/
RetroBoardNotFoundException.java

```java
package com.apress.myretro.exception;

public class RetroBoardNotFoundException extends RuntimeException{

    public RetroBoardNotFoundException(){
        super("RetroBoard Not Found");
    }

    public RetroBoardNotFoundException(String message) {
        super(String.format("RetroBoard Not Found: {}", message));
    }

    public RetroBoardNotFoundException(String message, Throwable cause) {
        super(String.format("RetroBoard Not Found: {}", message), cause);
    }
}
```

Listing 4-25. src/main/java/apress/com/myretro/exception/
RetroBoardResponseEntityExceptionHanlder.java

```java
package com.apress.myretro.exception;

import org.springframework.http.HttpHeaders;
import org.springframework.http.HttpStatus;
import org.springframework.http.ResponseEntity;
import org.springframework.web.bind.annotation.ControllerAdvice;
import org.springframework.web.bind.annotation.ExceptionHandler;
import org.springframework.web.context.request.WebRequest;
import org.springframework.web.servlet.mvc.method.annotation.
ResponseEntityExceptionHandler;
```

```
import java.time.LocalDateTime;
import java.time.format.DateTimeFormatter;
import java.util.HashMap;
import java.util.Map;

@ControllerAdvice
public class RetroBoardResponseEntityExceptionHandler extends
ResponseEntityExceptionHandler {

    @ExceptionHandler(value
            = { CardNotFoundException.class,RetroBoardNotFoundException.
            class })
    protected ResponseEntity<Object> handleNotFound(
            RuntimeException ex, WebRequest request) {

        Map<String, Object> response = new HashMap<>();

        response.put("msg","There is an error");
        response.put("code",HttpStatus.NOT_FOUND.value());
        response.put("time", LocalDateTime.now().format(DateTimeFormatter.
        ofPattern("yyyy-mm-dd HH:mm:ss")));

        Map<String, String> errors = new HashMap<>();
        errors.put("msg",ex.getMessage());
        response.put("errors",errors);

        return handleExceptionInternal(ex, response,
                new HttpHeaders(), HttpStatus.NOT_FOUND, request);
    }
}
```

The exception package is the same as in previous versions. Take a moment and review these classes, which are very straightforward.

Next, create the service package with the RetroBoardService class, as shown in Listing 4-26.

Listing 4-26. src/main/java/apress/com/myretro/service/
RetroBoardService.java

```
package com.apress.myretro.service;

import com.apress.myretro.board.Card;
import com.apress.myretro.board.RetroBoard;
import com.apress.myretro.persistence.RetroBoardRepository;
import com.apress.myretro.persistence.SimpleRepository;
import lombok.AllArgsConstructor;
import org.springframework.stereotype.Service;

import java.util.ArrayList;
import java.util.HashMap;
import java.util.List;
import java.util.UUID;

@AllArgsConstructor
@Service
public class RetroBoardService {

    SimpleRepository<RetroBoard,UUID> retroBoardRepository;

    public RetroBoard save(RetroBoard domain) {
        return this.retroBoardRepository.save(domain);
    }

    public RetroBoard findById(UUID uuid) {
        return this.retroBoardRepository.findById(uuid).get();
    }

    public Iterable<RetroBoard> findAll() {
        return this.retroBoardRepository.findAll();
    }

    public void delete(UUID uuid) {
        this.retroBoardRepository.deleteById(uuid);
    }
```

```java
public Iterable<Card> findAllCardsFromRetroBoard(UUID uuid) {
    return this.findById(uuid).getCards().values();
}

public Card addCardToRetroBoard(UUID uuid, Card card){
    RetroBoard retroBoard = this.findById(uuid);
    if (card.getId() == null) {
        card.setId(UUID.randomUUID());
    }
    retroBoard.getCards().put(card.getId(),card);
    this.save(retroBoard);
    return card;
}

public Card findCardByUUID(UUID  uuid,UUID uuidCard){
    RetroBoard retroBoard = this.findById(uuid);
    return retroBoard.getCards().get(uuidCard);
}

public Card saveCard(UUID  uuid,Card card){
    RetroBoard retroBoard = this.findById(uuid);
    retroBoard.getCards().put(card.getId(),card);
    this.save(retroBoard);
    return card;
}

public void removeCardByUUID(UUID uuid,UUID cardUUID){
    RetroBoard retroBoard = this.findById(uuid);
    retroBoard.getCards().remove(cardUUID);
    this.save(retroBoard);
}
}
```

The RetroBoardService class now contains additional methods that help to interact even more with the whole application, from the RetroBoard and its Cards. This class is very straightforward. What is important to notice is that this class has the @Service annotation to make it a Spring bean. This class will be injected into our web controller.

Next, create the web package and the RetroBoardController class as shown in Listing 4-27.

Listing 4-27. src/main/java/apress/com/myretro/web/
RetroBoardController.java

```
package com.apress.myretro.web;

import com.apress.myretro.board.Card;
import com.apress.myretro.board.RetroBoard;
import com.apress.myretro.service.RetroBoardService;
import jakarta.validation.Valid;
import lombok.AllArgsConstructor;
import org.springframework.http.HttpStatus;
import org.springframework.http.ResponseEntity;
import org.springframework.validation.FieldError;
import org.springframework.web.bind.MethodArgumentNotValidException;
import org.springframework.web.bind.annotation.*;
import org.springframework.web.servlet.support.ServletUriComponentsBuilder;

import java.net.URI;
import java.time.LocalDateTime;
import java.time.format.DateTimeFormatter;
import java.util.HashMap;
import java.util.Map;
import java.util.UUID;

@AllArgsConstructor
@RestController
@RequestMapping("/retros")
public class RetroBoardController {

    private RetroBoardService retroBoardService;

    @GetMapping
    public ResponseEntity<Iterable<RetroBoard>> getAllRetroBoards(){
        return ResponseEntity.ok(retroBoardService.findAll());
    }
```

```java
@PostMapping
public ResponseEntity<RetroBoard> saveRetroBoard(@Valid @RequestBody
RetroBoard retroBoard){
    RetroBoard result = retroBoardService.save(retroBoard);
    URI location = ServletUriComponentsBuilder
            .fromCurrentRequest()
            .path("/{uuid}")
            .buildAndExpand(result.getId().toString())
            .toUri();
    return ResponseEntity.created(location).body(result);
}

@GetMapping("/{uuid}")
public ResponseEntity<RetroBoard> findRetroBoardById(@PathVariable
UUID uuid){
    return ResponseEntity.ok(retroBoardService.findById(uuid));
}

@GetMapping("/{uuid}/cards")
public ResponseEntity<Iterable<Card>> getAllCardsFromBoard(@
PathVariable UUID uuid){
    return ResponseEntity.ok(retroBoardService.findAllCardsFromRetro
    Board(uuid));
}

@PutMapping("/{uuid}/cards")
public ResponseEntity<Card> addCardToRetroBoard(@PathVariable UUID
uuid,@Valid @RequestBody Card card){
    Card result = retroBoardService.addCardToRetroBoard(uuid,card);
    URI location = ServletUriComponentsBuilder
            .fromCurrentRequest()
            .path("/{uuid}/cards/{uuidCard}")
            .buildAndExpand(uuid.toString(),result.getId().toString())
            .toUri();
    return ResponseEntity.created(location).body(result);
}
```

```java
@GetMapping("/{uuid}/cards/{uuidCard}")
public ResponseEntity<Card> getCardByUUID(@PathVariable UUID uuid,
@PathVariable UUID uuidCard){
    return ResponseEntity.ok(retroBoardService.findCardByUUID(uuid,
    uuidCard));
}

@PutMapping("/{uuid}/cards/{uuidCard}")
public ResponseEntity<Card> updateCardByUUID(@PathVariable UUID uuid,
@PathVariable UUID uuidCard, @RequestBody Card card){
    return ResponseEntity.ok(retroBoardService.saveCard(uuid,card));
}

@ResponseStatus(HttpStatus.NO_CONTENT)
@DeleteMapping("/{uuid}/cards/{uuidCard}")
public void deleteCardFromRetroBoard(@PathVariable UUID uuid,
@PathVariable UUID uuidCard){
    retroBoardService.removeCardByUUID(uuid,uuidCard);
}

@ExceptionHandler(MethodArgumentNotValidException.class)
@ResponseStatus(HttpStatus.BAD_REQUEST)
public Map<String, Object> handleValidationExceptions(MethodArgumentNot
ValidException ex) {
    Map<String, Object> response = new HashMap<>();

    response.put("msg","There is an error");
    response.put("code",HttpStatus.BAD_REQUEST.value());
    response.put("time", LocalDateTime.now().format(DateTimeFormatter.
    ofPattern("yyyy-MM-dd HH:mm:ss")));

    Map<String, String> errors = new HashMap<>();
    ex.getBindingResult().getAllErrors().forEach((error) -> {
        String fieldName = ((FieldError) error).getField();
        String errorMessage = error.getDefaultMessage();
        errors.put(fieldName, errorMessage);
    });
```

```
        response.put("errors",errors);

        return response;
    }
}
```

The RetroBoardController class will serve the /retros, /retros/{uuid}/cards endpoints. As you can see, this class hasn't changed too much from the previous version. Analyze it and then continue.

Next, create the config package and the UsersProperties, MyRetroProperties, and MyRetroConfiguration classes as shown in Listings 4-28, 4-29, and 4-30, respectively.

Listing 4-28. src/main/java/apress/com/myretro/config/UsersProperties.java

```
package com.apress.myretro.config;

import lombok.Data;

@Data
public class UsersProperties {
    String server;
    Integer port;
    String username;
    String password;
}
```

Listing 4-29. src/main/java/apress/com/myretro/config/
MyRetroProperties.java

```
package com.apress.myretro.config;

import lombok.Data;
import org.springframework.boot.context.properties.ConfigurationProperties;

@ConfigurationProperties(prefix="service")
@Data
public class MyRetroProperties {
    UsersProperties users;
}
```

Listing 4-30. src/main/java/apress/com/myretro/config/
MyRetroConfiguration.java

```java
package com.apress.myretro.config;

import com.apress.myretro.board.Card;
import com.apress.myretro.board.CardType;
import com.apress.myretro.board.RetroBoard;
import com.apress.myretro.service.RetroBoardService;
import org.springframework.boot.context.event.ApplicationReadyEvent;
import org.springframework.boot.context.properties.
EnableConfigurationProperties;
import org.springframework.context.ApplicationListener;
import org.springframework.context.annotation.Bean;
import org.springframework.context.annotation.Configuration;

import java.util.ArrayList;
import java.util.UUID;

@EnableConfigurationProperties({MyRetroProperties.class})
@Configuration
public class MyRetroConfiguration {

    @Bean
    ApplicationListener<ApplicationReadyEvent> ready(RetroBoardService
    retroBoardService) {
        return applicationReadyEvent -> {
            UUID retroBoardId = UUID.fromString("9dc9b71b-a07e-418b-
            b972-40225449aff2");
            RetroBoard retroBoard = RetroBoard.builder()
                    .id(retroBoardId)
                    .name("Spring Boot Conference")
                        .card(UUID.fromString("bb2a80a5-a0f5-4180-
                        a6dc-80c84bc014c9"),Card.builder().
                        id(UUID.fromString("bb2a80a5-a0f5-4180-
                        a6dc-80c84bc014c9")).comment("Spring Boot
                        Rocks!").cardType(CardType.HAPPY).build())
```

```
                        .card(UUID.fromString("f9de7f11-5393-4b5b-
                        8e9d-10eca5f50189"),Card.builder().id(UUID.
                        randomUUID()).comment("Meet everyone in
                        person").cardType(CardType.HAPPY).build())
                        .card(UUID.fromString("6cdb30d6-43f2-42b7-
                        b0db-f3acbc53d467"),Card.builder().id(UUID.
                        randomUUID()).comment("When is the next one?").
                        cardType(CardType.MEH).build())
                        .card(UUID.fromString("9de1f7f9-2470-4c8d-8
                        6f2-371203620fcd"),Card.builder().id(UUID.
                        randomUUID()).comment("Not enough time to talk
                        to everyone").cardType(CardType.SAD).build())
                .build();
            retroBoardService.save(retroBoard);

        };
    }
}
```

Notice in Listing 4-28 that we've renamed the UsersConfiguration.java file from Chapter 3 to UsersProperties.java (this makes more sense). And in Listing 4-30 we added to MyRetroConfiguration some initial data using the ApplicationListener, and when the application is ready it will execute all the code in the ready() method. In this method (ready()) we have the RetroBoardService as a parameter. This will be injected by the Spring Framework.

In our previous version we have an Advice, so create the package advice and RetroBoardAdvice class, as shown in Listing 4-31.

Listing 4-31. src/main/java/apress/com/myretro/advice/RetroBoardAdvice.java

```
package com.apress.myretro.advice;

import com.apress.myretro.board.RetroBoard;
import com.apress.myretro.exception.RetroBoardNotFoundException;
import lombok.extern.slf4j.Slf4j;
import org.aspectj.lang.ProceedingJoinPoint;
import org.aspectj.lang.annotation.Around;
```

```
import org.aspectj.lang.annotation.Aspect;
import org.springframework.stereotype.Component;

import java.util.Optional;
import java.util.UUID;

@Slf4j
@Component
@Aspect
public class RetroBoardAdvice {

    @Around("execution(* com.apress.myretro.persistence.
    RetroBoardRepository.findById(..))")
    public Object checkFindRetroBoard(ProceedingJoinPoint
    proceedingJoinPoint) throws Throwable {
        log.info("[ADVICE] {}", proceedingJoinPoint.getSignature().getName());
        Optional<RetroBoard> retroBoard = (Optional<RetroBoard>)
        proceedingJoinPoint.proceed(new Object[]{
                UUID.fromString(proceedingJoinPoint.getArgs()[0].toString())
        });
        if (retroBoard.isEmpty())
            throw new RetroBoardNotFoundException();

        return retroBoard;
    }
}
```

Recall that the RetroBoardAdvice class will intercept the findById and, if it's not found, will throw the RetroBoardNotFoundException, which will be handled by our controller advice RetroBoardResponseEntityExceptionHandler class.

Next, open the application.properties file and add the content shown in Listing 4-32.

Listing 4-32. src/main/resource/application.properties

```
# DataSource
spring.datasource.generate-unique-name=false
spring.datasource.name=test-db

# SQL init
spring.sql.init.mode=always
```

Listing 4-32 shows that now we are using the `spring.sql.init.mode` so it reads the `schema.sql` file. Next, create the `schema.sql` file as shown in Listing 4-33.

Listing 4-33. src/main/resources/schema.sql

```sql
CREATE EXTENSION IF NOT EXISTS "uuid-ossp";
DROP TABLE IF EXISTS CARD CASCADE;
DROP TABLE IF EXISTS RETRO_BOARD CASCADE;
CREATE TABLE CARD
(
    ID              UUID DEFAULT uuid_generate_v4() NOT NULL,
    CARD_TYPE       VARCHAR(5)                      NOT NULL,
    COMMENT         VARCHAR(255),
    RETRO_BOARD_ID UUID,
    PRIMARY KEY (ID)
);
CREATE TABLE RETRO_BOARD
(
    ID   UUID DEFAULT uuid_generate_v4() NOT NULL,
    NAME VARCHAR(255),
    PRIMARY KEY (ID)
);
ALTER TABLE IF EXISTS CARD
    ADD CONSTRAINT RETRO_BOARD_CARD FOREIGN KEY (RETRO_BOARD_ID) REFERENCES
    RETRO_BOARD;
```

The first line of Listing 4-33 enables a function (uuid_generate_v4()) that generates the UUID that we required. We need two tables, one for the `RetroBoard`, and one for the `Card`, where we need to add the relationship, an one-to-many relationship.

Next, create in the root of the project a `docker-compose.yaml` file with the content shown in Listing 4-34.

Listing 4-34. docker-compose.yaml

```yaml
version: "3"
services:
  postgres:
```

```
image: postgres
restart: always
environment:
  POSTGRES_USER: postgres
  POSTGRES_PASSWORD: postgres
  POSTGRES_DB: test-db
ports:
  - 5432:5432
```

As you can see, this file is the same as in the Users App. We need to use the
PostgreSQL database, so we are all set.

Running the My Retro App

Now, it's time to run our application. You can run it by using your IDE or using the
following command:

```
./gradle bootRun
..
.s.b.d.c.l.DockerComposeLifecycleManager : Using Docker Compose file ...
..
..
```

There is new info about our docker-compose file, but what is happening?
More about this later.

If you go an open a browser (http://localhost:8080/retros) or execute a curl
command in another terminal window. The following code shows the use of the jq tool
(https://stedolan.github.io/jq/) to print the result of the curl command:

```
curl -s http://localhost:8080/retros | jq .

[
  {
    "id": "9dc9b71b-a07e-418b-b972-40225449aff2",
    "name": "Spring Boot Conference",
    "cards": {
      "2ca35157-63eb-4950-ac10-fc75ab828fcb": {
```

```
        "id": "2ca35157-63eb-4950-ac10-fc75ab828fcb",
        "comment": "Meet everyone in person",
        "cardType": "HAPPY",
        "retroBoardId": "9dc9b71b-a07e-418b-b972-40225449aff2"
      },
      "bb2a80a5-a0f5-4180-a6dc-80c84bc014c9": {
        "id": "bb2a80a5-a0f5-4180-a6dc-80c84bc014c9",
        "comment": "Spring Boot Rocks!",
        "cardType": "HAPPY",
        "retroBoardId": "9dc9b71b-a07e-418b-b972-40225449aff2"
      },
      "b0b993c7-83a3-4ab8-9a15-d9b160228da4": {
        "id": "b0b993c7-83a3-4ab8-9a15-d9b160228da4",
        "comment": "When is the next one?",
        "cardType": "MEH",
        "retroBoardId": "9dc9b71b-a07e-418b-b972-40225449aff2"
      },
      "394676ba-8609-4677-8ef1-420851139410": {
        "id": "394676ba-8609-4677-8ef1-420851139410",
        "comment": "Not enough time to talk to everyone",
        "cardType": "SAD",
        "retroBoardId": "9dc9b71b-a07e-418b-b972-40225449aff2"
      }
    }
  }
]
```

But wait... what... again?

Did you realize that we never specified the connection parameters for the DataSource? What happen is that we have the spring-boot-docker-compose dependency that auto-configured our application. It runs in the background docker image by executing docker compose up, and it creates the DataSource with the right parameters based on our docker-compose.yaml file. This feature is amazing! Of course, this only happens in development time, we don't need to do extra steps to create an infrastructure anymore. We are going to continue to use this development approach in the following chapters.

Summary

In this chapter you discovered how Spring Framework Data Access works with Spring Boot, and how Spring Boot helps to configure most of the Data Access components, such as the `DataSource` and the `JdbcTemplate`.

You learned the following in this chapter:

- You need to specify SQL statements and work directly with `ResultSets`.

- Spring Framework Data Access brings simplified classes like `JdbcTemplate` to remove all the boilerplate of dealing with connections, sessions, transactions, exceptions, and much more.

- Spring Boot helps you initialize embedded database engines such as H2, HSQL, and Derby without providing any connection parameters to set up the `DataSource`.

- You can have multiple drivers in your dependencies, but you need to provide the connection parameter to the driver you want to connect, as well as the initialization of the database through the `schema.sql` or `data.sql`.

- A new Spring Boot 3.1 feature allows you to add a `docker-compose.yaml` file and declare a service, and by adding the `spring-boot-docker-compose` starter dependency, Spring Boot will auto-configure all the necessary parameters to set up the `DataSource` based on the environment variables declared in `docker-compose.yaml`.

In Chapter 5 we'll explore the Spring Data project and how Spring Boot helps to configure everything.

CHAPTER 5

Spring Data with Spring Boot

This chapter introduces the Spring Data project and three of its subprojects, Spring Data JDBC, Spring Data JPA, and Spring Data REST. We'll examine all the Spring Data features and how Spring Boot can help us to use them in our two applications. Let's get started.

Spring Data

Chapter 4 introduced the features that the Spring Framework Data Access provides for creating persistent data applications. Another extension of the Spring Framework is the Spring Data technology, which offers even more features to make data persistence a breeze.

Spring Data brings the consistent way of programming model that do the heavy lifting when talking to persistence engines, from creating the connection and sessions, mapping your classes to a tables and rows, implement database operations, all the way to commit or rollback your data and show you a better error message if something happens during any operation, making it easier for developers. Spring Data makes data access easy not only for relational and nonrelational databases but also for MapReduce frameworks (such as Apache CouchDB, Apache Hadoop, Infinispan, and Riak, among others), reactive databases, and cloud-based data services. The Spring Data project (which includes the Spring Framework Data Access as its core) is an umbrella project that contains many other subprojects, such as the Spring Data JDBC, Spring Data JPA, Spring Data REST subprojects covered in this chapter.

Spring Data provides all the following features (and many more) that we are going to discuss in this chapter and the following chapters:

© Felipe Gutierrez 2024
F. Gutierrez, *Pro Spring Boot 3*, https://doi.org/10.1007/978-1-4842-9294-5_5

- An easy way to extend interfaces (using the domain model and Id's, Spring Data provides the Repository<T,ID>, CrudRepository<T,ID> interfaces) that gives you the base for CRUD actions

- Implementation of these interfaces with basic properties

- A query derivation based on the fields of the domain class

- Support for auditing based on events and callbacks (such as date created, last changed)

- A way to override any implementation from the interfaces

- JavaConfig support for configuration

- Experimental support for cross-store persistence, meaning that you can use the same model and be able to use multiple engines

So, let's start by exploring the features that Spring Data JDBC offers to developers.

Spring Data JDBC

Spring Data JDBC provides enhanced support for JDBC-based data access, making it a simple and opinionated Object Relational Mapping (ORM) tool. Spring Data JDBC is based on the *Aggregate Root* design pattern explained in the book *Domain Driven Design* by Eric Evans (Addison-Wesley Professional, 2003). These are some of the many features that you will find in Spring Data JDBC:

- A NamingStrategy implementation based on the CrudRepository that provides a specific table structure based on your domain classes. You can override this by using annotations such as @Table and @Column, among others.

- When you have a particular SQL query to execute that is not part of the default implementation, you can use the @Query annotation that allows you to customize your queries.

- Support for the legacy MyBatis framework.

- Support for callbacks and events.

- Support for JavaConfig classes where you configure your repositories by using the @EnableJdbcRepositories annotation. Normally this annotation is used (the same as @ComponentScan in Spring Boot) only in Spring applications. If you are using Spring Boot, you don't need to use the @EnabledJdbcRepositories annotation, because it gets activated from the auto-configuration.

- Entity state detection strategies. By default, Spring Data JDBC looks at the identifier; if it's null or 0, Spring Data JDBC recognizes that the entity is new.

- Default transactionality when extending the CrudRepository, so there's no need to mark your interface using the @Transactional annotation. You can override some methods, though, and add very specify parameters to the annotation.

- Auditing with annotations @CreatedBy, @LastModifiedBy, @CreatedDate, and @LastModifiedDate.

Spring Data JDBC with Spring Boot

When you use Spring Boot and add the spring-boot-starter-data-jdbc starter dependency to your application, the auto-configuration will add the @EnableJdbcRepositories annotation and configure everything depending on what driver you have and the properties you've set for your applications. The following sections demonstrate both the Users App and My Retro App with Spring Boot and Spring Data JDBC.

Users App with Spring Boot and Spring Data JDBC

It's time to see in action the Users App with Spring Boot and Spring Data JDBC. You can start from scratch by going to the Spring Initializr (https://start.spring.io) and selecting a base project. In the Project Metadata section, make sure to set the Group field to com.apress and the Artifact and Name fields to users. For dependencies, add Web, Validation, JDBC, H2, PostgreSQL, and Lombok. Download the project, unzip it, and import it into your favorite IDE. By the end of this section, you should have the structure shown in Figure 5-1.

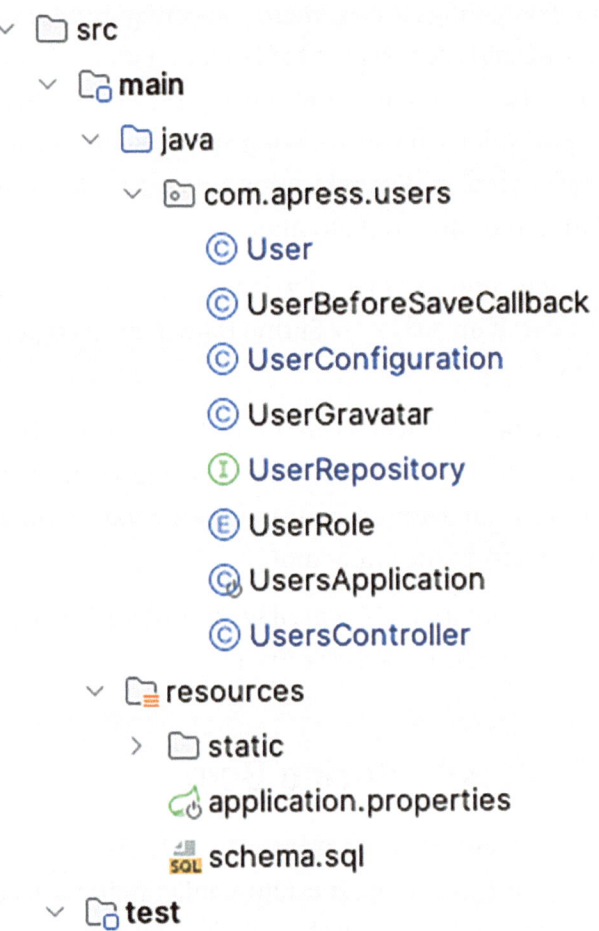

Figure 5-1. *Users App structure*

Tip   After clicking Add Dependencies in the Spring Initializr, you can easily locate each dependency from the long list of dependencies by typing its name in the field at the top of the panel.

Open the `build.gradle` file and add the `spring-boot-starter-data-jdbc` starter dependency as shown in Listing 5-1.

Listing 5-1. build.gradle

```
plugins {
    id 'java'
    id 'org.springframework.boot' version '3.2.3'
    id 'io.spring.dependency-management' version '1.1.4'
}

group = 'com.apress'
version = '0.0.1-SNAPSHOT'
sourceCompatibility = '17'

repositories {
    mavenCentral()
}

dependencies {
    implementation 'org.springframework.boot:spring-boot-starter-web'
    implementation 'org.springframework.boot:spring-boot-starter-
    validation'

    implementation 'org.springframework.boot:spring-boot-starter-data-jdbc'
    runtimeOnly 'com.h2database:h2'
    runtimeOnly 'org.postgresql:postgresql'

    compileOnly 'org.projectlombok:lombok'
    annotationProcessor 'org.projectlombok:lombok'

    // Web
    implementation 'org.webjars:bootstrap:5.2.3'

    testImplementation 'org.springframework.boot:spring-boot-starter-test'
}

tasks.named('test') {
    useJUnitPlatform()
}

test {
    testLogging {
```

```
        events "passed", "skipped", "failed"
        showExceptions true
        exceptionFormat "full"
        showCauses true
        showStackTraces true
        showStandardStreams = false
    }
}
```

The addition of spring-boot-starter-data-jdbc will bring all the Spring Data dependencies. Note that we have two drivers, h2 (embedded in-memory) and PostgreSQL. You already know from Chapter 4 what happens when Spring Boot sees two drivers and no connection parameters.

Next, create the UserRole enum (see Listing 5-2) and the User class (see Listing 5-3).

Listing 5-2. src/main/java/apress/com/users/UserRole.java

```
package com.apress.users;

public enum UserRole {
    USER, ADMIN, INFO
}
```

Listing 5-3. src/main/java/apress/com/users/User.java

```
package com.apress.users;

import jakarta.validation.constraints.NotBlank;
import jakarta.validation.constraints.Pattern;
import lombok.*;
import org.springframework.data.annotation.Id;
import org.springframework.data.relational.core.mapping.Table;

import java.util.List;

@Builder
@AllArgsConstructor
@NoArgsConstructor
@Data
@Table("USERS")
```

```java
public class User {

    @Id
    Long id;

    @NotBlank(message = "Email can not be empty")
    private String email;

    @NotBlank(message = "Name can not be empty")
    private String name;

    private String gravatarUrl;

    @Pattern(message = "Password must be at least 8 characters long and
    contain at least one number, one uppercase, one lowercase and one
    special character",
            regexp = "^(?=.*[0-9])(?=.*[a-z])(?=.*[A-Z])(?=.*[@#$%^&+=!])
            (?=\\S+$).{8,}$")
    private String password;

    @Singular("role")
    private List<UserRole> userRole;

    private boolean active;
}
```

As you can see in Listing 5-2, the UserRole enum is the same as in previous chapters. On the other hand, Listing 5-3 shows that the User class is different from it was in the previous chapters. As shown next, we need to extend from the CrudRepository<T,ID> interface, which requests the *entity* (T) and its *identifier* (ID). With this interface, Spring Data uses a NamingStrategy logic that uses the name of the entity as the name of the table and the identifier as the primary/indexed key. In this example, it will be USER as the table name and id (type Long) as the primary key. But in this case, that logic would result in USER as the table name, but in Listing 5-3 we marked the class with the @Table annotation and passed the name USERS as a parameter, overriding the default strategy so that the name of the table will be USERS. Also, we are using the @Id annotation that marks the Long type as our identifier, in this case our primary key in our database. An important point here is that both annotations come from the Spring Data package (org. springframework.data.*).

Next, create the UserRepository interface. See Listing 5-4.

Listing 5-4. src/main/java/apress/com/users/UserRepository.java

```
package com.apress.users;

import org.springframework.data.repository.CrudRepository;

import java.util.Optional;

public interface UserRepository extends CrudRepository<User,Long>{
    Optional<User> findByEmail(String email);
        void deleteByEmail(String email);
}
```

Note that the UserRepository interface is new. Let's analyze it:

- CrudRepository: Our interface extends the public interface
  CrudRepository<T,ID> extends Repository<T,ID> interface. The
  CrudRepository declares several methods:

  ```
  public interface CrudRepository<T, ID> extends Repository<T, ID> {

      <S extends T> S save(S entity);
      <S extends T> Iterable<S> saveAll(Iterable<S> entities);
      Optional<T> findById(ID id);
      boolean existsById(ID id);
      Iterable<T> findAll();
      Iterable<T> findAllById(Iterable<ID> ids);
      long count();
      void deleteById(ID id);
      void delete(T entity);
      void deleteAllById(Iterable<? extends ID> ids);
      void deleteAll(Iterable<? extends T> entities);
      void deleteAll();
  }
  ```

  One of the most impressive details here is that we don't need to
  implement anything—Spring Data will implement everything for
  us, so we can use this UserRepository interface directly in our
  Service or *Controller*.

- findBy and deleteBy: We are declaring these methods in the
  UserRepository. Using Spring Data, we can use the naming
  convention to create query methods that execute queries based on the
  name of the field and some keywords (that normally you would use in
  a SQL statement). In this case we are using the keyword findBy and we
  are adding the field name Email, in this case findByEmail and we have
  also the deleteBy and we are adding the field Email as well, then the
  Spring Data will generate the implementation to do something like:

```
select * from users where email = ?
delete from users where email = ?
```

  If you use something like findByMyAwesomeName, Spring Data
  won't do anything, because it must follow the fields declared in
  the entity, in other words, the fields must name-match the query
  methods you are creating. Additional keywords include After,
  GreaterThan, Before, In, NotIn, Like, Containing, IsTrue, and
  IsFalse, among others. To see all the available keywords, go to
  https://docs.spring.io/spring-data/jdbc/docs/current/
  reference/html/#jdbc.query-methods. If you have an IDE with
  IntelliSense, you should get all the possible combinations.

Spring Data also brings a *Query Lookup Strategy* in place, so you can extend some of the
default queries by using the @Query annotation that accepts a String (a SQL statement) as a
parameter. Sometimes you will find out that the default implementation of the CRUD doesn't
cover our requirements, so imagine that you only one to retrieve the Gravatar from the User,
then you could declare a method in the UserRepository interface like the following:

```
@Query("SELECT GRAVATAR_URL FROM USERS WHERE EMAIL = :email")
    String getGravatarByEmail(@Param("email") String email);
```

It's that simple!

The UserRepository interface is an amazing new feature for speeding up the
development process!

Next, create the UserGravatar and UserConfiguration classes. See Listings 5-5
and 5-6. Note that the UserGravatar class is the same as in Chapter 4 (Listing 4-3).

Listing 5-5. src/main/java/apress/com/users/UserGravatar.java

```java
package com.apress.users;

import java.io.UnsupportedEncodingException;
import java.security.MessageDigest;
import java.security.NoSuchAlgorithmException;

public class UserGravatar {

    public static String getGravatarUrlFromEmail(String email){
        return String.format("https://www.gravatar.com/
        avatar/%s?d=wavatar", md5Hex(email));
    }

    private static String hex(byte[] array) {
        StringBuffer sb = new StringBuffer();
        for (int i = 0; i < array.length; ++i) {
            sb.append(Integer.toHexString((array[i]
                    & 0xFF) | 0x100).substring(1, 3));
        }
        return sb.toString();
    }

    private static String md5Hex(String message) {
        try {
            MessageDigest md =
                    MessageDigest.getInstance("MD5");
            return hex(md.digest(message.getBytes("CP1252")));
        } catch (NoSuchAlgorithmException e) {
        } catch (UnsupportedEncodingException e) {
        }
        return "23bb62a7d0ca63c9a804908e57bf6bd4";
    }
}
```

Listing 5-6. src/main/java/apress/com/users/UserConfiguration.java

```java
package com.apress.users;

import org.springframework.boot.CommandLineRunner;
import org.springframework.boot.context.event.ApplicationReadyEvent;
import org.springframework.context.ApplicationListener;
import org.springframework.context.annotation.Bean;
import org.springframework.context.annotation.Configuration;

@Configuration
public class UserConfiguration {

    @Bean
    ApplicationListener<ApplicationReadyEvent> init(UserRepository
    userRepository) {
        return applicationReadyEvent -> {
            userRepository.save(User.builder()
                    .email("ximena@email.com")
                    .name("Ximena")
                    .password("aw2s0meR!")
                    .role(UserRole.USER)
                    .active(true)
                    .build());
            userRepository.save(User.builder()
                    .email("norma@email.com")
                    .name("Norma")
                    .password("aw2s0meR!")
                    .role(UserRole.USER)
                    .role(UserRole.ADMIN)
                    .active(true)
                    .build());
        };
    }

}
```

The UserGravatar class will create the Gravatar based on the email digest. Listing 5-6 shows the configuration. As you can see, the same as the previous chapter. The init method will be executed when the application is ready.

If you pay attention to the code (Listing 5-6), we are not setting the gravatarUrl field; how can we calculate it and add it to the User entity? In Chapter 4, we used the User as the record type, and we used the compact constructor to set the gravatarUrl field, so every time there is a new instance, we can get the Gravatar for that user. To solve this issue, we can use a callback feature from Spring Data. So, let's create the UserBeforeSaveCallback class to implement the BeforeSaveCallback<T>; see Listing 5-7.

Listing 5-7. src/main/java/apress/com/users/UserBeforeSaveCallback.java

```
package com.apress.users;

import org.springframework.data.relational.core.conversion.
MutableAggregateChange;
import org.springframework.data.relational.core.mapping.event.
BeforeSaveCallback;
import org.springframework.stereotype.Component;

import java.util.List;

@Component
public class UserBeforeSaveCallback implements BeforeSaveCallback<User> {
    @Override
    public User onBeforeSave(User aggregate, MutableAggregateChange<User>
    aggregateChange) {
        if (aggregate.getGravatarUrl()==null)
            aggregate.setGravatarUrl(UserGravatar.getGravatarUrlFromEmail(a
            ggregate.getEmail()));
        if (aggregate.getUserRole() == null)
            aggregate.setUserRole(List.of(UserRole.INFO));
        return aggregate;
    }
}
```

The UserBeforeSaveCallback class is marked as @Component so that Spring can find it and add it to the logic to execute a callback before the User entity is saved into the database. We are implementing the BeforeSaveCallback of type User, which allows us to implement the method onBeforeSave, with User and MutableAggreateChange as parameters. Here, we have access to our entity to add the defaults we want, in this case the gravatarUrl.

Spring Data offers more entity callbacks by process: BeforeDeleteCallback and AfterDeleteCallback for the delete process, BeforeConvertCallback, BeforeSaveCallback, and AfterSaveCallback for the save process, and AfterConvertCallback for the load process.

Next, create the UsersController class as shown in Listing 5-8.

Listing 5-8. src/main/java/apress/com/users/UsersController.java

```java
package com.apress.users;

import lombok.AllArgsConstructor;
import org.springframework.http.HttpStatus;
import org.springframework.http.ResponseEntity;
import org.springframework.validation.FieldError;
import org.springframework.web.bind.MethodArgumentNotValidException;
import org.springframework.web.bind.annotation.*;
import org.springframework.web.servlet.support.ServletUriComponentsBuilder;

import java.net.URI;
import java.util.HashMap;
import java.util.Map;

@AllArgsConstructor
@RestController
@RequestMapping("/users")
public class UsersController {

    private UserRepository userRepository;

    @GetMapping
    public ResponseEntity<Iterable<User>> getAll(){
        return ResponseEntity.ok(this.userRepository.findAll());
    }
```

```java
@GetMapping("/{email}")
public ResponseEntity<User> findUserByEmail(@PathVariable String email)
throws Throwable {
    return ResponseEntity.of(this.userRepository.findByEmail(email));
}

@RequestMapping(method = {RequestMethod.POST,RequestMethod.PUT})
public ResponseEntity<User> save(@RequestBody User user){
    this.userRepository.save(user);

    URI location = ServletUriComponentsBuilder
            .fromCurrentRequest()
            .path("/{email}")
            .buildAndExpand(user.getId())
            .toUri();
    return ResponseEntity.created(location).body(this.userRepository.
    findByEmail(user.getEmail()).get());
}

@DeleteMapping("/{email}")
@ResponseStatus(HttpStatus.NO_CONTENT)
public void delete(@PathVariable String email){
    this.userRepository.deleteByEmail(email);
}

@ExceptionHandler(MethodArgumentNotValidException.class)
@ResponseStatus(HttpStatus.BAD_REQUEST)
public Map<String, String> handleValidationExceptions(MethodArgumentNot
ValidException ex) {
    Map<String, String> errors = new HashMap<>();

    ex.getBindingResult().getAllErrors().forEach((error) -> {
        String fieldName = ((FieldError) error).getField();
        String errorMessage = error.getDefaultMessage();
        errors.put(fieldName, errorMessage);
    });
```

```
    return errors;
  }
}
```

As you can see, the `UsersController` class differs from in the previous chapter because we are using the methods `findByEmail` and `deleteByEmail` in our repository, and the most important part is that we are using the `UserRepository` interface directly.

Next, open the `application.properties` file and use the content shown in Listing 5-9. As you can see, it's the same as in Chapter 4.

Listing 5-9. src/main/resources/application.properties

```
spring.h2.console.enabled=true
spring.datasource.generate-unique-name=false
spring.datasource.name=test-db
```

Next, create the `schema.sql` file. See Listing 5-10.

Listing 5-10. src/main/resources/schema.sql

```
DROP TABLE IF EXISTS USERS CASCADE;
CREATE TABLE USERS (
    ID LONG NOT NULL AUTO_INCREMENT,
    EMAIL VARCHAR(255) NOT NULL UNIQUE,
    NAME VARCHAR(100) NOT NULL,
    GRAVATAR_URL VARCHAR(255) NOT NULL,
    PASSWORD VARCHAR(255) NOT NULL,
    USER_ROLE VARCHAR(5) ARRAY NOT NULL DEFAULT ARRAY['INFO'],
    ACTIVE BOOLEAN NOT NULL,
    PRIMARY KEY (ID));
```

The `schema.sql` file will be executed when our app starts—this is the power of Spring Boot. Remember that this happens only with embedded databases such as H2, Derby, or HSQL, and because we haven't specified any connection parameters, Spring Boot will set up H2 as the in-memory database. So, we are all set!

Testing the Users App

To test the Users App, create the `UsersHttpRequestTests` class in the `test` folder structure. See Listing 5-11.

Listing 5-11. src/test/java/apress/com/users/UsersHttpRequestTests.java

```java
package com.apress.users;

import org.junit.jupiter.api.Test;
import org.springframework.beans.factory.annotation.Autowired;
import org.springframework.beans.factory.annotation.Value;
import org.springframework.boot.test.context.SpringBootTest;
import org.springframework.boot.test.web.client.TestRestTemplate;

import java.util.Collection;

import static org.assertj.core.api.Assertions.assertThat;

@SpringBootTest(webEnvironment = SpringBootTest.WebEnvironment.RANDOM_PORT)
public class UsersHttpRequestTests {

    @Value("${local.server.port}")
    private int port;

    private final String BASE_URL = "http://localhost:";
    private final String USERS_PATH = "/users";

    @Autowired
    private TestRestTemplate restTemplate;

    @Test
    public void indexPageShouldReturnHeaderOneContent() throws Exception {
        assertThat(this.restTemplate.getForObject(BASE_URL + port,
                String.class)).contains("Simple Users Rest Application");
    }

    @Test
    public void usersEndPointShouldReturnCollectionWithTwoUsers() throws
    Exception {
        Collection<User> response = this.restTemplate.
                getForObject(BASE_URL + port + USERS_PATH,
                Collection.class);

        assertThat(response.size()).isGreaterThanOrEqualTo(2);
    }
```

```java
@Test
public void userEndPointPostNewUserShouldReturnUser() throws Exception {

    User user =  User.builder()
            .email("dummy@email.com")
            .name("Dummy")
            .gravatarUrl("https://www.gravatar.com/avatar/23bb62a7d0ca6
            3c9a804908e57bf6bd4?d=wavatar")
            .password("aw2sOmeR!")
            .role(UserRole.USER)
            .active(true)
            .build();

    User response =  this.restTemplate.postForObject(BASE_URL + port +
    USERS_PATH,user,User.class);

    assertThat(response).isNotNull();
    assertThat(response.getEmail()).isEqualTo(user.getEmail());

    Collection<User> users = this.restTemplate.
            getForObject(BASE_URL + port + USERS_PATH, Collection.class);

    assertThat(users.size()).isGreaterThanOrEqualTo(2);

}

@Test
public void userEndPointDeleteUserShouldReturnVoid() throws Exception {
    this.restTemplate.delete(BASE_URL + port + USERS_PATH + "/norma@
    email.com");

    Collection<User> users = this.restTemplate.
            getForObject(BASE_URL + port + USERS_PATH,
            Collection.class);

    assertThat(users.size()).isLessThanOrEqualTo(2);
}

@Test
public void userEndPointFindUserShouldReturnUser() throws Exception{
```

```java
        User user = this.restTemplate.getForObject(BASE_URL + port + USERS_
        PATH + "/ximena@email.com",User.class);

        assertThat(user).isNotNull();
        assertThat(user.getEmail()).isEqualTo("ximena@email.com");
    }
}
```

You can run the test class in you IDE or you can use the following command:

```
./gradlew clean test
```

Running the Users App

You can run the Users App in your IDE or you can use the following command:

```
./gradlew clean bootRun
```

and you can either open a browser and go to http://localhost:8080/users or execute the following curl command at the command line:

```
curl -s http://localhost:8080/users | jq .
```

```json
[
  {
    "id": 1,
    "email": "ximena@email.com",
    "name": "Ximena",
    "gravatarUrl": "https://www.gravatar.com/avatar/f07f7e553264c9710105ede
    be6c465e7?d=wavatar",
    "password": "aw2s0meR!",
    "userRole": [
      "USER"
    ],
    "active": true
  },
  {
    "id": 2,
    "email": "norma@email.com",
```

```
    "name": "Norma",
    "gravatarUrl": "https://www.gravatar.com/avatar/23bb62a7d0ca63c9a80490
    8e57bf6bd4?d=wavatar",
    "password": "aw2s0meR!",
    "userRole": [
      "USER",
      "ADMIN"
    ],
    "active": true
  }
]
```

Note   You can find all the source code at the Apress website: `https://www.apress.com/gp/services/source-code.` or `https://github.com/felipeg48/pro-spring-boot-3rd`

My Retro App with Spring Boot and Spring Data JDBC

As with the Users App, to configure the My Retro App with Spring Boot and Spring Data JDBC, you can start from scratch by going to the Spring Initializr (`https://start.spring.io`). Set the Group field to `com.apress` and the Artifact and Name fields to `myretro`. For dependencies, add Web, Validation, JDBC, Docker Compose, and Lombok. Download the project, unzip it, and import it into your favorite IDE. By the end of this section, you should have the structure shown in Figure 5-2.

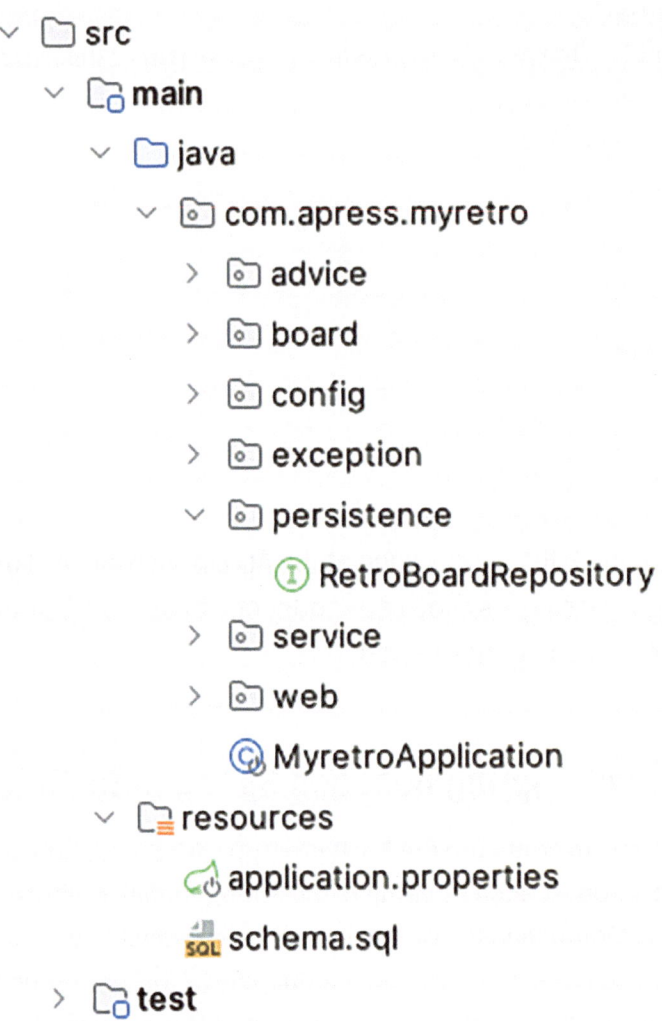

Figure 5-2. *My Retro App structure*

Note that the structure is very similar to that of the previous version (see Figure 4-5) and we are going to reuse almost the same code, so I'll describe only the most important changes here. And if you look and compare the persistence package, we have only one interface! That is the power of Spring Data and Spring Boot.

Open the build.gradle file and replace its content with the content shown in Listing 5-12.

Listing 5-12. build.gradle

```
plugins {
    id 'java'
    id 'org.springframework.boot' version '3.2.3'
    id 'io.spring.dependency-management' version '1.1.4'
}

group = 'com.apress'
version = '0.0.1-SNAPSHOT'
sourceCompatibility = '17'

repositories {
    mavenCentral()
}

dependencies {
    implementation 'org.springframework.boot:spring-boot-starter-web'
    implementation 'org.springframework.boot:spring-boot-starter-validation'
    implementation 'org.springframework.boot:spring-boot-starter-aop'

    implementation 'org.springframework.boot:spring-boot-starter-data-jdbc'
developmentOnly 'org.springframework.boot:spring-boot-docker-compose'

    annotationProcessor 'org.springframework.boot:spring-boot-
    configuration-processor'

    runtimeOnly 'org.postgresql:postgresql'

    compileOnly 'org.projectlombok:lombok'
    annotationProcessor 'org.projectlombok:lombok'

     // Web
    implementation 'org.webjars:bootstrap:5.2.3'

    testImplementation 'org.springframework.boot:spring-boot-starter-test'
}

tasks.named('test') {
    useJUnitPlatform()
}
```

Note in Listing 5-12 that we are using the dependencies `spring-boot-starter-data-jdbc` starter and `spring-boot-docker-compose` (as development only; remember that you need to add the connection properties if you go to production).

Note   Again, we are going to work only on the classes that changed using Spring Data JDBC. You can find the complete code by going to `https://www.apress.com/gp/services/source-code`.

Next, create/view the board package and create/open the `Card` and `RetroBoard` classes, as shown in Listing 5-13 and Listing 5-14, respectively.

Listing 5-13. src/main/java/apress/com/myretro/board/Card.java

```
package com.apress.myretro.board;

import com.fasterxml.jackson.annotation.JsonIgnore;
import jakarta.validation.constraints.NotBlank;
import jakarta.validation.constraints.NotNull;
import lombok.AllArgsConstructor;
import lombok.Builder;
import lombok.Data;
import lombok.NoArgsConstructor;
import org.springframework.data.annotation.Id;
import org.springframework.data.relational.core.mapping.Table;

import java.util.UUID;

@Builder
@AllArgsConstructor
@NoArgsConstructor
@Data
@Table
public class Card {

    @Id
    private UUID id;

    @NotBlank
```

```
    private String comment;

    @NotNull
    private CardType cardType;

    @JsonIgnore
    private UUID retroBoardId;
}
```

Notice in the Card class that we are using the @Table and @Id annotation (from the Spring Data package), plus one special annotation, @JsonIgnore, which is more for the web endpoint (this annotation will not render the retroBoardId value when requested). We'll discuss @JsonIgnore in more detail later in the chapter.

Listing 5-14. src/main/java/apress/com/myretro/board/RetroBoard.java

```
package com.apress.myretro.board;

import jakarta.validation.constraints.NotBlank;
import lombok.AllArgsConstructor;
import lombok.Builder;
import lombok.Data;
import lombok.NoArgsConstructor;
import org.springframework.data.annotation.Id;
import org.springframework.data.relational.core.mapping.MappedCollection;
import org.springframework.data.relational.core.mapping.Table;

import java.util.HashMap;
import java.util.Map;
import java.util.UUID;

@Builder
@AllArgsConstructor
@NoArgsConstructor
@Data
@Table
public class RetroBoard {

    @Id
    private UUID id;
```

```
@NotBlank(message = "A name must be provided")
private String name;

@MappedCollection(idColumn = "retro_board_id",keyColumn = "id")
    private Map<UUID,Card> cards = new HashMap<>();
}
```

The RetroBoard class will be our entity class. Note that we are using the @Table and @Id annotations. The important code is the @MappedCollection annotation, which uses two parameters, the idColumn and the keyColumn. It's important to know that Spring Data JDBC has support for the following types (among others):

- enum: Get mapped to their name.

- int, flow, Integer, Float, and more: All primitive types.

- String, java.util.Date, java.time.LocalDate, java.time.
  LocalDateTime, java.time.LocalTime:

- Arrays and Collections are mapped if your selected database engine
  that has array types.

- Set, Map: This is normally good for *one-to-many relationships*.

If you want to see the complete list of types, go to https://docs.spring.io/ spring-data/jdbc/docs/current/reference/html/#jdbc.entity-persistence. types. It's important to mention that reference is limited, and this is based on what we talked before (about the *Aggregate Root pattern*). In our case, we have a relationship of *one-to-many*. So, *many-to-one* or *many-to-many* relationships need to be handled using the AggregateReference interface. This interface is a wrapper around an id that is the reference of a different aggregate you are mapping.

Also notice in the @Table annotation that we didn't use the name parameters (like @Table("RETROBOARD")); we are going to let the NamingStrategy logic do its job, which uses *snake name convention*, instead of camel case, so our RetroBoard class will be retro_board (snake name convention) as the table name. And in the Card class, the field retroBoardId will become retro_board_id as the column for the card table.

Next, open/create the persistence package and open/create the RetroBoardRepository interface. See Listing 5-15.

Listing 5-15. src/main/java/apress/com/myretro/persistence/
RetroBoardRepository.java

```
package com.apress.myretro.persistence;

import com.apress.myretro.board.RetroBoard;
import org.springframework.data.repository.CrudRepository;

import java.util.UUID;

public interface RetroBoardRepository extends
CrudRepository<RetroBoard,UUID> {
}
```

The `RetroBoardRepository` interface extends from the `CrudRepository` interface. Remember that this interface brings several methods (previously discussed in the "Users App with Spring Boot and Spring Data JDBC" section).

Next, open/create the `MyRetroConfiguration` class. See Listing 5-16.

Listing 5-16. src/main/java/apress/com/myretro/config/
MyRetroConfiguration.java

```
package com.apress.myretro.config;

import com.apress.myretro.board.Card;
import com.apress.myretro.board.CardType;
import com.apress.myretro.board.RetroBoard;
import com.apress.myretro.service.RetroBoardService;
import org.springframework.boot.context.event.ApplicationReadyEvent;
import org.springframework.boot.context.properties.
EnableConfigurationProperties;
import org.springframework.context.ApplicationListener;
import org.springframework.context.annotation.Bean;
import org.springframework.context.annotation.Configuration;

@EnableConfigurationProperties({MyRetroProperties.class})
@Configuration
public class MyRetroConfiguration {

    @Bean
```

```
ApplicationListener<ApplicationReadyEvent> ready(RetroBoardService
retroBoardService) {
    return applicationReadyEvent -> {
        RetroBoard retroBoard = retroBoardService.save(RetroBoard.builder()
                .name("Spring Boot Conference")
                .build());

        retroBoardService.addCardToRetroBoard(retroBoard.getId(), Card.
        builder().comment("Spring Boot Rocks!").cardType(CardType.
        HAPPY).retroBoardId(retroBoard.getId()).build());
        retroBoardService.addCardToRetroBoard(retroBoard.getId(), Card.
        builder().comment("Meet everyone in person").cardType(CardType.
        HAPPY).retroBoardId(retroBoard.getId()).build());
        retroBoardService.addCardToRetroBoard(retroBoard.getId(), Card.
        builder().comment("When is the next one?").cardType(CardType.
        MEH).retroBoardId(retroBoard.getId()).build());
        retroBoardService.addCardToRetroBoard(retroBoard.getId(), Card.
        builder().comment("Not enough time to talk to everyone").
        cardType(CardType.SAD).retroBoardId(retroBoard.getId()).
        build());

    };
}
}
```

The MyRetroConfiguration class changed in the way we are saving some initial data. We are using the RetroBoardService to do the job, and as you can imagine the RetroBoardService uses the RetroBoardRepository interface (as a field is injected in the constructor).

Next, open/create the application.properties file. See Listing 5-17.

Listing 5-17. src/main/resources/application.properties

```
# DataSource
spring.datasource.generate-unique-name=false
spring.datasource.name=test-db

# SQL init
```

spring.sql.init.mode=always

logging.level.org.springframework.jdbc.core.JdbcTemplate=DEBUG

The only new addition to `application.properties` from the previous chapter is the logging, and we are going to do a DEBUG over the `JdbcTemplate` class. Why? Because at the end, the `JdbcTemplate` is the one executing all the SQL statements. So, when you start the app, you should see some of the SQL queries being executed (which is a way to see if we need to improve how we are modeling the classes or do a better job in normalizing some of the relationships).

The `schema.sql` file and the `advice.*`, `exception.*`, `service.*`, and `web.*` packages and classes are the same as in Chapter 4, so there's no need to add them here.

Running the My Retro App

You can run the My Retro App in your IDE or you can execute the following command:

`./gradlew clean bootRun`

In the output, you should see that Docker Compose will start up the PostgreSQL service and you should see part of the SQL queries being executed, thanks to the logging property.

```
...
Executing SQL update and returning generated keys
DEBUG [main] o.s.jdbc.core.JdbcTemplate         : Executing prepared SQL
statement [INSERT INTO "retro_board" ("name") VALUES (?)]
 INFO [main] c.a.myretro.advice.RetroBoardAdvice: [ADVICE] findById
DEBUG [main] o.s.jdbc.core.JdbcTemplate         : Executing prepared
SQL query
DEBUG [main] o.s.jdbc.core.JdbcTemplate         : Executing prepared SQL
statement [SELECT "retro_board"."id" AS "id", "retro_board"."name" AS
"name" FROM "retro_board" WHERE "retro_board"."id" = ?]
DEBUG [main] o.s.jdbc.core.JdbcTemplate         : Executing prepared
SQL query
DEBUG [main] o.s.jdbc.core.JdbcTemplate         : Executing prepared SQL
statement [SELECT "card"."id" AS "id", "card"."comment" AS "comment",
"card"."card_type" AS "card_type", "card"."retro_board_id" AS "retro_board_
id", "card"."id" AS "id" FROM "card" WHERE "card"."retro_board_id" = ?]
```

```
DEBUG [main] o.s.jdbc.core.JdbcTemplate          : Executing prepared
SQL update
...
```

Now, you can either open your browser and to to http://localhost:8080/retros or execute the following command:

```
curl -s  http://localhost:8080/retros | jq .

[
  {
    "id": "8e2d85d3-2521-45d0-975c-2753bac964c5",
    "name": "Spring Boot Conference",
    "cards": {
      "9aa89e63-d85d-41bc-af5e-ad6fd41d9447": {
        "id": "9aa89e63-d85d-41bc-af5e-ad6fd41d9447",
        "comment": "Spring Boot Rocks!",
        "cardType": "HAPPY"
      },
      "57e89fec-5720-4207-8de6-d2c11470b422": {
        "id": "57e89fec-5720-4207-8de6-d2c11470b422",
        "comment": "Not enough time to talk to everyone",
        "cardType": "SAD"
      },
      "f0ba172b-396e-471b-bff3-84eb5fd15934": {
        "id": "f0ba172b-396e-471b-bff3-84eb5fd15934",
        "comment": "When is the next one?",
        "cardType": "MEH"
      },
      "c29d3809-688b-4102-a65a-7fdfc2062875": {
        "id": "c29d3809-688b-4102-a65a-7fdfc2062875",
        "comment": "Meet everyone in person",
        "cardType": "HAPPY"
      }
    }
  }
]
```

Note that in the `Card` element the `retroBoardId` is not printed out, and this is because of the `@JsonIgnore` annotation in the `Card` class.

Now, you know how Spring Boot uses the Spring Data technology: it auto-configures all the `DataSource`, `JdbcTemplate`, and everything underneath, initializes the database by finding a `schema.sql` or `data.sql` (it's important to use `spring.sql.init.mode=always` to allow the database initialization to kick in), and it enables the repositories (`@EnableJdbcRepositories`) without telling where to find them.

Next, let's review how Spring Boot relies on Spring Data JPA to create awesome apps using this technology.

Spring Data JPA

The *Jakarta Persistence API (JPA)* (formerly known as *Java Persistence API*) is not a tool or framework, but a set of specifications that define concepts for any implementer. Like Spring Data JDBC, Spring Data JPA can be used as an ORM tool. And now, with the JPA 3 specification, JPA can be extended to be used for NoSQL databases. The difference between JDBC and JPA is that in JPA you need to think not in terms of relationships but in terms of the rules of persistence of the Java code, and in JDBC you need to manually translate your code to relational tables, columns, and keys back-and-forth.

If you use plain JPA in a non-Spring application, you still need to do a lot of coding and a lot of configuration in your app, such as handling connection and session management, handling transaction setup and management, defining mapping and persistence units with XML files, defining the JPA implementation to be used, and much more.

Spring Data JPA eliminates most of this effort by auto-configuring the implementation of the data access layers. The main feature of Spring Data JPA is that developers only need to write repository interfaces and, if needed, any custom finder methods. Spring Data JPA takes care of the implementation. Here are some of its other features:

- Spring programming model

- The choice of using either the `CrudRepository` interface or the `JpaRepository` interfaces

- Spring Data Extension with the support for Querydsl extension predicates and thus type-safe JPA queries

- Repository query-based methods, based on several keywords
  (findBy<field-name>, deleteBy<field-name>, findBy<field-
  name>And<other-field-name>, etc.) together with the field names of
  the domain classes

- Transparent auditing of domain classes

- Pagination support, dynamic query execution, and the ability to
  integrate custom data access code through several interfaces with the
  PagingAndSortingRepository

- Callbacks and events

- Customization of individual repositories

- Validation of @Query annotated queries at bootstrap time, providing
  control of customization to what you really need

- JavaConfig-based repository configuration using the
  @EnableJpaRepositories annotation

- Powerful extensions to extend Spring Data JPA, such as the use of
  Hibernate Envers for auditing/revisions

Spring Data JPA with Spring Boot

Using Spring Data JPA in a Spring app (no Spring Boot) requires adding
some configuration (either JavaConfig or XML), declaring the DataSource,
PlatformTransactionManager, and LocalContainerEntityManagerFactoryBean, and
using the @EnableJpaRepositories and @EnableTransactionManagement annotations in
the class where we are using the @Configuration annotation.

Thanks to its Spring Boot auto-configuration, we don't need to do anything.
It works just by adding the spring-boot-starter-data-jpa starter dependency.
Spring Boot allows us to overwrite the defaults by providing spring.jpa.* properties.
Another cool feature that Spring Boot provides is the default registration of the
OpenEntityManagerInViewInterceptor that applies the *Open EntityManager in View*
pattern to allow for lazy loading in web views.

By default, Spring Boot uses *Hibernate JPA* implementation, and that also brings an
enterprise-grade level for data applications.

In the following sections we'll see in action how we can use Spring Data JPA with Spring Boot in our applications.

Users App with Spring Boot and Spring Data JPA

Remember that you have access to the code, so if you want to start from scratch, you can go to the Spring Initializr (`https://start.spring.io`) and accept the default settings. Set the Group field to `com.apress` and the Artifact and Name fields to `users`. For dependencies, add Web, Validation, JPA, Lombok, H2, and PostgreSQL. Download the project, unzip it, and import it to you favorite IDE. This section covers only the classes that changed from previous sections. By the end of this section, you should have the structure shown in Figure 5-3.

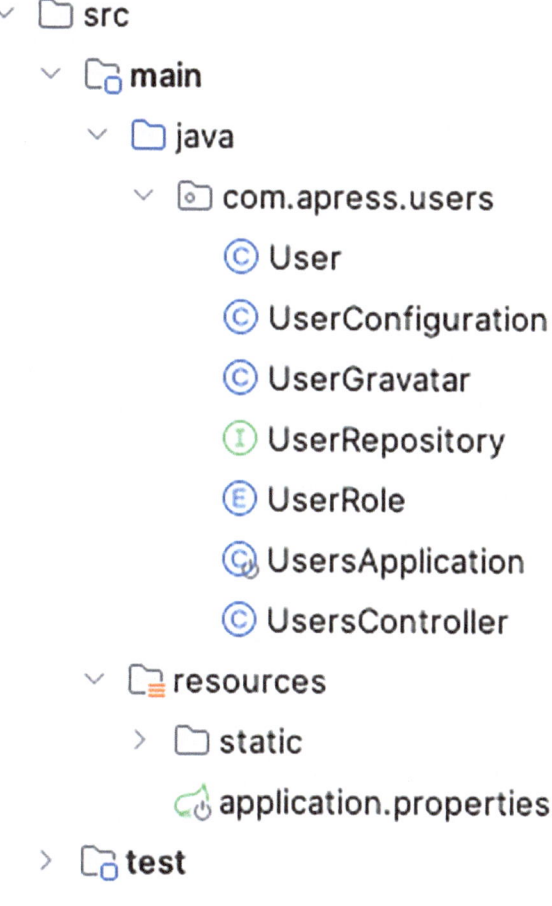

Figure 5-3. *Users App with Spring Data JPA*

Importantly, we no longer have the schema.sql file. Spring Boot with Spring Data JPA will take care of creating out database tables, as described in the following sections.

Start by opening the build.gradle file and replacing its content with the content shown in Listing 5-18.

Listing 5-18. build.gradle

```
plugins {
    id 'java'
    id 'org.springframework.boot' version '3.2.3'
    id 'io.spring.dependency-management' version '1.1.4'
}

group = 'com.apress'
version = '0.0.1-SNAPSHOT'
sourceCompatibility = '17'

repositories {
    mavenCentral()
}

dependencies {
    implementation 'org.springframework.boot:spring-boot-starter-web'
    implementation 'org.springframework.boot:spring-boot-starter-
    validation'

    implementation 'org.springframework.boot:spring-boot-starter-data-jpa'
    runtimeOnly 'com.h2database:h2'
    runtimeOnly 'org.postgresql:postgresql'

    compileOnly 'org.projectlombok:lombok'
    annotationProcessor 'org.projectlombok:lombok'

    // Web
    implementation 'org.webjars:bootstrap:5.2.3'

    testImplementation 'org.springframework.boot:spring-boot-starter-test'
}

tasks.named('test') {
```

```
        useJUnitPlatform()
}

test {
    testLogging {
        events "passed", "skipped", "failed"
        showExceptions true
        exceptionFormat "full"
        showCauses true
        showStackTraces true
        showStandardStreams = false
    }
}
```

As you can see, we are using the spring-boot-starter-data-jpa starter dependency and two drivers.

Next, let's look at the domain class. Open/Create the User class. See Listing 5-19.

Listing 5-19. src/main/java/apress/com/users/User.java

```
package com.apress.users;

import jakarta.persistence.*;
import jakarta.validation.constraints.NotBlank;
import jakarta.validation.constraints.Pattern;
import lombok.*;

import java.util.Collections;
import java.util.List;

@Builder
@AllArgsConstructor
@NoArgsConstructor
@Data
@Entity(name="USERS")
public class User {

    @GeneratedValue(strategy = GenerationType.IDENTITY)
    @Id
```

```
    private Long id;

    @NotBlank(message = "Email can not be empty")
    private String email;

    @NotBlank(message = "Name can not be empty")
    private String name;

    private String gravatarUrl;

    @Pattern(message = "Password must be at least 8 characters long and
    contain at least one number, one uppercase, one lowercase and one
    special character",
            regexp = "^(?=.*[0-9])(?=.*[a-z])(?=.*[A-Z])(?=.*[@#$%^&+=!])
            (?=\\S+$).{8,}$")
    private String password;

    @Singular("role")
    private List<UserRole> userRole;

    private boolean active;

    @PrePersist
        private void prePersist(){
          if (this.gravatarUrl == null)
                this.gravatarUrl = UserGravatar.
                getGravatarUrlFromEmail(this.email);
          if(this.userRole == null)
                this.userRole = Collections.
                singletonList(UserRole.INFO);
        }
}
```

Listing 5-19 shows that the User class includes the following annotations (all of which are from the jakarta.persistence.* package):

- @Entity: This annotation marks our class as an entity and accepts a parameter that will be used to set the name of the table—in this case, the USERS table.

- @Id: This annotation is required to set the class identity, which is the primary key in the database.

- @GeneratedValue: This annotation is used to set the Id generation strategy. It accepts as a parameter which strategy to use; this choice will depend on your database engine. If your database doesn't support an identity column, you can use the strategy GenerationType.SEQUENCE. If your database does support an indentity column, then you can use any of these: GenerationType. TABLE (indicates that the persistence provider must assign primary keys for the entity using an underlying database table to ensure uniqueness), GenerationType.UUID (indicates that the persistence provider must assign primary keys for the entity by generating an RFC 4122 Universally Unique Identifier), GenerationType.AUTO (indicates that the persistence provider should pick an appropriate strategy for the database).

- @PrePersist: This annotation marks the method as a callback for the corresponding life cycle event. There are many more of these callback annotation markers, such as @PreUpdate, @PreRemove, @PostUpddate, @PostRemove, @PostPersist, and @PostLoad.

Next, open/create the UserRepository interface. See Listing 5-20.

Listing 5-20. src/main/java/apress/com/users/UserRepository.java

```
package com.apress.users;

import org.springframework.data.repository.CrudRepository;
import org.springframework.transaction.annotation.Transactional;

import java.util.Optional;

public interface UserRepository extends CrudRepository<User,Long>{
    Optional<User> findByEmail(String email);

    @Transactional
    void deleteByEmail(String email);
}
```

The `UserRepository` interface extends from the `CrudRepository`...wait...what? We used the same interface in Listing 5-4. We could use the JPA-related interface, but we will use it in the My Retro App project. Remember that the `CrudRepository` interface has several methods declared and that Spring Data will take care of the implementation. Also note that we are adding the `@Transactional` annotation, which allows us to have a thread-safe call.

Next, open the `application.properties` file and replace its content with the content shown in Listing 5-21.

Listing 5-21. src/main/resources/application.properties

```
# H2
spring.h2.console.enabled=true

# DataSource
spring.datasource.generate-unique-name=false
spring.datasource.name=test-db

# JPA
spring.jpa.show-sql=true
spring.jpa.generate-ddl=true
spring.jpa.hibernate.ddl-auto=update
spring.jpa.properties.hibernate.format_sql=true
```

The new additions to the `application.properties` file are the two `spring.jpa.*` properties, the first of which will show every SQL statement executed and the second of which will help to auto-generate the DDL for our database—this is awesome, because we no longer have to worry about creating a `schema.sql` file!

The `spring.jpa.hibernate.ddl-auto` property provides the following values:

- `update`: Updates the schema if necessary.

- `create`: Creates the schema and destroys previous data.

- `create-drop`: Creates the schema and then destroys the schema at the end of the session.

- `none`: Disables the DDL handling.

- `validate`: Validates the schema and makes no changes to the database.

As you can see, the DDL generation support covers different scenarios. The other classes, UserConfiguration, UserGravatar, UserRole, and UsersController, and the test UsersHttpRequestTests remain the same.

Testing and Running the Users App

You can run the tests in your IDE or by using the following command:

```
./gradlew clean test
```

```
UsersHttpRequestTests > userEndPointFindUserShouldReturnUser() PASSED
UsersHttpRequestTests > userEndPointDeleteUserShouldReturnVoid() PASSED
UsersHttpRequestTests > indexPageShouldReturnHeaderOneContent() PASSED
UsersHttpRequestTests > userEndPointPostNewUserShouldReturnUser() PASSED
UsersHttpRequestTests >
usersEndPointShouldReturnCollectionWithTwoUsers() PASSED
```

You can run the Users App in your IDE or by using the following command:

```
./gradlew bootRun
```

```
...
Hibernate:
    create table users (
        id bigint generated by default as identity,
        active boolean not null,
        email varchar(255),
        gravatar_url varchar(255),
        name varchar(255),
        password varchar(255),
        user_role tinyint array,
        primary key (id)
    )
...
```

This output shows that the JPA/Hibernate SQL statements are being executed when saving the data (in the UserConfiguration class).

My Retro App Using Spring Boot and Spring Data JPA

Next, let's look at what we need to do to use Spring Data JPA in the My Retro App. I'll add all the classes that are important to review, remember that you have access to the source code. But if you want to start from scratch, go to the Spring Initializr (`https://start.spring.io`) and start an empty project. Set the Group field to `com.apress` and the Artifact and Name fields to `myretro`. For dependencies, add Web, Validation, JPA, Docker Compose, Processor, and PostgreSQL. Download the project, unzip it, and import it into your favorite IDE. By the end of this section, you should have the structure shown in Figure 5-4.

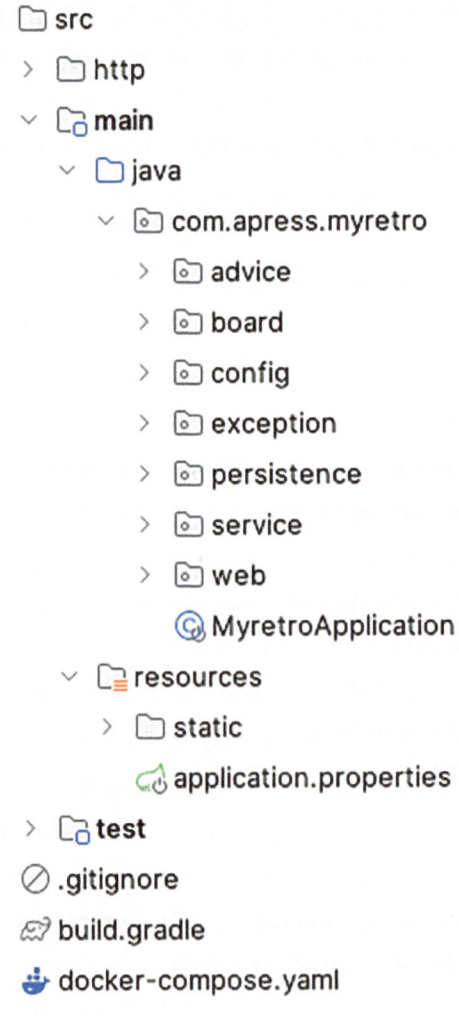

Figure 5-4. *My Retro App with Spring Data JPA*

As you can see, we no longer need schema.sql. Spring Boot with Spring Data JPA takes care of creating our database's tables.

Open the build.gradle file and replace its content with the content shown in Listing 5-22.

Listing 5-22. build.gradle

```
plugins {
    id 'java'
    id 'org.springframework.boot' version '3.2.3'
    id 'io.spring.dependency-management' version '1.1.4'
}

group = 'com.apress'
version = '0.0.1-SNAPSHOT'
sourceCompatibility = '17'

repositories {
    mavenCentral()
}

dependencies {
    implementation 'org.springframework.boot:spring-boot-starter-web'
    implementation 'org.springframework.boot:spring-boot-starter-
    validation'
    implementation 'org.springframework.boot:spring-boot-starter-aop'

    implementation 'org.springframework.boot:spring-boot-starter-data-jpa'
    developmentOnly 'org.springframework.boot:spring-boot-docker-compose'

    annotationProcessor 'org.springframework.boot:spring-boot-
    configuration-processor'

    runtimeOnly 'org.postgresql:postgresql'

    compileOnly 'org.projectlombok:lombok'
    annotationProcessor 'org.projectlombok:lombok'

    // Web
    implementation 'org.webjars:bootstrap:5.2.3'
```

```
    testImplementation 'org.springframework.boot:spring-boot-starter-test'
}

tasks.named('test') {
    useJUnitPlatform()
}
```

As you can see, we are including the spring-boot-starter-data-jpa and spring-boot-docker-compose starter dependencies.

Next, open/create the board package and the Card and RetroBoard classes. Listing 5-23 shows the Card class.

Listing 5-23. src/main/java/apress/com/myretro/board/Card.java

```
package com.apress.myretro.board;

import com.fasterxml.jackson.annotation.JsonIgnore;
import jakarta.persistence.*;
import jakarta.validation.constraints.NotBlank;
import jakarta.validation.constraints.NotNull;
import lombok.AllArgsConstructor;
import lombok.Builder;
import lombok.Data;
import lombok.NoArgsConstructor;

import java.util.UUID;

@Builder
@AllArgsConstructor
@NoArgsConstructor
@Data
@Entity
public class Card {

        @GeneratedValue(strategy = GenerationType.UUID)
    @Id
    private UUID id;

    @NotBlank
    private String comment;
```

```
@NotNull
@Enumerated(EnumType.STRING)
private CardType cardType;

@ManyToOne
    @JoinColumn(name = "retro_board_id")
@JsonIgnore
RetroBoard retroBoard;
}
```

Let's analyze the Card class annotations and see what changed from previous versions:

- @Entity: This annotation marks our class as an entity. This annotation comes from the jakarta.persistence.* package. By default, the name of the table will be CARD.

- @GeneratedValue: This annotation is used to set the Id generation strategy. It accepts as a parameter which strategy to use; this will depend on your database engine. In this case we are using the UUID.

- @Id: This annotation is required to set the class identity, which is the primary key in the database. This annotation also comes from the jakarta.persistence.* package.

- @Enumerated: This annotation specifies that a property or field should be persisted as an enumerated type. It accepts a value of EnumType, ORDINAL, or STRING. With this setting, it will generate the following SQL statement (this is just a snippet):

```
create table card (
        id uuid not null,
        retro_board_id uuid,
        card_type varchar(255) not null check (card_type in
        ('HAPPY','MEH','SAD')),
        comment varchar(255),
        primary key (id)
    )
```

- @ManyToOne: This annotation specifies a single-valued association to another entity class that has many-to-one multiplicity. There is also the @ManyToMany, @OneToOne, @OneToMany.

- @JoinColumn: This annotation specifies a column for joining an entity association or element collection. In this case we are identifying which column to join, retro_board_id. Remember that the NamingStrategy takes place, using snake case to create the fields. You can see that in the previous snippet.

As previously mentioned, with JPA we need to take care of the objects and the dependencies. In this case we have a dependency of the RetroBoard class, shown in Listing 5-24.

Listing 5-24. src/main/java/apress/com/myretro/board/RetroBoard.java

```
package com.apress.myretro.board;

import jakarta.persistence.*;
import jakarta.validation.constraints.NotBlank;
import lombok.*;

import java.util.ArrayList;
import java.util.List;
import java.util.UUID;

@Builder
@AllArgsConstructor
@NoArgsConstructor
@Data
@Entity
public class RetroBoard {

    @GeneratedValue(strategy = GenerationType.UUID)
        @Id
    private UUID id;

    @NotBlank(message = "A name must be provided")
    private String name;
```

```
@Singular
@OneToMany(mappedBy = "retroBoard")
private List<Card> cards = new ArrayList<>();
}
```

Note in Listing 5-24 that we are still using the required annotations, @Entity and @Id. We also are using the new @OneToMany annotation, which knows how to get all the Cards based on the RetroBoard id. The @OneToMany annotation specifies a many-valued association with one-to-many multiplicity.

Next, open/create the persistence package and the CardRepository and RetroBoardRepository classes, as shown in Listing 5-25 and 5-26, respectively.

Listing 5-25. src/main/java/apress/com/myretro/persistence/ CardRepository.java

```
package com.apress.myretro.persistence;

import com.apress.myretro.board.Card;
import org.springframework.data.jpa.repository.JpaRepository;

import java.util.UUID;

public interface CardRepository extends JpaRepository<Card, UUID> {
}
```

Listing 5-26. src/main/java/apress/com/myretro/persistence/ RetroBoardRepository.java

```
package com.apress.myretro.persistence;

import com.apress.myretro.board.RetroBoard;
import org.springframework.data.jpa.repository.JpaRepository;

import java.util.UUID;

public interface RetroBoardRepository extends JpaRepository<RetroBoard,UUID> {
}
```

Listings 5-25 and 5-26 show that the CardRepository and RetroBoardRepository interfaces both are extending from the JpaRepository interface, which is all about JPA. take a look at the following snippet:

```
public interface JpaRepository<T, ID> extends ListCrudRepository<T, ID>,
ListPagingAndSortingRepository<T, ID>, QueryByExampleExecutor<T> {

    void flush();

    <S extends T> S saveAndFlush(S entity);

    <S extends T> List<S> saveAllAndFlush(Iterable<S> entities);

    void deleteAllInBatch(Iterable<T> entities);

    void deleteAllByIdInBatch(Iterable<ID> ids);

    void deleteAllInBatch();

    T getReferenceById(ID id);

    @Override
    <S extends T> List<S> findAll(Example<S> example);

    @Override
    <S extends T> List<S> findAll(Example<S> example, Sort sort);
}
```

As you can see, the JpaRepository extends from other interfaces that allow paging and sorting, and from the CrudRepository, which brings the methods you already know.

Next, open/create the service package and the RetroBoardService class. See Listing 5-27.

Listing 5-27. src/main/java/apress/com/myretro/service/
RetroBoardService.java

```
package com.apress.myretro.service;

import com.apress.myretro.board.Card;
import com.apress.myretro.board.RetroBoard;
import com.apress.myretro.exception.CardNotFoundException;
```

```
import com.apress.myretro.exception.RetroBoardNotFoundException;
import com.apress.myretro.persistence.CardRepository;
import com.apress.myretro.persistence.RetroBoardRepository;
import lombok.AllArgsConstructor;
import org.springframework.stereotype.Service;

import java.util.List;
import java.util.Optional;
import java.util.UUID;

@AllArgsConstructor
@Service
public class RetroBoardService {

    RetroBoardRepository retroBoardRepository;
        CardRepository cardRepository;

    public RetroBoard save(RetroBoard domain) {
        return this.retroBoardRepository.save(domain);
    }

    public RetroBoard findById(UUID uuid) {
        return this.retroBoardRepository.findById(uuid).get();
    }

    public Iterable<RetroBoard> findAll() {
        return this.retroBoardRepository.findAll();
    }

    public void delete(UUID uuid) {
        this.retroBoardRepository.deleteById(uuid);
    }

    public Iterable<Card> findAllCardsFromRetroBoard(UUID uuid) {
        return this.findById(uuid).getCards();
    }

    public Card addCardToRetroBoard(UUID uuid, Card card){
        Card result = retroBoardRepository.findById(uuid).
        map(retroBoard -> {
```

```java
            card.setRetroBoard(retroBoard);
            return cardRepository.save(card);
        }).orElseThrow(() -> new RetroBoardNotFoundException());
        return result;
    }

    public void addMultipleCardsToRetroBoard(UUID uuid, List<Card> cards){
        RetroBoard retroBoard = this.findById(uuid);
        cards.forEach(card -> card.setRetroBoard(retroBoard));
        cardRepository.saveAll(cards);
    }

    public Card findCardByUUID(UUID uuidCard){
        Optional<Card> result = cardRepository.findById(uuidCard);
        if(result.isPresent()){
            return result.get();
        }else{
            throw new CardNotFoundException();
        }
    }

    public Card saveCard(Card card){
        return cardRepository.save(card);
    }

    public void removeCardByUUID(UUID cardUUID){
        cardRepository.deleteById(cardUUID);
    }
}
```

As you can see, the RetroBoardService class is very straightforward, and in this case we are using two repositories, the CardRepository and RetroBoardRepository interfaces, as part of the dependencies that will be injected by Spring.

Next, open the application.properties file and replace its content with the content shown in Listing 5-28.

Listing 5-28. src/main/resources/application.properties

```
# DataSource
spring.datasource.generate-unique-name=false
spring.datasource.name=test-db

# JPA
spring.jpa.show-sql=true
spring.jpa.generate-ddl=true
spring.jpa.hibernate.ddl-auto=create-drop
spring.jpa.properties.hibernate.format_sql=true
spring.jpa.properties.hibernate.dialect=org.hibernate.dialect.PostgreSQLDialect
```

The only new property is `hibernate.dialect`, just to show that you can specify a dialect, but this is not necessary, because when the auto-configuration kicks in, it already knows that you are using PostgreSQL.

The `advice`, `config`, `exception`, and `web` packages and their classes are the same from the previous sections and chapters. As previously noted, you can access the source code from the Apress website (`https://www.apress.com/gp/services/source-code`).

Running the My Retro App

You can run My Retro App in your IDE or you can execute the following command:

```
./gradlew clean bootRun
```

In the output you should see that Docker Compose will start up the PostgreSQL service and you should see part of the SQL queries being executed. Note that the JPA generates three tables, `card`, `retro_board`, and `retro_board_card`, which is the result of how we create our classes and relationships.

Now, you can either open your browser and go to `http://localhost:8080/retros` or execute the following command:

```
curl -s  http://localhost:8080/retros | jq .

[
  {
    "id": "57964a9a-9b56-453d-925a-64f63b502a48",
    "name": "Spring Boot Conference",
```

```
  "cards": [
    {
      "id": "080d4feb-8f84-4fc7-b6c3-9da741291846",
      "comment": "Spring Boot Rocks!",
      "cardType": "HAPPY"
    },
    {
      "id": "6a642199-ca5f-4d25-9242-1fe3301cf49d",
      "comment": "Meet everyone in person",
      "cardType": "HAPPY"
    },
    {
      "id": "5b4fd29f-89a8-4842-9381-6a0ed4cddb4f",
      "comment": "When is the next one?",
      "cardType": "MEH"
    },
    {
      "id": "91a49f48-a99b-4163-a0c7-230aa0142dc2",
      "comment": "Not enough time to talk to everyone",
      "cardType": "SAD"
    }
  ]
  }
]
```

As you can see, the Card is not printing out the retroBoard instance due to the @JsonIgnore annotation.

That's it, you're now aware of the power of JPA. But is there something simpler?

Spring Data REST

Spring Data REST, another project under the umbrella of Spring Data, provides an easy way to build hypermedia-driven REST web services using the interface repositories. When you include Spring Data REST, it will analyze all your domain models and create all the necessary HTTP resources for aggregates contained in your models; in other words, it will generate the REST controllers (with the Hypertext Application Language as the media type) having access to HTTP methods (such as POST, GET, PUT, PATCH, and so on) for your application.

The following are some of the most common features of Spring Data REST:

- Exposes a discoverable REST API based on your domain model

- Uses HAL+JSON by default (Hypertext Application Language)

- Exposes collection, item and any association resources that represent your model

- Support pagination via navigational links

- Creates search resources for query methods defined in your repositories

- Exposes metadata about the model discovered as Application-Level Profile Semantics (ALPS) and JSON Schema

- Brings a HAL Explorer to exposed metadata

- Currently supports JPA, MongoDB, Neo4j, Solr, Cassandra, and Gemfire

- Allows customizations of the default resources exposed

Spring Data REST with Spring Boot

If you want to use Spring Data REST with Spring Boot, you need to add the spring-boot-starter-data-rest starter along with the supported technologies. All the auto-configuration will set up everything for you to use the REST API from your models. There's no need to do anything else—it's that simple.

Users App Using Spring Boot and Spring Data REST

Again, you have access to the code, so if you want to start from scratch, you can go to the Spring Initializr (https://start.spring.io) and accept the defaults. Set the Group field to com.apress and the Artifact and Name fields to users. For dependencies, add Web, Validation, JPA, Data REST, H2, and Lombok. Download the project, unzip it, and import it to your favorite IDE. I'll present only the classes that change from previous sections. By the end of this section, you should have the structure shown in Figure 5-5.

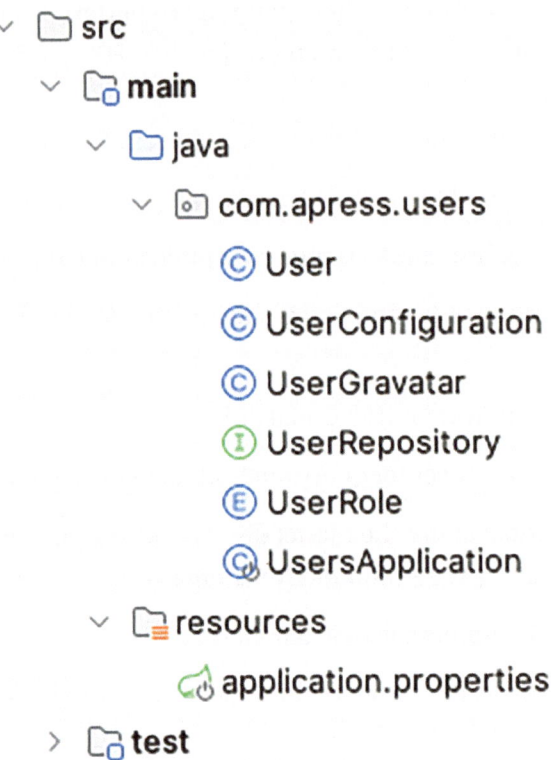

Figure 5-5. *Users App with JPA and Data REST*

Notice in particular that it no longer includes the UsersController class or static content!

Start by opening the build.gradle file and replacing its content with the content shown in Listing 5-29.

Listing 5-29. build.gradle

```
plugins {
    id 'java'
    id 'org.springframework.boot' version '3.2.3'
    id 'io.spring.dependency-management' version '1.1.4'
}

group = 'com.apress'
version = '0.0.1-SNAPSHOT'
sourceCompatibility = '17'
```

```
repositories {
    mavenCentral()
}

dependencies {
    implementation 'org.springframework.boot:spring-boot-starter-web'
    implementation 'org.springframework.boot:spring-boot-starter-validation'

    implementation 'org.springframework.boot:spring-boot-starter-data-jpa'

        implementation 'org.springframework.boot:spring-boot-starter-
        data-rest'
        implementation 'org.springframework.data:spring-data-rest-hal-
        explorer'

    runtimeOnly 'com.h2database:h2'

    compileOnly 'org.projectlombok:lombok'
    annotationProcessor 'org.projectlombok:lombok'

    // Web
    implementation 'org.webjars:bootstrap:5.2.3'

    testImplementation 'org.springframework.boot:spring-boot-starter-test'
}

tasks.named('test') {
    useJUnitPlatform()
}

test {
    testLogging {
        events "passed", "skipped", "failed"
        showExceptions true
        exceptionFormat "full"
        showCauses true
        showStackTraces true
        showStandardStreams = false
    }
}
```

We've added the `spring-boot-starter-data-rest`, `spring-data-rest-hal-explorer`, and, of course, `spring-boot-starter-data-jpa` starter dependencies.

Next, open/create the UserRepository interface. See Listing 5-30.

Listing 5-30. src/main/java/apress/com/users/UserRepository.java

```java
package com.apress.users;

import org.springframework.data.repository.CrudRepository;
import org.springframework.data.repository.PagingAndSortingRepository;

import java.util.Optional;

public interface UserRepository extends CrudRepository<User,Long>, PagingAndSortingRepository<User,Long> {
    Optional<User> findByEmail(String email);
}
```

We've only removed the delete method and extended the PaginAndSortingRespository interface. This interface is also part of Spring Data (the core) and can be used along with JPA.

The User, UserConfiguration, UserGravatar, and UserRole classes haven't changed from the Spring Data JPA project, nor has the application.properties file (and, yes, we removed the UsersController class). Modifying the UserRepository wasn't necessary, but I wanted to show you the effects of having pagination and sorting. And that's it, just a minimal change!

Note   When you run the Users App, Spring Data REST inspects your application and gets your repositories. Based on the domain classes, it will create several endpoints that accept any HTTP requests and more, the access will be using the domain model in this case User class, and it will generate the /users (lowercase and plural, thanks to the Evo Inflector library) endpoint. The response will be always based on **HAL+JSON media type/content-type**, so you need to know how to handle those responses.

Running the Users App

Normally we test our app before running it, but I want to run it first to show you the result of including the `spring-data-rest-hal-explorer` starter dependency and discuss the benefits that it brings to the app.

Either run the application in your IDE or run it with the following command:

```
./gradlew clean bootRun
```

Open your browser and go to `http://localhost:8080` and you should see the HAL Explorer, as shown in Figure 5-6.

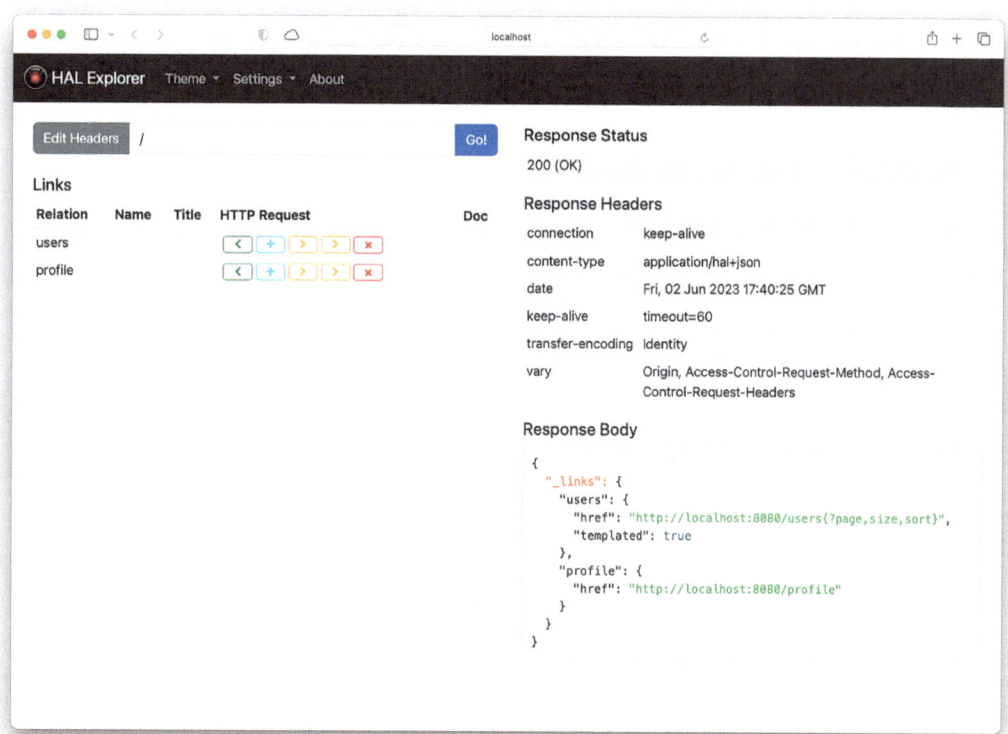

Figure 5-6. *Users App -* `http://localhost:8080` *- HAL Explorer*

By default, you have access to the HAL Explorer in the root endpoint /. If you click the < symbol in the `users` row (in the left pane below the HTTP Request), you will see a dialog box asking for parameters, as shown in Figure 5-7.

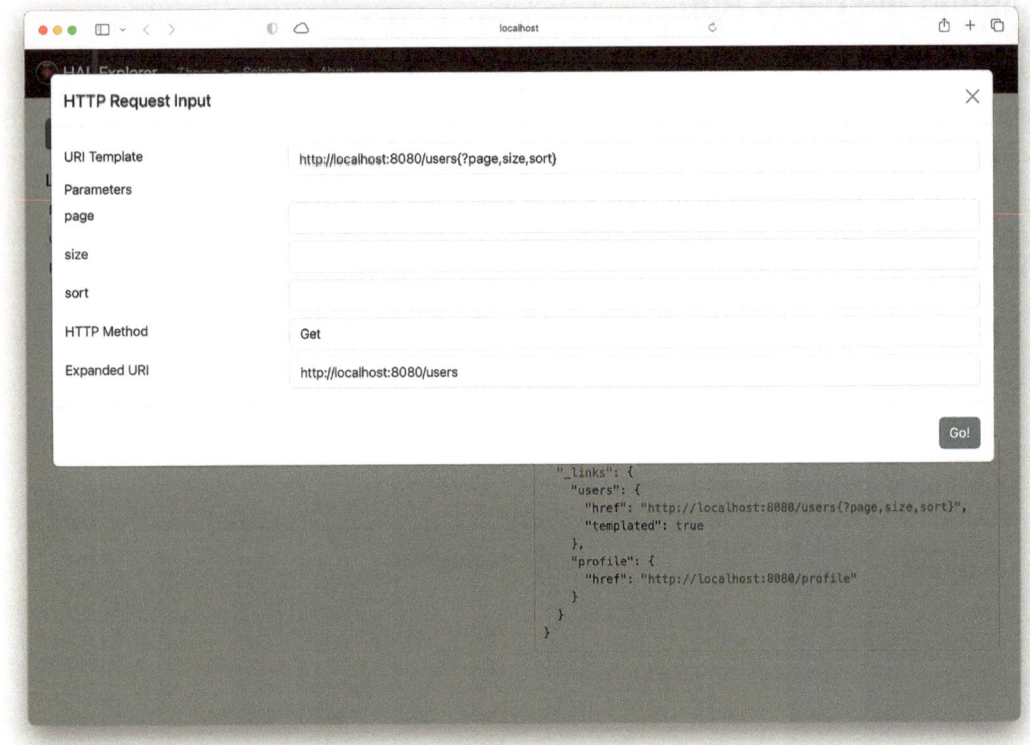

Figure 5-7. *Users App - HAL Explorer - users Dialog*

If you click the Go! button, you will see the updated HAL Explorer window shown in Figure 5-8.

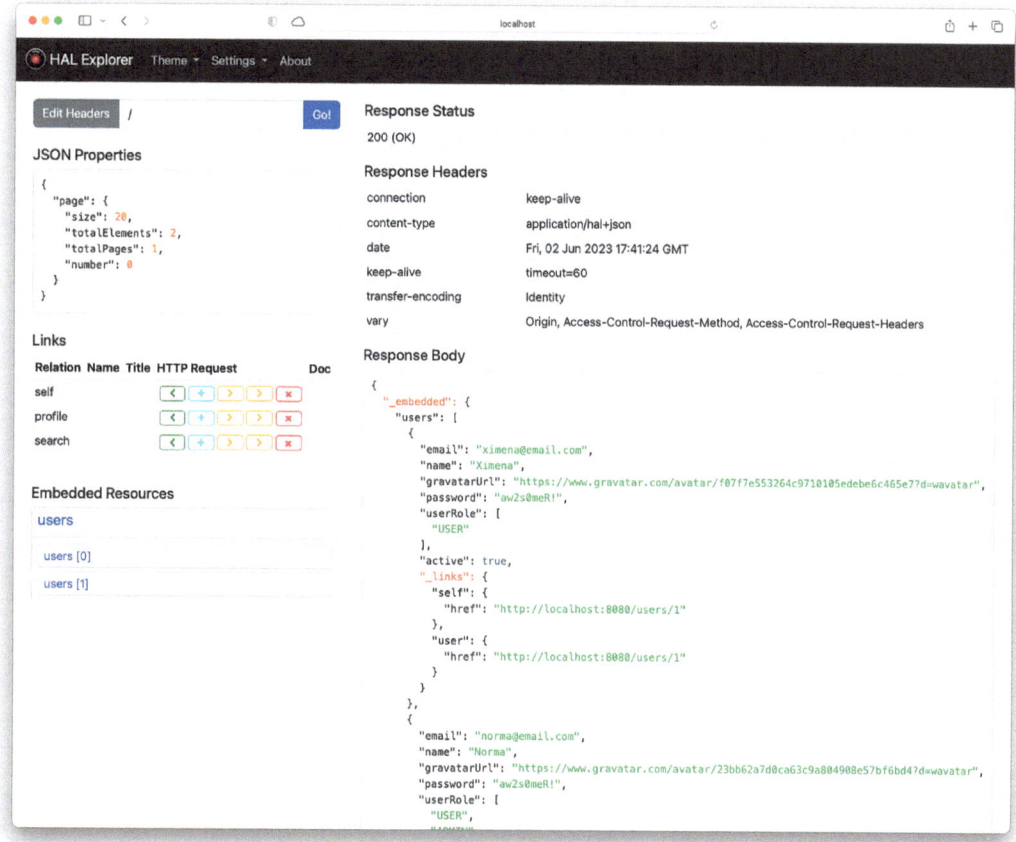

Figure 5-8. *Users App - HAL Explorer - users HAL hyperlinks*

Notice that the response now has more metadata, with new keywords; that's the HAL. response. If you click `users[0]` in the left pane under Embedded Resources, you should see the window shown in Figure 5-9.

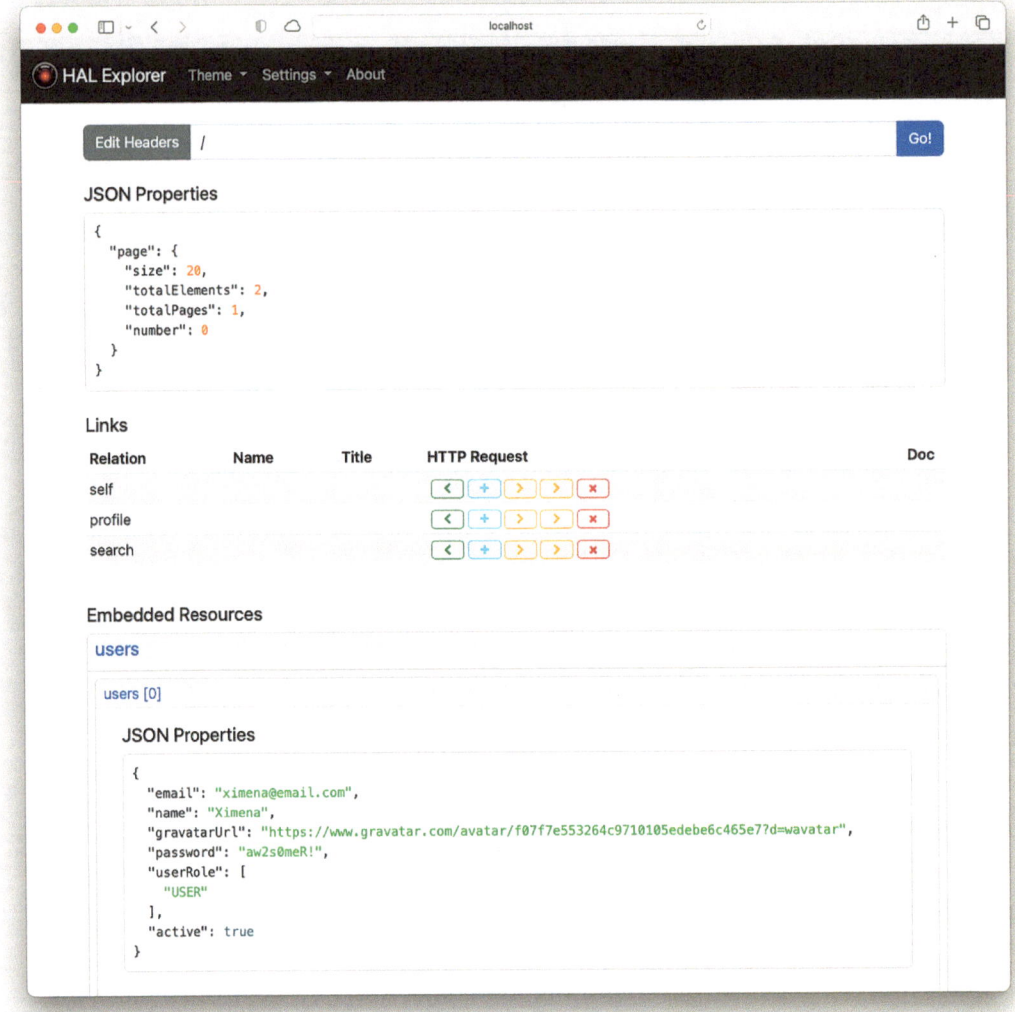

Figure 5-9. *Users App - HAL Explorer - Embedded Resources - users[0]*

This is awesome, because now you have an API that is discoverable by other clients. Take a moment to click around and look at the HAL format. If you want to take a look at the command line, open a terminal and execute the following:

```
curl -s  http://localhost:8080/users | jq .

{
  "_embedded": {
    "users": [
```

```json
{
  "email": "ximena@email.com",
  "name": "Ximena",
  "gravatarUrl": "https://www.gravatar.com/avatar/f07f7e553264c971010
  5edebe6c465e7?d=wavatar",
  "password": "aw2sOmeR!",
  "userRole": [
    "USER"
  ],
  "active": true,
  "_links": {
    "self": {
      "href": "http://localhost:8080/users/1"
    },
    "user": {
      "href": "http://localhost:8080/users/1"
    }
  }
},
{
  "email": "norma@email.com",
  "name": "Norma",
  "gravatarUrl": "https://www.gravatar.com/avatar/23bb62a7d0ca63c
  9a804908e57bf6bd4?d=wavatar",
  "password": "aw2sOmeR!",
  "userRole": [
    "USER",
    "ADMIN"
  ],
  "active": true,
  "_links": {
    "self": {
      "href": "http://localhost:8080/users/2"
    },
    "user": {
```

```
            "href": "http://localhost:8080/users/2"
          }
        }
      }
    ]
  },
  "_links": {
    "self": {
      "href": "http://localhost:8080/users?page=0&size=20"
    },
    "profile": {
      "href": "http://localhost:8080/profile/users"
    },
    "search": {
      "href": "http://localhost:8080/users/search"
    }
  },
  "page": {
    "size": 20,
    "totalElements": 2,
    "totalPages": 1,
    "number": 0
  }
}
```

Before you continue, analyze the response and all the elements. In particular, notice that there is a "_links" element and that it declares the "search" endpoint. This endpoint was created thanks to the findByEmail method declaration in the UserRepository, so you can search for an email using the following command:

curl -s "http://localhost:8080/users/search/findByEmail?email=ximena@email.com" | jq .

```
{
  "email": "ximena@email.com",
  "name": "Ximena",
```

```
"gravatarUrl": "https://www.gravatar.com/avatar/f07f7e553264c9710105edebe
6c465e7?d=wavatar",
"password": "aw2sOmeR!",
"userRole": [
  "USER"
],
"active": true,
"_links": {
  "self": {
    "href": "http://localhost:8080/users/1"
  },
  "user": {
    "href": "http://localhost:8080/users/1"
  }
}
}
```

It will return the user and self-links (that help to navigate to an individual record). And if you are in a browser, you can access to the http://localhost:8080/users and get all the users records. And if you install the JSON Viewer plugin, you can click any of the links displayed in the page. See Figure 5-10.

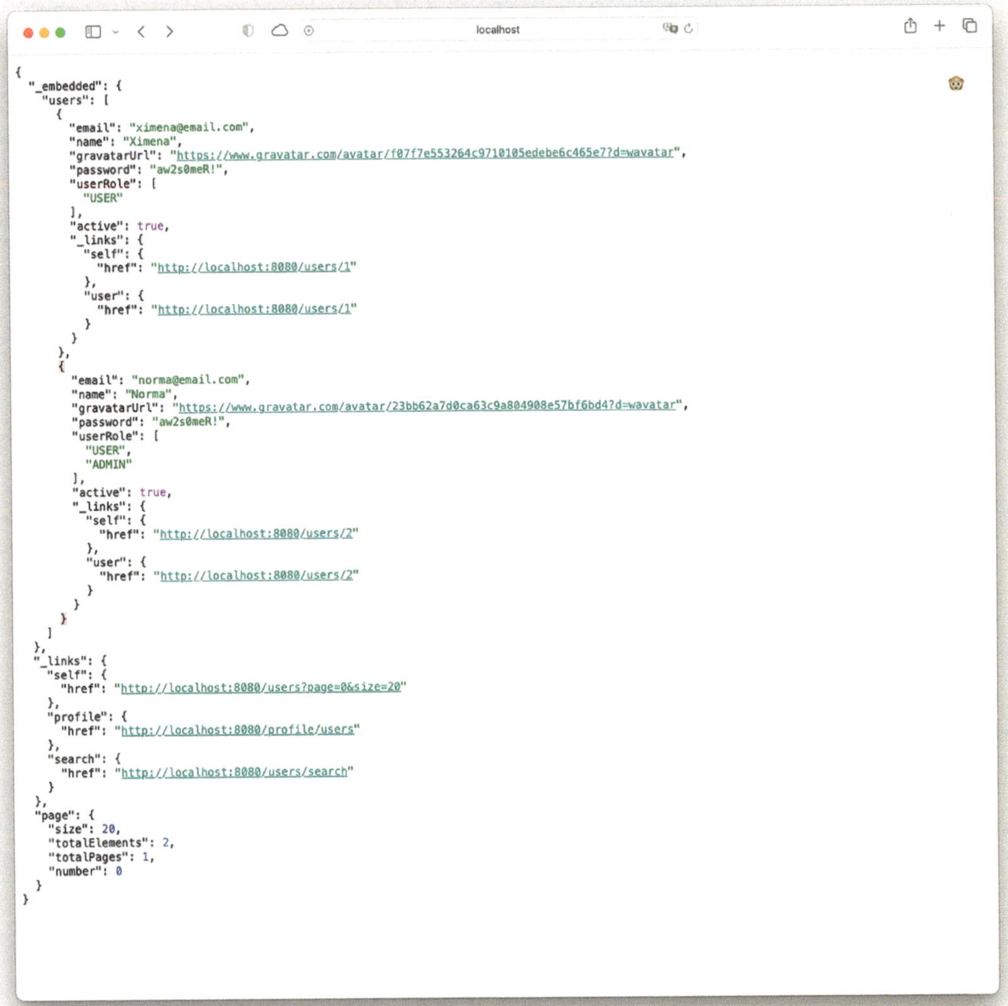

Figure 5-10. *Users App -* `http://localhost:8080/users` *- JSON Viewer Plugin*

Testing the Users App

Now, it's time to test our app using integration testing. In this case it's a little different from what we have seen in previous sections and chapters. We are going to use new classes that support the HAL/HATEOAS media type to do the testing.

So, open/create the `UsersHttpRequestTests` class and replace/add the content as shown in Listing 5-31.

Listing 5-31. src/test/java/apress/com/users/UsersHttpRequestTests.java

```
package com.apress.users;

import com.fasterxml.jackson.core.JsonProcessingException;
import com.fasterxml.jackson.databind.ObjectMapper;
import org.junit.jupiter.api.BeforeEach;
import org.junit.jupiter.api.Test;
import org.springframework.beans.factory.annotation.Autowired;
import org.springframework.boot.test.context.SpringBootTest;
import org.springframework.boot.test.web.client.TestRestTemplate;
import org.springframework.core.ParameterizedTypeReference;
import org.springframework.hateoas.CollectionModel;
import org.springframework.hateoas.EntityModel;
import org.springframework.hateoas.MediaTypes;
import org.springframework.http.*;

import static org.assertj.core.api.Assertions.assertThat;

@SpringBootTest(webEnvironment = SpringBootTest.WebEnvironment.RANDOM_PORT)
public class UsersHttpRequestTests {

    private String baseUrl;

    @Autowired
    private TestRestTemplate restTemplate;

    @BeforeEach
    public void setUp() throws Exception {
        baseUrl = "/users";
    }

    @Test
    public void usersEndPointShouldReturnCollectionWithTwoUsers() throws
    Exception {
        ResponseEntity<CollectionModel<EntityModel<User>>> response =
                restTemplate.exchange(baseUrl, HttpMethod.GET, null, new
                ParameterizedTypeReference<CollectionModel<EntityModel<Us
                er>>>() {});
```

```
    assertThat(response).isNotNull();
    assertThat(response.getStatusCode()).isEqualTo(HttpStatus.OK);
    assertThat(response.getBody()).isNotNull();
    assertThat(response.getHeaders().getContentType()).
    isEqualTo(MediaTypes.HAL_JSON);
    assertThat(response.getBody().getContent().size()).
    isGreaterThanOrEqualTo(2);
}

@Test
public void userEndPointPostNewUserShouldReturnUser() throws
Exception {
    HttpHeaders createHeaders = new HttpHeaders();
    createHeaders.setContentType(MediaTypes.HAL_JSON);

    User user =  User.builder()
            .email("dummy@email.com")
            .name("Dummy")
            .gravatarUrl("https://www.gravatar.com/avatar/23bb62a7d0ca6
            3c9a804908e57bf6bd4?d=wavatar")
            .password("aw2sOmeR!")
            .role(UserRole.USER)
            .active(true)
            .build();

    HttpEntity<String> createRequest = new HttpEntity<>(convertToJson(u
    ser), createHeaders);
    ResponseEntity<EntityModel<User>> response =  this.restTemplate.
    exchange(baseUrl, HttpMethod.POST, createRequest, new Parameterized
    TypeReference<EntityModel<User>>() {});

    assertThat(response).isNotNull();
    assertThat(response.getStatusCode()).isEqualTo(HttpStatus.CREATED);

    EntityModel<User> userResponse = response.getBody();
    assertThat(userResponse).isNotNull();
    assertThat(userResponse.getContent()).isNotNull();
    assertThat(userResponse.getLink("self")).isNotNull();
```

```
    assertThat(userResponse.getContent().getEmail()).isEqualTo(user.
    getEmail());

}

@Test
public void userEndPointDeleteUserShouldReturnVoid() throws Exception {

    this.restTemplate.delete(baseUrl + "/1");

    ResponseEntity<CollectionModel<EntityModel<User>>> response =
            restTemplate.exchange(baseUrl, HttpMethod.GET, null, new
            ParameterizedTypeReference<CollectionModel<EntityModel<Us
            er>>>() {});

    assertThat(response).isNotNull();
    assertThat(response.getStatusCode()).isEqualTo(HttpStatus.OK);
    assertThat(response.getBody()).isNotNull();
    assertThat(response.getHeaders().getContentType()).
    isEqualTo(MediaTypes.HAL_JSON);
    assertThat(response.getBody().getContent().size()).
    isGreaterThanOrEqualTo(1);

}

@Test
public void userEndPointFindUserShouldReturnUser() throws Exception{
    String email = "ximena@email.com";
    ResponseEntity<EntityModel<User>> response = restTemplate.
    exchange(baseUrl + "/search/findByEmail?email={email}", HttpMethod.
    GET, null, new ParameterizedTypeReference<EntityModel<User>>()
    {}, email);
    assertThat(response.getStatusCode()).isEqualTo(HttpStatus.OK);
    EntityModel<User> users = response.getBody();
    assertThat(users).isNotNull();
    assertThat(users.getContent().getEmail()).isEqualTo(email);
}
```

```
private String convertToJson(User user) throws
JsonProcessingException {
    ObjectMapper objectMapper = new ObjectMapper();
    return objectMapper.writeValueAsString(user);
}
}
```

The UsersHttpRequestTests test class now is using classes from the org.
springframework.hateoas.* package. The spring-boot-starter-data-rest starter
dependency that we added includes the spring-hateoas library, which contains the
necessary classes to handle the HAL+JSON type: CollectionModel and EntityModel. Note
that we can get the references, the links, and all the declaration that the response has.

You can run the test using your IDE or with the following command:

```
./gradlew clean test
```

```
UsersHttpRequestTests > userEndPointFindUserShouldReturnUser() PASSED
UsersHttpRequestTests > userEndPointDeleteUserShouldReturnVoid() PASSED
UsersHttpRequestTests > userEndPointPostNewUserShouldReturnUser() PASSED
UsersHttpRequestTests >
usersEndPointShouldReturnCollectionWithTwoUsers() PASSED
```

My Retro App Using Spring Boot and Spring Data REST

Now it's the turn of My Retro App to benefit from Spring Data REST. Again, you have
the code or you can start from scratch by going to the Spring Initializr (https://start.
spring.io). You can use an empty project, but again, make sure to set the Group field to
com.apress and the Artifact and Name fields to myretro. For dependencies, add Web,
Validation, JPA, Docker Compose, Data REST, PostgreSQL, and Lombok. Download
the project, unzip it, and import it into your favorite IDE. By the end of this section, you
should have the structure shown in Figure 5-11.

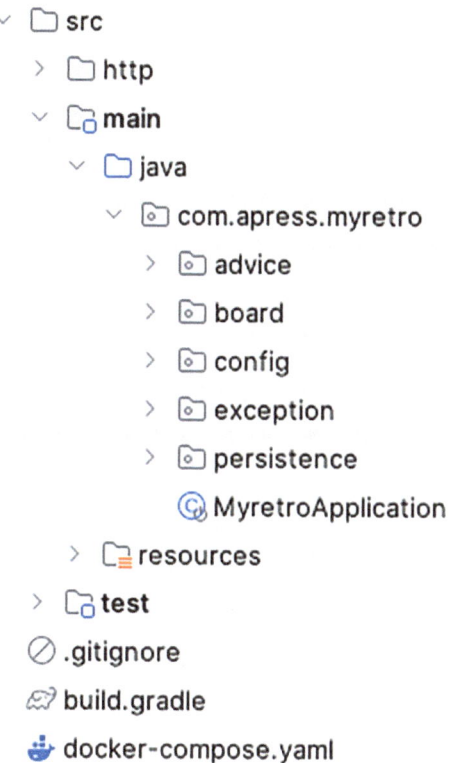

Figure 5-11. *My Retro App with Spring Data JPA and REST*

You already know what's missing, right? Yes, the web and service packages and classes are gone! We don't need them anymore. So, what changed?

Open the build.gradle file and replace its content with the content shown in Listing 5-32.

Listing 5-32. build.gradle

```
plugins {
    id 'java'
    id 'org.springframework.boot' version '3.2.3'
    id 'io.spring.dependency-management' version '1.1.4'
}

group = 'com.apress'
version = '0.0.1-SNAPSHOT'
sourceCompatibility = '17'
```

```
repositories {
    mavenCentral()
}

dependencies {
    implementation 'org.springframework.boot:spring-boot-starter-web'
    implementation 'org.springframework.boot:spring-boot-starter-
    validation'
    implementation 'org.springframework.boot:spring-boot-starter-aop'

    implementation 'org.springframework.boot:spring-boot-starter-data-jpa'
    implementation 'org.springframework.boot:spring-boot-starter-data-rest'

    developmentOnly 'org.springframework.boot:spring-boot-docker-compose'

    annotationProcessor 'org.springframework.boot:spring-boot-
    configuration-processor'

    runtimeOnly 'org.postgresql:postgresql'

    compileOnly 'org.projectlombok:lombok'
    annotationProcessor 'org.projectlombok:lombok'

    // Web
    implementation 'org.webjars:bootstrap:5.2.3'

    testImplementation 'org.springframework.boot:spring-boot-starter-test'
}

tasks.named('test') {
    useJUnitPlatform()
}
```

The only addition is the `spring-boot-starter-data-rest` starter dependency. The advice, board, config, exception, and persistence packages and their classes remain the same. Of course, we removed the service and web packages and their classes.

Run the My Retro App

To run the My Retro App, use your IDE or use the following command:

`./gradlew clean bootRun`

Now, what happens if you point your browser to `http://localhost:8080/retros`? You will get a 404 - Not Found error, but why? We have been using the `/retros` as an endpoint, so what happened? Recall that Spring Data REST reviews your repositories and changes the names of the domain classes to lowercase and plural to create the endpoints, so in this case we have the endpoints `/retroBoards` and `/cards`.

If you go now to `http://localhost:8080/retroBoards` you will get all the RetroBoard records, or you can use the following command:

```
curl -s http://localhost:8080/retroBoards | jq .
{
  "_embedded": {
    "retroBoards": [
      {
        "name": "Spring Boot Conference",
        "_links": {
          "self": {
            "href": "http://localhost:8080/retroBoards/fc873a23-15ee-42f6-
            b9f4-bfbab4537ff4"
          },
          "retroBoard": {
            "href": "http://localhost:8080/retroBoards/fc873a23-15ee-42f6-
            b9f4-bfbab4537ff4"
          },
          "cards": {
            "href": "http://localhost:8080/retroBoards/fc873a23-15ee-42f6-
            b9f4-bfbab4537ff4/cards"
          }
        }
      }
    ]
  },
```

```
  "_links": {
    "self": {
      "href": "http://localhost:8080/retroBoards?page=0&size=20"
    },
    "profile": {
      "href": "http://localhost:8080/profile/retroBoards"
    }
  },
  "page": {
    "size": 20,
    "totalElements": 1,
    "totalPages": 1,
    "number": 0
  }
}
```

Look at the _embedded and _links data, which refer, respectively, to retroBoards
(plural) and retroBoard (singular). What happen if I want to still have my /retros so my
client doesn't change just because now I'm using a HAL response; in other words, when
using HAL by default, The HAL auto-configuration will pluralized the class and create an
endpoint of the form /retroBoards (camel case naming convention), but my clients are
connect to a /retros, so how can I switch it back? Well, there is a solution to that. Open
the RetroBoardRepository interface and add the code shown in Listing 5-33.

Listing 5-33. src/main/java/apress/com/myretro/persistence/
RetroBoardRepository.java

```
package com.apress.myretro.persistence;

import com.apress.myretro.board.RetroBoard;
import org.springframework.data.jpa.repository.JpaRepository;
import org.springframework.data.rest.core.annotation.
RepositoryRestResource;

import java.util.UUID;

@RepositoryRestResource(path = "retros",itemResourceRel = "retros",
collectionResourceRel = "retros")
```

```
public interface RetroBoardRepository extends
JpaRepository<RetroBoard,UUID> {
}
```

The modified RetroBoardRepository interface now includes the
@RepositoryRestResource annotation. You can see that it takes three parameters:

- path: This parameter creates the endpoint we want, /retros.

- itemResourceRel: This parameter changes the name in the "_links"
  to "retros" (instead of "retroBoard").

- collectionResourceRel: This parameter changes the name in the
  "_embedded" from "retroBoards" to "retros".

Now, rerun the app either by going to http://localhost:8080/retros or by using
the following command:

```
curl -s http://localhost:8080/retros |  jq .

{
  "_embedded": {
    "retros": [
      {
        "name": "Spring Boot Conference",
        "_links": {
          "self": {
            "href": "http://localhost:8080/retros/e800d409-9295-4565-bf4b-
            f3b95ff32eff"
          },
          "retros": {
            "href": "http://localhost:8080/retros/e800d409-9295-4565-bf4b-
            f3b95ff32eff"
          },
          "cards": {
            "href": "http://localhost:8080/retros/e800d409-9295-4565-bf4b-
            f3b95ff32eff/cards"
          }
        }
      }
```

```
      }
    ]
  },
  "_links": {
    "self": {
      "href": "http://localhost:8080/retros?page=0&size=20"
    },
    "profile": {
      "href": "http://localhost:8080/profile/retros"
    }
  },
  "page": {
    "size": 20,
    "totalElements": 1,
    "totalPages": 1,
    "number": 0
  }
}
```

Before continue, and if you are in the browser, you can take a look at the /profile/ retros endpoint in your browser or command line, that will give you the ALPS (Application-Level Profile Semantics) and JSON Schema:

```
curl -s -H 'Accept:application/schema+json' http://localhost:8080/profile/
retros | jq .
{
  "title": "Retro board",
  "properties": {
    "cards": {
      "title": "Cards",
      "readOnly": false,
      "type": "string",
      "format": "uri"
    },
    "name": {
```

```
        "title": "Name",
        "readOnly": false,
        "type": "string"
      }
   },
   "definitions": {},
   "type": "object",
   "$schema": "http://json-schema.org/draft-04/schema#"
}
```

With these options, you can expose your API to the public.

Summary

In this chapter we covered Spring Data JDBC, Spring Data JPA, and Spring Data REST, and how Spring Boot help us to remove all the extra configuration by using the auto-configuration based on the dependencies we used. You learned the following:

- Spring Data (the core) uses repository interfaces, such as CrudRepository and PagingAndSortingRepository (others include ListCrudRepository, ListPagingAndSortingRepository, Repository, NoRepositoryBean, and RepositoryDefinition).

- In Spring Data JDBC, you need to focus on relationships, but in Spring Data JPA, you need to focus on object composition.

- JDBC is limited to get many-to-many relationships, but you can still perform the job using SQL statements.

- You learned that JDBC can initialize the database, and this is thanks to Spring Boot that can look at the class path and if it finds the schema.sql or data.sql can create the table and add data.

- JPA can generate tables based on the spring.jpa.* properties, and thanks to the Spring Boot auto-configuration, all the interface repositories are enabled.

- You also learned about a REST implementation based on Spring Data REST, and how Spring Data REST with Spring Boot helps to configure everything so that it looks like just plug and play.

- With Spring Data REST, you can have out-of-the-box web controllers that expose many endpoints.

In Chapter 6 we are going to cover the NoSQL databases.

CHAPTER 6

Spring Data NoSQL with Spring Boot

In this chapter we discuss NoSQL databases, specifically *MongoDB* and *Redis*. There are many NoSQL database options available, but MongoDB and Redis currently are the most commonly used in the IT industry. Spring Framework Data Access and the Spring Data project support several NoSQL databases, and you can access a dedicated page for each of these technologies, including SQL and NoSQL, at the Spring Data site: `https://spring.io/projects/spring-data`.

Spring Data MongoDB

Spring Data MongoDB implements the Spring programming model for the MongoDB document type database. One of the most important features of Spring Data MongoDB is that it offers the POJO (plain old Java object) model for interacting with collections by using the well-known repository interfaces.

The following are some of the many features of Spring Data MongoDB:

- Full configuration through JavaConfig classes or XML configuration files

- The well-known data exception management translation from Spring Data Access

- Life-cycle callbacks and events

- Implementation of the `Repository`, `CrudRepository`, and `MongoRepository` interfaces

- Custom query methods and Querydsl integration

© Felipe Gutierrez 2024
F. Gutierrez, *Pro Spring Boot 3*, https://doi.org/10.1007/978-1-4842-9294-5_6

- MapReduce integration.

- Annotation-based mapping metadata (using the @Document annotation to specify your domain class and @Id annotation to specify the identifier or key)

- A MongoTemplate and MongoOperations classes to help with all the boilerplate to connect and execute actions on MongoDB

- Low-level mapping using MongoReader and MongoWriter abstractions

Spring Data MongoDB with Spring Boot

If you want to use MongoDB with Spring Boot, you need to include the spring-boot-starter-data-mongodb starter dependency. The Spring Boot auto-configuration will set up all the necessary defaults to connect to a MongoDB database and will identify all the repositories and domain classes that are required for your application.

This auto-configuration will set up the MongoDatabaseFactory with the default URI: mongodb://localhost/test. You can use an injected MongoTemplate and MongoOperations classes, and if you want to change these defaults, you can modify this behavior by using the spring.data.mongodb.* properties.

Users App with Spring Data MongoDB and Spring Boot

If you are following along, you can reuse some of the code from previous versions. Or you can start from scratch by going to the Spring Initializr (https://start.spring.io) and generating a base project (no dependencies), but make sure to set the Group field to com.apress and the Artifact and Name Fields to users. Download the project, unzip it, and import it into your favorite IDE. By the end of this section, you should have the structure shown in Figure 6-1.

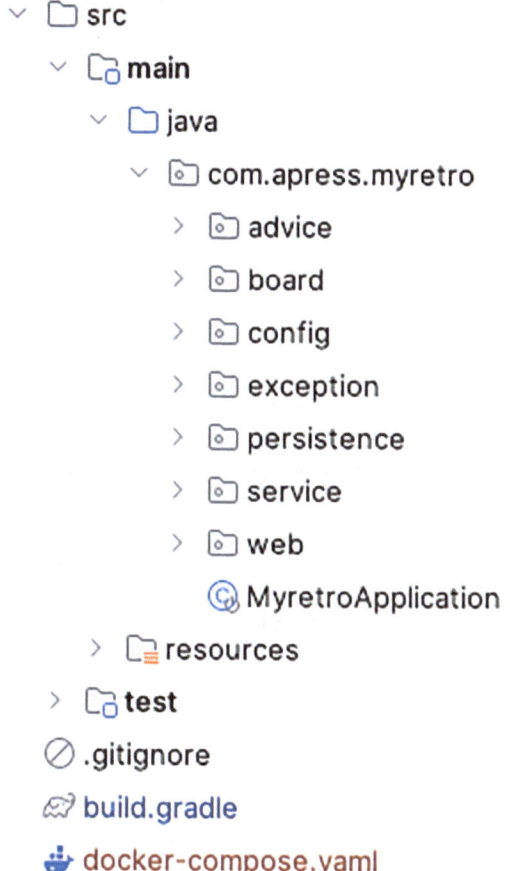

Figure 6-1. *Users App with Spring Data MongoDB with Spring Boot*

Open the build.gradle file and replace its content with the content shown in Listing 6-1.

Listing 6-1. build.gradle

```
plugins {
    id 'java'
    id 'org.springframework.boot' version '3.2.3'
    id 'io.spring.dependency-management' version '1.1.4'
}
```

```
group = 'com.apress'
version = '0.0.1-SNAPSHOT'
sourceCompatibility = '17'

repositories {
    mavenCentral()
}

dependencies {

    implementation 'org.springframework.boot:spring-boot-starter-web'
    implementation 'org.springframework.boot:spring-boot-starter-validation'

    implementation 'org.springframework.boot:spring-boot-starter-
    data-mongodb'
    developmentOnly 'org.springframework.boot:spring-boot-docker-compose'

    compileOnly 'org.projectlombok:lombok'
    annotationProcessor 'org.projectlombok:lombok'

    // Web
    implementation 'org.webjars:bootstrap:5.2.3'

    testImplementation 'org.springframework.boot:spring-boot-starter-test'
}

tasks.named('test') {
    useJUnitPlatform()
}

test {
    testLogging {
        events "passed", "skipped", "failed"
        showExceptions true
        exceptionFormat "full"
        showCauses true
        showStackTraces true
        showStandardStreams = false
    }
}
```

Note that we are including the `spring-boot-starter-data-mongodb` and `spring-boot-docker-compose` starter dependencies. The Spring Boot auto-configuration feature will use the defaults to configure everything to connect to the MongoDB engine and perform the necessary operations and interaction.

Next, create/open the User class; it should look like Listing 6-2.

Listing 6-2. src/main/java/apress/com/users/User.java

```java
package com.apress.users;

import jakarta.validation.constraints.NotBlank;
import jakarta.validation.constraints.Pattern;
import lombok.*;
import org.springframework.data.annotation.Id;
import org.springframework.data.mongodb.core.mapping.Document;

import java.util.Collection;

@Builder
@AllArgsConstructor
@NoArgsConstructor
@Data
@Document
public class User {

    @Id
    @NotBlank(message = "Email can not be empty")
    private String email;

    @NotBlank(message = "Name can not be empty")
    private String name;

    private String gravatarUrl;

    @Pattern(message = "Password must be at least 8 characters long and
    contain at least one number, one uppercase, one lowercase and one
    special character",
            regexp = "^(?=.*[0-9])(?=.*[a-z])(?=.*[A-Z])(?=.*[@#$%^&+=!])
            (?=\\S+$).{8,}$")
    private String password;
```

```
@Singular("role")
private Collection<UserRole> userRole;

private boolean active;
}
```

Do you know what changed compared to the previous version of the User class (in JPA, shown in Listing 5-19)? We now use the @Document annotation (instead of @Entity) that belongs to the org.springframework.data.mongo.* package. The @Document annotation identifies the class as a domain object so that it can be persisted into the MongoDB database. And practically that's it.

Next, create/open the UserRepository interface. See Listing 6-3.

Listing 6-3. src/main/java/apress/com/users/UserRepository.java

```
package com.apress.users;

import org.springframework.data.repository.CrudRepository;

public interface UserRepository extends CrudRepository<User,String>{}
```

As you can see, this class hasn't changed at all, we still extend from the CrudRepository interface. There is the MongoRepository, but we are going to use that in the My Retro App project. The objective here is to show you that you can still use your base without too much effort when migrating from one technology to another.

Next, create/open the UsersController class. See Listing 6-4.

Listing 6-4. src/main/java/apress/com/users/UsersController.java

```
package com.apress.users;

import lombok.AllArgsConstructor;
import org.springframework.http.HttpStatus;
import org.springframework.http.ResponseEntity;
import org.springframework.validation.FieldError;
import org.springframework.web.bind.MethodArgumentNotValidException;
import org.springframework.web.bind.annotation.*;
import org.springframework.web.servlet.support.ServletUriComponentsBuilder;
```

```java
import java.net.URI;
import java.util.HashMap;
import java.util.Map;

@AllArgsConstructor
@RestController
@RequestMapping("/users")
public class UsersController {

    private UserRepository userRepository;

    @GetMapping
    public ResponseEntity<Iterable<User>> getAll(){
        return ResponseEntity.ok(this.userRepository.findAll());
    }

    @GetMapping("/{email}")
    public ResponseEntity<User> findUserById(@PathVariable String email)
    throws Throwable {
        return ResponseEntity.of(this.userRepository.findById(email));
    }

    @RequestMapping(method = {RequestMethod.POST,RequestMethod.PUT})
    public ResponseEntity<User> save(@RequestBody User user){
        this.userRepository.save(user);

        URI location = ServletUriComponentsBuilder
                .fromCurrentRequest()
                .path("/{email}")
                .buildAndExpand(user.getEmail())
                .toUri();
        return ResponseEntity.created(location).body(user);
    }

    @DeleteMapping("/{email}")
    @ResponseStatus(HttpStatus.NO_CONTENT)
    public void save(@PathVariable String email){
        this.userRepository.deleteById(email);
    }
```

```
@ExceptionHandler(MethodArgumentNotValidException.class)
@ResponseStatus(HttpStatus.BAD_REQUEST)
public Map<String, String> handleValidationExceptions(MethodArgumentNot
ValidException ex) {
    Map<String, String> errors = new HashMap<>();

    ex.getBindingResult().getAllErrors().forEach((error) -> {
        String fieldName = ((FieldError) error).getField();
        String errorMessage = error.getDefaultMessage();
        errors.put(fieldName, errorMessage);
    });

    return errors;
    }
}
```

The UsersController class hasn't change either. It's the same from previous versions. But, not that we can use directly the UserRepository that will be injected by Spring (as part of the constructor, thanks to the @AllArgsConstructor annotation from Lombok).

Next, create/open the UserConfiguration class. See Listing 6-5.

Listing 6-5. src/main/java/apress/com/users/UserConfiguration.java

```
package com.apress.users;

import org.springframework.boot.context.event.ApplicationReadyEvent;
import org.springframework.context.ApplicationListener;
import org.springframework.context.annotation.Bean;
import org.springframework.context.annotation.Configuration;
import org.springframework.data.mongodb.core.mapping.event.
BeforeConvertCallback;

import java.util.Arrays;

@Configuration
public class UserConfiguration implements BeforeConvertCallback<User> {

    @Bean
    ApplicationListener<ApplicationReadyEvent> init(UserRepository
    userRepository){
```

```
        return applicationReadyEvent -> {
            userRepository.save(User.builder()
                    .email("ximena@email.com")
                    .name("Ximena")
                    .password("aw2s0meR!")
                    .role(UserRole.USER)
                    .active(true)
                    .build());
            userRepository.save(User.builder()
                    .email("norma@email.com")
                    .name("Norma")
                    .password("aw2s0meR!")
                    .role(UserRole.USER)
                    .role(UserRole.ADMIN)
                    .active(true)
                    .build());
        };
    }

    @Override
    public User onBeforeConvert(User entity, String collection) {
        if (entity.getGravatarUrl()==null)
            entity.setGravatarUrl(UserGravatar.
            getGravatarUrlFromEmail(entity.getEmail()));

        if (entity.getUserRole() == null)
            entity.setUserRole(Arrays.asList(UserRole.INFO));

        return entity;
    }
}
```

The UserConfiguration class is implementing the BeforeConvertCallback interface, where you can add your logic before the entity gets saved into the MongoDB database. You need to implement the onBeforeConvert method. Note that instead of creating another class for the callback event, you can use your configuration class to add the necessary implementation. Also note that this class is initializing with some documents (as in previous versions) using the ApplicationReadyEvent and using the UserRepository interface.

Next, create/open the docker-compose.yaml file. See Listing 6-6.

Listing 6-6. docker-compose.yaml

```
version: "3.1"
services:
  mongo:
    image: mongo
    restart: always
    environment:
      MONGO_INITDB_DATABASE: retrodb
    ports:
      - "27017:27017"
```

The docker-compose.yaml file ensures your MongoDB database is up and running when executing the application. Remember that this file is used only in development (not testing).

We are finished here. The UserGravatar and UserRole classes have no changes. The application.properties file must be empty; we don't need to specify anything there.

Testing the Users App

To test the Users App, create/open the UsersHttpRequestTests class. See Listing 6-7.

Listing 6-7. src/test/java/apress/com/users/UsersHttpRequestTests.java

```
package com.apress.users;

import org.junit.jupiter.api.Test;
import org.springframework.beans.factory.annotation.Autowired;
import org.springframework.beans.factory.annotation.Value;
import org.springframework.boot.test.context.SpringBootTest;
import org.springframework.boot.test.web.client.TestRestTemplate;

import java.util.Collection;

import static org.assertj.core.api.Assertions.assertThat;
```

```java
@SpringBootTest(webEnvironment = SpringBootTest.WebEnvironment.RANDOM_PORT,
        properties = {"spring.data.mongodb.database=retrodb"})
public class UsersHttpRequestTests {

    @Value("${local.server.port}")
    private int port;

    private final String BASE_URL = "http://localhost:";
    private final String USERS_PATH = "/users";

    @Autowired
    private TestRestTemplate restTemplate;

    @Test
    public void indexPageShouldReturnHeaderOneContent(){
        assertThat(this.restTemplate.getForObject(BASE_URL + port,
                String.class)).contains("Simple Users Rest Application");
    }

    @Test
    public void usersEndPointShouldReturnCollectionWithTwoUsers() {
        Collection response = this.restTemplate.
                getForObject(BASE_URL + port + USERS_PATH,
                Collection.class);

        assertThat(response.size()).isEqualTo(2);
    }

    @Test
    public void userEndPointPostNewUserShouldReturnUser(){
        User user =  User.builder()
                .email("dummy@email.com")
                .name("Dummy")
                .gravatarUrl("https://www.gravatar.com/avatar/23bb62a7d0ca6
                3c9a804908e57bf6bd4?d=wavatar")
                .password("aw2sOmeR!")
                .role(UserRole.USER)
                .active(true)
```

```
            .build();
    User response =  this.restTemplate.postForObject(BASE_URL + port +
    USERS_PATH,user,User.class);

    assertThat(response).isNotNull();
    assertThat(response.getEmail()).isEqualTo(user.getEmail());

    Collection users = this.restTemplate.
            getForObject(BASE_URL + port + USERS_PATH, Collection.class);

    assertThat(users.size()).isGreaterThanOrEqualTo(2);

}

@Test
public void userEndPointDeleteUserShouldReturnVoid() {
    this.restTemplate.delete(BASE_URL + port + USERS_PATH + "/norma@
    email.com");

    Collection users = this.restTemplate.
            getForObject(BASE_URL + port + USERS_PATH, Collection.class);

    assertThat(users.size()).isLessThanOrEqualTo(2);
}

@Test
public void userEndPointFindUserShouldReturnUser() {
    User user = this.restTemplate.getForObject(BASE_URL + port + USERS_
    PATH + "/ximena@email.com",User.class);

    assertThat(user).isNotNull();
    assertThat(user.getEmail()).isEqualTo("ximena@email.com");
}
}
```

The UsersHttpRequestTests class is the same as the previous versions, but there is an interesting point to note here. In the previous section we left the application. properties file empty. Well, the spring-boot-docker-compose dependency we added works for development only, so to use most of these defaults from Spring Boot, the minimum requirement is to add the name of the database (retrodb) we are going to use; otherwise, the name will be test.

To run the tests, you first need to start MongoDB, either using your IDE (both VS Code and IntelliJ have plugins for running docker-compose files) or through the command line:

```
docker compose up -d
```

This command will use the docker-compose dependency and start the mongodb service in the background. Next, you can run the test with

```
./gradlew clean test
```

and you should get the following output:

```
UsersHttpRequestTests > userEndPointFindUserShouldReturnUser() PASSED
UsersHttpRequestTests > userEndPointDeleteUserShouldReturnVoid() PASSED
UsersHttpRequestTests > indexPageShouldReturnHeaderOneContent() PASSED
UsersHttpRequestTests > userEndPointPostNewUserShouldReturnUser() PASSED
UsersHttpRequestTests >
usersEndPointShouldReturnCollectionWithTwoUsers() PASSED
```

You can now stop the docker compose with

```
docker compose down
```

It would be nice to have this docker compose feature working as test support, right? You'll see how to do that in the testing chapter (Chapter 8) with *Test Containers*.

Running the Users App

To run the Users App, make sure you don't have the docker compose running—the Users App will run this for you. Use the following command:

```
./gradlew clean bootRun

...
INFO -[           main] ..eLifecycleManager : Using Docker Compose file ..
INFO -[utReader-stderr] ...core.DockerCli   : Container users-mongo-1  Created
INFO -[utReader-stderr] ...core.DockerCli   : Container users-mongo-1  Starting
INFO -[utReader-stderr] ...core.DockerCli   : Container users-mongo-1  Started
...
```

The output shows that the container just started up. Now, open another terminal and use the following command:

```
curl -s http://localhost:8080/users  |  jq .
  {
    "email": "ximena@email.com",
    "name": "Ximena",
    "gravatarUrl": "https://www.gravatar.com/avatar/f07f7e553264c9710105ede
    be6c465e7?d=wavatar",
    "password": "aw2sOmeR!",
    "userRole": [
      "USER"
    ],
    "active": true
  },
  {
    "email": "dummy@email.com",
    "name": "Dummy",
    "gravatarUrl": "https://www.gravatar.com/avatar/23bb62a7d0ca63c9a80490
    8e57bf6bd4?d=wavatar",
    "password": "aw2sOmeR!",
    "userRole": [
      "USER"
    ],
    "active": true
  },
  {
    "email": "norma@email.com",
    "name": "Norma",
    "gravatarUrl": "https://www.gravatar.com/avatar/23bb62a7d0ca63c9a80490
    8e57bf6bd4?d=wavatar",
    "password": "aw2sOmeR!",
    "userRole": [
      "USER",
      "ADMIN"
```

```
  ],
    "active": true
  }
]
```

Based on our previous versions and how the Users App works, switching from JPA to MongoDB was very straightforward, with minimal change necessary. And in this version, we used the `docker-compose` dependency to use MongoDB.

If you want to run this app using a remote MongoDB database, you need to add the `spring.data.mongo.*` properties in the `application.properties` file similar to the following:

```
# MongoDB
spring.data.mongodb.uri=mongodb://retroadmin:aw2s0me@other-server:27017/ret
rodb?directConnection=true&serverSelectionTimeoutMS=2000&authSource=admin&a
ppName=mongosh+1.7.1
spring.data.mongodb.database=retrodb
```

where the mongodb URI looks like this:

mongodb://<username>:<password>@<remote-server>:<port>/<database>[?...]

In the connection URL you must specified the username and password as one of the parameters is the authorization source, that is `admin` as default value. Just check your URI details with your administrator.

My Retro App with Spring Data MongoDB Using Spring Boot

Let's switch to the My Retro App and see what we need to do to use the MongoDB persistence. And if you are following along, you can reuse some of the code from previous versions. Or you can start from scratch by going to the Spring Initializr (`https://start.spring.io`), generating a base project (no dependencies), and setting the Group field to `com.apress` and the Artifact and Name fields to `myretro`. Download the project, unzip it, and import it into your favorite IDE. By the end of this section, you should have the structure shown in Figure 6-2.

Figure 6-2. *My Retro App with Spring Data MongoDB with Spring Boot*

As you can see, some of the packages and their classes remain the same from previous versions. I want to show you the persistence package that contains the RetroBoardPersistenceCallback (as a separate) class that implements the BeforeConvertCallback to perform an operation before the document gets persisted in MongoDB.

Open the build.gradle file and replace the content with the content shown in Listing 6-8.

Listing 6-8. build.gradle

```
plugins {
    id 'java'
    id 'org.springframework.boot' version '3.2.3'
    id 'io.spring.dependency-management' version '1.1.4'
}

group = 'com.apress'
version = '0.0.1-SNAPSHOT'
sourceCompatibility = '17'

repositories {
    mavenCentral()
}

dependencies {
    implementation 'org.springframework.boot:spring-boot-starter-web'
    implementation 'org.springframework.boot:spring-boot-starter-validation'
    implementation 'org.springframework.boot:spring-boot-starter-aop'
    implementation 'com.fasterxml.uuid:java-uuid-generator:4.0.1'

    implementation 'org.springframework.boot:spring-boot-starter-data-mongodb'
    developmentOnly 'org.springframework.boot:spring-boot-docker-compose'

    annotationProcessor 'org.springframework.boot:spring-boot-configuration-processor'

    compileOnly 'org.projectlombok:lombok'
    annotationProcessor 'org.projectlombok:lombok'

    // Web
    implementation 'org.webjars:bootstrap:5.2.3'

    testImplementation 'org.springframework.boot:spring-boot-starter-test'
}

tasks.named('test') {
    useJUnitPlatform()
}
```

As you can see, the build.gradle file uses the spring-boot-starter-data-mongodb starter dependency.

Next, create/open the board package and the Card class and CardType enum as shown in Listings 6-9 and 6-10, respectively.

Listing 6-9. src/main/java/apress/com/myretro/board/Card.java

```java
package com.apress.myretro.board;

import jakarta.validation.constraints.NotBlank;
import jakarta.validation.constraints.NotNull;
import lombok.AllArgsConstructor;
import lombok.Builder;
import lombok.Data;
import lombok.NoArgsConstructor;

import java.util.UUID;

@Builder
@AllArgsConstructor
@NoArgsConstructor
@Data
public class Card {

    @NotNull
    private UUID id;

    @NotBlank
    private String comment;

    @NotNull
    private CardType cardType;

}
```

Listing 6-10. src/main/java/apress/com/myretro/board/CardType.java

```
package com.apress.myretro.board;

public enum CardType {
    HAPPY,MEH,SAD
}
```

As you can see, the Card and CardType classes haven't changed from the base version. Here, we are not using any persistence mechanism or annotation.

Next, create/open the RetroBoard class. See Listing 6-11.

Listing 6-11. src/main/java/apress/com/myretro/board/RetroBoard.java

```
package com.apress.myretro.board;

import jakarta.validation.constraints.NotBlank;
import jakarta.validation.constraints.NotNull;
import lombok.*;
import org.springframework.data.annotation.Id;
import org.springframework.data.mongodb.core.mapping.Document;

import java.util.ArrayList;
import java.util.List;
import java.util.UUID;

@Builder
@AllArgsConstructor
@NoArgsConstructor
@Data
@Document
public class RetroBoard {

    @Id
    @NotNull
    private UUID id;

    @NotBlank(message = "A name must be provided")
    private String name;
```

```
@Singular("card")
private List<Card> cards;

public void addCard(Card card){
    if (this.cards == null)
        this.cards = new ArrayList<>();
    this.cards.add(card);
}

public void addCards(List<Card> cards){
    if (this.cards == null)
        this.cards = new ArrayList<>();
    this.cards.addAll(cards);
}

}
```

The @Document annotation makes the RetroBoard class ready for MongoDB, and the @Id annotation creates an identifier, which in this case is UUID type. Listing 6-11 also adds some helper methods for adding cards, the addCard and addCards methods.

Next, create/open the persistence package and the RetroBoardPersistenceCallback class and RetroBoardRepository interface. Listing 6-12 shows the RetroBoardPersistenceCallback class.

Listing 6-12. src/main/java/apress/com/myretro/persistence/RetroBoardPersistenceCallback.java

```
package com.apress.myretro.persistence;

import com.apress.myretro.board.RetroBoard;
import org.springframework.data.mongodb.core.mapping.event.
BeforeConvertCallback;
import org.springframework.stereotype.Component;

@Component
public class RetroBoardPersistenceCallback implements BeforeConvertCallback
<RetroBoard> {
```

```
    @Override
    public RetroBoard onBeforeConvert(RetroBoard retroBoard, String s) {
        if (retroBoard.getId() == null)
            retroBoard.setId(java.util.UUID.randomUUID());
        return retroBoard;
    }
}
```

Listing 6-12 shows you the RetroBoardPersistenceCallback class, see that we are implementing the BeforeConvertCallback interface. Did you remember this in the Users App? We declared in the same java config (UserConfiguration class; see Listing 6-5). The difference is that here we are creating a separate class. Which is better? Whichever makes sense for your application, really. And, of course, for this class to work, we need to mark it as a Spring bean by using the @Component annotation.

Listing 6-13 shows the RetroBoardRepository interface.

Listing 6-13. src/main/java/apress/com/myretro/persistence/ RetroBoardRepository.java

```
package com.apress.myretro.persistence;

import com.apress.myretro.board.RetroBoard;
import com.apress.myretro.exception.RetroBoardNotFoundException;
import org.springframework.data.mongodb.repository.MongoRepository;
import org.springframework.data.mongodb.repository.Query;

import java.util.Optional;
import java.util.UUID;

public interface RetroBoardRepository extends MongoRepository<RetroBoard,
UUID> {

    @Query("{'id': ?0}")
    Optional<RetroBoard> findById(UUID id);

    @Query("{}, { cards: { $elemMatch: { _id: ?0 } } }")
    Optional<RetroBoard> findRetroBoardByIdAndCardId(UUID cardId);
```

```
default void removeCardFromRetroBoard(UUID retroBoardId, UUID cardId) {
    Optional<RetroBoard> retroBoard = findById(retroBoardId);
    if (retroBoard.isPresent()) {
        retroBoard.get().getCards().removeIf(card -> card.getId().
        equals(cardId));
        save(retroBoard.get());
    }else {
        throw new RetroBoardNotFoundException();
    }
}
```

}

In Listing 6-13, we are following the *Spring Data application model*, the use of repositories, so that we don't need to worry about the implementation. In this case, we are using a specific repository, MongoRepository, which changed the signatures from the traditional Repository or CrudRepository. It's defined like this:

```
public interface MongoRepository<T, ID>
        extends ListCrudRepository<T, ID>, ListPagingAndSortingRepository
        <T, ID>, QueryByExampleExecutor<T> {

    <S extends T> S insert(S entity);
    <S extends T> List<S> insert(Iterable<S> entities);
    <S extends T> List<S> findAll(Example<S> example);
    <S extends T> List<S> findAll(Example<S> example, Sort sort);
}
```

And guess what? Because the RetroBoardRepository interface extends the ListCrudRepository interface, which in turn extends from CrudRepository, we have access to the methods that we are accustomed to using. The RetroBoardRepository interface is defining the findById method but is using the @Query annotation that accepts a String value in which you can add the JavaScript code to query MongoDB. The RetroBoardRepository interface also is declaring the findRetroBoardByIdAndCardId method with its @Query annotation and the query to be executed. As you can see, we need this type of method because we use the Card as an aggregated to the RetroBoard. Again, this is just an example of what we can do to get the relationship from the RetroBoard

and Card classes applying the MongoDB context. Finally, we are declaring a default
implementation for the removeCardFromRetroBoard method, which is necessary to deal
with the Card instances.

Next, create/open the service package and the RetroBoardService class, as shown
in Listing 6-14.

Listing 6-14. src/main/java/apress/com/myretro/service/
RetroBoardService.java

```java
package com.apress.myretro.service;

import com.apress.myretro.board.Card;
import com.apress.myretro.board.RetroBoard;
import com.apress.myretro.exception.CardNotFoundException;
import com.apress.myretro.persistence.RetroBoardRepository;
import lombok.AllArgsConstructor;
import org.springframework.stereotype.Service;

import java.util.ArrayList;
import java.util.List;
import java.util.Optional;
import java.util.UUID;

@AllArgsConstructor
@Service
public class RetroBoardService {

    RetroBoardRepository retroBoardRepository;

    public RetroBoard save(RetroBoard domain) {
        if (domain.getCards() == null)
            domain.setCards(new ArrayList<>());
        return this.retroBoardRepository.save(domain);
    }

    public RetroBoard findById(UUID uuid) {
        return this.retroBoardRepository.findById(uuid).get();
    }
```

```java
public Iterable<RetroBoard> findAll() {
    return this.retroBoardRepository.findAll();
}

public void delete(UUID uuid) {
    this.retroBoardRepository.deleteById(uuid);
}

public Iterable<Card> findAllCardsFromRetroBoard(UUID uuid) {
    return this.findById(uuid).getCards();
}

public Card addCardToRetroBoard(UUID uuid, Card card){
    if (card.getId() == null)
        card.setId(UUID.randomUUID());
    RetroBoard retroBoard = this.findById(uuid);
    retroBoard.addCard(card);
    retroBoardRepository.save(retroBoard);
    return card;
}

public void addMultipleCardsToRetroBoard(UUID uuid, List<Card> cards){
    RetroBoard retroBoard = this.findById(uuid);
    retroBoard.addCards(cards);
    retroBoardRepository.save(retroBoard);
}

public Card findCardByUUID(UUID uuidCard){
    Optional<RetroBoard> result = retroBoardRepository.findRetroBoardBy
    IdAndCardId(uuidCard);
    if(result.isPresent() && result.get().getCards().size() > 0
            && result.get().getCards().get(0).getId().equals(uuidCard)){
        return result.get().getCards().get(0);
    }else{
        throw new CardNotFoundException();
    }
}
```

```
    public void removeCardByUUID(UUID uuid,UUID cardUUID){
        retroBoardRepository.removeCardFromRetroBoard(uuid,cardUUID);
    }
}
```

As you can see, the `RetroBoardService` class uses the `RetroBoardRepository` and is using the declared methods in the interface. Also note that this time we are taking care of the UUID in the `save` and `addCardToRetroBoard` methods. Again, this is another example demostrating that the UUID can be managed in the callbacks or events, but it's up to you to decide where it makes more sense to have this logic.

Next, create/open the `config` package and the `MyRetroConfiguration` class. See Listing 6-15.

Listing 6-15. src/main/java/apress/com/myretro/config/
MyRetroConfiguration.java

```
package com.apress.myretro.config;

import com.apress.myretro.board.Card;
import com.apress.myretro.board.CardType;
import com.apress.myretro.board.RetroBoard;
import com.apress.myretro.service.RetroBoardService;
import org.springframework.boot.context.event.ApplicationReadyEvent;
import org.springframework.boot.context.properties.
EnableConfigurationProperties;
import org.springframework.context.ApplicationListener;
import org.springframework.context.annotation.Bean;
import org.springframework.context.annotation.Configuration;

import java.util.UUID;

@EnableConfigurationProperties({MyRetroProperties.class})
@Configuration
public class MyRetroConfiguration {

    @Bean
    ApplicationListener<ApplicationReadyEvent> ready(RetroBoardService
    retroBoardService) {
        return applicationReadyEvent -> {
```

```
            UUID retroBoardId = UUID.fromString("9dc9b71b-a07e-418b-
            b972-40225449aff2");
            retroBoardService.save(RetroBoard.builder()
                    .id(retroBoardId)
                    .name("Spring Boot Conference")
                    .card(Card.builder().id(UUID.fromString("bb2a80a5-
                    a0f5-4180-a6dc-80c84bc014c9")).comment("Spring Boot
                    Rocks!").cardType(CardType.HAPPY).build())
                    .card(Card.builder().id(UUID.randomUUID()).
                    comment("Meet everyone in person").cardType(CardType.
                    HAPPY).build())
                    .card(Card.builder().id(UUID.randomUUID()).
                    comment("When is the next one?").cardType(CardType.
                    MEH).build())
                    .card(Card.builder().id(UUID.randomUUID()).comment("Not
                    enough time to talk to everyone").cardType(CardType.
                    SAD).build())
                    .build());
        };
    }
}
```

Note in Listing 6-15 that we are using the ready event and adding some data using the RetroBoardService.

The advice, exception, and web packages and classes remain the same as in previous versions. The rest of the config package classes also remain the same (MyRetroProperties and UsersConfiguration classes).

Next, create/open the docker-compose.yaml file that uses the mongo service, as shown in Listing 6-16.

Listing 6-16. docker-compose.yaml

```
version: "3.1"
services:
  mongo:
    image: mongo
    restart: always
```

```
environment:
  MONGO_INITDB_DATABASE: retrodb
ports:
  - "27017:27017"
```

As you can see, this file is the same as in the Users App.

At this point, the `application.properties` file remains empty, but we need to add a property after we run the app, as described in the following section.

Running My Retro App

To run the application, use your IDE or run it from the terminal with the following command:

```
./gradlew clean bootRun
```

If you direct your browser to `http://localhost:8080/retros` or execute the following `curl` command line, you'll get an error:

```
curl -s http://localhost:8080/retros | jq .
{
  "timestamp": "2023-06-06T21:10:08.737+00:00",
  "status": 500,
  "error": "Internal Server Error",
  "path": "/retros"
}
```

If you look at the console (where the My Retro App is running), you'll see this error:

```
org.springframework.core.convert.ConverterNotFoundException: No converter
found capable of converting from type [org.bson.types.Binary] to type
[java.util.UUID]
```

Before we solve this issue, let's connect to the running container with another mongo container and use the mongo client (`mongosh`). You can execute the following command:

```
docker run -it --rm --network myretro_default mongo mongosh --host
mongo retrodb
```

First, we need to identify the network where our MongoDB container is running. If you are using the docker compose, you can get the network information with the following command:

```
docker network ls
```

```
NETWORK ID     NAME               DRIVER    SCOPE
cee592df7cc9   bridge             bridge    local
2ac86d937732   educates           bridge    local
087927a8ef8c   host               host      local
83b09c37d02e   kind               bridge    local
3e5ded7b4095   minikube           bridge    local
757602f101d2   myretro_default    bridge    local
38898bbc5491   none               null      local
3c0967bf62b8   test-scdf_default  bridge    local
71d99f2d6934   users_default      bridge    local
```

You can see that the name of the network is **myretro_default**. Then, we are using the command mongosh pointing to the mongo host (this comes from the docker-compose.yaml) and the name of the database, retrodb. Within the mongosh client if you:

```
retrodb> db.retroBoard.find({});
```

```
[{
    _id: Binary(Buffer.from("8b417ea01bb7c99df2af4954224072b9", "hex"), 3),
    name: 'Spring Boot Conference',
    cards: [
      {
        _id: Binary(Buffer.from("8041f5a0a5802abbc914c04bc880dca6",
        "hex"), 3),
        comment: 'Spring Boot Rocks!',
        cardType: 'HAPPY'
      },
      {
        _id: Binary(Buffer.from("dd448b729c5fa1a0c6ab3919ffc48b84",
        "hex"), 3),
        comment: 'Meet everyone in person',
        cardType: 'HAPPY'
```

```
  },
  {
    _id: Binary(Buffer.from("094924fee8d3ef9d60bef876439eeeb4",
    "hex"), 3),
    comment: 'When is the next one?',
    cardType: 'MEH'
  },
  {
    _id: Binary(Buffer.from("a6401bca1b9f6cda0a1ff95bd2dc61b9",
    "hex"), 3),
    comment: 'Not enough time to talk to everyone',
    cardType: 'SAD'
  }
],
_class: 'com.apress.myretro.board.RetroBoard'
}
]
```

The output indicates that the MongoDB representation of the UUID is the Binary(Buffer.from) object and that Spring doesn't know about it or how to habndle it. But there is an easy fix. You can add the property shown in Listing 6-17 to the application.properties file.

Listing 6-17. src/main/resource/application.properties

```
# MongoDB
spring.data.mongodb.uuid-representation=standard
```

One of the main formats that MongoDB uses is Binary JSON (BSON), which extends the capabilities of JSON by adding additional data types and binary encodings. And in this case, it tries to use its own "hex" binary representation for the UUID, but with this property, we can use the standard UUID format.

Before you rerun the My Retro App, you need to drop the values from the collection with

```
retrofb> db.retroBoard.drop({});

true
```

This is necessary because we are using docker compose and it creates a volume that is reused every time we run the app and the container is started up. Now, if you rerun the My Retro App after adding this property to the `application.properties` file, you can execute the `curl` command:

```
curl -s http://localhost:8080/retros |  jq .
```

```
[
  {
    "id": "9dc9b71b-a07e-418b-b972-40225449aff2",
    "name": "Spring Boot Conference",
    "cards": [
      {
        "id": "bb2a80a5-a0f5-4180-a6dc-80c84bc014c9",
        "comment": "Spring Boot Rocks!",
        "cardType": "HAPPY"
      },
      {
        "id": "bf2e263e-b698-43a9-adc7-bec07e94c8fd",
        "comment": "Meet everyone in person",
        "cardType": "HAPPY"
      },
      {
        "id": "130441b7-6b77-465e-b879-006163de5279",
        "comment": "When is the next one?",
        "cardType": "MEH"
      },
      {
        "id": "92c7d841-9e2a-40b1-847c-362fb5fe53cc",
        "comment": "Not enough time to talk to everyone",
        "cardType": "SAD"
      }
    ]
  }
]
```

If you are curious about how it looks in Mongo, execute (in the mongo client that is connected to the mongo database) the following command:

```
retrodb> db.retroBoard.find({});

[
  {
    _id: new UUID("9dc9b71b-a07e-418b-b972-40225449aff2"),
    name: 'Spring Boot Conference',
    cards: [
      {
        _id: new UUID("bb2a80a5-a0f5-4180-a6dc-80c84bc014c9"),
        comment: 'Spring Boot Rocks!',
        cardType: 'HAPPY'
      },
      {
        _id: new UUID("bf2e263e-b698-43a9-adc7-bec07e94c8fd"),
        comment: 'Meet everyone in person',
        cardType: 'HAPPY'
      },
      {
        _id: new UUID("130441b7-6b77-465e-b879-006163de5279"),
        comment: 'When is the next one?',
        cardType: 'MEH'
      },
      {
        _id: new UUID("92c7d841-9e2a-40b1-847c-362fb5fe53cc"),
        comment: 'Not enough time to talk to everyone',
        cardType: 'SAD'
      }
    ],
    _class: 'com.apress.myretro.board.RetroBoard'
  }
]
```

Now, you can see the data using the UUID. Did you notice the _class element? This is provided by Spring as metadata, so mapping between Mongo documents and classes is easy.

Note   You can find the complete source code for this book at the Apress website. https://www.apress.com/gp/services/source-code.

Spring Data Redis

Spring Data Redis is another NoSQL in-memory data structure data store, with support for key/value maps, lists, sets, sorted sets, bitmaps, hyperloglogs, and much more. Redis can also be used for pub/sub messaging, because it is fast. It has various additional features such as scalability and high availability. Some of its uses include session management, real-time analytics, task queues, caching, chat applications, leaderboard, etc.

Spring Data Redis provides a lot of low-level and high-level abstractions to interact with Redis from Spring applications. These are some of its many features:

- Spring programming model.

- RedisTemplate, a class based on template implementation that provides high-level abstraction for performing various Redis operations, exception translation, and serialization. It also provides the StringRedisTemplate, which allows string-focused operations.

- Easy connection across multiple Redis drivers(Lettuce and Jedis).

- Exception translation to Spring's portable Data Access exception hierarchy.

- Pub/sub support (such as a MessageListenerContainer for message-driven POJOs).

- Support for Redis Cluster and Redis Sentinel.

- Reactive API support for the Lettuce driver.

- JDK, String, JSON, and Spring Object/XML mapping serializers.

- JDK Collection implementations on top of Redis.

- Redis implementation for Spring 3.1 cache abstraction.

- Repository interfaces including support for custom query methods using @EnableRedisRepositories.

- Stream support.

Spring Data Redis with Spring Boot

By using Spring Data Redis with Spring Boot, you get the auto-configuration for Lettuce and Jedis clients and all the abstraction provided by Spring Data Redis. If you want to use Spring Data Redis in your Spring Boot app, you need to add the `spring-boot-starter-data-redis` starter dependency. If you intend to use Spring Data Redis only, without Spring Boot, and you are using the repository programming model, you need to use the `@EnableRedisRepositories` annotation in your JavaConfig class; of course, if you are using Spring Boot, there's no need for this annotation because Spring Boot takes care of the configuration.

Let's see what we need to do in our projects to use Redis as data persistence, starting with the Users App.

Users App with Spring Data Redis Using Spring Boot

Again, you can reuse some of the code from previous versions or you can start from scratch by going to the Spring Initializr (`https://start.spring.io`) and generating a base project (no dependencies); as always, make sure to set the Group field to `com.apress` and the Artifact and Name fields to `users`. Download the project, unzip it, and import it into your favorite IDE. By the end of this section, you should have the structure shown in Figure 6-3.

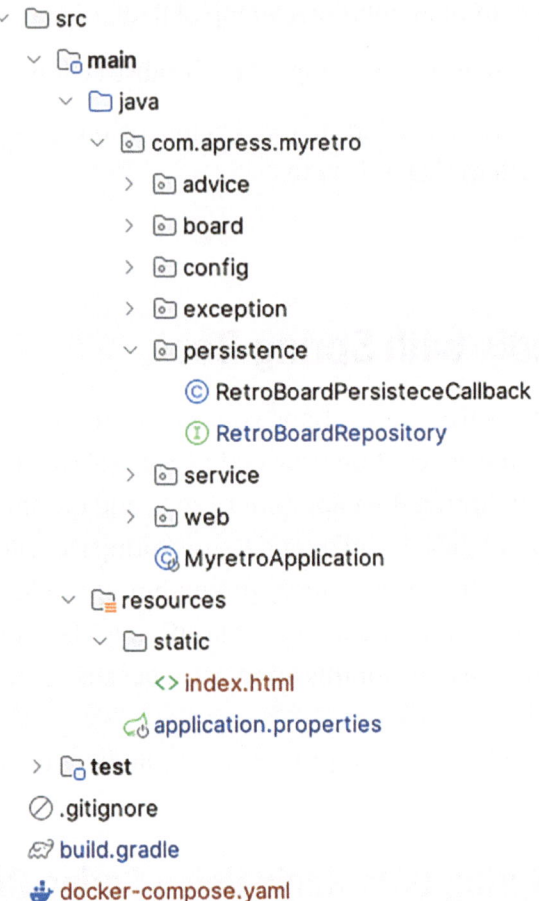

Figure 6-3. *Users App with Spring Data Redis with Spring Boot*

As you can see, we are going to use docker compose with Redis so that it's easy to test and run our application.

Start by opening the `build.gradle` file and replacing the content with the content shown in Listing 6-18.

Listing 6-18. build.gradle

```
plugins {
    id 'java'
    id 'org.springframework.boot' version '3.2.3'
    id 'io.spring.dependency-management' version '1.1.4'
}
```

```
group = 'com.apress'
version = '0.0.1-SNAPSHOT'
sourceCompatibility = '17'

repositories {
    mavenCentral()
}

dependencies {
    implementation 'org.springframework.boot:spring-boot-starter-web'
    implementation 'org.springframework.boot:spring-boot-starter-validation'

    implementation 'org.springframework.boot:spring-boot-starter-data-redis'
    developmentOnly 'org.springframework.boot:spring-boot-docker-compose'

    compileOnly 'org.projectlombok:lombok'
    annotationProcessor 'org.projectlombok:lombok'

    // Web
    implementation 'org.webjars:bootstrap:5.2.3'

    testImplementation 'org.springframework.boot:spring-boot-starter-test'
}

tasks.named('test') {
    useJUnitPlatform()
}

test {
    testLogging {
        events "passed", "skipped", "failed"
        showExceptions true
        exceptionFormat "full"
        showCauses true
        showStackTraces true
        showStandardStreams = false
    }
}
```

We are adding the `spring-boot-starter-data-redis` and the `spring-boot-docker-compose` starter dependencies, which will allow Spring Boot to use the auto-configuration feature to set up everything about Redis, from connections and all the necessary classes to do any interaction.

Next, create/open the User class. See Listing 6-19.

Listing 6-19. src/main/java/apress/com/users/User.java

```java
package com.apress.users;

import jakarta.validation.constraints.NotBlank;
import lombok.*;
import org.springframework.data.annotation.Id;
import org.springframework.data.redis.core.RedisHash;

import java.util.Collection;

@Builder
@AllArgsConstructor
@NoArgsConstructor
@Data
@RedisHash("USERS")
public class User {

    @Id
    @NotBlank(message = "Email can not be empty")
    private String email;

    @NotBlank(message = "Name can not be empty")
    private String name;

    private String gravatarUrl;

    @NotBlank(message = "Password can not be empty")
    private String password;

    @Singular("role")
    private Collection<UserRole> userRole;

    private boolean active;

}
```

The User class now is marked with the @RedisHashMark annotation, which accepts a String value that will be the prefix that distinguishes it from other domain classes. This annotation marks objects as aggregate roots to be stored in the Redis hash.

Next, create/open the UserRepository interface, as shown in Listing 6-20.

Listing 6-20. src/main/java/apress/com/users/UserRepository.java

```
package com.apress.users;

import org.springframework.data.repository.CrudRepository;

public interface UserRepository extends CrudRepository<User,String>{ }
```

As you can see, the UserRepository interface is the same from the previous versions, extending from the CrudRepository interface, which (as you already know) comes with several methods that will be implemented by Spring Data core and the Redis abstractions.

Next, create/open the UsersController class. See Listing 6-21.

Listing 6-21. src/main/java/apress/com/users/UsersController.java

```
package com.apress.users;

import lombok.AllArgsConstructor;
import org.springframework.http.HttpStatus;
import org.springframework.http.ResponseEntity;
import org.springframework.validation.FieldError;
import org.springframework.web.bind.MethodArgumentNotValidException;
import org.springframework.web.bind.annotation.*;
import org.springframework.web.servlet.support.ServletUriComponentsBuilder;

import java.net.URI;
import java.util.Collections;
import java.util.HashMap;
import java.util.Map;

@AllArgsConstructor
@RestController
@RequestMapping("/users")
public class UsersController {
```

```
private UserRepository userRepository;

@GetMapping
public ResponseEntity<Iterable<User>> getAll(){
    return ResponseEntity.ok(this.userRepository.findAll());
}

@GetMapping("/{email}")
public ResponseEntity<User> findUserById(@PathVariable String email)
throws Throwable {
    return ResponseEntity.of(this.userRepository.findById(email));
}

@RequestMapping(method = {RequestMethod.POST,RequestMethod.PUT})
public ResponseEntity<User> save(@RequestBody User user){

    if (user.getGravatarUrl()==null)
        user.setGravatarUrl(UserGravatar.getGravatarUrlFromEmail(user.
        getEmail()));
    if (user.getUserRole()==null)
        user.setUserRole(Collections.singleton(UserRole.INFO));

    this.userRepository.save(user);

    URI location = ServletUriComponentsBuilder
            .fromCurrentRequest()
            .path("/{email}")
            .buildAndExpand(user.getEmail())
            .toUri();
    return ResponseEntity.created(location).body(user);
}

@DeleteMapping("/{email}")
@ResponseStatus(HttpStatus.NO_CONTENT)
public void delete(@PathVariable String email){
    this.userRepository.deleteById(email);
}
```

```
@ExceptionHandler(MethodArgumentNotValidException.class)
@ResponseStatus(HttpStatus.BAD_REQUEST)
public Map<String, String> handleValidationExceptions(MethodArgumentNot
ValidException ex) {
    Map<String, String> errors = new HashMap<>();

    ex.getBindingResult().getAllErrors().forEach((error) -> {
        String fieldName = ((FieldError) error).getField();
        String errorMessage = error.getDefaultMessage();
        errors.put(fieldName, errorMessage);
    });

    return errors;
    }
}
```

The UsersController class also is the same from previous versions. Note that the UserRepository is used, and it will be injected (by constructor injection) when the application starts.

Next, create/open the UserConfiguration class. See Listing 6-22.

Listing 6-22. src/main/java/apress/com/users/UserConfiguration.java

```
package com.apress.users;

import lombok.extern.slf4j.Slf4j;
import org.springframework.boot.context.event.ApplicationReadyEvent;
import org.springframework.context.ApplicationListener;
import org.springframework.context.annotation.Bean;
import org.springframework.context.annotation.Configuration;

@Slf4j
@Configuration
public class UserConfiguration{

    @Bean
    ApplicationListener<ApplicationReadyEvent> init(UserRepository
    userRepository) {
        return applicationReadyEvent -> {
```

```
            userRepository.save(User.builder()
                    .email("ximena@email.com")
                    .name("Ximena")
                    .gravatarUrl("https://www.gravatar.com/avatar/23bb62a7d
                    0ca63c9a804908e57bf6bd4?d=wavatar")
                    .password("aw2sOmeR!")
                    .role(UserRole.USER)
                    .active(true)
                    .build());
            userRepository.save(User.builder()
                    .email("norma@email.com")
                    .name("Norma")
                    .gravatarUrl("https://www.gravatar.com/avatar/f07f
                    7e553264c9710105edebe6c465e7?d=wavatar")
                    .password("aw2sOmeR!")
                    .role(UserRole.USER)
                    .role(UserRole.ADMIN)
                    .active(true)
                    .build());
        };
    }

}
```

In the UserConfiguration class, we are adding some users, and in this case we are also adding the gravatarUrl field value. In Spring Data Redis, there are no callbacks or events before persisting any object, so we can add the gravatarUrl field value either manually in the controller when receiving the object by using a record type and the custom constructor or by creating a service that takes care of it before saving. So, as you can see, we have options, and for now we can add it here, in the configuration.

The UserGravatar and UserRole classes remain the same as previous versions. The application.properties file is empty; no properties. Remember that the auto-configuration will use the default connection parameters. If you want to override them, you can set the spring.data.redis.* properties as environment variables, either from the command line or in the application.properties file.

Testing the Users App

To test the Users App, create/open the UsersHttpRequestTests class. See Listing 6-23.

Listing 6-23. src/test/java/apress/com/users/UsersHttpRequestTests.java

```
package com.apress.users;

import org.junit.jupiter.api.Test;
import org.springframework.beans.factory.annotation.Autowired;
import org.springframework.beans.factory.annotation.Value;
import org.springframework.boot.test.context.SpringBootTest;
import org.springframework.boot.test.web.client.TestRestTemplate;

import java.util.Collection;

import static org.assertj.core.api.Assertions.assertThat;

@SpringBootTest(webEnvironment = SpringBootTest.WebEnvironment.RANDOM_PORT)
public class UsersHttpRequestTests {

    @Value("${local.server.port}")
    private int port;

    private final String BASE_URL = "http://localhost:";
    private final String USERS_PATH = "/users";

    @Autowired
    private TestRestTemplate restTemplate;

    @Test
    public void indexPageShouldReturnHeaderOneContent() throws Exception {
        assertThat(this.restTemplate.getForObject(BASE_URL + port,
                String.class)).contains("Simple Users Rest Application");
    }

    @Test
    public void usersEndPointShouldReturnCollectionWithTwoUsers() throws
    Exception {
        Collection<User> response = this.restTemplate.
                getForObject(BASE_URL + port + USERS_PATH, Collection.class);
```

```java
        assertThat(response.size()).isEqualTo(2);
    }

    @Test
    public void userEndPointPostNewUserShouldReturnUser() throws Exception {
        User user =  User.builder()
                    .email("dummy@email.com")
                    .name("Dummy")
                    .gravatarUrl("https://www.gravatar.com/avatar/23bb62a7d0ca6
                    3c9a804908e57bf6bd4?d=wavatar")
                    .password("aw2sOmeR!")
                    .role(UserRole.USER)
                    .active(true)
                    .build();

        User response =  this.restTemplate.postForObject(BASE_URL + port +
        USERS_PATH,user,User.class);

        assertThat(response).isNotNull();
        assertThat(response.getEmail()).isEqualTo(user.getEmail());

        Collection<User> users = this.restTemplate.
                    getForObject(BASE_URL + port + USERS_PATH, Collection.class);

        assertThat(users.size()).isGreaterThanOrEqualTo(2);

    }

    @Test
    public void userEndPointDeleteUserShouldReturnVoid() throws Exception {
        this.restTemplate.delete(BASE_URL + port + USERS_PATH + "/norma@
        email.com");

        Collection<User> users = this.restTemplate.
                    getForObject(BASE_URL + port + USERS_PATH, Collection.class);

        assertThat(users.size()).isLessThanOrEqualTo(2);
    }
```

```
    @Test
    public void userEndPointFindUserShouldReturnUser() throws Exception{
        User user = this.restTemplate.getForObject(BASE_URL + port + USERS_
        PATH + "/ximena@email.com",User.class);

        assertThat(user).isNotNull();
        assertThat(user.getEmail()).isEqualTo("ximena@email.com");
    }
}
```

As you can see, the UsersHttpRequestTests class has not changed. To begin the test, you need to have Redis up and running. You can use your IDE to run the docker-compose.yaml service or execute the following in the command line:

```
docker compose up -d
```

Once Redis is up and running, you can run the tests by using either the IDE or the following command:

```
./gradlew clean test
```

```
UsersHttpRequestTests > userEndPointFindUserShouldReturnUser() PASSED
UsersHttpRequestTests > userEndPointDeleteUserShouldReturnVoid() PASSED
UsersHttpRequestTests > indexPageShouldReturnHeaderOneContent() PASSED
UsersHttpRequestTests > userEndPointPostNewUserShouldReturnUser() PASSED
UsersHttpRequestTests > usersEndPointShouldReturnCollectionWithTwoUsers()
PASSED
```

Running the Users App

Before you run the application, make sure Redis (from docker compose) is stopped. Then you can run the app by using your IDE or the following command:

```
./gradlew clean bootRun
```

First docker compose starts, then the application. Next, you can direct your browser to http://localhost:8080/users or using the following command:

```
curl -s http://localhost:8080/users | jq .
```

```
[
  {
    "email": "ximena@email.com",
    "name": "Ximena",
    "gravatarUrl": "https://www.gravatar.com/avatar/23bb62a7d0ca63c9a80490
    8e57bf6bd4?d=wavatar",
    "password": "aw2s0meR!",
    "userRole": [
      "USER"
    ],
    "active": true
  },
  {
    "email": "norma@email.com",
    "name": "Norma",
    "gravatarUrl": "https://www.gravatar.com/avatar/f07f7e553264c9710105ede
    be6c465e7?d=wavatar",
    "password": "aw2s0meR!",
    "userRole": [
      "USER",
      "ADMIN"
    ],
    "active": true
  }
]
```

If you are curious, you can use the Redis CLI and take a peek at how Spring Data Redis saves the data. You can execute the following command:

```
docker run -it --rm --network users_default redis:alpine redis-cli -h redis
```

This command is using the network users_default and is using the image redis:alpine by passing the execution to redis-cli with the -h (host) flag and redis (the name of the service, declared in docker-compose.yaml). After the command has executed, you should get the redis:6379> prompt. Then you can take a look at the keys by executing the KEYS command:

```
redis:6379> KEYS *
1) "USERS"
2) "USERS:norma@email.com"
3) "USERS:ximena@email.com"
```

you can get the type, for example:

```
redis:6379> TYPE "USERS:norma@email.com"
hash
```

You can get the keys of the hash with the following command:

```
redis:6379> HKEYS "USERS:norma@email.com"
1) "email"
2) "active"
3) "_class"
4) "userRole.[1]"
5) "password"
6) "userRole.[0]"
7) "gravatarUrl"
8) "name"
```

You can get the values of that hash with

```
redis:6379> HVALS "USERS:norma@email.com"
1) "norma@email.com"
2) "1"
3) "com.apress.users.User"
4) "ADMIN"
5) "aw2s0meR!"
```

```
6) "USER"
7) "https://www.gravatar.com/avatar/f07f7e553264c9710105edebe6c465e7?
d=wavatar"
8) "Norma"
```

And you can get a specific field with

```
redis:6379> HGET "USERS:norma@email.com" name
"Norma"
```

Play around with the commands in the Redis client (you can view the documentation at `https://redis.io/commands/`).

Next, let's look at using Spring Data Redis in the My Retro App project.

My Retro App with Spring Data Redis Using Spring Boot

If you are following along, you can reuse some of the code from previous versions. Or you can start from scratch by going to the Spring Initializr (`https://start.spring.io`) and generating a base project (no dependencies); just make sure you have set the Group field to `com.apress` and the Artifact and Name fields to `myretro`. Download the project, unzip it, and import it into your favorite IDE. By the end of this section, you should have the structure shown in Figure 6-4.

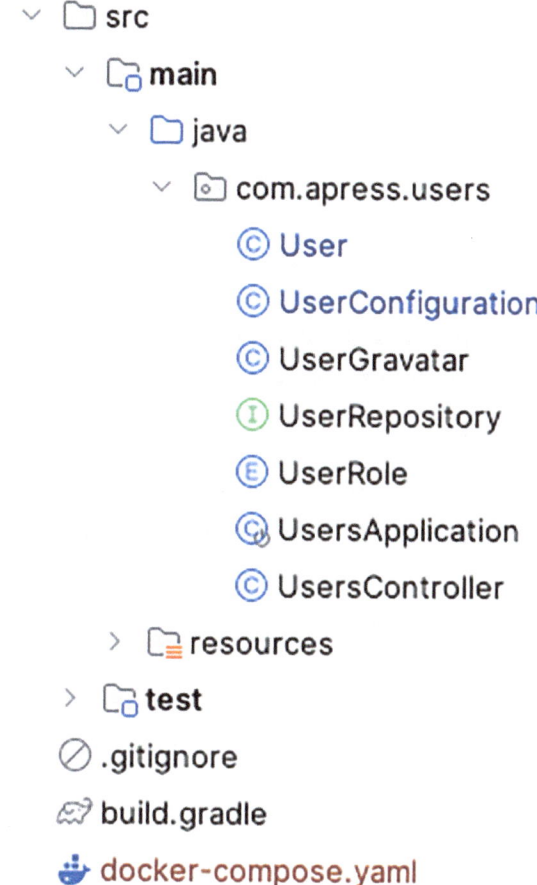

Figure 6-4. *My Retro App with Spring Data Redis with Spring Boot*

Open the build.gradle file and replace the content with the content shown in Listing 6-24.

Listing 6-24. build.gradle

```
plugins {
    id 'java'
    id 'org.springframework.boot' version '3.2.3'
    id 'io.spring.dependency-management' version '1.1.4'
}
```

```
group = 'com.apress'
version = '0.0.1-SNAPSHOT'
sourceCompatibility = '17'

repositories {
    mavenCentral()
}

dependencies {
    implementation 'org.springframework.boot:spring-boot-starter-web'
    implementation 'org.springframework.boot:spring-boot-starter-validation'
    implementation 'org.springframework.boot:spring-boot-starter-aop'

    implementation 'org.springframework.boot:spring-boot-starter-data-redis'
    developmentOnly 'org.springframework.boot:spring-boot-docker-compose'

    annotationProcessor 'org.springframework.boot:spring-boot-
    configuration-processor'

    compileOnly 'org.projectlombok:lombok'
    annotationProcessor 'org.projectlombok:lombok'

    // Web
    implementation 'org.webjars:bootstrap:5.2.3'

    testImplementation 'org.springframework.boot:spring-boot-starter-test'
}

tasks.named('test') {
    useJUnitPlatform()
}
```

Listing 6-24 shows that we included the spring-boot-starter-data-redis starter dependency and the docker compose.

Next, create/open the board package and the RetroBoard class. See Listing 6-25.

Listing 6-25. src/main/java/apress/com/myretro/board/RetroBoard.java

```java
package com.apress.myretro.board;

import jakarta.validation.constraints.NotBlank;
import jakarta.validation.constraints.NotNull;
import lombok.*;
import org.springframework.data.annotation.Id;
import org.springframework.data.redis.core.RedisHash;

import java.util.List;
import java.util.UUID;

@Builder
@AllArgsConstructor
@NoArgsConstructor
@Data
@RedisHash("RETRO_BOARD")
public class RetroBoard {

    @Id
    @NotNull
    private UUID id;

    @NotBlank(message = "A name must be provided")
    private String name;

    @Singular
    private List<Card> cards;
}
```

Listing 6-25 shows you the RetroBoard class, see that is marked using the @RedisHash annotation that using the parameter RETRO_BOARD, this will identify the hash for Redis. Also, we are using the @Id annotation to mark an identifier, which is used (as in the Users App) to establish the key for that hash value.

Next, create/open the persistence package and the RetroBoardRepository interface. See Listing 6-26.

Listing 6-26. src/main/java/apress/com/myretro/persistence/
RetroBoardRepository.java

```
package com.apress.myretro.persistence;
```

```
import com.apress.myretro.board.RetroBoard;
import org.springframework.data.repository.CrudRepository;
```

```
import java.util.UUID;
```

```
public interface RetroBoardRepository extends CrudRepository<RetroBoard,
UUID> { }
```

We are using the `CrudRepository`, which is the same as in other versions, and we are using the Spring Data core to implement this interface on our behalf using all the Redis operations.

The `advice`, `config`, `exception`, `service`, and `web` packages and all their classes remain the same from previous versions, and in the `board` package, the `Card` and `CardType` remain the same. Also, the `application.properties` file is empty. So, to sum up, the only significant change was the addition of the `@RedisHash` annotation to the `RetroBoard` class.

Running the My Retro App

To run the application, you can use the IDE or use the following command:

```
./gradlew clean bootRun
```

Then, point your browser to `http://localhost:8080/retros` or execute the following command:

```
curl -s http://localhost:8080/retros | jq .

[
  {
    "id": "9dc9b71b-a07e-418b-b972-40225449aff2",
    "name": "Spring Boot Conference",
    "cards": [
      {
        "id": "bb2a80a5-a0f5-4180-a6dc-80c84bc014c9",
```

```
      "comment": "Spring Boot Rocks!",
      "cardType": "HAPPY"
    },
    {
      "id": "e39cd241-4c75-41c8-9750-8b01e8225774",
      "comment": "Meet everyone in person",
      "cardType": "HAPPY"
    },
    {
      "id": "e8947c01-fd30-4f78-81cf-31acf5cfd46b",
      "comment": "When is the next one?",
      "cardType": "MEH"
    },
    {
      "id": "ce228247-d414-4cf9-8356-dbccb89f9370",
      "comment": "Not enough time to talk to everyone",
      "cardType": "SAD"
    }
  ]
  }
]
```

If you want to know how it is represented in Redis, you can run the following docker container client to run the redis-cli:

```
docker run -it --rm --network myretro_default redis:alpine redis-cli -h redis
```

Notice that we are using the network myretro_default that was generated when the application ran.

Then you can review with the redis commands:
you can review the keys with the following Redis commands:

```
redis:6379> keys *
1) "RETRO_BOARD"
2) "RETRO_BOARD:9dc9b71b-a07e-418b-b972-40225449aff2"
```

You can review the values with

```
redis:6379> hvals "RETRO_BOARD:9dc9b71b-a07e-418b-b972-40225449aff2"
 1) "com.apress.myretro.board.RetroBoard"
 2) "HAPPY"
 3) "Spring Boot Rocks!"
 4) "bb2a80a5-a0f5-4180-a6dc-80c84bc014c9"
 5) "HAPPY"
 6) "Meet everyone in person"
 7) "08821ce0-79ed-4dd4-ad8d-c6d0e841738f"
 8) "MEH"
 9) "When is the next one?"
10) "edf4dcc7-c800-48e2-bf2d-0af41b5c3a23"
11) "SAD"
12) "Not enough time to talk to everyone"
13) "7545e981-4042-4f9f-8986-78ff92242c3c"
14) "9dc9b71b-a07e-418b-b972-40225449aff2"
15) "Spring Boot Conference"
```

You can view the keys with

```
redis:6379> Hkeys "RETRO_BOARD:9dc9b71b-a07e-418b-b972-40225449aff2"
 1) "_class"
 2) "cards.[0].cardType"
 3) "cards.[0].comment"
 4) "cards.[0].id"
 5) "cards.[1].cardType"
 6) "cards.[1].comment"
 7) "cards.[1].id"
 8) "cards.[2].cardType"
 9) "cards.[2].comment"
10) "cards.[2].id"
11) "cards.[3].cardType"
12) "cards.[3].comment"
13) "cards.[3].id"
14) "id"
15) "name"
```

And you can view the field values with

```
redis:6379> hget "RETRO_BOARD:9dc9b71b-a07e-418b-b972-40225449aff2"
"cards.[3].cardType"
"SAD"
```

Note that our `Card` domain class is set as keys using a composition, the name of the instance pluralized to `cards`, the array notation, and the `Card` field name.

So, that's how you can run your applications using Spring Data Redis and Spring Boot. Of course, here we are using the repository programming model, but you can use directly the `RedisTemplate`, `StringRedisTemplate,` or any other operation classes such as `HashOperation`, `ListOperations`, `ZSetOperations`, etc.

Summary

In this chapter we covered MongoDB and Redis as NoSQL databases, and you learned how they are based in the Spring Data core project and how they expose the same repository programming model, making this easier with a minimal changes just in the domain classes.

You learned that Spring Boot uses the auto-configuration feature to set the default connection parameters and enable extra features for NoSQL databases, and the only thing you are required to do is add these dependencies.

You also learned with both projects (Users App and My Retro App) what to do to mark the classes with the right annotation (`@Document`, `@RedisHash`) and even use the core such as `@Id` annotation and the `CrudRepository` interface. Again, only minimal changes are needed if you let Spring Data do the work for you.

In Chapter 7 we are going to explore Spring Reactor and how to create Reactive applications.

CHAPTER 7

Spring Boot Reactive

In this chapter we are going to talk about how the Spring Framework utilizes the power of Project Reactor (`https://projectreactor.io/`) for building data and web reactive applications, and how Spring Boot helps with its auto-configuration feature to wire everything up to create awesome Reactive apps with ease. The Spring Framework introduced the Reactive technology with WebFlux in version 5.0 and further integrated 6.x to offer several improvements.

Spring Web Flux and Spring Reactive Data are fully nonblocking frameworks that rely on Project Reactor, which supports Reactive Streams back pressure (flow control) and runs on servers such as Netty and Undertow and servlet containers. Let's explore reactive systems and how Project Reactor implements them.

Reactive Systems

In the past decade, the software industry has responded to the demands of mobile and cloud computing by improving software development processes to produce software that is more stable, more robust, more resilient, and more flexible to accept even more modern demands, not only by users (using the desktop or the Web) but by a variety of devices (mobile phones, sensors, etc.). Accomodating these new workloads has many challenges, which is why a group of organizations worked together to produce a manifesto to cover many aspects of today's data demands.

The Reactive Manifesto

The Reactive Manifesto (`https://www.reactivemanifesto.org/`) was published on September 16, 2014. It defines the characteristics of reactive systems. Reactive systems are flexible, loosely coupled, and scalable. They are more tolerant to failure, and when a failure occurs, they deal with it by applying patterns to avoid disaster.

© Felipe Gutierrez 2024
F. Gutierrez, *Pro Spring Boot 3*, https://doi.org/10.1007/978-1-4842-9294-5_7

To paraphrase the Reactive Manifesto, Reactive systems have the following characteristics:

- *Responsive*: Reactive systems respond in a timely manner, if possible. They focus on providing fast and consistent response times and establishing reliable upper bounds for delivering a consistent quality of service.

- *Resilient*: Reactive systems apply replication, containment, isolation, and delegation patterns to achieve resilience. They contain failures through isolation so that failures do not affect other parts of the system or other systems. Recovery must be from another system, so that high availability (HA) is ensured.

- *Elastic*: Reactive systems are responsive under any kind of workload. They can react to changes in the input rate by increasing or decreasing the resources allocated to service these inputs. They should not have any bottlenecks, meaning they have the capability to shard or replicate components and distributed inputs to them. Reactive systems must support predictive algorithms, which ensures cost-effective elasticity on commodity hardware.

- *Message driven*: Reactive systems rely on asynchronous messaging to establish a boundary between components, making sure that the systems are loosely coupled, isolated, and location transparent. Reactive Systems must support load management, elasticity, and flow control by providing a back-pressure pattern when needed. The communication must be nonblocking to allow recepients to consume resources while active, which leads to lower system overhead.

After the publication of the Reactive Manifesto, different initiatives started to emerge and implement frameworks and libraries that help many developers around the world. Reactive Streams (`https://www.reactive-streams.org/`) is a specification that defines four simple interfaces: `Publisher<T>`, a provider of an unbounded number of sequenced elements, publishing them according to the demand of the subscriber; `Subscriber<T>`, which subscribes to the publisher; `Subscription`, which represents the one-to-one life cycle of a subscriber subscribing to a publisher; and `Processor`, which is the processing stage for both *Subscriber* and *Publisher*. Reactive Streams also has different implementations, such as *ReactiveX RXJava* (`https://github.com/ReactiveX/`

RxJava), *Akka Streams* (https://akka.io), *Ratpack* (https://ratpack.io), *Vert.x* (https://vertx.io), *Slick* (https://scala-slick.org/), *Project Reactor* (https://projectreactor.io), *Reactive Relational Database Connectivity (R2DBC)* (https://r2dbc.io/), and many more.

The Reactive Streams API has its own implementation in the Java 9 SDK release; in other words, as of December 2017, Reactive Streams version 1.0.2 is part of JDK9.

Project Reactor

Project Reactor 3.5.x is a library that is built around the Reactive Streams specification, bringing the reactive programming paradigm to JVM. Reactive programming is a paradigm that is an event-based model, where data is pushed to the consumer as it becomes available; it deals with asynchronous sequences of events. Offering fully asynchronous and nonblocking patterns, reactive programming is an alternative to the limited ways of doing async code in the JDK (callbacks, APIs, and the Future<V> interface).

Reactor is a full, nonblocking reactive programming framework that manages back pressure and integrates interaction with the Java 8+ functional APIs (CompletableFuture, Stream, and Duration). Reactor provides two reactive composable asynchronous APIs: Flux [N] (for N elements) and Mono [0|1] (for 0 or 1 elements). Reactor can be used to develop microservices architectures because it offers interprocess communication (IPC) with reactor-ipc components and backpressure-ready network engines for HTTP (including *WebSockets*, *TCP*, and *UDP*), and reactive encoding and decoding is fully supported.

Mono<T>, an Asynchronous [0|1] Result

Mono<T> is a specialized Publisher<T> interface that emits one item, and it can optionally terminate with onComplete or onError signals. You can apply operators to manipulate the item (see Figure 7-1).

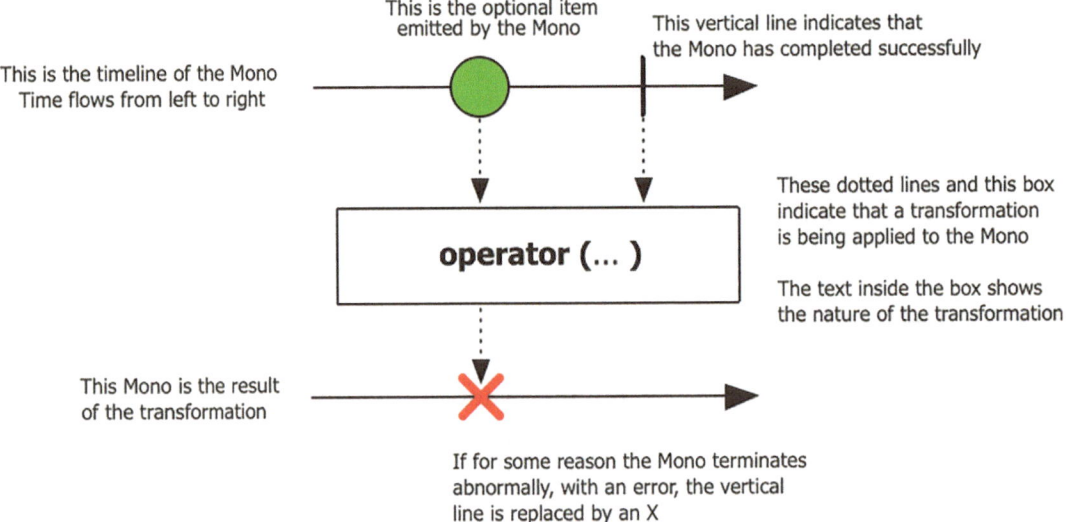

Figure 7-1. *Mono [0|1] (Source: https://projectreactor.io documentation)*

Flux<T>: An Asynchronous Sequence of [0|N] Items

Flux<T> is a Publisher<T> that represents an asynchronous sequence of 0 to N emitted items that can optionally terminate by using onComplete or an onError signals (see Figure 7-2).

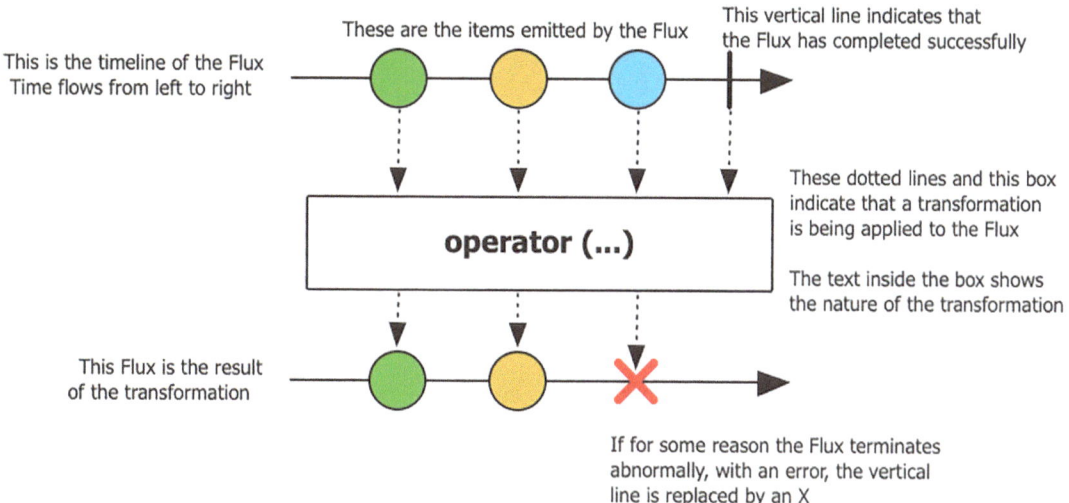

Figure 7-2. *Flux [0|N] (Source: https://projectreactor.io documentation)*

Project Reactor provides processors, operators, and timers that can sustain a high throughput rate on tens of millions of messages per second with a low memory footprint.

Note   If you want to know more about Project Reactor, peruse its documentation at https://projectreactor.io/docs/core/release/reference/.

Reactive Web and Data Applications with Spring Boot

By using Spring Boot, creating your Reactive apps is simpler. Spring Boot provides auto-configuration of all your web and data reactive apps using Web Flux, by adding the spring-boot-starter-webflux starter and any reactive data dependencies. When you create a project in the Spring Initializr (https://start.spring.io), if you click Add Dependencies and type **reactive**, you get a list of dependencies similar to the list shown in Figure 7-3.

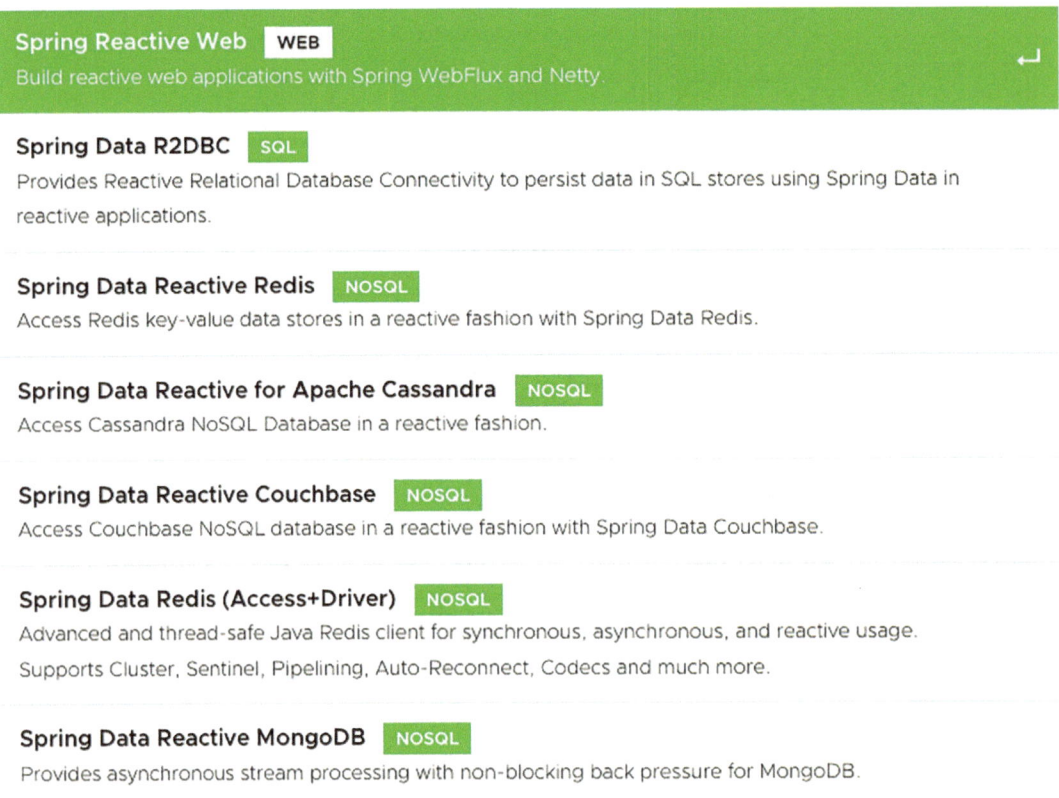

Figure 7-3. *Spring Initializr Reactive dependencies*

from Web to any SQL or NoSQL Data reactive. You can use any programming style model, from using annotations to functional programming, and use the reactive types of Mono<T> or Flux<T> classes for your responses. When you are using Spring Boot with Spring Reactive Web (a.k.a. WebFlux), the application container used is *Netty* by default (no more Tomcat). You can switch to use *Undertow* instead.

Figure 7-4 depicts the respective programming models when you are using Spring MVC or Spring WebFlux.

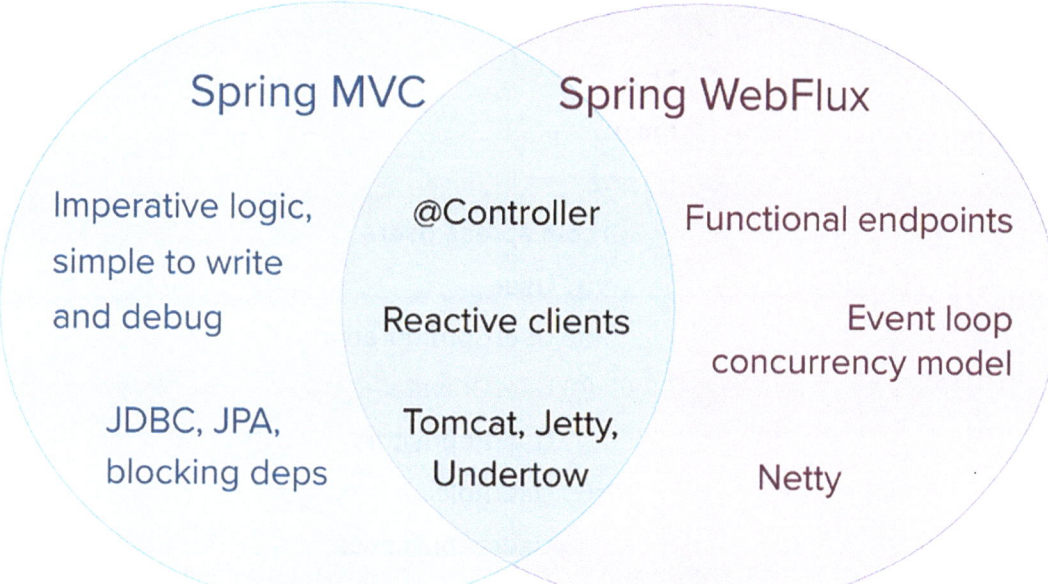

Figure 7-4. *Applicability - Spring MVC WebFlux (Source: `https://docs.spring.io/spring-framework/reference/web/webflux/new-framework.html`)*

As you can see, we have options to choose, from using functional to annotated endpoints.

Users App with Spring Boot Reactive

For our Users App, we are going to use WebFlux and R2DBC with H2 and functional programming for our endpoints, which will be a little different from previous versions. You can start from scratch by going to the Spring Initializr (`https://start.spring.io`) and generating a base project (no dependencies). Make sure to set the group field to `com.apress` and the Artifact and Name fields to `users`. Download the project, unzip it, and import it into your favorite IDE. By the end of this section, you should have the structure shown in Figure 7-5.

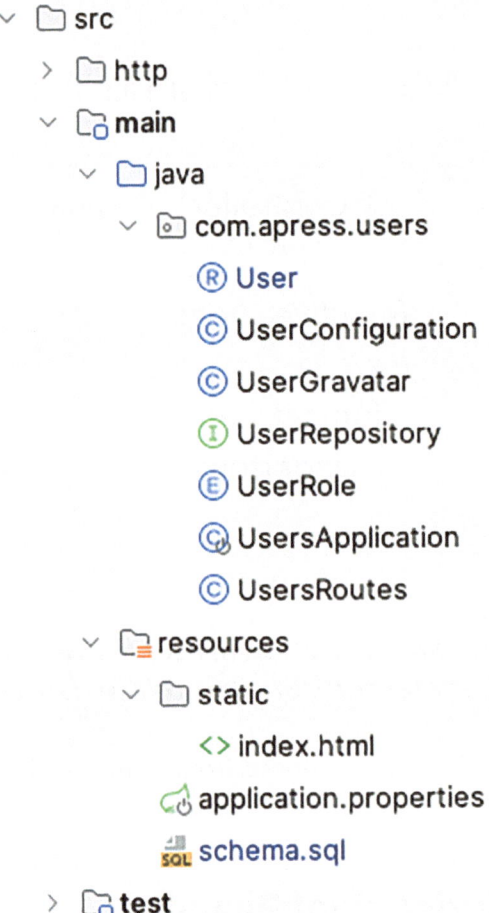

Figure 7-5. *Users App with Web Flux and R2DBC*

Next, open the build.gradle file and replace the content with the content shown in Listing 7-1.

Listing 7-1. build.gradle

```
plugins {
    id 'java'
    id 'org.springframework.boot' version '3.2.3'
    id 'io.spring.dependency-management' version '1.1.4'
}
```

```
group = 'com.apress'
version = '0.0.1-SNAPSHOT'
sourceCompatibility = '17'

repositories {
    mavenCentral()
}

dependencies {
    implementation 'org.springframework.boot:spring-boot-starter-webflux'
    implementation 'org.springframework.boot:spring-boot-starter-
    validation'

    implementation 'org.springframework.boot:spring-boot-starter-
    data-r2dbc'
            runtimeOnly 'io.r2dbc:r2dbc-h2'

            // Reactive Postgres
            //runtimeOnly 'org.postgresql:r2dbc-postgresql'

    compileOnly 'org.projectlombok:lombok'
    annotationProcessor 'org.projectlombok:lombok'

    // Web
    implementation 'org.webjars:bootstrap:5.2.3'

    testImplementation 'org.springframework.boot:spring-boot-starter-test'
    testImplementation 'io.projectreactor:reactor-test'
}


tasks.named('test') {
    useJUnitPlatform()
}

test {
    testLogging {
        events "passed", "skipped", "failed"
        showExceptions true
        exceptionFormat "full"
```

```
        showCauses true
        showStackTraces true
        showStandardStreams = false
    }
}
```

Listing 7-1 shows the necessary dependencies to use WebFlux and R2DBC. Note that we need to include the DB driver (in this example, will use the H2 Database) and add the reactive driver (in this case, the r2dbc-h2 driver). If H2 is not a good option for you, you can switch it to the Reactive PostgreSQL r2dbc-postgresql dependency. Also notice that we are using the reactor-test dependency that allows us to test new types such as Flux and Mono.

Next, create/open the User class. See Listing 7-2.

Listing 7-2. src/main/java/apress/com/users/User.java

```
package com.apress.users;

import jakarta.validation.constraints.NotBlank;
import org.springframework.data.annotation.Id;
import org.springframework.data.relational.core.mapping.Table;

import java.util.Collection;
import java.util.UUID;

@Table("PEOPLE")
public record User(

        @Id
        UUID id,

        @NotBlank(message = "Email can not be empty")
        String email,

        @NotBlank(message = "Name can not be empty")
        String name,

        String gravatarUrl,

        @NotBlank(message = "Password can not be empty")
        String password,
```

```
    Collection<UserRole> userRole,

    boolean active
) {

    public User withGravatarUrl(String email) {
        String url = UserGravatar.getGravatarUrlFromEmail(email);
        return new User(UUID.randomUUID(), email, name, url, password,
        userRole, active);
    }

}
```

Listing 7-2 shows the User class of type record! Yes, we can return to using the record type introduced in Chapter 4, but this time as simply as possible. We are creating a custom method that will be our factory that will build a new User (just remember about record type immutability). We are using the @Table annotation with the value PEOPLE, which will be the name of the table. We are also using the @Id annotation. Note that here we are using the *Spring Data core*, a common use case for our domain classes.

Next, create/open the UserRepository interface. See Listing 7-3.

Listing 7-3. src/main/java/apress/com/users/UserRepository.java

```
package com.apress.users;

import org.springframework.data.repository.reactive.ReactiveCrudRepository;
import reactor.core.publisher.Mono;

import java.util.UUID;

public interface UserRepository extends
ReactiveCrudRepository<User, UUID> {
    Mono<User> findByEmail(String email);
    Mono<Void> deleteByEmail(String email);
}
```

In Listing 7-3, we are using a new extended interface, ReactiveCrudRepository, which takes the domain class (in this case the User) and the identifier (in this case a UUID).

Next, create/open the UserRoutes class. See Listing 7-4.

Listing 7-4. src/main/java/apress/com/users/UserRoutes.java

```java
package com.apress.users;

import lombok.RequiredArgsConstructor;
import org.springframework.context.annotation.Bean;
import org.springframework.context.annotation.Configuration;
import org.springframework.web.reactive.function.server.RouterFunction;
import org.springframework.web.reactive.function.server.ServerResponse;

import static org.springframework.web.reactive.function.
BodyExtractors.toMono;
import static org.springframework.web.reactive.function.server.
RequestPredicates.*;
import static org.springframework.web.reactive.function.server.
RouterFunctions.route;

@RequiredArgsConstructor
@Configuration
public class UsersRoutes {

    private final UserRepository userRepository;

    @Bean
    public RouterFunction<ServerResponse> getUsersRoute() {
        return route(GET("/users"), request -> ServerResponse.ok()
                .body(this.userRepository.findAll(),User.class));
    }

    @Bean
    public RouterFunction<ServerResponse> postUserRoute() {
        return route(POST("/users"), request -> request
                .body(toMono(User.class))
                .flatMap(this.userRepository::save)
                .then(ServerResponse.ok().build()));
    }

    @Bean
    public RouterFunction<ServerResponse> findUserByEmail(){
        return route(GET("/users/{email}"), request -> ServerResponse.ok()
```

```
                    .body(this.userRepository.findByEmail(request.
                    pathVariable("email")),User.class));
        }

        @Bean
        public RouterFunction<ServerResponse> deleteUserByEmail(){
            return route(DELETE("/users/{email}"), request -> {
                this.userRepository.deleteByEmail(request.
                pathVariable("email"));
                return ServerResponse.noContent().build();
            });
        }

    }
```

In Listing 7-4 we are defining our routes for the endpoints and configuring the server responses, all in the UserRoutes class (this is different from the previous versions, in which we have two separated files for defining routes and another for the responses). Note that this is a JavaConfig class and that we are marking it with the @Configuration annotation, meaning that we need to declare beans with the @Bean annotation. We are using the RouterFunction with the ServerResponse type and we are using the route configuration to accept a request and return a ServerResponse. This all is very straightforward because we are based on the core, the *Spring Data core* with the repositories programming model (meaning that Spring Data is doing the implementation of the ReactiveCrudRepository interface on our behalf).

Because we are using the ServerResponse this wraps all the responses either Flux<T> or Mono<T> classes depending on the response from the reactive database.

If we were just using Project Reactor to access the database (without Spring R2DBC), we would need to write something like the following code snippet:

```
ConnectionFactory connectionFactory = ConnectionFactories
  .get("r2dbc:h2:mem:///testdb");

Mono.from(connectionFactory.create())
  .flatMapMany(connection -> connection
    .createStatement("SELECT name FROM PEOPLE WHERE email = $1")
    .bind("$1", "norma@email.com")
    .execute())
```

```
.flatMap(result -> result
  .map((row, rowMetadata) -> row.get("name", String.class)))
.doOnNext(System.out::println)
.subscribe();
```

Next, create/open the UserConfiguration class. See Listing 7-5.

Listing 7-5. src/main/java/apress/com/users/UserConfiguration.java

```
package com.apress.users;

import org.springframework.boot.CommandLineRunner;
import org.springframework.context.annotation.Bean;
import org.springframework.context.annotation.Configuration;
import org.springframework.data.r2dbc.mapping.event.BeforeConvertCallback;
import org.springframework.data.relational.core.sql.SqlIdentifier;
import org.springframework.stereotype.Component;
import reactor.core.publisher.Mono;

import java.time.Duration;
import java.util.Arrays;

@Configuration
public class UserConfiguration {

    @Component
    class GravatarUrlGeneratingCallback implements
    BeforeConvertCallback<User>{

        @Override
        public Mono<User> onBeforeConvert(User user, SqlIdentifier
        sqlIdentifier) {

            if (user.id() == null && (user.gravatarUrl() == null || user.
            gravatarUrl().isEmpty())) {
                return Mono.just(user.withGravatarUrl(user.email()));
            }
            return Mono.just(user);
        }
    }
```

```
@Bean
CommandLineRunner init(UserRepository userRepository){
    return args -> {

        userRepository.saveAll(Arrays.asList(new User(null,"ximena@
        email.com","Ximena",null,"aw2s0me", Arrays.asList(UserRole.
        USER),true)
        ,new User(null,"norma@email.com","Norma" ,null, "aw2s0me",
        Arrays.asList(UserRole.USER, UserRole.ADMIN),true)))
                .blockLast(Duration.ofSeconds(10));
    };
}

}
```

Let's analyze the UserConfiguration class:

- BeforeConvertCallback: We are declaring this inner class that implements the BeforeConvertCallback interface. We have been implementing it in previous versions (for the data persistence apps), and in this case we are declaring it as an inner class instead of having a separate class (which option to use is up to you and where you want to have certain logic). In this method we are checking for the gravatarUrl field and setting it if it's null.

- CommandLineRunner: Instead of using the ApplicationListener (that will listen for the ApplicationReadyEvent), we are using the CommanLineRunner interface just as a reminder that we have options for how we execute code when the application is ready.

Next, create/open the schema.sql file, shown in Listing 7-6.

Listing 7-6. src/main/resources/schema.sql

```
drop table if exists people cascade;
create table people (
    id uuid default random_uuid() not null,
    email varchar(255) not null,
    active boolean not null,
    gravatar_url varchar(255),
```

```
name varchar(255),
password varchar(255),
user_role VARCHAR(100) array,
primary key (id));
```

If you want to use PostgreSQL, you need to change to the uuid_generate_v4() and enable the plugin that handles those functions:

```
CREATE EXTENSION IF NOT EXISTS "uuid-ossp";
```

The UserGravatar and UserRole classes remain the same as in previous versions. The application.properties file is empty. Remember that Spring Boot will do all the auto-configuration, which means that it will initialize the database with the SQL statements found in the schema.sql file and it will set all the default values needed for the connection.

Testing the Users App

To test the Users App, we need to use a different client that accepts the Flux<T> or Mono<T> types. Specifically, we need the reactor-test dependency in the build.gradle file (Listing 7-1). So, create/open the UsersHttpRequestTests class. See Listing 7-7.

Listing 7-7. src/test/java/apress/com/users/UsersHttpRequestTests.java

```
package com.apress.users;

import org.junit.jupiter.api.Test;
import org.springframework.beans.factory.annotation.Autowired;
import org.springframework.boot.test.context.SpringBootTest;
import org.springframework.test.web.reactive.server.WebTestClient;
import reactor.core.publisher.Mono;

import java.util.Arrays;
import java.util.Collection;

import static org.assertj.core.api.Assertions.assertThat;

@SpringBootTest(webEnvironment = SpringBootTest.WebEnvironment.RANDOM_PORT)
public class UsersHttpRequestTests {

    @Autowired
    private WebTestClient webTestClient;
```

```java
@Test
public void indexPageShouldReturnHeaderOneContent() throws Exception {
    webTestClient.get().uri("/")
                    .exchange()
                            .expectStatus().isOk()
                    .expectBody(String.class).value( value -> {
                assertThat(value).contains("Simple Users Rest
                Application");
            });
}

@Test
public void usersEndPointShouldReturnCollectionWithTwoUsers() throws
Exception {
    webTestClient.get().uri("/users")
            .exchange().expectStatus().isOk()
            .expectBody(Collection.class).value( collection -> {
                assertThat(collection.size()).isGreaterThanOrEqualTo(3);
            });
}

@Test
public void userEndPointPostNewUserShouldReturnUser() throws
Exception {
    webTestClient.post().uri("/users")
            .body(Mono.just(new User(null,"dummy@email.com","Dummy",
            null,"aw2s0me", Arrays.asList(UserRole.USER),true)),
            User.class)
            .exchange().expectStatus().isOk();
}

@Test
public void userEndPointDeleteUserShouldReturnVoid() throws Exception {
    webTestClient.delete().uri("/users/norma@email.com")
            .exchange().expectStatus().isNoContent();
}
```

```
@Test
public void userEndPointFindUserShouldReturnUser() throws Exception{
    webTestClient.get().uri("/users/ximena@email.com")
            .exchange().expectStatus().isOk()
            .expectBody(User.class).value( user -> {
                assertThat(user).isNotNull();
                assertThat(user.email()).isEqualTo("ximena@email.com");
            });
}
}
```

Listing 7-7 shows that we are using the WebTestClient class instead of the TestRestTemplate class. The WebTestClient class is a client for testing web servers that uses the WebClient class internally to perform requests. The WebTestClient class provides a fluent API, using get(), post(), and delete() methods that allow you to easily configure your tests. This client can connect not only to any server over HTTP, but to any WebFlux app.

To run your test, you can use your IDE or use the following command:

```
./gradlew clean test
```

```
UsersHttpRequestTests > userEndPointFindUserShouldReturnUser() PASSED
UsersHttpRequestTests > userEndPointDeleteUserShouldReturnVoid() PASSED
UsersHttpRequestTests > indexPageShouldReturnHeaderOneContent() PASSED
UsersHttpRequestTests > userEndPointPostNewUserShouldReturnUser() PASSED
UsersHttpRequestTests > usersEndPointShouldReturnCollectionWith
                        TwoUsers() PASSED
```

Running the Users App

Run the Users App either by using your IDE or by executing the following command:

```
./gradlew clean bootRun
```

Next, either point your browser to http://localhost:8080/users or open another terminal and use the following curl command:

```
curl -s http://localhost:8080/users |  jq .

{
    "id": "9ca8bbbe-4814-4223-9dc9-1c6aec783e43",
    "email": "ximena@email.com",
    "name": "Ximena",
    "gravatarUrl": "https://www.gravatar.com/avatar/f07f7e553264c9710105
    edebe6c465e7?d=wavatar",
    "password": "aw2sOme",
    "userRole": [
      "USER"
    ],
    "active": true
  },
  {
    "id": "f1e5570d-15f4-41e6-bb87-de333871232c",
    "email": "norma@email.com",
    "name": "Norma",
    "gravatarUrl": "https://www.gravatar.com/avatar/23bb62a7d0ca63c9a804
    908e57bf6bd4?d=wavatar",
    "password": "aw2sOme",
    "userRole": [
      "USER",
      "ADMIN"
    ],
    "active": true
  }
]
```

Note   If you want more information about Spring R2DBC, visit the reference
documentation at https://docs.spring.io/spring-data/r2dbc/docs/
current/reference/html/#reference.

My Retro App with Spring Boot Reactive

Switching to the My Retro App, we are going to use Reactive MongoDB persistence and the annotation-based controller endpoints. And if you are following along, you can reuse some of the code from previous versions. Or you can start from scratch by going to the Spring Initializr (https://start.spring.io) and generating a base project (no dependencies). Make sure to set the Group field to com.apress and the Artifact and Name fields to myretro. Download the project, unzip it, and import it into your favorite IDE. By the end of this section, you should have the structure shown in Figure 7-6.

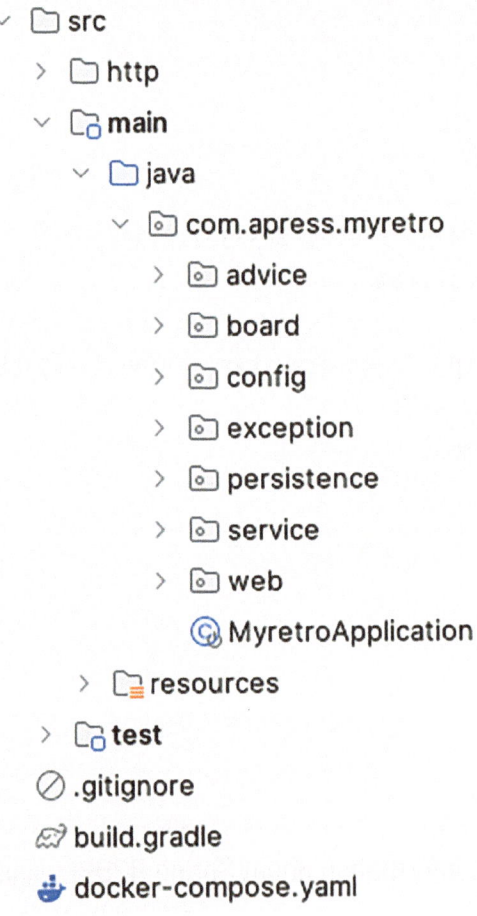

Figure 7-6. *My Retro with WebFlux / MongoDB*

The My Retro App final structure is very similar to the MongoDB version. Open the build.gradle file. See Listing 7-8.

Listing 7-8. build.gradle

```
plugins {
    id 'java'
    id 'org.springframework.boot' version '3.2.3'
    id 'io.spring.dependency-management' version '1.1.4'
}

group = 'com.apress'
version = '0.0.1-SNAPSHOT'
sourceCompatibility = '17'

repositories {
    mavenCentral()
}

dependencies {
    implementation 'org.springframework.boot:spring-boot-starter-webflux'
    implementation 'org.springframework.boot:spring-boot-starter-
    validation'
    implementation 'org.springframework.boot:spring-boot-starter-aop'
    implementation 'com.fasterxml.uuid:java-uuid-generator:4.0.1'

    implementation 'org.springframework.boot:spring-boot-starter-data-
    mongodb-reactive'
    developmentOnly 'org.springframework.boot:spring-boot-docker-compose'

    annotationProcessor 'org.springframework.boot:spring-boot-
    configuration-processor'

    compileOnly 'org.projectlombok:lombok'
    annotationProcessor 'org.projectlombok:lombok'

    // Web
    implementation 'org.webjars:bootstrap:5.2.3'
```

```
    testImplementation 'org.springframework.boot:spring-boot-starter-test'
    testImplementation 'io.projectreactor:reactor-test'
}

tasks.named('test') {
    useJUnitPlatform()
}
```

Listing 7-8 shows the build.gradle dependencies, including the new spring-boot-starter-data-mongodb-reactive starter dependency added to this file.

Create/open the board package and the Card, CardType, and RetroBoard classes. See Listings 7-9, 7-10, and 7-11, respectively.

Listing 7-9. src/main/java/apress/com/myretro/board/Card.java

```
package com.apress.myretro.board;

import lombok.AllArgsConstructor;
import lombok.Builder;
import lombok.Data;
import lombok.NoArgsConstructor;

import java.util.UUID;

@Builder
@AllArgsConstructor
@NoArgsConstructor
@Data
public class Card {

    private UUID id;

    private String comment;

    private CardType cardType;

}
```

Listing 7-10. src/main/java/apress/com/myretro/board/CardType.java

```
package com.apress.myretro.board;

public enum CardType {
    HAPPY,MEH,SAD
}
```

Listing 7-11. src/main/java/apress/com/myretro/board/RetroBoard.java

```
package com.apress.myretro.board;

import lombok.*;
import org.springframework.data.annotation.Id;
import org.springframework.data.mongodb.core.mapping.Document;

import java.util.ArrayList;
import java.util.List;
import java.util.UUID;

@Builder
@AllArgsConstructor
@NoArgsConstructor
@Data
@Document
public class RetroBoard {

    @Id
    private UUID id;

    private String name;

    @Singular("card")
    private List<Card> cards;

    public void addCard(Card card){
        if (this.cards == null)
            this.cards = new ArrayList<>();
        this.cards.add(card);
    }
```

```
    public void addCards(List<Card> cards){
        if (this.cards == null)
            this.cards = new ArrayList<>();
        this.cards.addAll(cards);
    }

}
```

The Card class and CardType enum remain the same, but the RetroBoard class now includes the @Document and @Id annotations. The @Document annotation marks the class as a persistence domain for the MongoDB, and the @Id annotation adds an identifier for our document, in this case the UUID.

Next, create/open the persistence package and the RetroBoardRepository interface and the RetroBoardPersistenceCallback class. Listing 7-12 shows the RetroBoardRepository interface.

Listing 7-12. src/main/java/apress/com/myretro/persistence/
RetroBoardRepository.java

```
package com.apress.myretro.persistence;

import com.apress.myretro.board.RetroBoard;
import org.springframework.data.mongodb.repository.Query;
import org.springframework.data.mongodb.repository.ReactiveMongoRepository;
import reactor.core.publisher.Mono;

import java.util.UUID;

public interface RetroBoardRepository extends ReactiveMongoRepository
<RetroBoard,UUID> {

    @Query("{'id': ?0}")
    Mono<RetroBoard> findById(UUID id);

    @Query("{}, { cards: { $elemMatch: { _id: ?0 } } }")
    Mono<RetroBoard> findRetroBoardByIdAndCardId(UUID cardId);

    default Mono<Void> removeCardFromRetroBoard(UUID retroBoardId,
    UUID cardId) {
        findById(retroBoardId)
```

```
        .doOnNext(retroBoard -> retroBoard.getCards().removeIf
        (card -> card.getId().equals(cardId)))
        .flatMap(this::save)
        .subscribe();
    return Mono.empty();
  }
}
```

Listing 7-11 shows that the RetroBoardRepository interface includes the following:

- ReactiveMongoRepository: This is a Mongo-specific interface with reactive support. It extends from the ReactiveCrudRepository, ReactiveSortingRepository, and ReactiveQueryByExampleExecutor interfaces. It brings all the usual methods (insert, save, find*, etc.) but in this case using the Flux and Mono types. Spring Data and Project Reactor will implement these interfaces using all the reactive support.

- Mono<RetroBoard>: As in previous versions, we can define our own find* methods and even use our own queries, in this case using the MongoDB query language. We are using the @Query annotation to define a particular query that will find a Card by its UUID among all the retroBoard objects. Of course, this approach is not the best because we can narrow our search if we facilitate the retroBoard UUID or do a different map reduce option. The idea here is that you can extend, use, and create your own custom queries.

- Mono<Void>: We are creating a default method, removeCardFromRetroBoard; take a moment and analyze it. First, we are using the findById method (in this case the retroBoard), which will emit a Mono<RetroBoard> entity if found. Then, we can remove it using the doOnNext method, which will emit a modified Mono<RetroBoard> entity (because we are removing the Card by UUID if found). Finally, we can apply a transformation using the flatMap method (this transformation will emit a Mono<RetroBoard>); the transformation will be just to save it. Remember that we are dealing with reactive entities, so it's necessary to subscribe() to the flow. Finally, we return just an empty Mono entity.

Listing 7-13 shows the RetroBoardPersistenceCallback class.

Listing 7-13. src/main/java/apress/com/myretro/persistence/
RetroBoardPersistenceCallback.java

```java
package com.apress.myretro.persistence;

import com.apress.myretro.board.RetroBoard;
import lombok.extern.slf4j.Slf4j;
import org.reactivestreams.Publisher;
import org.springframework.data.mongodb.core.mapping.event.
ReactiveBeforeConvertCallback;
import org.springframework.stereotype.Component;
import reactor.core.publisher.Mono;

@Slf4j
@Component
public class RetroBoardPersisteceCallback implements ReactiveBeforeConvert
Callback<RetroBoard> {

    @Override
    public Publisher<RetroBoard> onBeforeConvert(RetroBoard entity, String
    collection) {
        if (entity.getId() == null)
            entity.setId(java.util.UUID.randomUUID());
        if (entity.getCards() == null)
            entity.setCards(new java.util.ArrayList<>());
        log.info("[CALLBACK] onBeforeConvert {}", entity);
        return Mono.just(entity);
          }
}
```

Notice in the RetroBoardPersistenceCallback class that now we are using
the ReactiveBeforeConvertCallback interface. This interface has a method,
onBeforeConvert, that expects to return a Publisher. This method will be used to add
a UUID when a RetroBoard entity doesn't have one. Also note that we are returning a
Mono, in this case with the Mono.just() method. Again, this is one way to implement
this business rule, but you can add logic to the service if you prefer. With the code in
Listing 7-13, we are making our service as clean as possible.

Next, open/create the advice package and the RetroBoardAdvice class. See
Listing 7-14.

Listing 7-14. src/main/java/apress/com/myretro/advice/RetroBoardAdvice.java

```java
package com.apress.myretro.advice;

import com.apress.myretro.board.RetroBoard;
import com.apress.myretro.exception.RetroBoardNotFoundException;
import lombok.extern.slf4j.Slf4j;
import org.aspectj.lang.ProceedingJoinPoint;
import org.aspectj.lang.annotation.Around;
import org.aspectj.lang.annotation.Aspect;
import org.springframework.stereotype.Component;
import reactor.core.publisher.Mono;

import java.util.UUID;

@Slf4j
@Component
@Aspect
public class RetroBoardAdvice {

    @Around("execution(* com.apress.myretro.persistence.
    RetroBoardRepository.findById(..))")
    public Object checkFindRetroBoard(ProceedingJoinPoint
    proceedingJoinPoint) throws Throwable {
        log.info("[ADVICE] {}", proceedingJoinPoint.getSignature().
        getName());
        Mono<RetroBoard> retroBoard = (Mono<RetroBoard>)
        proceedingJoinPoint.proceed(new Object[]{
                UUID.fromString(proceedingJoinPoint.getArgs()[0].
                toString())
        });
        if (retroBoard == null)
            throw new RetroBoardNotFoundException();

        return retroBoard;
    }
}
```

The RetroBoardAdvice class will help to intercept the calls to avoid any other errors (we are removing some concerns from our main code, to avoid code tangling and scattering). Because we are using Reactive programming, we need to use either the Flux type or Mono type, and in this case we are using the Mono type.

Next, create/open the service package with the RetroBoardService class. See Listing 7-15.

Listing 7-15. src/main/java/apress/com/myretro/service/RetroBoardService.java

```java
package com.apress.myretro.service;

import com.apress.myretro.board.Card;
import com.apress.myretro.board.RetroBoard;
import com.apress.myretro.persistence.RetroBoardRepository;
import lombok.AllArgsConstructor;
import org.springframework.stereotype.Service;
import reactor.core.publisher.Flux;
import reactor.core.publisher.Mono;

import java.util.UUID;

@AllArgsConstructor
@Service
public class RetroBoardService {

    RetroBoardRepository retroBoardRepository;

    public Mono<RetroBoard> save(RetroBoard domain) {
        return this.retroBoardRepository.save(domain);
    }

    public Mono<RetroBoard> findById(UUID uuid) {
        return this.retroBoardRepository.findById(uuid);
    }

    public Flux<RetroBoard> findAll() {
        return this.retroBoardRepository.findAll();
    }

    public Mono<Void> delete(UUID uuid) {
```

```java
        return this.retroBoardRepository.deleteById(uuid);
    }

    public Flux<Card> findAllCardsFromRetroBoard(UUID uuid) {
        return this.findById(uuid).flatMapIterable(RetroBoard::getCards);
    }

    public Mono<Card> addCardToRetroBoard(UUID uuid, Card card) {
        return this.findById(uuid).flatMap(retroBoard -> {
            if (card.getId() == null)
                card.setId(UUID.randomUUID());
            retroBoard.getCards().add(card);
            return this.save(retroBoard).thenReturn(card);
        });
    }

    public Mono<Card> findCardByUUID(UUID uuidCard) {
        Mono<RetroBoard> result = retroBoardRepository.findRetroBoardByIdAn
        dCardId(uuidCard);
        return result.flatMapIterable(RetroBoard::getCards).filter(card ->
        card.getId().equals(uuidCard)).next();
    }

    public Mono<Void> removeCardByUUID(UUID uuid, UUID cardUUID) {
        return retroBoardRepository.removeCardFromRetroBoard(uuid,
        cardUUID);
    }
}
```

In Listing 7-15 we are using the RetroBoardRepository interface that will be injected through the constructor (thanks to the @AllArgsConstructor annotation from Lombok). We are using either the Flux type or Mono type depending on the service method. Every method will emit an entity type (either Flux or Mono), and somehow we need to subscribe to them. But how or where? Some of the methods interact with (i.e., are transformed by) the entities using flatMap or flatMapIterable, and even so, we are emitting these types.

Next, create/open the web package and the RetroBoardController class, shown in Listing 7-16.

Listing 7-16. src/main/java/apress/com/myretro/web/
RetroBoardController.java

```java
package com.apress.myretro.web;

import com.apress.myretro.board.Card;
import com.apress.myretro.board.RetroBoard;
import com.apress.myretro.service.RetroBoardService;
import lombok.AllArgsConstructor;
import org.springframework.http.HttpStatus;
import org.springframework.web.bind.annotation.*;
import reactor.core.publisher.Flux;
import reactor.core.publisher.Mono;

import java.util.UUID;

@AllArgsConstructor
@RestController
@RequestMapping("/retros")
public class RetroBoardController {

    private RetroBoardService retroBoardService;

    @GetMapping
    public Flux<RetroBoard> getAllRetroBoards() {
        return retroBoardService.findAll();
    }

    @PostMapping
    public Mono<RetroBoard> saveRetroBoard(@RequestBody RetroBoard
    retroBoard) {
        return retroBoardService.save(retroBoard);
    }

    @GetMapping("/{uuid}")
    public Mono<RetroBoard> findRetroBoardById(@PathVariable UUID uuid) {
        return retroBoardService.findById(uuid);
    }
```

```java
@GetMapping("/{uuid}/cards")
public Flux<Card> getAllCardsFromBoard(@PathVariable UUID uuid) {
    return retroBoardService.findAllCardsFromRetroBoard(uuid);
}

@PutMapping("/{uuid}/cards")
public Mono<Card> addCardToRetroBoard(@PathVariable UUID uuid,
@RequestBody Card card) {
    return retroBoardService.addCardToRetroBoard(uuid, card);
}

@GetMapping("/cards/{uuidCard}")
public Mono<Card> getCardByUUID(@PathVariable UUID uuidCard) {
    return retroBoardService.findCardByUUID(uuidCard);
}

@ResponseStatus(HttpStatus.NO_CONTENT)
@DeleteMapping("/{uuid}/cards/{uuidCard}")
public Mono<Void> deleteCardFromRetroBoard(@PathVariable UUID uuid,
@PathVariable UUID uuidCard) {
    return retroBoardService.removeCardByUUID(uuid, uuidCard);
}
}
```

In the RetroBoardController class, we are using *annotation-based programming* to create the web endpoints, and in this case every endpoint will expose either Flux or Mono types. This class uses the RetroBoardService and see that we still emitting these types without subscription, and the reason is that Spring WebFlux takes care of the subscription for our reactive controllers.

Next, create/open the config package and the MyRetroConfiguration class, shown in Listing 7-17.

Listing 7-17. src/main/java/apress/com/myretro/config/
MyRetroConfiguration.java

```java
package com.apress.myretro.config;

import com.apress.myretro.board.Card;
import com.apress.myretro.board.CardType;
```

```java
import com.apress.myretro.board.RetroBoard;
import com.apress.myretro.service.RetroBoardService;
import lombok.extern.slf4j.Slf4j;
import org.springframework.boot.context.event.ApplicationReadyEvent;
import org.springframework.boot.context.properties.
EnableConfigurationProperties;
import org.springframework.context.ApplicationListener;
import org.springframework.context.annotation.Bean;
import org.springframework.context.annotation.Configuration;

import java.util.UUID;

@Slf4j
@EnableConfigurationProperties({MyRetroProperties.class})
@Configuration
public class MyRetroConfiguration {

    @Bean
    ApplicationListener<ApplicationReadyEvent> ready(RetroBoardService
    retroBoardService) {
        return applicationReadyEvent -> {
            log.info("Application Ready Event");
            UUID retroBoardId = UUID.fromString("9dc9b71b-a07e-418b-
            b972-40225449aff2");
            retroBoardService.save(RetroBoard.builder()
                    .id(retroBoardId)
                    .name("Spring Boot Conference 2023")
                    .card(Card.builder().id(UUID.fromString("bb2a80a5-
                    a0f5-4180-a6dc-80c84bc014c9")).comment("Spring Boot
                    Rocks!").cardType(CardType.HAPPY).build())
                    .card(Card.builder().id(UUID.randomUUID()).
                    comment("Meet everyone in person").cardType(CardType.
                    HAPPY).build())
                    .card(Card.builder().id(UUID.randomUUID()).comment
                    ("When is the next one?").cardType(CardType.MEH).
                    build())
```

```
                .card(Card.builder().id(UUID.randomUUID()).comment
                ("Not enough time to talk to everyone").cardType
                (CardType.SAD).build())
                .build()).subscribe();
        };
    }
}
```

The MyRetroConfiguration class is using the RetroBoardService, and we are saving a retroBoard. Note that the save method emits an entity, and for us to save it, we need to subscribe, and that's why we are using the subscribe() method. Again, in our reactive controllers, we don't need to subscribe because Spring WebFlux will take care of it.

The package exception is like the other versions; the MyRetroProperties and UsersConfiguration haven't changed. You can reuse these classes.

Next, create/open the application.properties file. See Listing 7-18.

Listing 7-18. src/main/resources/application.properties

```
# MongoDB
spring.data.mongodb.uuid-representation=standard
spring.data.mongodb.database=retrodb
spring.data.mongodb.repositories.type=reactive
```

The only property in application.properties that is important to us here is spring.data.mongodb.uuid-representation. Recall that it is important because MongoDB uses its own implementation to deal with UUID objects, and if we want just to use our UUID, then we need to use this property. Also notice that we have specified the name of the database; if you don't set the name, by default (Spring Boot auto-configuration), it will use the name test, but you need to change it accordingly in your database setup (docker-compose.yaml). Also, we are using a property to set the repositories to reactive, even though this is set by Spring Boot auto-configuration.

Next, create/open the docker-compose.yaml file. See Listing 7-19.

Listing 7-19. docker-compose.yaml

```
version: "3.1"
services:
  mongo:
    image: mongo
    restart: always
    environment:
      MONGO_INITDB_DATABASE: retrodb
    ports:
      - "27017:27017"
```

The `docker-compose.yaml` file is the same as the previous version—nothing different is needed to use reactive programming.

Running My Retro App

You can run the My Retro App by using the IDE or by using the following command:

```
./gradlew clean bootRun
```

You should see the docker containers starting up and the Netty server ready to accept requests. You can do some testing using either VS Code (and the REST Client plugin - `humao.rest-client`) or if you are using the IntelliJ Enterprise Edition, you have it by default.

Create/open the `myretro.http` file. See Listing 7-20.

Listing 7-20. src/http/myretro.http

```
### Get All Retro Boards
GET http://localhost:8080/retros
Content-Type: application/json

### Get Retro Board
GET http://localhost:8080/retros/9dc9b71b-a07e-418b-b972-40225449aff2
Content-Type: application/json

### Get All Cards from Retro Board
GET http://localhost:8080/retros/9dc9b71b-a07e-418b-b972-40225449aff2/cards
Content-Type: application/json
```

```
### Get Single Card No Retro Board
GET http://localhost:8080/retros/cards/bb2a80a5-a0f5-4180-a6dc-80c84bc014c9
Content-Type: application/json

### Create a Retro Board
POST http://localhost:8080/retros
Content-Type: application/json

{
  "name": "Spring Boot Videos 2024"
}

### Add Card to Retro
PUT http://localhost:8080/retros/9dc9b71b-a07e-418b-b972-40225449aff2/cards
Content-Type: application/json

{
  "comment": "We are back in business",
  "cardType": "HAPPY"
}

### Delete Card from Retro
DELETE http://localhost:8080/retros/9dc9b71b-a07e-418b-b972-40225449aff2/
cards/bb2a80a5-a0f5-4180-a6dc-80c84bc014c9
Content-Type: application/json
```

The myretro.http file allows you to execute REST calls to different endpoints. The "Testing the Users App" section already covered the WebClient and WebTestClient classes; you can create your own integration test.

Summary

Now, you know what to do when you need to create a Reactive application with Spring Boot and Project Reactor. In this chapter you learned about Reactive programming and the many implementations from different frameworks.

You learned about Project Reactor and how it exposes the new Flux and Mono types to make it easier for your application to emit new flows of streaming data.

You saw the different ways to create a web API using either functional or annotation-based reactive applications where you emit the results using `Flux` or `Mono`. Although we didn't mention it, you can still use something like `Mono<ResponseEntity<RetroBoard>>` to return a reponse that can have custom HTTP Headers for example.

Chapter 8 covers how to test with Spring Boot and what bring to the Integration and Unit testing for your applications.

CHAPTER 8

Spring Boot Testing

In this chapter we are going to talk about how Spring Boot uses the power of the Spring Testing Framework to facilitate development by providing powerful tools for creating unit and integration testing with ease. In previous chapters we have done some testing of our two applications, but we haven't covered other important features of the Spring Boot Testing Framework, so let's start talking about the Spring Testing Framework, which is the basis for Spring Boot testing.

Spring Testing Framework

One of the main ideas of the Spring Framework is to encourage developers to create simple and loosely coupled classes and to program to interfaces, making the software more robust and extensible. The Spring Framework provides the tools for making unit and integration testing easy (actually, you don't need Spring to test the functionality of your system if you really program to interfaces); in other words, your application should be testable using either the JUnit or TestNG test engine with objects (by simple instantiation using the new operator)—without Spring or any other container.

The Spring Framework has several testing packages that help create unit and integration testing for applications. It offers unit testing by providing several mock objects (`Environment`, `PropertySource`, JNDI, Servlet; Reactive: `ServerHttpRequest` and `ServerHttpResponse` test utilities) that help you to test your code in isolation.

One of the most commonly used testing features of the Spring Framework is integration testing. Its primary's goals are

- Managing the Spring IoC container caching between test execution
- Transaction management
- Dependency injection of test fixture instances
- Spring-specific base classes

© Felipe Gutierrez 2024
F. Gutierrez, *Pro Spring Boot 3*, https://doi.org/10.1007/978-1-4842-9294-5_8

The Spring Framework provides an easy way to do testing by integrating the ApplicationContext in the tests. The Spring testing module offers several ways to use the ApplicationContext, programmatically and through annotations:

- BootstrapWith: A class-level annotation to configure how the Spring TestContext Framework is bootstrapped.

- @ContextConfiguration: Defines a class-level metadata to determine how to load and configure an ApplicationContext for integration tests. This is a must-have annotation for your classes because that's where the ApplicationContext loads all your bean definitions.

- @WebAppConfiguration: A class-level annotation to declare that the ApplicationContext that loads for an integration test should be a WebApplicationContext.

- @ActiveProfile: A class-level annotation to declare which bean definition profile(s) should be active when loading an ApplicationContext for an integration test.

- @TestPropertySource: A class-level annotation to configure the locations of properties files and inline properties to be added to the set of PropertySources in the Environment for an ApplicationContext loaded for an integration test.

- @DirtiesContext: Indicates that the underlying Spring ApplicationContext has been dirtied during the execution of a test (modified or corrupted, for example, by changing the state of a singleton bean) and should be closed.

The Spring Framework offers many more annotations, including @TestExecutionListeners, @Commit, @Rollback, @ BeforeTransaction, @AfterTransaction, @Sql, @SqlGroup, @SqlConfig, @Timed, @Repeat, @IfProfileValue, and so forth.

As you can see, there are a lot of choices when you test with the Spring Framework. Normally, you always use the @RunWith annotation that wires up all the test framework goodies. For example, the following code shows how you can do unit/integration testing using just Spring:

```
@RunWith(SpringRunner.class)
@ContextConfiguration({"/app-config.xml", "/test-data-access-config.xml"})

@ActiveProfiles("dev")
@Transactional
public class UsersTests {
    @Test
    public void userPersistenceTest(){
//...
}
}
```

Now, let's explore the Spring Boot set of features that enable you to create better unit, integration, and isolation tests (per layer) with ease.

Spring Boot Testing Framework

Spring Boot uses the power of the Spring Testing Framework by enhancing and adding new annotations and features that make testing easier for developers.

If you want to start using all the testing features provided by Spring Boot, you only need to add the `spring-boot-starter-test` dependency with scope test to your application. This dependency is already in place if you used the Spring Initializr (`https://start.spring.io`) to create your projects.

The `spring-boot-starter-test` dependency provides several test frameworks that play very well with all the Spring Boot testing features: Junit 5, AssertJ, Hamcrest, Mockito, JSONassert, JsonPath, and the Spring Test and Spring Boot Test utilities and integration support for Spring Boot applications. If you are using a different test framework, it likely will play very nicely with the Spring Boot Test module; you only need to include those dependencies manually.

Spring Boot provides the `@SpringBootTest` annotation that simplifies the way you can test Spring apps. Normally, with Spring testing, you are required to add several annotations to test a particular feature or functionality of your app, but not in Spring Boot. The `@SpringBootTest` annotation has parameters that are useful when testing a web app with the parameter `webEnvironment` which accepts values such as a `RANDOM_PORT` or `DEFINED_PORT`, `MOCK`, and `NONE`. The `@SpringBootTest` annotation also defines `properties` (properties that you want to test for different values), `args` (for arguments you normally passed in the command line, such @SpringBootTest(args = "--app. name=Users")), classes (a list of classes in which you declared your beans using the @Configuration annotation), and useMainMethod, with values such as ALWAYS and WHEN_AVAILABLE (meaning that it will use the main method of the application to the set the ApplicationContext to run the tests).

In the following sections I'll show you some of the main test classes from the two different projects that we have been developing so far.

Note   In this chapter we are going to use the code in the folder 08-testing. You have complete access to the code at the Apress website: https://www.apress.com/gp/services/source-code.

Testing Web Apps with a Mock Environment

The @SpringBootTest annotation by default uses a mock environment, meaning that it doesn't start the server and it's ready for testing web endpoints. To use this feature, we need to use the MockMvc class along with the @AutoConfigurationMockMvc annotation as a marker for the test class. This annotation will configure the MockMvc class and all its dependencies, such as filters, security (if any), and so forth. If we were using just Spring (no Spring Boot), we could use the MockMvc class, but we would need to do some extra steps because it relies on the WebApplicationContext. Fortunately, with Spring Boot, we only need to inject it with the @Autowired annotation.

Let's look at the Users App project in the test folder. The name of the class is UserMockMvcTests. See Listing 8-1.

Listing 8-1. src/test/java/apress/com/users/UserMockMvcTests.java

```
package com.apress.users;

import org.junit.jupiter.api.Test;
import org.springframework.beans.factory.annotation.Autowired;
import org.springframework.boot.test.autoconfigure.web.servlet.
AutoConfigureMockMvc;
import org.springframework.boot.test.context.SpringBootTest;
import org.springframework.test.context.ActiveProfiles;
```

```
import org.springframework.test.web.servlet.MockMvc;

import static org.hamcrest.Matchers.hasItem;
import static org.hamcrest.Matchers.hasSize;
import static org.springframework.test.web.servlet.request.
MockMvcRequestBuilders.get;
import static org.springframework.test.web.servlet.request.
MockMvcRequestBuilders.post;
import static org.springframework.test.web.servlet.result.
MockMvcResultMatchers.*;

@SpringBootTest
@AutoConfigureMockMvc
@ActiveProfiles("mockMvc")
public class UserMockMvcTests {

    @Autowired
    MockMvc mockMvc;

    @Test
    void createUserTests() throws Exception {
        String location = mockMvc.perform(post("/users")
                    .contentType("application/json")
                    .content("""
                    {
                        "email": "dummy@email.com",
                        "name": "Dummy",
                        "password": "aw2s0meR!",
                        "gravatarUrl": "https://www.gravatar.com/
                        avatar/fb651279f4712e209991e05610dfb03a?d
                        =wavatar",
                        "userRole": ["USER"],
                        "active": true
                    }
                    """))
                .andExpect(status().isCreated())
                .andExpect(header().exists("Location"))
                .andReturn().getResponse().getHeader("Location");
```

367

```
    mockMvc.perform(get(location))
            .andExpect(status().isOk())
            .andExpect(jsonPath("$.email").exists())
            .andExpect(jsonPath("$.active").value(true));
}

@Test
void getAllUsersTests() throws Exception {
    mockMvc.perform(get("/users"))
            .andExpect(status().isOk())
            .andExpect(jsonPath("$[0].name").value("Dummy"))
            .andExpect(jsonPath("$..active").value(hasItem(true)))
            .andExpect(jsonPath("$[*]").value(hasSize(1)));
}
}
```

Let's analyze the UserMockMvcTests class:

- **@SpringBootTest**: This annotation is essential for testing with Spring Boot. It provides the SpringBootContextLoader as the default ContextLoader. It searches for all the @Configuration classes, it allows custom Environment properties, and it can register either a TestRestTemplate (which we have been using in previous chapters when running our tests) or a WebTestClient (for Reactive apps, such as MongoDB). As I've already mentioned, it also accepts parameters where you can add the type of web environment—by default, MOCK–arguments, and properties.

- **@AutoConfigureMockMvc**: This annotation creates the MockMvc bean that can be injected to perform all the server-side testing, among other features, such as filters, security, etc.

- **@ActiveProfiles**: This annotation helps to run only the defined beans under the name set (in this case mockMvc). This is helpful when you have a lot of testing and you require certain behavior. This annotation will activate the mockMvc profile. Normally, in every configuration class, you can add the @Profile({"<profile-name>"}) annotation to specify which beans are needed for such profile(s).

- MockMvc: This bean class is the entry point for all the server-side testing, and it supports request builders and result matchers that can be combined with different test libraries such as Mockito, Hamcrest, and AssertJ, among others.

- MockMvcRequestBuilders: The MockMvc class has a perform method that accepts a request builder, and the MockMvcRequestBuilders class provides a fluent API where you can perform the get, post, put, etc. These request builders accept the URI in the form of a path (it's not necessary to add the complete URL), and you can build upon whatever request you need, like adding headers, content type, content, etc., and this is because it returns a MockHttpServletRequestBuilder class that brings a lot of customization for the request.

- Hamcrest and MockMvcResultMatchers: After calling the perform method (from MockMvc), you have access to a ResultsActions interface where you can add all the expectations on the result of the executed request; it includes the andExpect(ResultMatcher), andExpectAll(ResultMatcher...), andDo(ResultHandler), and andReturn() methods. The andReturn() method returns a MvcResult interface that is also a fluent API where you can get anything that brings the request (content, headers, etc.). And because some of these methods require a ResultMatcher interface, you can find implementation using the Hamcrest library (such as contains, hasSize, etc.). In our tests, we are using the jsonPath result matchers that allow us to interact with the JSON response.

It's important to mention that these tests need something extra that allows them to use the repositories, and we are going to talk about that in the following sections.

Using Mocking and Spying Beans

Spring Boot testing includes two annotations that allow you to mock Spring beans, which is helpful when some of these beans depend on external services, such a third-party REST endpoints, database connections, and so forth. If one of these services is unavailable, these annotations can help. Let's first look at the UserMockBeanTests class. See Listing 8-2.

Listing 8-2. src/test/java/apress/com/users/UserMockBeanTests.java

```
package com.apress.users;

import org.junit.jupiter.api.Test;
import org.mockito.Mockito;
import org.springframework.beans.factory.annotation.Autowired;
import org.springframework.boot.test.autoconfigure.web.servlet.
AutoConfigureMockMvc;
import org.springframework.boot.test.context.SpringBootTest;
import org.springframework.boot.test.mock.mockito.MockBean;
import org.springframework.http.MediaType;
import org.springframework.test.context.ActiveProfiles;
import org.springframework.test.web.servlet.MockMvc;

import static org.mockito.Mockito.*;
import static org.springframework.test.web.servlet.request.
MockMvcRequestBuilders.post;
import static org.springframework.test.web.servlet.result.
MockMvcResultMatchers.jsonPath;
import static org.springframework.test.web.servlet.result.
MockMvcResultMatchers.status;

@SpringBootTest
@AutoConfigureMockMvc
@ActiveProfiles("mockBean")
public class UserMockBeanTests {

    @Autowired
    private MockMvc mockMvc;

    @MockBean
    UserRepository userRepository;

    @Test
    void saveUsers() throws Exception {
        var user = UserBuilder.createUser()
                .withName("Dummy")
                .withEmail("dummy@email.com")
```

```
            .active()
            .withRoles(UserRole.USER)
            .withPassword("aw3sOm3R!")
            .build();

    when(userRepository.save(any())).thenReturn(user);

    mockMvc.perform(post("/users")
            .contentType(MediaType.APPLICATION_JSON)
            .content("""
    {
                    "email": "dummy@email.com",
                    "name": "Dummy",
                    "password": "aw2sOmeR!",
                    "gravatarUrl": "https://www.gravatar.com/
                    avatar/fb651279f4712e209991e05610dfb03a?d
                    =wavatar",
                    "userRole": ["USER"],
                    "active": true
        }
    """))
            .andExpect(status().isCreated())
            .andExpect(jsonPath("$.name").value(user.getName()))
            .andExpect(jsonPath("$.email").value(user.getEmail()))
            .andExpect(jsonPath("$.userRole").isArray())
            .andExpect(jsonPath("$.userRole[0]").value("USER"));

    verify(userRepository, times(1)).save(Mockito.any(User.class));
    }
}
```

The UserMockBeanTests class includes the following:

- @MockBean: This annotation mocks the object marked, enabling
  you to add the necessary behavior and outcome for that object. It
  replaces the object implementation, so it can be easily customized.

To use this bean, you need to use a framework such as Mockito. In our test in Listing 8-2, we are mocking the when() UserRepository, meaning that we don't need any connection or anything related to the database; basically, we are mocking its behavior.

- Mockito.*: The Mockito library help us to add the behavior to our mock bean (in this case, the UserRepository) with the thenReturn() methods. Then we can verify the behavior after we performed the call with the verify() method. Note that the Mockito library has a lot of useful fluent APIs. In our test, we are specifying that when the UserRepository uses the method save (with any value), it will return the user we specified, then at the end we verify that our mock bean was call once when saving the User object.

Remember, the @MockBean will replace the actual bean with a mock implementation that is easy to modify to get the desired behavior.

Next, let's go to the UserSpyBeanTests class, shown in Listing 8-3.

Listing 8-3. src/test/java/apress/com/users/UserSpyBeanTests.java

```
package com.apress.users;

import org.junit.jupiter.api.Test;
import org.springframework.beans.factory.annotation.Autowired;
import org.springframework.boot.test.autoconfigure.web.servlet.
AutoConfigureMockMvc;
import org.springframework.boot.test.context.SpringBootTest;
import org.springframework.boot.test.mock.mockito.SpyBean;
import org.springframework.test.context.ActiveProfiles;
import org.springframework.test.web.servlet.MockMvc;

import java.util.Arrays;
import java.util.List;

import static org.mockito.Mockito.doReturn;
import static org.mockito.Mockito.verify;
import static org.springframework.test.web.servlet.request.
MockMvcRequestBuilders.get;
```

```
import static org.springframework.test.web.servlet.result.
MockMvcResultMatchers.jsonPath;
import static org.springframework.test.web.servlet.result.
MockMvcResultMatchers.status;

@SpringBootTest
@AutoConfigureMockMvc
@ActiveProfiles("spyBean")
public class UserSpyBeanTests {

    @Autowired
    private MockMvc mockMvc;

    @SpyBean
    private UserRepository userRepository;

    @Test
    public void testGetAllUsers() throws Exception {

        List<User> mockUsers = Arrays.asList(
                UserBuilder.createUser()
                        .withName("Ximena")
                        .withEmail("ximena@email.com")
                        .build(),
                UserBuilder.createUser()
                        .withName("Norma")
                        .withEmail("norma@email.com")
                        .build()
        );
        doReturn(mockUsers).when(userRepository).findAll();

        mockMvc.perform(get("/users"))
                .andExpect(status().isOk())
                .andExpect(jsonPath("$[0].name").value("Ximena"))
                .andExpect(jsonPath("$[1].name").value("Norma"));

        verify(userRepository).findAll();
    }
}
```

Let's analyze the `UserSpyBeanTests` class:

> @SpyBean: This annotation creates a spy bean. A spy bean is
> similar to a mock object, but it retains the original behavior of the
> real bean. It allows you to intercept and verify method invocations
> and allows you to specify specific behavior for selected methods
> while keeping the rest of the methods intact. In our example in
> Listing 8-3 we are using the Mockito library to use the `doReturn()`
> method when the `UserRepository` instance is used, and when we
> call the `findAll()` method. At the end, we are verifying that the
> actual `findAll()` method was called.

Testcontainers

Testcontainers (`https://testcontainers.com/`) is an open source framework that
allows you to run some services in a docker container through unit testing frameworks.
Spring Boot now has the capability to use Testcontainers with ease, without any
configuration; you simply include two dependencies in your Maven or Gradle build
files: `org.springframework.boot:spring-boot-testcontainers` and
`org.testcontainers:junit-jupiter` (with `scope test` for Maven or
`testImplementation` for Gradle).

Previous chapters demonstrated the use of the `spring-boot-docker-compose`
feature, which enables to read a `docker-compose.yaml` file provided by you (the
developer) and allows you to run your app by creating the necessary environment for
your app. Recall that this feature works only when running the app, not when executing
the tests. So, the solution is Testcontainers!

With Testcontainers, Spring Boot introduces the following annotations:

- @Testcontainers: Starts and stops the containers; this annotation
  looks for every @Container annotation defined.

- @Container: Sets up all the necessary configuration that is required
  for the Testcontainers framework to initialize.

- @ServiceConnection: Takes care of creating the default connection details to be used with your application. And in this case, it will depend on which technology you will use. For example, in this chapter we are using Postgres for the Users App project and MongoDB for the My Retro App project, so it will be necessary to include the necessary Testcontainers dependency, `org.testcontainers:postgresql` or `org.testcontainers:mongodb`, respectively.

You will find out that with Testcontainers in Spring Boot, normally it will take some time to pull down the image (if it's not there) and start the testing. For this behavior, the Spring Boot team also created the `@RestartScope` annotation that allows you to recreate the container (keep it running) when your app is restarted if you are using Spring Boot Dev Tools. One of the main features of Dev Tools is that you can use it in your favorite IDE and it will restart your app in any new file modification when it's saved. The normal steps without Dev Tools are to modify your app, save it, and then either stop or restart the app; with Spring Boot Dev Tools, you don't need to do this, because it automatically restarts the app when the file is saved.

So, let's look at how to use Testcontainers with Spring Boot. The following snippet shows you how to use the `@Testcontainers`, `@Container`, and `@ServiceConnection` annotations and how to enable the PostgreSQL container:

```
@SpringBootTest
@Testcontainers
public class UserTests {

    @Container
    @ServiceConnection
    static PostgreSQLContainer<?> postgreSQLContainer =
        new PostgreSQLContainer<>("postgres:latest");

    // Your test here ...
}
```

In the following sections, we'll see this as part of the test.

Spring Boot Testing Slices

So far, we have seen integration testing because normally we require the whole server to do such tests, in some other chapters we used this a lot, and we use the `TestRestTemplate` to perform some calls to the web application using the `SpringBootTest.WebEnvironment.RANDOM_PORT` to get the port and such.

Spring Boot offers a way to test only the layers we need to test instead of testing the whole environment. So, suppose we only need to parse the JSON response and verify that the serialization was done correctly, the values of the field match, and so forth. With Spring Boot, we don't need the whole environment for that. The Spring Boot slice testing feature allows us to test every layer separately, including the controller layer, the data layer, and the domain layer.

- `@WebMvcTest`

  Purpose: Tests Spring MVC controllers in isolation.

  Includes: Web layer components (controllers, filters, view resolvers, etc.)

  Excludes: Service layer, repository layer, and other non-web components.

  Suitable For: Testing the behavior of controllers, request mappings, validation, and rendering views (if applicable).

- `@DataJpaTest`

  Purpose: Tests Spring Data JPA repositories in isolation.

  Includes: JPA repositories, entity classes, and related configuration.

  Excludes: Service layer, web layer, and other non-JPA components.

  Suitable For: Testing repository CRUD operations, queries, and custom repository methods.

- **@JdbcTest**

  Purpose: Tests data access code that uses plain JDBC (without Spring Data JPA).

  Includes: DataSource configuration, JDBC templates, and SQL scripts.

  Excludes: JPA repositories, web layer, and other non-JDBC components.

  Suitable For: Testing low-level JDBC operations, SQL scripts, and data access logic not using JPA.

- **@JsonTest**

  Purpose: Tests JSON serialization and deserialization.

  Includes: Jackson or Gson configurations and custom serializers/deserializers.

  Excludes: Web layer, data access layers, and other non-JSON components.

  Suitable For: Verifying correct serialization and deserialization of your objects to and from JSON.

- **@RestClientTest**

  Purpose: Tests Spring's RestTemplate or WebClient based REST clients.

  Includes: REST client beans, error handling configurations, and related components.

  Excludes: Web layer (server-side), data access layers, and other non-REST client components.

  Suitable For: Verifying REST client configurations, request building, response handling, and error scenarios.

- @DataMongoTest

  Purpose: Tests Spring Data MongoDB repositories in isolation.

  Includes: MongoDB repositories, entity classes, and related configuration.

  Excludes: Web layer, JPA components, and other non-MongoDB components.

  Suitable For: Testing repository operations with MongoDB, queries, and custom repository methods.

- @SpringBootTest (Special Case)

  Purpose: Not strictly a "slice," but provides a way to test the full application context.

  Includes: All Spring beans and configurations.

  Suitable For: End-to-end tests or integration tests that need the complete application context.

Let's start with the @JsonTest annotation.

@JsonTest

@JsonTest is an annotation that allows you to test your JSON serialization and deserialization. This annotation also auto-configures the right mapper support if it finds any of these libraries in your classpath: Jackson, Gson, or Jsonb.

Listing 8-4 shows the UserJsonTest class.

Listing 8-4. src/test/java/apress/com/users/UserJsonTest.java

```
package com.apress.users;

import jakarta.validation.ConstraintViolationException;
import jakarta.validation.Validation;
import org.junit.jupiter.api.Test;
import org.springframework.beans.factory.annotation.Autowired;
import org.springframework.boot.test.autoconfigure.json.JsonTest;
import org.springframework.boot.test.json.JacksonTester;
import org.springframework.boot.test.json.JsonContent;
```

```java
import java.io.IOException;

import static org.assertj.core.api.Assertions.assertThat;
import static org.assertj.core.api.Assertions.assertThatExceptionOfType;
import static org.junit.jupiter.api.Assertions.assertThrows;

@JsonTest
public class UserJsonTests {

    @Autowired
    private JacksonTester<User> jacksonTester;

    @Test
    void serializeUserJsonTest() throws IOException{
        User user = UserBuilder.createUser(Validation.
        buildDefaultValidatorFactory().getValidator())
                .withEmail("dummy@email.com")
                .withPassword("aw2sOme")
                .withName("Dummy")
                .withRoles(UserRole.USER)
                .active().build();

        JsonContent<User> json =  jacksonTester.write(user);

        assertThat(json).extractingJsonPathValue("$.email").
        isEqualTo("dummy@email.com");
        assertThat(json).extractingJsonPathArrayValue("$.userRole").size().
        isEqualTo(1);
        assertThat(json).extractingJsonPathBooleanValue("$.active").
        isTrue();
        assertThat(json).extractingJsonPathValue("$.gravatarUrl").
        isNotNull();
        assertThat(json).extractingJsonPathValue("$.gravatarUrl").
        isEqualTo(UserGravatar.getGravatarUrlFromEmail(user.getEmail())));
    }
```

```java
@Test
void serializeUserJsonFileTest() throws IOException{
    User user = UserBuilder.createUser(Validation.
    buildDefaultValidatorFactory().getValidator())
            .withEmail("dummy@email.com")
            .withPassword("aw2s0me")
            .withName("Dummy")
            .withRoles(UserRole.USER)
            .active().build();

    System.out.println(user);

    JsonContent<User> json =  jacksonTester.write(user);

            assertThat(json).isEqualToJson("user.json");

}

@Test
void deserializeUserJsonTest() throws Exception{
    String userJson = """
            {
               "email": "dummy@email.com",
               "name": "Dummy",
               "password": "aw2s0me",
               "userRole": ["USER"],
               "active": true
            }
            """;
    User user = this.jacksonTester.parseObject(userJson);

    assertThat(user.getEmail()).isEqualTo("dummy@email.com");
    assertThat(user.getPassword()).isEqualTo("aw2s0me");
    assertThat(user.isActive()).isTrue();

}

@Test
void userValidationTest(){
    assertThatExceptionOfType(ConstraintViolationException.class)
```

```
        .isThrownBy( () -> UserBuilder.createUser(Validation.
        buildDefaultValidatorFactory().getValidator())
                .withEmail("dummy@email.com")
                .withName("Dummy")
                .withRoles(UserRole.USER)
                .active().build());

// Junit 5
Exception exception = assertThrows(ConstraintViolationException.
class, () -> {
    UserBuilder.createUser(Validation.
    buildDefaultValidatorFactory().getValidator())
            .withName("Dummy")
            .withRoles(UserRole.USER)
            .active().build();
});

String expectedMessage = "email: Email can not be empty";
assertThat(exception.getMessage()).contains(expectedMessage);

    }
}
```

The UserJsonTest class includes the following:

- @JsonTest: This annotation marks this class as only JSON serialization. It auto-configures the test based on what is in the classpath on what dependency library are you using, either Jackson (comes with spring-boot-starter-web as default), Jsonb, or Gson.

- JacksonTester: This class takes care of the *serialization* and *deserialization* of the domain class. It uses the Jackson library (the default) and uses the ObjectMapper class to do the serialization.

- JsonContent: This class includes the content from a JSON tester, which is useful to get the values from the serialization.

- **UserBuilder**: This class receives a `Validation` class that helps to inspect and validate the values of fields marked with `@NotBlank` or `@NotNull` annotations.

- **Validation**: This class belongs to the Jakarta validation package that helps to review and validate the fields marked with annotations such as @NotBlank, @NoNull, etc.

- **assertThat**: We are using the AssertJ package to use the assertations for our serialization and deserialization in this class.

- **assertThatExceptionOfType**: This class helps to identify the type of any error or exception thrown during testing. In Listing 8-4, we are creating a `User` class that doesn't comply with the `@NotBlank` annotation, and because of that, this class identifies it as a `ConstraintViolationException` exception.

- **assertThrows**: And we can do the same with this `assertThrows` call. Different way to accomplish the same asserts.

As you can see, doing tests about your domain gets simpler that even, more waiting for some response or a particular service, you can test your own domain schemas by doing serialization and deserialization of your classes. Also, note that we are allowed to test against JSON files, in this case the `user.json` file. This file must be placed in the `src/test/resources/com/apress/users` folder in order to be picked up by the test framework.

@WebMvcTest

If we can do stand-alone domain/schemas tests, can we do them over our web controllers? Yes, we can! We can use the `@WebMvcTest` annotation to test only the web endpoint—our controllers. The `@WebMvcTest` annotation auto-configures all the Spring MVC infrastructure, but it will be limited just to the `@RestController`, annotated classes, `Filter` interface implementations, and all other web-related classes, and by default it sets the `webEnvironment` to `MOCK`, which means that this annotation inherits from the `@AutoConfigureMockMvc`, which in turn means that you can use the `MockMvc` class.

Let's look at the `UserControllerTests` class. See Listing 8-5.

Listing 8-5. src/main/java/apress/com/users/UserControllerTests.java

```
package com.apress.users;

import com.fasterxml.jackson.databind.ObjectMapper;
import org.junit.jupiter.api.Test;
import org.springframework.beans.factory.annotation.Autowired;
import org.springframework.boot.test.autoconfigure.web.servlet.WebMvcTest;
import org.springframework.boot.test.mock.mockito.MockBean;
import org.springframework.http.MediaType;
import org.springframework.test.web.servlet.MockMvc;

import java.util.Arrays;
import java.util.Optional;

import static org.mockito.Mockito.doNothing;
import static org.mockito.Mockito.when;
import static org.springframework.test.web.servlet.request.
MockMvcRequestBuilders.*;
import static org.springframework.test.web.servlet.result.
MockMvcResultHandlers.print;
import static org.springframework.test.web.servlet.result.
MockMvcResultMatchers.*;

@WebMvcTest(controllers = { UsersController.class })
public class UserControllerTests {

  @Autowired
    private MockMvc mockMvc;

    @MockBean
    private UserRepository userRepository;

    @Test
    void getAllUsersTest() throws Exception {
        when(userRepository.findAll()).thenReturn(Arrays.asList(
                UserBuilder.createUser()
```

```
                        .withName("Ximena")
                        .withEmail("ximena@email.com")
                        .active()
                        .withRoles(UserRole.USER, UserRole.ADMIN)
                        .withPassword("aw3s0m3R!")
                        .build(),
                UserBuilder.createUser()
                        .withName("Norma")
                        .withEmail("norma@email.com")
                        .active()
                        .withRoles(UserRole.USER)
                        .withPassword("aw3s0m3R!")
                        .build()
    ));

    mockMvc.perform(get("/users"))
            .andDo(print())
            .andExpect(status().isOk())
  .andExpect(content().contentType(MediaType.APPLICATION_JSON))
            .andExpect(jsonPath("$[0].active").value(true));
}

@Test
void newUserTest() throws Exception {
    User user = UserBuilder.createUser()
            .withName("Dummy")
            .withEmail("dummy@email.com")
            .active()
            .withRoles(UserRole.USER, UserRole.ADMIN)
            .withPassword("aw3s0m3R!")
            .build();
    when(userRepository.save(user)).thenReturn(user);

    mockMvc.perform(post("/users")
                    .content(toJson(user))
                    .contentType(MediaType.APPLICATION_JSON))
            .andDo(print())
```

```
                .andExpect(status().isCreated())
                .andExpect(content().contentType(MediaType.
                APPLICATION_JSON))
                .andExpect(jsonPath("$.email").value("dummy@email.com"));
}

@Test
void findUserByEmailTest() throws Exception {
    User user = UserBuilder.createUser()
                .withName("Dummy")
                .withEmail("dummy@email.com")
                .active()
                .withRoles(UserRole.USER, UserRole.ADMIN)
                .withPassword("aw3sOm3R!")
                .build();
    when(userRepository.findById(user.getEmail())).thenReturn(Optional.
    of(user));

    mockMvc.perform(get("/users/{email}",user.getEmail())
                .contentType(MediaType.APPLICATION_JSON))
                .andDo(print())
                .andExpect(status().isOk())
                .andExpect(content().contentType(MediaType.
                APPLICATION_JSON))
                .andExpect(jsonPath("$.email").value("dummy@email.com"));
}

@Test
void deleteUserByEmailTest() throws Exception{
    User user = UserBuilder.createUser()
                .withEmail("dummy@email.com")
                .build();
    doNothing().when(userRepository).deleteById(user.getEmail());

    mockMvc.perform(delete("/users/{email}",user.getEmail()))
                .andExpect(status().isNoContent());

}
```

```
    private static String toJson(final Object obj) {
        try {
            return new ObjectMapper().writeValueAsString(obj);
        } catch (Exception e) {
            throw new RuntimeException(e);
        }
    }
}
```

The `UserControllerTests` class includes the following:

- `@WebMvcvTest`: This annotation inherits the `@AutoConfiguraMockMvc` annotation, so it's ready to use the `MockMvc` class, which also means that the `webEnvironment` is a `MOCK`. Thanks to this, we can test only our web controllers without starting up the complete server to do some testing. You can declare more controllers in this annotation.

- `MockMvc`: As you already know, this class is the entry point for the server side for the Spring MVC test support. In this case, it allows us to perform HTTP method requests to our controllers.

- `@MockBean`: This annotation will mock the bean behavior. This is helpful when you don't want to wait for an external service to be ready, because you can mock the behavior. In this case, we are mocking the `UserRepository`.

- `Mockito.*`: We are using the `when().thenReturn` to prepare our call and then perform the calls and expect the results. Also, we are using the `doNothing().when()` calls. As you can see, the Mockito library offer a nice fluent API that you apply for these scenarios.

Before you continue to the next section, take time to review the `@WebMvcTest` annotation and `UserControllerTests` class in depth.

@DataJpaTest

With this annotation, you can test everything about the JPA technology. It auto-configures all the repositories and the needed entities for you to test without starting a web server or any other dependency; in other words, `@DataJpaTest` is dedicated to testing the data layer. It also sets the `spring.jpa.show-sql` property to `true` so that you can see what the queries are when the tests are been executed. Let's look at the `UserJpaRepositoryTests` class. See Listing 8-6.

Listing 8-6. src/test/java/apress/com/users/UserJpaRepositoryTests.java

```
package com.apress.users;

import org.junit.jupiter.api.Test;
import org.springframework.beans.factory.annotation.Autowired;
import org.springframework.boot.test.autoconfigure.jdbc.
AutoConfigureTestDatabase;
import org.springframework.boot.test.autoconfigure.orm.jpa.DataJpaTest;
import org.springframework.boot.testcontainers.service.connection.
ServiceConnection;
import org.springframework.context.annotation.Import;
import org.testcontainers.containers.PostgreSQLContainer;
import org.testcontainers.junit.jupiter.Container;
import org.testcontainers.junit.jupiter.Testcontainers;

import static org.assertj.core.api.Assertions.assertThat;

@Import({UserConfiguration.class})
@Testcontainers
@AutoConfigureTestDatabase(replace = AutoConfigureTestDatabase.Replace.NONE)
@DataJpaTest
public class UserJpaRepositoryTests {

    @Container
    @ServiceConnection
    static PostgreSQLContainer<?> postgreSQLContainer = new PostgreSQL
    Container<>("postgres:latest");

    @Autowired
    UserRepository userRepository;

    @Test
    void findAllTest(){
        var expectedUsers = userRepository.findAll();
        assertThat(expectedUsers).isNotEmpty();
        assertThat(expectedUsers).isInstanceOf(Iterable.class);
```

```java
        assertThat(expectedUsers).element(0).isInstanceOf(User.class);
        assertThat(expectedUsers).element(0).matches( user -> user.
        isActive());
    }

    @Test
    void saveTest(){
        var dummyUser =  UserBuilder.createUser()
                .withName("Dummy")
                .withEmail("dummy@email.com")
                .active()
                .withRoles(UserRole.INFO)
                .withPassword("aw3sOm3R!")
                .build();
        var expectedUser = userRepository.save(dummyUser);
        assertThat(expectedUser).isNotNull();
        assertThat(expectedUser).isInstanceOf(User.class);
        assertThat(expectedUser).hasNoNullFieldsOrProperties();
        assertThat(expectedUser.isActive()).isTrue();
    }

    @Test
    void findByIdTest(){
        var expectedUser = userRepository.findById("norma@email.com");

        assertThat(expectedUser).isNotNull();
        assertThat(expectedUser.get()).isInstanceOf(User.class);
        assertThat(expectedUser.get().isActive()).isTrue();
        assertThat(expectedUser.get().getName()).isEqualTo("Norma");
    }

    @Test
    void deleteByIdTest(){
        var expectedUser = userRepository.findById("ximena@email.com");
        assertThat(expectedUser).isNotNull();
        assertThat(expectedUser.get()).isInstanceOf(User.class);
        assertThat(expectedUser.get().isActive()).isTrue();
        assertThat(expectedUser.get().getName()).isEqualTo("Ximena");
```

```
userRepository.deleteById("ximena@email.com");

expectedUser = userRepository.findById("ximena@email.com");
assertThat(expectedUser).isNotNull();
assertThat(expectedUser).isEmpty();
    }

}
```

The UserJpaRepositoryTests class includes the following:

- @DataJpaTest: This annotation auto-configures everything related to the JPA, from the repositories to the EntityManager (required as part of the JPA implementation), and even inherits from the @Transactional annotation, meaning that you are making sure your tests are completely transactional. With this annotation, you don't need any web layer to perform any of this action against your data persistence.

- @Import: One of the cool features of Spring is that you can import specific configurations with this annotation. In this case, we are importing the UserConfiguration where we are adding some users.

- @Testcontainers: As previously described, this annotation starts and stops the containers; this annotation looks for every @Container annotation defined. In this case we are marking this class as Testcontainers, and this will look for any @Container annotation and set up the environment for running the container image specified.

- @AutoConfigureTestDatabase: When testing the data layer, normally it is better to use an in-memory database, because it's fast and efficient; but sometimes you are required to do testing over a real database, and in this case the @AutoConfigureTestDatabase annotation will not use the embedded auto-configuration, but instead will follow what you specified with the @Container and @ServerConnection.

- `@Container`, `@ServerConnection`, `PostgreSQLContainer`: You already know these annotations. All these annotations will auto-configure the start and stop of the PostgreSQL container, the `DataSource` interface that required connection parameters such as `username`, `password`, `url`, `dialect`, `driverClass`, etc.

- `Assertions.*`: In this class we are using AssertJ, which comes with a fluent API to do assertions.

Spring Boot not only has support for the JPA, but also has the following annotations (and many more) that follow the same pattern, isolate your tests based on the technology you need to test for:

- `@JdbcTest`: For tests related to `JdbcTemplate` programming, covered in Chapters 4 and 5

- `@DataJdbcTest`: For tests on the Spring Data repositories

- `@JooqTest`: For all the jOOQ-related tests

- `@DataNeo4jTest`: For Spring Data Neo4J–related tests

- `@DataRedisTest`: For all Spring Data Redis–related tests

- `@DataLdapTest`: For LDAP tests

@WebFluxTest

If you have any `Reactive` web app with Spring Boot WebFlux, then you can test it using the `@WebFluxTest` annotation, which will auto-configure all the web-related beans and annotations for a WebFlux application. It auto-configures the `WebTestClient` interface to perform any exchange/request to the WebFlux endpoints, and it plays well with the `@MockBean` annotation for any services or repositories.

Let's look at the `RetroBoardWebFluxTests` class. See Listing 8-7.

Listing 8-7.   src/test/java/apress/com/myretro/ RetroBoardWebFluxTests.java

```
package com.apress.myretro;

import com.apress.myretro.board.Card;
import com.apress.myretro.board.CardType;
import com.apress.myretro.board.RetroBoard;
```

```
import com.apress.myretro.service.RetroBoardService;
import com.apress.myretro.web.RetroBoardController;
import org.junit.jupiter.api.Test;
import org.mockito.Mockito;
import org.springframework.beans.factory.annotation.Autowired;
import org.springframework.boot.test.autoconfigure.web.reactive.
WebFluxTest;
import org.springframework.boot.test.mock.mockito.MockBean;
import org.springframework.http.HttpHeaders;
import org.springframework.http.MediaType;
import org.springframework.test.web.reactive.server.WebTestClient;
import org.springframework.web.reactive.function.BodyInserters;
import reactor.core.publisher.Flux;
import reactor.core.publisher.Mono;

import java.util.Arrays;
import java.util.UUID;

@WebFluxTest(controllers = {RetroBoardController.class})
public class RetroBoardWebFluxTests {

    @MockBean
    RetroBoardService retroBoardService;

    @Autowired
    private WebTestClient webClient;

    @Test
    void getAllRetroBoardTest(){

        Mockito.when(retroBoardService.findAll()).thenReturn(Flux.just(
                new RetroBoard(UUID.randomUUID(),"Simple Retro",
                Arrays.asList(
                        new Card(UUID.randomUUID(),"Happy to be here",
                        CardType.HAPPY),
                        new Card(UUID.randomUUID(),"Meetings everywhere",
                        CardType.SAD),
                        new Card(UUID.randomUUID(),"Vacations?",
                        CardType.MEH),
```

```
                new Card(UUID.randomUUID(),"Awesome Discounts",
                CardType.HAPPY),
                new Card(UUID.randomUUID(),"Missed my train",
                CardType.SAD)
        ))
    ));

    webClient.get()
            .uri("/retros")
            .accept(MediaType.APPLICATION_JSON)
            .exchange()
            .expectStatus().isOk()
            .expectBody().jsonPath("$[0].name").isEqualTo("Simple
            Retro");

    Mockito.verify(retroBoardService,Mockito.times(1)).findAll();
}

@Test
void findRetroBoardByIdTest(){
    UUID uuid = UUID.randomUUID();
    Mockito.when(retroBoardService.findById(uuid)).
    thenReturn(Mono.just(
            new RetroBoard(uuid,"Simple Retro", Arrays.asList(
                    new Card(UUID.randomUUID(),"Happy to be here",
                    CardType.HAPPY),
                    new Card(UUID.randomUUID(),"Meetings everywhere",
                    CardType.SAD),
                    new Card(UUID.randomUUID(),"Vacations?",
                    CardType.MEH),
                    new Card(UUID.randomUUID(),"Awesome Discounts",
                    CardType.HAPPY),
                    new Card(UUID.randomUUID(),"Missed my train",
                    CardType.SAD)
            ))
    ));
```

```java
    webClient.get()
            .uri("/retros/{uuid}",uuid.toString())
            .header(HttpHeaders.ACCEPT, MediaType.APPLICATION_
            JSON_VALUE)
            .exchange()
            .expectStatus().isOk()
            .expectBody(RetroBoard.class);

    Mockito.verify(retroBoardService,Mockito.times(1)).findById(uuid);
}

@Test
void saveRetroBoardTest(){
    RetroBoard retroBoard = new RetroBoard();
    retroBoard.setName("Simple Retro");

    Mockito.when(retroBoardService.save(retroBoard))
            .thenReturn(Mono.just(retroBoard));

    webClient.post()
            .uri("/retros")
            .contentType(MediaType.APPLICATION_JSON)
            .body(BodyInserters.fromValue(retroBoard))
            .exchange()
            .expectStatus().isOk();

    Mockito.verify(retroBoardService,Mockito.times(1)).
    save(retroBoard);
}

@Test
void deleteRetroBoardTest(){
    UUID uuid = UUID.randomUUID();
    Mockito.when(retroBoardService.delete(uuid)).thenReturn(Mono.
    empty());

    webClient.delete()
            .uri("/retros/{uuid}",uuid.toString())
```

```
          .exchange()
          .expectStatus().isOk();

     Mockito.verify(retroBoardService,Mockito.times(1)).delete(uuid);
  }

}
```

The RetroBoardWebFluxTests class includes the following:

- @WebFluxTests: This annotation configures everything related to a WebFlux application, looking for @Controller, Filter, etc. And you can add the controller you want to test. In this case, it is the RetroBoardController class. See that this test isolates from the data layer, only testing the web layer.

- WebTestClient: Recall from Chapter 7 that this class is used for testing HTTP and WebFlux endpoints and returns mocks of the response, making it easier to do the assertions and test our classes.

- Mockito.*: Again, we are using the Mockito library not only to prepare our call but also to verify that the call was executed the times we said we did.

@DataMongoTest

Similar to its support for SQL testing, Spring Boot provides testing support for NoSQL databases, such as the @DataMongoTest annotation to test only the MongoDB data layer.

Let's look at the RetroBoardMongoTests class. See Listing 8-8.

Listing 8-8. src/main/java/apress/com/myretro/RetroBoardMongoTests.java

```
package com.apress.myretro;

import com.apress.myretro.board.RetroBoard;
import com.apress.myretro.persistence.RetroBoardRepository;
import org.junit.jupiter.api.Test;
import org.springframework.beans.factory.annotation.Autowired;
import org.springframework.boot.test.autoconfigure.data.mongo.
DataMongoTest;
```

```
import org.springframework.test.context.ActiveProfiles;
import org.springframework.test.context.DynamicPropertyRegistry;
import org.springframework.test.context.DynamicPropertySource;
import org.testcontainers.containers.MongoDBContainer;
import org.testcontainers.junit.jupiter.Container;
import reactor.core.publisher.Mono;
import reactor.test.StepVerifier;

import java.util.UUID;

import static org.assertj.core.api.Assertions.assertThat;

@ActiveProfiles("mongoTest")
@DataMongoTest
public class RetroBoardMongoTests {

    @Container
    static MongoDBContainer mongoDBContainer = new MongoDBContainer("mongo
    :latest");

    static {
        mongoDBContainer.start();
    }

    @DynamicPropertySource
    static void setProperties(DynamicPropertyRegistry registry) {
        registry.add("spring.data.mongodb.uri", mongoDBContainer::getRepli
        caSetUrl);
    }

    @Autowired
    RetroBoardRepository retroBoardRepository;

    @Test
    void saveRetroTest(){
        var name = "Spring Boot 3 Retro";
        RetroBoard retroBoard = new RetroBoard();
        retroBoard.setId(UUID.randomUUID());
        retroBoard.setName(name);
```

```java
        var retroBoardResult = this.retroBoardRepository.
        insert(retroBoard).block();

        assertThat(retroBoardResult).isNotNull();
        assertThat(retroBoardResult.getId()).isInstanceOf(UUID.class);
        assertThat(retroBoardResult.getName()).isEqualTo(name);
    }

    @Test
    void findRetroBoardById(){
        RetroBoard retroBoard = new RetroBoard();
        retroBoard.setId(UUID.randomUUID());
        retroBoard.setName("Migration Retro");

        var retroBoardResult = this.retroBoardRepository.
        insert(retroBoard).block();
        assertThat(retroBoardResult).isNotNull();
        assertThat(retroBoardResult.getId()).isInstanceOf(UUID.class);

        Mono<RetroBoard> retroBoardMono = this.retroBoardRepository.
        findById(retroBoardResult.getId());

        StepVerifier
                .create(retroBoardMono)
                .assertNext( retro -> {
                    assertThat(retro).isNotNull();
                    assertThat(retro).isInstanceOf(RetroBoard.class);
                    assertThat(retro.getName()).isEqualTo("Migration
                    Retro");
                })
                .expectComplete()
                .verify();
    }
}
```

The `RetroBoardMongoTests` class includes the following:

- `@DataMongoTest`: This annotation auto-configures all the mongo layer, where it should find all related to Mongo data layer, from domain classes to repositories.

- `@ActiveProfiles`: You know this annotation will set the profile for a `mongoTest`. Useful to isolate tests.

- `@Container`: You also know this annotation, which will start and stop the container. In this case we statically declared the MongoDB container and omitted the `@Testcontainers` annotation, simply to demonstrate another way to achieve the same result.

- `@DynamcPropertiesSource`: With the `@ServiceConnection` annotation (introduced earlier in the chapter), the test will set up with the right mongo connection properties (the defaults), but in this case we are using another approach, a dynamic way to set the connection properties using the `@DynamicPropertiesSource` annotation. With this annotation, you can override all the defaults.

Using Testcontainers to Run Your Spring Boot Applications

So far, we have been using the `spring-boot-docker-compose` dependency to run the apps and the `spring-boot-testcontainers` dependency only for testing. With Testcontainers, Spring Boot provides a way to run your app without the need to use the `docker-compose,` directly from a container. How? It's very easy to do. Open the `RetroBoardTestConfiguration` class. See Listing 8-9.

Listing 8-9. src/main/java/apress/com/myretro/
RetroBoardTestConfiguration.java

```
package com.apress.myretro;

import org.springframework.boot.SpringApplication;
import org.springframework.boot.devtools.restart.RestartScope;
import org.springframework.boot.testcontainers.service.connection.
ServiceConnection;
import org.springframework.context.annotation.Bean;
import org.springframework.context.annotation.Configuration;
import org.springframework.context.annotation.Profile;
import org.testcontainers.containers.MongoDBContainer;

@Profile({"!mongoTest"})
@Configuration
public class RetroBoardTestConfiguration {

        @Bean
        @RestartScope
        @ServiceConnection
        public MongoDBContainer mongoDBContainer(){
                return new MongoDBContainer("mongo:latest");
        }

        public static void main(String[] args) {
                SpringApplication.from(MyretroApplication::main).run(args);
        }
}
```

Let's analyze the RetroBoardTestConfiguration class:

- @RestartScope: Every time run your application the container
  resttart, and one way to avoid this situation is to use the spring-
  boot-devtools dependency, then using the @RestartScope will
  prevent of a container to restart for every test method you have
  in your Tests. The time required to run the tests will be shorter!
  because you don't need to wait for the container to restart.

- `@ServiceConnection`: As you already know, this annotation sets up all the connection parameters from the container you are trying to run, so you don't need to worry about these parameters.

- `SpringApplication.from`: This is very important! Notice that we are using a main method (yes, another entry point for our application), and in this case, instead of using `SpringApplication.run()` method, we are using the **`SpringApplication.from()`** method that is pointing to our main class (in this case, **`MyretroApplication::main`**).

Summary

In this chapter you learned all about the unit and integration testing support in Spring Boot. You discovered that Spring Boot testing is based on the Spring Framework Testing, and that Spring Boot auto-configuration helps to configure a lot of the testing.

You also learned that Spring Boot testing includes slice testing, which enables you to test all the layers in isolation and by technology—from your domain classes with the `@JsonTest` annotation, to the web controllers with `@WebMvcTest`, all the way to the data layer with `@DataJpaTest` annotations.

This chapter also introduced you to Testcontainers and how it can help not only for testing but also for running your application. You saw many ways to do testing thanks to the Spring Boot Testing Framework that offers the choice of AssertJ, Mockito, Hamcrest, and many more libraries for your tests.

PART II

More on Spring Boot 3

CHAPTER 9

Spring Boot Security

In this chapter we are going to review how Spring Security works and how you can create secure apps with Spring Boot. Of course, Spring Boot helps to auto-configure all the necessary infrastructure that you need when implementing security for your applications.

Spring Security

The Spring Security framework provides *authentication*, *authorization*, and ways to prevent common attacks such cross-site request forgery (CSRF) and cross-origin resource sharing (CORS), among others. It also supports integration with different technologies, such as Spring Data, Cryptography, Concurrency, Spring MVC, Spring Web Flux, WebSockets, Observability, and much more. The Spring Security team have been working very hard to create ways to secure your applications with ease, such as by simplifying how you expose security to your code and, recently, by launching Spring Authorization Server, a framework that supports implementation of OAuth 2 and similar specifications.

One of the most important features of the Spring Security framework is that implements and makes use of the *chain-of-responsibility design pattern*, which allows you to easily add your own custom logic for securing your apps.

Before we get into more details, let's review the main Spring Security architecture, depicted in Figure 9-1, to understand how it works and how you can use it to secure your apps.

© Felipe Gutierrez 2024
F. Gutierrez, *Pro Spring Boot 3*, https://doi.org/10.1007/978-1-4842-9294-5_9

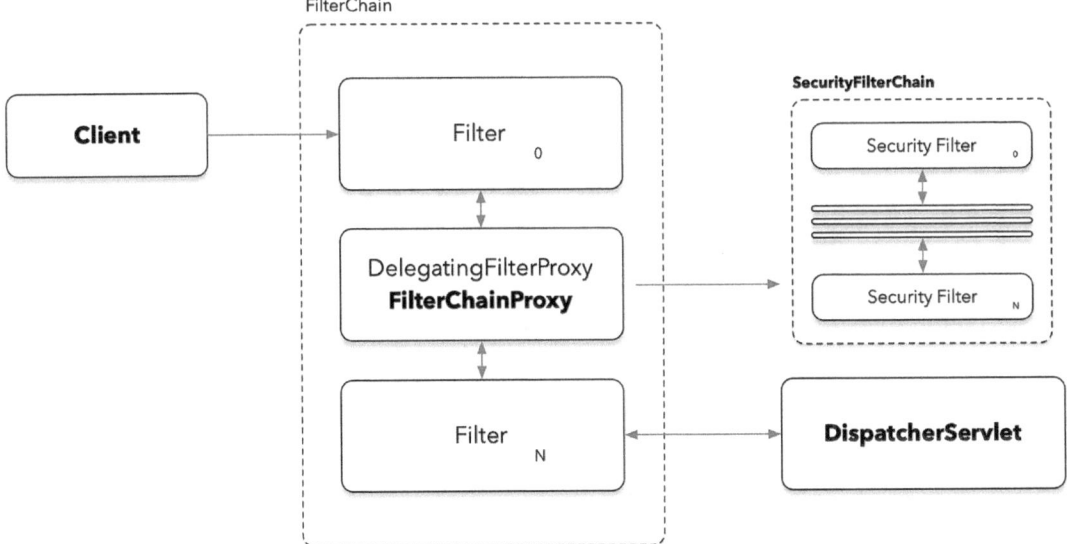

Figure 9-1. *Spring Security architecture*

Let's review each part of the Spring Security architecture:

- Client: This is any client, from a web-based browser to any device that can use HTTP, WebSocket, etc.

- FilterChain: The Spring Security servlet is based on servlet filters, so when a client makes a request, the Spring container will create a FilterChain that contains Filter instances and the servlet instance (DispatcherServlet) to accept any incoming requests (HttpServletRequest) and returns a response (HttpServletResponse). You will have multiple filters with different purposes not only for your security but for your own business logic.

- DelegatingFilterProxy > FilterChainProxy: Spring provides a Filter named DelegatingFilterProxy that delegates an instance of FilterChainProxy, which is the main class that Spring Security provides not only to apply default security features to your application but also to add your own SecurityFilterChain with your own custom rules for authentication and authorization.

- SecurityFilterChain: The SecurityFilterChain interface is used by the FilterChainProxy class and is where you specify your own custom rules for authentication and authrorization.

This is where the Spring Security `Filter` will create the necessary instances that allow your custom security to be placed. The `SecurityFilterChain` is where you provide what to secure by HTTP method (`GET`, `POST`, `DELETE`, `PATCH`, etc.), by URL pattern (e.g., `/users/**`, `/retros/admin/**`, etc.), and much more. You can add as many filters as you need. In fact, there are already some defaults, such as `CsrfFilter` (provided with the `HttpSecurity.csrf` call), `BasicAuthenticationFilter` (provided with `HttpSecurity.httpBasic`), and `AuthorizationFilter` (provided when the `HttpSecurity.authorizationHttpRequests` is called). We are going to revisit these filters very soon and see how to configure them. Another important filter in the `SecurityFilterChain` is the `ExceptionTranslationFilter` management; it allows the translation of the `AccessDeniedException` and `AuthenticationException` responses.

- `DispatcherServlet`: is the core component that handles all incoming web requests. It acts as the "front controller," receiving requests, determining the appropriate handler (controller), and delegating request processing to that handler. Think of it as the central traffic controller directing incoming web requests to their correct destinations within your Spring application. When a web request arrives, Spring Security's filters intercept it before it reaches the `DispatcherServlet`. These filters handle authentication (verifying user identity) and authorization (checking permissions), ensuring that only authorized requests are forwarded to the `DispatcherServlet` for further processing. Once security checks pass, the `DispatcherServlet` takes over, finding the correct controller method to handle the request and generating the response.

Now that you know a little more about how Spring Security helps to secure your applications, let's start with the Users App and My Retro App with Spring Boot.

Spring Security with Spring Boot

In this chapter we will cover the MVC (Servlet) and Reactive integrations with our two projects and how they will interact with each other. This section provides a brief overview of how Spring Security works within Spring Boot and what you need to do if you want to add security to an existing application.

Start by opening your `build.gradle` file and adding the security dependencies as shown in the following snippet:

build.gradle - snippet

```
// ...
implementation 'org.springframework.boot:spring-boot-starter-security'
testImplementation 'org.springframework.security:spring-security-test'
//...
```

When Spring Boot runs, it will find these dependencies and auto-configure everything necessary to secure your apps. If there is no custom configuration (no `SecurityFilterChain` declared), the auto-configuration provides several default beans that will make every access to your application secure, requiring a username and password. By default, Spring Security auto-configuration creates the `UserDetailsService` bean that provides the user details for authentication; it generates the username `user` and randomly generates a password that will be printed out in the console logs, something similar to the following snippet:

```
...
Using generated security password: 2a569843-122a-4559-a245-60f5ab2b6c51
...
```

Also, the auto-configuration creates the `HttpSecurity` bean that, by default, requires all requests to be authenticated, and it also sets the `PasswordEncoder` to use the BCrypt (algorithm) password encoder as the default. So, if you try to access your application either by using an external client or through a web browser, you will be prompted for a username (`user`) and password (from the console output).

This only covers authentication, but what about authorization? Don't worry, I will cover it in the "Adding Security to the My Retro App" section.

In the following section, we'll start using a UI application that is written using jQuery and (of course) HTML.

Adding Security to the Users App

In this section we are going to not only secure the Users App but also connect it to a UI app. This UI app is written using HTML and jQuery, something simple and quick. Of course, you are welcome to add any other UI framework, such as Angular, VueJS, React, etc.

First, let's talk about the back end, our Users App. If you are following along, you can open the Users App project in your favorite IDE, open the build.gradle file, and add the security dependencies needed. See Listing 9-1.

Listing 9-1. build.gradle

```
plugins {
    id 'java'
    id 'org.springframework.boot' version '3.2.3'
    id 'io.spring.dependency-management' version '1.1.4'
}

group = 'com.apress'
version = '0.0.1-SNAPSHOT'
sourceCompatibility = '17'

repositories {
    mavenCentral()
}

dependencies {
    implementation 'org.springframework.boot:spring-boot-starter-web'
    implementation 'org.springframework.boot:spring-boot-starter-
    validation'
    implementation 'org.springframework.boot:spring-boot-starter-security'
    implementation 'org.springframework.boot:spring-boot-starter-data-jpa'

    runtimeOnly 'com.h2database:h2'
    runtimeOnly 'org.postgresql:postgresql'

    compileOnly 'org.projectlombok:lombok'
    annotationProcessor 'org.projectlombok:lombok'

    // Web
    implementation 'org.webjars:bootstrap:5.2.3'
```

```
    testImplementation 'org.springframework.boot:spring-boot-starter-test'
    testImplementation 'org.springframework.security:spring-security-test'
}

tasks.named('test') {
    useJUnitPlatform()
}

test {
    testLogging {
        events "passed", "skipped", "failed" //, "standardOut",
        "standardError"

        showExceptions true
        exceptionFormat "full"
        showCauses true
        showStackTraces true

        // Change to `true` for more verbose test output
        showStandardStreams = false
    }
}
```

Note in Listing 9-1 that we have both postgresql and h2. By default, we are going to use the H2 database. In this project, we are going to skip the default behavior for the Spring Security auto-configuration with Spring Boot. We are going to create a custom authentication for all the requests with a few users that will be persistent in-memory only. Of course, this is just one way to do it, but the reality is that you will find that authentication normally involves a persistence mechanism, and we are going to cover that too, so don't worry about it too much at this point.

Next, open/create the UserSecurityConfig class, shown in Listing 9-2.

Listing 9-2. src/main/java/com/apress/users/security/UserSecurityConfig.java

```
package com.apress.users.security;
import org.springframework.context.annotation.Bean;
import org.springframework.context.annotation.Configuration;
import org.springframework.security.config.Customizer;
```

```
import org.springframework.security.config.annotation.web.builders.
HttpSecurity;
import org.springframework.security.core.userdetails.UserDetails;
import org.springframework.security.crypto.bcrypt.BCryptPasswordEncoder;
import org.springframework.security.crypto.password.PasswordEncoder;
import org.springframework.security.provisioning.InMemoryUserDetailsManager;
import org.springframework.security.provisioning.UserDetailsManager;
import org.springframework.security.web.SecurityFilterChain;
import org.springframework.security.core.userdetails.User;

@Configuration
public class UserSecurityConfig {

    @Bean
    SecurityFilterChain filterChain(HttpSecurity http) throws Exception {
        http
            .csrf(csrf -> csrf.disable())
                .authorizeHttpRequests( auth -> auth.anyRequest().
                authenticated())
                .httpBasic(Customizer.withDefaults());

        return http.build();
    }

    @Bean
    UserDetailsManager userDetailsManager(PasswordEncoder passwordEncoder){
        UserDetails admin = User
                .builder()
                .username("admin")
                .password(passwordEncoder.encode("admin"))
                .roles("ADMIN","USER")
                .build();

        UserDetails manager = User
                .builder()
                .username("manager@email.com")
                .password(passwordEncoder.encode("aw2sOmeR!"))
                .roles("ADMIN","USER")
                .build();
```

```
    UserDetails user = User
            .builder()
            .username("user@email.com")
            .password(passwordEncoder.encode("aw2sOmeR!"))
            .roles("USER")
            .build();

    return new InMemoryUserDetailsManager(manager,user,admin);
}

@Bean
PasswordEncoder passwordEncoder(){
    return new BCryptPasswordEncoder();
}
}
```

The UserSecurityConfig class includes the following:

- SecurityFilterChain: We are declaring a SecurityFilterChain
  bean, which will provide all the configuration for our application.
  This is where we can declare what default security filters to use and
  even create our own custom filter.

- HttpSecurity: This class allows the configuration of all the web-
  based security for specific HTTP requests. If nothing is provided
  (default), the security will be applied to all requests. It can be
  configured to specific requests using RequestMatcher patterns,
  custom filters implementations, and much more.

- csrf: In our code, we are declaring that the CsrfFilter (with
  csrf() method call) needs to be disabled (this is just for now; but
  for production purposes, it always should be enabled to avoid any
  surprises!). In Spring Security, CSRF (Cross-Site Request Forgery)
  protection is enabled by default. However, there are valid reasons
  why you might want to disable it in certain scenarios:

  - If your API is stateless (doesn't rely on server-side sessions) and
    uses token-based authentication (like JWT), then CSRF attacks are
    not a concern. Since each request carries its own authentication
    token, there's no session cookie for an attacker to exploit.

- If your API is primarily consumed by non-browser clients (e.g., mobile apps, other backend services), then CSRF attacks are also not relevant. CSRF exploits browser behavior, and non-browser clients typically don't send cookies with requests.

- You might have specific endpoints in your application that don't change state (e.g., GET requests that only retrieve data). These endpoints are less susceptible to CSRF attacks, and disabling CSRF protection for them can simplify development and testing. Disabling CSRF protection should be done with caution. If your application has stateful endpoints that are accessible via browsers, disabling CSRF opens it up to potential CSRF attacks.

- `authorizeHttpRequests`: this method provides a way to configure the `AuthorizationManagerRequestMatcherRegister`, a `Customizer` that allows you to declare endpoint match patterns, specific HTTP methods requests (`GET`, `POST`, `PUT`, `DELETE`, `PATCH`, `OPTIONS`, etc.), and a combination of both to secure your application. If nothing is provided, by default (as we saw earlier) it will secure all incoming requests. With the usage of `authorizedHttpRequests`, we are configuring the `AuthorizationFilter`.

- `httpBasic`: We are calling this method to configure the `BasicAuthenticationFilter`, where either the username/password or authentication token is required for authentication.

So, that's how we configure the `SecurityFilterChain`, by providing some default or custom behavior to secure our app. It's important to remember that here we are only using authentication, not authorization yet. Next, we have two more beans declarations:

- `UserDetailsManager`/`InMemoryUserDetailsManager`: This interface allows you to create new users and update existing ones. This interface extends from the `UserDetailsService`, which (in my opinion) is the heart of the authentication and authorization because you can add any implementation here, from in-memory (with `InMemoryUserDetailsManager`) to JDBC, LDAP, and much more. In our code we are using the `UserDetails` to create a user with the `User.builder()` method call and provide the username, password, and roles (you can add more properties, such as `accountExpired`, `credentialsExpired`, `disabled`, `accountLocked` and `PasswordEncoder`).

411

- `PasswordEncoder/BCryptPasswordEncoder`: This bean declares how the password should be persisted, by declaring an encoder algorithm, in this case the usage of the `BCryptPasswordEncoder` class, where the *log round* (controls the amount of work required to hash a password) to use is set to 10 by default (but can configured anywhere between 4 and 31). The higher the log round, the more work is required, and the more secure the hash will be.

That's it! That's how we can add authentication to the Users App.

Users App Web Controller

If you are interested in reviewing the source code, you can find it in the `chapter 09-security` folder; there is a new method at the end that will respond to the `/{email}/gravatar` endpoint, returning the URL based on the user's username/email. We can calculate the hash needed to provide the ID for the gravatar, but I want to make an external call when we integrate the My Retro App with the Users App later in this chapter.

Note   You have access to all the source code at the Apress website or here: `https://github.com/felipeg48/pro-spring-boot-3rd`.

Let's test the Users App security!

Testing Security in the Users App

Spring Security provides different ways to test your code. We'll start with an integration test, using the well-known `TestRestTemplate` class from previous chapters.

Open/create the `UsersHttpRequestTests` class. See Listing 9-3.

Listing 9-3. src/test/java/com/apress/users/UsersHttpRequestTests.java

```
package com.apress.users;

import com.apress.users.model.User;
import com.apress.users.model.UserRole;
import org.junit.jupiter.api.Test;
import org.springframework.beans.factory.annotation.Autowired;
```

```java
import org.springframework.boot.test.context.SpringBootTest;
import org.springframework.boot.test.web.client.TestRestTemplate;

import java.util.Arrays;
import java.util.Collection;

import static org.assertj.core.api.Assertions.assertThat;

@SpringBootTest(webEnvironment = SpringBootTest.WebEnvironment.RANDOM_PORT)
public class UsersHttpRequestTests {
    private final String USERS_PATH = "/users";

    @Autowired
    private TestRestTemplate restTemplate;

    @Test
    public void indexPageShouldReturnHeaderOneContent() throws Exception {
        String html = this.restTemplate.withBasicAuth("manager@email.
        com","aw2sOmeR!").getForObject("/", String.class);
        assertThat(html).contains("Simple Users Rest Application");
    }

    @Test
    public void usersEndPointShouldReturnCollectionWithTwoUsers() throws
    Exception {
        Collection<User> response = this.restTemplate.
        withBasicAuth("manager@email.com","aw2sOmeR!").
                getForObject(USERS_PATH, Collection.class);

        assertThat(response.size()).isGreaterThan(1);
    }

    @Test
    public void userEndPointPostNewUserShouldReturnUser() throws
    Exception {
        User user =  new User("dummy@email.com","Dummy",
                "https://www.gravatar.com/avatar/23bb62a7d0ca63c9a804908e57
                bf6bd4?d=wavatar",
                "SomeOtherAw2sOmeR!", Arrays.asList(UserRole.USER),true);
```

```
        User response = this.restTemplate.withBasicAuth("manager@email.
    com","aw2sOmeR!").postForObject(USERS_PATH,user, User.class);

        assertThat(response).isNotNull();
        assertThat(response.getEmail()).isEqualTo(user.getEmail());

        Collection<User> users = this.restTemplate.withBasicAuth("manager@
    email.com","aw2sOmeR!").
                getForObject(USERS_PATH, Collection.class);

        assertThat(users.size()).isGreaterThanOrEqualTo(2);

    }

    @Test
    public void userEndPointDeleteUserShouldReturnVoid() throws Exception {
        this.restTemplate.withBasicAuth("manager@email.com","aw2sOmeR!").
        delete(USERS_PATH + "/norma@email.com");

        Collection<User> users = this.restTemplate.withBasicAuth(
        "manager@email.com","aw2sOmeR!").
                getForObject(USERS_PATH, Collection.class);

        assertThat(users.size()).isLessThanOrEqualTo(2);
    }

    @Test
    public void userEndPointFindUserShouldReturnUser() throws Exception{
        User user = this.restTemplate.withBasicAuth("manager@email.
        com","aw2sOmeR!").getForObject(USERS_PATH + "/ximena@email.
        com",User.class);

        assertThat(user).isNotNull();
        assertThat(user.getEmail()).isEqualTo("ximena@email.com");
    }
}
```

Listing 9-3 shows the `Integration  Test` when adding security to your app. The only difference from previous chapters is that now we are using the `TestRestTemplate` with the `withBasicAuth` method that receives the *username* and *password* as parameters. And because the Users App is configured with basic authentication, this method is exactly what we need here. If you run the tests, they will pass.

Mocking Security Tests

Mocking is another way to test security when we need to isolate the behavior of a particular object. This way, we can simulate the behavior of the real object, in this case security, and just focus on part of our business logic.

Open/create the `UserMockMvcTests` class. See Listing 9-4.

Listing 9-4. src/test/java/com/apress/users/UserMockMvcTests.java

```
package com.apress.users;

import org.junit.jupiter.api.Test;
import org.springframework.beans.factory.annotation.Autowired;
import org.springframework.boot.test.autoconfigure.web.servlet.
AutoConfigureMockMvc;
import org.springframework.boot.test.context.SpringBootTest;
import org.springframework.security.test.context.support.WithMockUser;
import org.springframework.test.web.servlet.MockMvc;

import static org.hamcrest.Matchers.*;
import static org.springframework.test.web.servlet.request.
MockMvcRequestBuilders.get;
import static org.springframework.test.web.servlet.request.
MockMvcRequestBuilders.post;
import static org.springframework.test.web.servlet.result.
MockMvcResultMatchers.*;

@SpringBootTest
@WithMockUser
@AutoConfigureMockMvc
public class UserMockMvcTests {
```

```java
@Autowired
MockMvc mockMvc;

@Test
void createUserTests() throws Exception {
    String location = mockMvc.perform(post("/users")
                    .contentType("application/json")
                    .content("""
                    {
                        "email": "dummy@email.com",
                        "name": "Dummy",
                        "password": "aw2s0meR!",
                        "gravatarUrl": "https://www.gravatar.com/
                        avatar/fb651279f4712e209991e05610dfb03a?d
                        =wavatar",
                        "userRole": ["USER"],
                        "active": true
                    }
                    """))
            .andExpect(status().isCreated())
            .andExpect(header().exists("Location"))
            .andReturn().getResponse().getHeader("Location");

    mockMvc.perform(get(location))
            .andExpect(status().isOk())
            .andExpect(jsonPath("$.email").exists())
            .andExpect(jsonPath("$.active").value(true));
}

@Test
void getAllUsersTests() throws Exception {
    mockMvc.perform(get("/users"))
            .andExpect(status().isOk())
            .andExpect(jsonPath("$[0].name").value("Ximena"))
            .andExpect(jsonPath("$..active").value(hasItem(true)));
}
}
```

The `UserMockMvcTests` class includes the following:

- `@AutoConfigureMockMvc`: As you already know, this annotation configures the `MockMvc` object for testing web apps.

- `MockMvc`: Also as you already know, this object simulates any HTTP requests and responses without using the real web server. In this case, we are using the `POST` and `GET` HTTP requests.

- `@WithMockUser`: This annotation comes with the `spring-security-test` dependency that allows us to mock the user in the security context, and in this case, we can test our functionality without worrying about getting authenticated. Of course, if you remove this annotation, the tests will fail. The `@WithMockUser` annotation accepts parameters; for example, you can use it like this: `@WithMockUser(username = "user", password = "password", roles = "USER")`, where you can make sure that the user is authenticated or has the permission if needed. And it can live as a class annotation, affecting every test method, or by method if you require more control in your mocking.

Now, you know that you have options for testing security, either unit or integration testing.

Using Persistence for Security Authentication in the Users App

We just saw that we can do security authentication with users in-memory, but in a real-world scenario, the users are in a database. Fortunately, Spring Security provides an easy way to use a standard JDBC implementation for the `UserDetailsService`. It requires you to provide the username, password, account status (enable or disable), and roles. The SQL schema shown in Listing 9-5 follows this standard implementation.

Listing 9-5. SQL Schema for Spring Security

```
create table users(
    username varchar_ignorecase(50) not null primary key,
    password varchar_ignorecase(50) not null,
    enabled boolean not null
);

create table authorities (
    username varchar_ignorecase(50) not null,
    authority varchar_ignorecase(50) not null,
    constraint fk_authorities_users foreign key(username) references
    users(username)
);
create unique index ix_auth_username on authorities (username,authority);
```

Spring Security provides a dedicated schema depending on the engine used (PostgreSQL, Oracle, MySQL, etc.). You can find more details here: `https://docs. spring.io/spring-security/reference/servlet/appendix/database-schema.html`.

If you want to use the default schema, you simply need to change the return type of `UserDetailsManager` to `JdbcUserDetailsManager` in the `UserSecurityConfig` class.

Open the `UserSecurityConfig` class and replace all the code with the content shown in Listing 9-6.

Listing 9-6. src/main/java/com/apress/users/security/UserSecurityConfig.java

```
package com.apress.users.security;

import org.springframework.context.annotation.Bean;
import org.springframework.context.annotation.Configuration;
import org.springframework.jdbc.datasource.embedded.
EmbeddedDatabaseBuilder;
import org.springframework.security.config.Customizer;
import org.springframework.security.config.annotation.web.builders.
HttpSecurity;
import org.springframework.security.core.userdetails.UserDetails;
import org.springframework.security.core.userdetails.jdbc.JdbcDaoImpl;
```

```
import org.springframework.security.crypto.bcrypt.BCryptPasswordEncoder;
import org.springframework.security.crypto.password.PasswordEncoder;
import org.springframework.security.provisioning.JdbcUserDetailsManager;
import org.springframework.security.provisioning.UserDetailsManager;
import org.springframework.security.web.SecurityFilterChain;
import org.springframework.security.core.userdetails.User;

import javax.sql.DataSource;
import static org.springframework.jdbc.datasource.embedded.
EmbeddedDatabaseType.H2;

@Configuration
public class UserSecurityConfig {

    @Bean
    SecurityFilterChain filterChain(HttpSecurity http) throws Exception {
        http
                .csrf(csrf -> csrf.disable())
                .authorizeHttpRequests( auth -> auth.anyRequest().
                authenticated())
                .httpBasic(Customizer.withDefaults());

        return http.build();
    }

    @Bean
    DataSource dataSource() {
        return new EmbeddedDatabaseBuilder()
                .setType(H2)
                .addScript(JdbcDaoImpl.DEFAULT_USER_SCHEMA_DDL_LOCATION)
                .build();
    }

    @Bean
    UserDetailsManager userDetailsManager(PasswordEncoder passwordEncoder,
    DataSource dataSource){
        UserDetails admin = User
```

```
                .builder()
                .username("admin")
                .password(passwordEncoder.encode("admin"))
                .roles("ADMIN","USER")
                .build();

        UserDetails manager = User
                .builder()
                .username("manager@email.com")
                .password(passwordEncoder.encode("aw2sOmeR!"))
                .roles("ADMIN","USER")
                .build();

        UserDetails user = User
                .builder()
                .username("user@email.com")
                .password(passwordEncoder.encode("aw2sOmeR!"))
                .roles("USER")
                .build();

        JdbcUserDetailsManager users = new JdbcUserDetailsManager(
        dataSource);
        users.createUser(admin);
        users.createUser(manager);
        users.createUser(user);
        return users;
    }

    @Bean
    PasswordEncoder passwordEncoder(){
        return new BCryptPasswordEncoder();
    }

}
```

Listing 9-6 shows the new UserSecurityConfig class. Note that our SecurityFilterChain didn't change; only the DataSource and JdbcUserDetailsManager changed.

That's it. If you run the same unit and integration tests, they should pass without any issues.

Using Custom Persistence for Your Authentication Security

In this section I want to show you that you can use your own custom models to supply part of the authentication and authorization for your apps, and in this case show that the Users App already comes with the required info that Spring Security needs. Of course, our model brings extra information that is valuable just for our domain, but it can be used for security (authentication and authorization) purposes.

To use our model, we need to replace all code in the UserSecurityConfig class with the code shown in Listing 9-7.

Listing 9-7. src/main/java/com/apress/users/security/UserSecurityConfig.java

```
package com.apress.users.security;

import com.apress.users.repository.UserRepository;
import org.springframework.context.annotation.Bean;
import org.springframework.context.annotation.Configuration;
import org.springframework.security.authentication.AuthenticationProvider;
import org.springframework.security.authentication.dao.
DaoAuthenticationProvider;
import org.springframework.security.config.Customizer;
import org.springframework.security.config.annotation.web.builders.
HttpSecurity;
import org.springframework.security.web.SecurityFilterChain;

@Configuration
public class UserSecurityConfig {

    @Bean
    SecurityFilterChain filterChain(HttpSecurity http,
    AuthenticationProvider authenticationProvider) throws Exception {
        http
                .csrf(csrf -> csrf.disable())
                .authorizeHttpRequests( auth -> auth.anyRequest().
                authenticated())
```

```
            .authenticationProvider(authenticationProvider)
            .httpBasic(Customizer.withDefaults());

        return http.build();
    }

    @Bean
    public AuthenticationProvider authenticationProvider(UserRepository
    userRepository){
        DaoAuthenticationProvider provider = new
        DaoAuthenticationProvider();
        provider.setUserDetailsService(new UserSecurityDetailsService(
        userRepository));
        return provider;
    }
}
```

Let's review the modified UserSecurityConfig class:

- AuthenticationProvider: In this example we are using our own
  database that will serve data for authentication and authorization,
  and in this case, we need to specify a provider that will do this type of
  authentication. You can provide multiple AuthenticationProvider
  instances that are managed by the ProviderManager. In fact,
  Spring Security already has the DaoAuthenticationProvider, the
  JwtAuthenticationProvider (if you want to handle JWT tokens),
  LdapAuthenticationProvider, CasAuthenticationProvider, and
  more. Of course, you can create your own custom authentication
  provider.

- DaoAuthenticationProvider: One of the main benefits of using an
  AuthenticationProvider is that it already implements what you
  require, and you only need to provide the logic to get the data, in this
  case the username/password and authorities/roles. In this example
  we are setting up the UserSecurityDetailsService that uses the
  UserRepository where our data lives.

- UserSecurityDetailsService/UserRepository: In this example we
  are declaring a new instance of the UserSecurityDetailsService
  class that requires the UserRepository. This class implements the
  UserDetailsService that will hold the data for the authentication
  process, in this case the username, the password, the authorities
  or roles, and other properties that can be useful when the
  authentication is happening.

Next, open/create the UserSecurityDetailsService class. See Listing 9-8.

Listing 9-8. src/main/java/com/apress/users/security/
UserSecurityDetailsService.java

```
package com.apress.users.security;

import com.apress.users.exception.UserNotFoundException;
import com.apress.users.model.User;
import com.apress.users.model.UserRole;
import com.apress.users.repository.UserRepository;
import lombok.AllArgsConstructor;
import org.springframework.security.core.userdetails.UserDetails;
import org.springframework.security.core.userdetails.UserDetailsService;
import org.springframework.security.core.userdetails.
UsernameNotFoundException;
import org.springframework.security.crypto.factory.PasswordEncoderFactories;
import org.springframework.security.crypto.password.PasswordEncoder;

@AllArgsConstructor
public class UserSecurityDetailsService implements UserDetailsService {

    private UserRepository userRepository;

    @Override
    public UserDetails loadUserByUsername(String username) throws
    UsernameNotFoundException {
        User user = userRepository.findById(username)
                .orElseThrow(UserNotFoundException::new);
        PasswordEncoder passwordEncoder = PasswordEncoderFactories.
        createDelegatingPasswordEncoder();
```

```
    return org.springframework.security.core.userdetails.User
            .withUsername(username)
            .roles(user.getUserRole().stream().map(UserRole::toString).
            toArray(String[]::new))
            .password(passwordEncoder.encode(user.getPassword()))
            .accountExpired(!user.isActive())
            .build();
    }
}
```

In the UserSecurityDetailsService class, we are implementing the
UserDetailsService that we'll use to retrieve the username, the password, and other
useful properties for authentication. In this case, we are using the User.isActive
method to set the accountExpired property as an example, and we are using the
UserRole instance to collect a list of roles.

Now you know that there are different ways to configure authentication with
Spring Security, from the default with the username user and the randomly generated
password in the console, to the in-memory and persistence (using the default schema or
a custom schema), and even LDAP. If you implement the UserDetails, you can plug in
anything you want to provide authentication and authorization.

Connecting a UI to the Users App

Now it's time to connect a UI to the Users App. In the source code, you will find a folder
named users-ui that includes all the HTML, JavaScript (jQuery), and necessary assets to
have a UI for the Users App. Before continuing with the UI and back-end code, take some
time to review it and analyze what is happening.

Let's first look at the login.js file and review part of its code, shown in Listing 9-9.

Listing 9-9. Snippet from users-ui/assets/js/login.js

```
...
$('#btn__login').click(function(e) {
    e.preventDefault();
    let user_email = $('input[name="user_email"]').val();
    let user_password = $('input[name="user_password"]').val();
```

```
$.ajax({
  url: 'http://localhost:8080/users/'+user_email,
  method: 'GET',
  headers: {
    'Authorization': 'Basic ' + btoa(user_email + ':' + user_password),
    'Content-Type': 'application/json'
  }
})
.done(function( response, textStatus, xhr ) {

  // ....
}
```

Note that in the $.ajax call, for the url value we are using the /users/{email} endpoint. Perhaps you are thinking that this is unusual, because it could be / authenticate or something similar, and yes, we can do that, but for the purpose of illustrating how this UI works, we are going to use that endpoint as part of the authentication.

Another important part of the code is the headers value, where we are using the HTTP Authorization header and the btoa() method that allows us to encode to *base64* the string we are sending, in this case the username and password. When we receive the response from the back end, we get it in the done function.

Now, switch to the Users App (back end) and modify/create the UserSecurityConfig class. See Listing 9-10.

Listing 9-10. src/main/java/com/apress/users/security/UserSecurityConfig.java

```
package com.apress.users.security;

import com.apress.users.repository.UserRepository;
import org.springframework.context.annotation.Bean;
import org.springframework.context.annotation.Configuration;
import org.springframework.security.config.Customizer;
import org.springframework.security.config.annotation.web.builders.
HttpSecurity;
import org.springframework.security.authentication.AuthenticationProvider;
import org.springframework.security.authentication.dao.
DaoAuthenticationProvider;
```

```java
import org.springframework.security.web.SecurityFilterChain;
import org.springframework.web.cors.CorsConfiguration;
import org.springframework.web.cors.CorsConfigurationSource;
import org.springframework.web.cors.UrlBasedCorsConfigurationSource;

import java.util.Arrays;

@Configuration
public class UserSecurityConfig {

    @Bean
    SecurityFilterChain filterChain(HttpSecurity http,
    AuthenticationProvider authenticationProvider,
                                    CorsConfigurationSource
                                    corsConfigurationSource) throws
                                    Exception {

        http
                .csrf(csrf -> csrf.disable())
                .cors(cors -> cors.configurationSource(
                corsConfigurationSource))
                .authorizeHttpRequests( auth -> auth.anyRequest().
                authenticated())
                .authenticationProvider(authenticationProvider)
                .formLogin(Customizer.withDefaults())
                .httpBasic(Customizer.withDefaults());

        return http.build();
    }

    @Bean
    CorsConfigurationSource corsConfigurationSource() {
        CorsConfiguration configuration = new CorsConfiguration();

        configuration.setAllowedOrigins(Arrays.asList("*"));
        configuration.setAllowedMethods(Arrays.asList("*"));
        configuration.setAllowedHeaders(Arrays.asList("*"));

        UrlBasedCorsConfigurationSource source = new
        UrlBasedCorsConfigurationSource();
```

```
    source.registerCorsConfiguration("/**", configuration);
    return source;
}

@Bean
public AuthenticationProvider authenticationProvider(UserRepository
userRepository){
    DaoAuthenticationProvider provider = new DaoAuthenticationProvider();
    provider.setUserDetailsService(new UserSecurityDetailsService(
    userRepository));
    return provider;
}
}
```

Listing 9-10 shows the modified UserSecurityConfig class that works with the UI. Let's review it:

- cors: As mentioned earlier in the chapter, Spring Security provides support for cross-origin resource sharing, which blocks any script from connecting or making calls using AJAX to a different origin. In this case, the requests won't be happening in the same origin, because they will be made using the UI (which lives in a different server), so we need to provide access to it (knowing, of course, the client), and we need to declare some rules by using a CorsConfigurationSource.

- formLogin: We are also adding a new way to authenticate as part of the SecurityFilterChain, in this case as formLogin with its defaults. So, if we try to get directly into the endpoint /users, Spring Security will respond with a Login page requesting a username and password.

- CorsConfigurationSource/CorsConfiguration/UrlBasedCors ConfigurationSource: We are creating the CorsConfigurationSource bean to declare the origins, headers, and methods of the requests to identify where they are coming from. In this case, we are enabling everything just for demonstration purposes, but when there are clients embedded in our apps, we must restrain and configure our security in such way that we can avoid any hacking.

Users Front End/Back End: Let's Give It a Try!

Now, let's run our two apps in the UI. First, make sure the Users App (back end) is up and running. Then, in your `users-ui`, you can use Python or any other app that can serve those HTML files. If you are using an IDE for the UI, it should have a way to run your apps. If you are using the JetBrains WebStorm IDE, for example, simply right-click the `index.html` file and select Run. If you are using VS Code, you can use the Live Server plugin and click (in the status bar) the Go Live button, which will launch a browser window in port 5500.

Using Python, go to the root of the `users-ui` folder and execute the following command:

```
python3 -m http.server
```

This opens the 8000 port. If you open your browser and go to `http://localhost:8000`, you will see the UI shown in Figure 9-2.

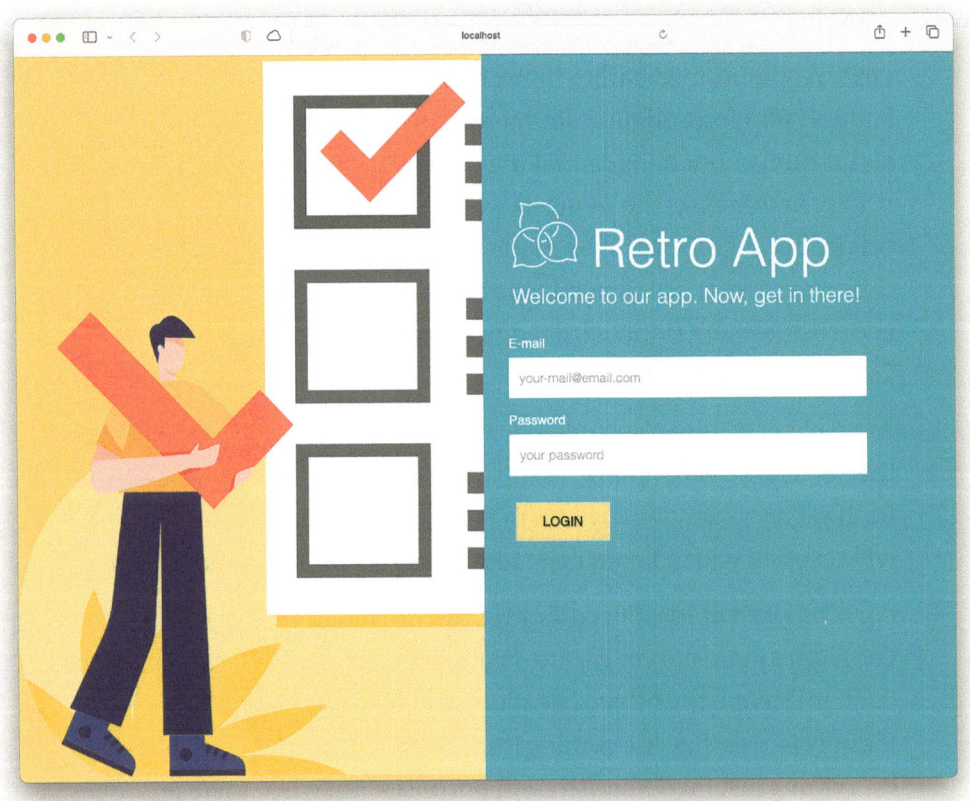

Figure 9-2. `http://localhost:8000` - *Users/Retro App*

Figure 9-2 shows the Users/Retro App, right now we are using the Retro App view to see the users, so it's not an error. Then, you can log in with the following credentials:

`manager@email.com/aw2sOmeR!`

After a successful login, you should see the Users screen shown in Figure 9-3.

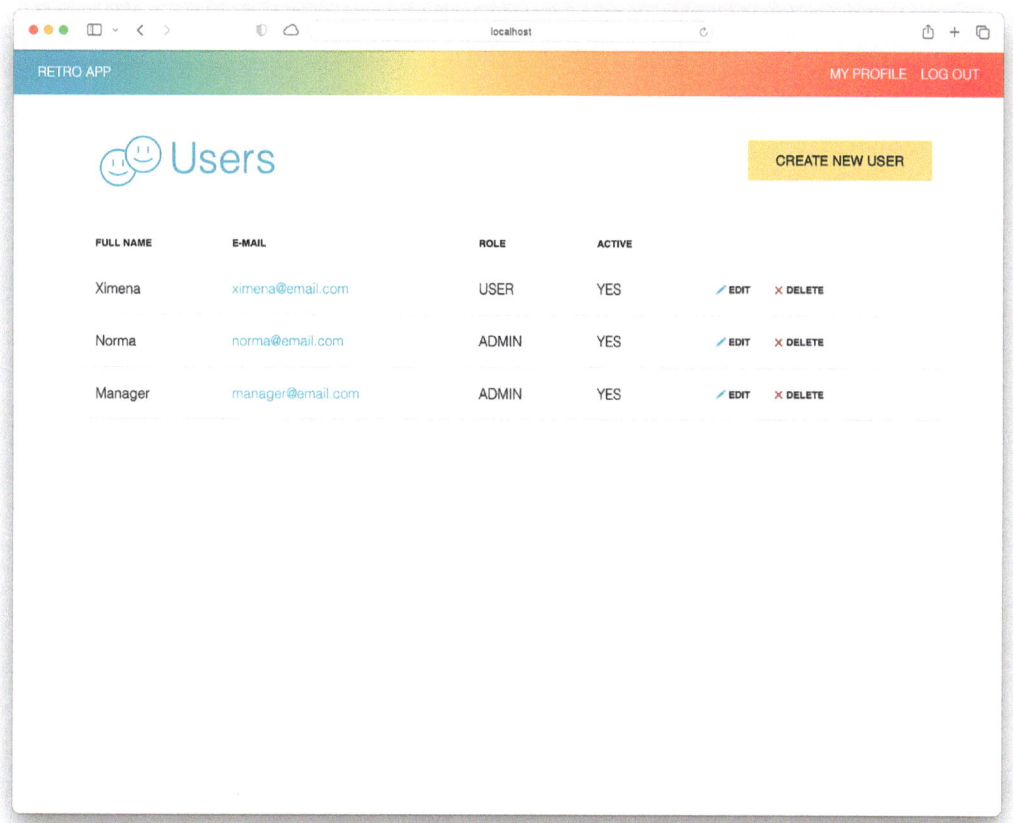

Figure 9-3. *Users page*

If you want to test its functionality, go ahead and experiment. The users listed are the default that we set in the back end in the `UserConfig` class.

Note   If you find a bug/issue in the UI or have a suggestion for an improvement, let me know; you can write a pull request here: `https://github.com/ felipeg48/pro-spring-boot-3rd`.

Now, if you try to access directly the back end, `http://localhost:8080`, you will see the login page shown in Figure 9-4.

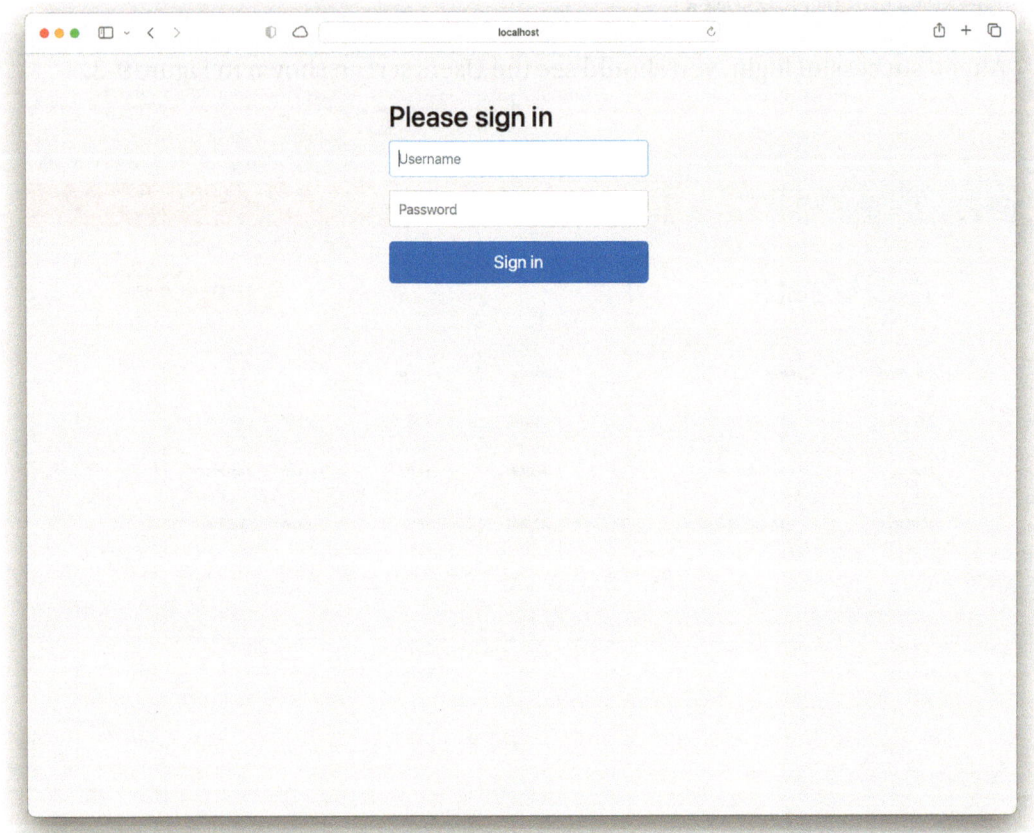

Figure 9-4. *Users App back end (`http://localhost:8080`)*

Figure 9-4 shows the default implementation of the `formLogin` declaration we set in the `SecurityFilterChain`. If you provide the credentials, you will see a list of users in JSON format, as shown in Figure 9-5.

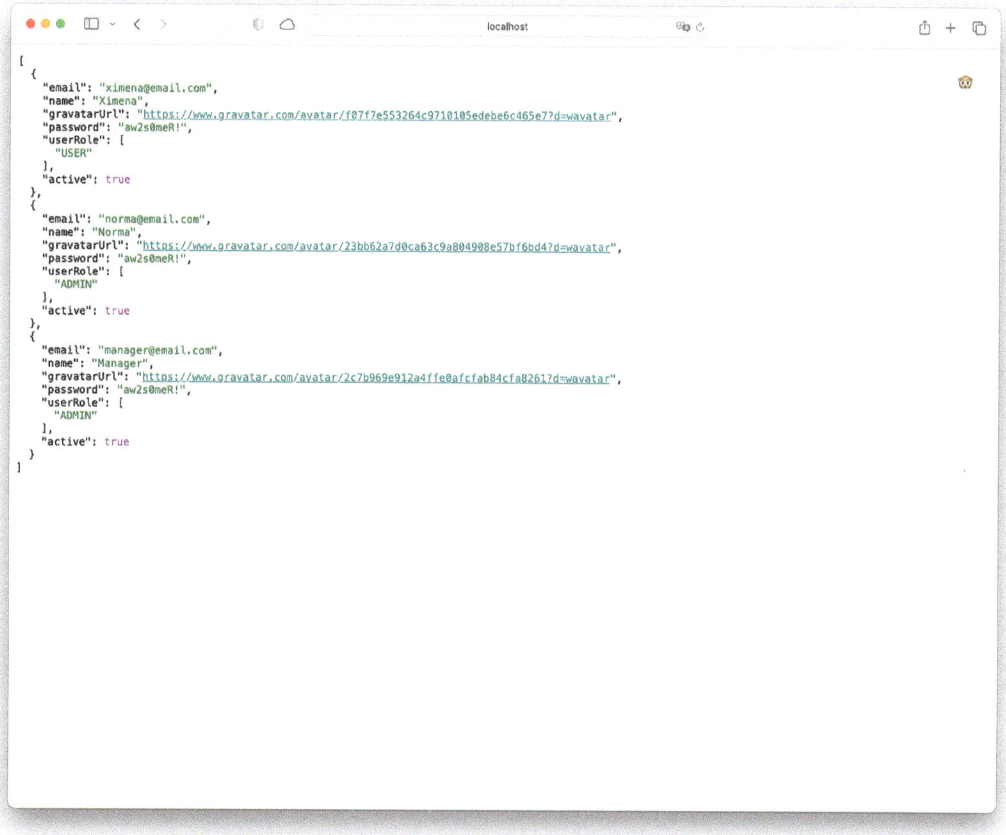

Figure 9-5. *List of users in JSON format, displayed after authenticating*

Now you know how to integrate an external UI to your Users App using Spring Security with Spring Boot! Before you move on to the next section, keep in mind that we discussed only *authentication* in this section. In the following sections, we'll be covering *authorization* also.

Adding Security to the My Retro App

The My Retro App is an integration of a Reactive technology, and Spring Security with Spring Boot provides the similar auto-configuration when the dependencies are based on WebFlux. The main idea here is that we are going to connect both apps: the Users App will provide user information lookup, and the My Retro App will secure that information (User info) based on the security configuration that will be defining in this section. In the My Retro App, we'll define not only authentication but authorization as well, as depicted in Figure 9-6.

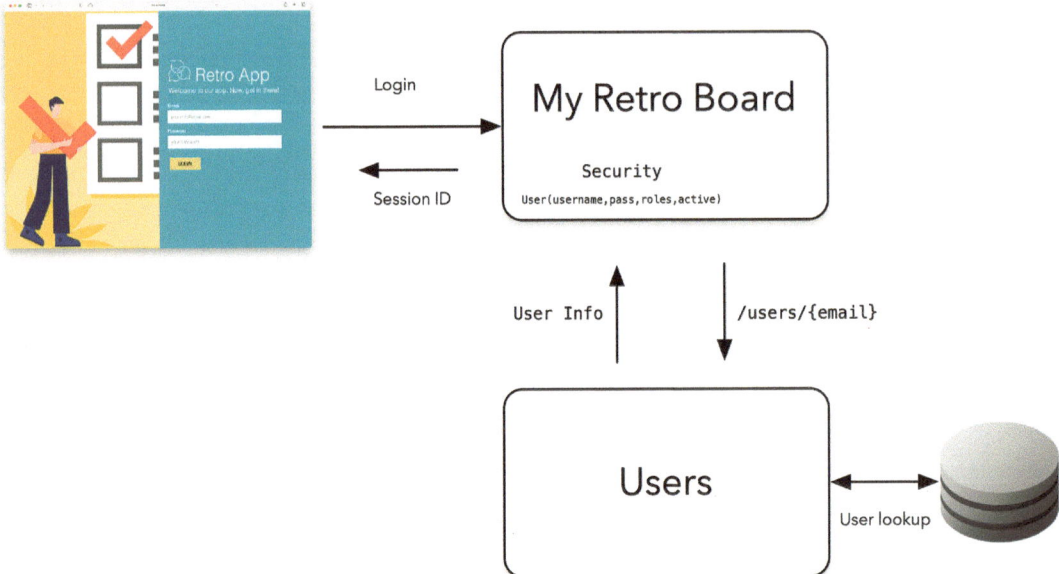

Figure 9-6. *My Retro App security flow*

Next, we are going to review the source code, which you can find in the 09-security/myretro folder. If you open the build.gradle file, you will see the code shown in Listing 9-11.

Listing 9-11. build.gradle

```
plugins {
    id 'java'
    id 'org.springframework.boot' version '3.2.3'
    id 'io.spring.dependency-management' version '1.1.4'
}

group = 'com.apress'
version = '0.0.1-SNAPSHOT'
sourceCompatibility = '17'
```

```
repositories {
    mavenCentral()
}

dependencies {
    implementation 'org.springframework.boot:spring-boot-starter-webflux'
    implementation 'org.springframework.boot:spring-boot-starter-
    validation'
    implementation 'org.springframework.boot:spring-boot-starter-security'
    implementation 'org.springframework.boot:spring-boot-starter-data-
    mongodb-reactive'

    annotationProcessor 'org.springframework.boot:spring-boot-
    configuration-processor'

    compileOnly 'org.projectlombok:lombok'
    annotationProcessor 'org.projectlombok:lombok'

    developmentOnly 'org.springframework.boot:spring-boot-docker-compose'

    // Web
    implementation 'org.webjars:bootstrap:5.2.3'
    implementation 'org.webjars:webjars-locator-core'

    // DevTools
    implementation 'org.springframework.boot:spring-boot-devtools'
    testImplementation 'org.springframework.boot:spring-boot-
    testcontainers'
    testImplementation 'org.testcontainers:junit-jupiter'
    testImplementation 'org.testcontainers:mongodb'

    testImplementation 'org.springframework.boot:spring-boot-starter-test'
    testImplementation 'org.springframework.security:spring-security-test'
    testImplementation 'io.projectreactor:reactor-test'
}

tasks.named('test') {
    useJUnitPlatform()
}
```

```
test {
    testLogging {
        events "passed", "skipped", "failed" //, "standardOut",
        "standardError"

        showExceptions true
        exceptionFormat "full"
        showCauses true
        showStackTraces true

        // Change to `true` for more verbose test output
        showStandardStreams = false
    }
}
```

Listing 9-11 shows the spring-boot-starter-security dependencies added to the build.gradle file.

Next, open/create the UserClient class. See Listing 9-12.

Listing 9-12. src/main/java/apress/com/myretro/client/UserClient.java

```
package com.apress.myretro.client;

import com.apress.myretro.config.MyRetroProperties;
import org.springframework.stereotype.Component;
import org.springframework.web.reactive.function.client.WebClient;
import reactor.core.publisher.Mono;

@Component
public class UserClient {
    WebClient webClient;

    public UserClient(WebClient.Builder webClientBuilder, MyRetroProperties
    props) {
        this.webClient = webClientBuilder
                .baseUrl(props.getUsers().getServer())
                .defaultHeaders(headers -> headers.setBasicAuth(props.
                getUsers().getUsername(), props.getUsers().getPassword()))
                .build();
    }
```

```
public Mono<User> getUserInfo(String email){
    return webClient.get()
            .uri("/users/{email}",email)
            .retrieve()
            .bodyToMono(User.class);
}

public Mono<String> getUserGravatar(String email){
    return webClient.get()
            .uri("/users/{email}/gravatar",email)
            .retrieve()
            .bodyToMono(String.class);
}
}
```

The UserClient class is marked as a Spring bean with the @Component annotation, which will make it available when needed. This class is like the RestTemplate class but for Reactive applications. It has its own way to deal with the WebFlux components, such as Flux and Mono types, and it can be used in non-reactive apps as well. The UserClient class has two methods that use the WebClient class:

- The getUserInfo method will use the /users/{email} endpoint to retrieve user information by sending an email (as a parameter of the method). In the constructor, we are using the *basic authentication* required by the Users App (or Users Service in this case) to access its API. We are using the MyRetroProperties class that is constructed by Spring and is expecting the server, port, username, and password properties. (You can look at the MyRetroProperties class to review these properties, and the application.properties file.) In other words, to secure the My Retro App, we need to have some users, and the Users App holds such information, so we are using the Users App as an authentication and authorization mechanism for the My Retro App, and to authenticate we are using the WebClient class that will access the Users Service.

- The getUserGravatar method will be used to retrieve the image (based on the email) and is using the WebClient to communicate to the Users Service.

Notice that both methods are calling a non-reactive app, and they are transforming the response into a Mono type.

Next, open/create the RetroBoardSecurityConfig class. See Listing 9-13.

Listing 9-13. src/main/java/apress/com/myretro/security/
RetroBoardSecurityConfig.java

```java
package com.apress.myretro.security;

import com.apress.myretro.client.User;
import com.apress.myretro.client.UserClient;
import com.apress.myretro.client.UserRole;
import lombok.extern.slf4j.Slf4j;
import org.springframework.context.annotation.Bean;
import org.springframework.context.annotation.Configuration;
import org.springframework.core.io.buffer.DataBuffer;
import org.springframework.http.HttpMethod;
import org.springframework.http.HttpStatusCode;
import org.springframework.http.ReactiveHttpOutputMessage;
import org.springframework.security.authentication.BadCredentialsException;
import org.springframework.security.authentication.
ReactiveAuthenticationManager;
import org.springframework.security.authentication.
UsernamePasswordAuthenticationToken;
import org.springframework.security.config.Customizer;
import org.springframework.security.config.web.server.ServerHttpSecurity;
import org.springframework.security.core.GrantedAuthority;
import org.springframework.security.core.authority.SimpleGrantedAuthority;
import org.springframework.security.web.server.SecurityWebFilterChain;
import org.springframework.security.web.server.authentication.
ServerAuthenticationSuccessHandler;
import org.springframework.web.cors.CorsConfiguration;
import org.springframework.web.cors.reactive.CorsConfigurationSource;
import org.springframework.web.cors.reactive.
UrlBasedCorsConfigurationSource;
import reactor.core.publisher.Flux;
import reactor.core.publisher.Mono;
```

```java
import java.util.Arrays;
import java.util.List;
import java.util.stream.Collectors;

@Slf4j
@Configuration
public class RetroBoardSecurityConfig {

    @Bean
    SecurityWebFilterChain securityWebFilterChain(ServerHttpSecurity http,
                            ReactiveAuthenticationManager
                            reactiveAuthenticationManager,
                            CorsConfigurationSource corsConfigurationSource)
                            throws Exception {
        http
                .csrf(csrf -> csrf.disable())
                .cors(cors -> cors.configurationSource(
                corsConfigurationSource))
                .authorizeExchange( auth -> auth

                        .pathMatchers(HttpMethod.POST,"/retros/**").
                        hasRole("ADMIN")
                        .pathMatchers(HttpMethod.DELETE,"/retros/**").
                        hasRole("ADMIN")
                        .pathMatchers("/retros/**").
                        hasAnyRole("USER","ADMIN")
                        .pathMatchers("/","/webjars/**").permitAll()
                )
                .authenticationManager(reactiveAuthenticationManager)
                .formLogin(formLoginSpec -> formLoginSpec.authentication
                SuccessHandler(serverAuthenticationSuccessHandler()))
                .httpBasic(Customizer.withDefaults());

        return http.build();
    }

    @Bean
    ServerAuthenticationSuccessHandler serverAuthentication
    SuccessHandler(){
```

```java
        return (webFilterExchange, authentication) -> {
            return webFilterExchange.getExchange().getSession()
                    .flatMap(session -> {

User user = (User) authentication.getDetails();

                    var body = """
                {
                    "email": "%s",
                    "name": "%s",
                    "password": "%s",
                    "userRole": "%s",
                    "gravatarUrl": "%s",
                    "active": %s
                }
                """.formatted(
                            user.email(),
                            user.name(),
                            user.password(),
                            authentication.getAuthorities().stream()
                                    .map(GrantedAuthority::getAuthority)
                                    .map(role -> role.replace("ROLE_",""))
                                    .collect(Collectors.joining(",")),
                            user.gravatarUrl(),
                            true);
                        webFilterExchange.getExchange().getResponse().
                        setStatusCode(HttpStatusCode.valueOf(200));
                        ReactiveHttpOutputMessage response =
                        webFilterExchange.getExchange().getResponse();
                        response.getHeaders().add("Content-Type",
                        "application/json");
                        response.getHeaders().add("X-MYRETRO",
                        "SESSION="+session.getId()+"; Path=/; HttpOnly;
                        SameSite=Lax");
                        DataBuffer dataBufferPublisher =
                        response.bufferFactory().wrap(body.getBytes());
```

```
        return response.writeAndFlushWith( Flux.
        just(dataBufferPublisher).windowUntilChanged());
    });
};
}

@Bean
CorsConfigurationSource corsConfigurationSource() {
    CorsConfiguration configuration = new CorsConfiguration();

    //configuration.setAllowedOrigins(Arrays.asList("*"));
    configuration.setAllowedMethods(Arrays.asList("GET","POST","PUT",
    "DELETE","OPTIONS"));
    configuration.setAllowedHeaders(Arrays.asList("x-ijt","Set-Cookie",
    "Cookie","Content-Type","X-MYRETRO","Allow","Authorization",
    "Access-Control-Allow-Origin","Access-Control-Allow-
    Credentials","Access-Control-Allow-Headers","Access-Control-
    Allow-Methods","Access-Control-Expose-Headers","Access-Control-
    Max-Age","Access-Control-Request-Headers","Access-Control-
    Request-Method","Origin","X-Requested-With","Accept","Accept-
    Encoding","Accept-Language","Host","Referer","Connection",
    "User-Agent"));
    configuration.setExposedHeaders(Arrays.asList("x-ijt",
    "Set-Cookie","Cookie","Content-Type","X-MYRETRO","Allow","Autho
    rization","Access-Control-Allow-Origin","Access-Control-Allow-
    Credentials","Access-Control-Allow-Headers","Access-Control-
    Allow-Methods","Access-Control-Expose-Headers","Access-Control-
    Max-Age","Access-Control-Request-Headers","Access-Control-
    Request-Method","Origin","X-Requested-With","Accept","Accept-
    Encoding","Accept-Language","Host","Referer","Connection",
    "User-Agent"));
    configuration.setAllowedOriginPatterns(Arrays.asList("http://
    localhost:*"));
    configuration.setAllowCredentials(true);

    UrlBasedCorsConfigurationSource source = new
    UrlBasedCorsConfigurationSource();
```

```java
        source.registerCorsConfiguration("/**", configuration);
        return source;
    }

    @Bean
    public ReactiveAuthenticationManager reactiveAuthenticationManager
    (UserClient userClient){
        return authentication -> {
            String username = authentication.getName();
            String password = authentication.getCredentials().toString();

            Mono<User> userResult = userClient.getUserInfo(username);

            return userResult.flatMap( user -> {
                if(user.password().equals(password)){
                    List<GrantedAuthority> grantedAuthorities = user.
                    userRole().stream()
                            .map(UserRole::name)
                            .map("ROLE_"::concat)
                            .map(SimpleGrantedAuthority::new)
                            .collect(Collectors.toList());

                    UsernamePasswordAuthenticationToken authentication
                    Token = new UsernamePasswordAuthenticationToken(
                    username,password,grantedAuthorities);
                    authenticationToken.setDetails(user);
                    return Mono.just(authenticationToken);

                }else{
                    return Mono.error(new BadCredentialsException("Invalid
                    username or password"));
                }});

        };

    }

}
```

The `RetroBoardSecurityConfig` class includes the following:

- `SecurityWebFilterChain`: This interface is quite familiar, right? In the Users App, we use the `SecurityFilterChain` interface for *Servlet web apps*, and in this case, we are using *WebFlux (Reactive web apps)*, so we need to use its counterpart, the `SecurityWebFilterChain` interface, which deals with `Mono` and `Flux` types. It has the same behavior as `SecurityFilterChain`; it will help to declare any custom filters we need. In our example, we also require the `ServerHttpSecurity`, `ReactiveAuthenticationManager,` and `CorsConfigurationSource` beans as parameters of the method.

- `ServerHttpSecurity`: Very similar to its counterpart (`HttpSecurity` class), the `ServerHttpSecurity` class configures all the security for specific requests. By default (if not used) the security will be applied for all requests (then will get the username user defines and the password printed out in the logs), and in this case we are defining certain rules:

  - We are disabling the CSRF filter.

  - We are configuring the CORS filter.

  - We are defining security access for the `/retros/*` endpoints requests. We are adding the methods `POST` and `DELETE` with role `ADMIN`, and any of these `USER` or `ADMIN` roles for the other Http request, such as `GET`. We are also permitting any / and / `webjars/**` endpoints that correspond to the `index.html` page and resources such as images, CSS, JS scripts. Also, we are using a `ReactiveAuthenticationManager,` and we are defining both *form* and *basic* authentication.

- `ReactiveAuthenticationManager`: This bean helps to contact the Users App. An important detail here is that we are using `Mono` and `Flux` types, so we need to use some of these reactive methods and do the appropriate handling to get the needed value. `ReactiveAuthenticationManager` is a functional interface that defines the `Mono<Authentication> authenticate(Authentication`

authentication) method; and when we get the request, we
can get the information sent, in this case the username and
password, then we can use our UserClient to get the User
info and return the Mono<Authentication> by creating the
UsernamePasswordAuthenticationToken or by throwing an
exception with BadCredentialsException. This is part of the
authentication process. Also, it's important to know that the
UsernamePasswordAuthenticationToken class extends the
AbstractAuthenticationToken abstract class that has the Object
details, where we are going to add the User info (retrieved by the
UserClient class) and use it in the authenticationSuccessHandler
as part of its configuration.

- CorsConfigurationSource: This interface defines the
  CorsConfiguration getCorsConfiguration(ServerWebExchange
  exchange) method, which we can configure by adding the headers
  and origins and allowing credentials for a specific endpoint(s), and in
  this case we are using everything with /** syntax.

- ServerAuthenticationSuccessHandler: We are going to use
  an external Login page (see Figure 9-6), which requires that
  we add some configuration for it to work, so we need to define
  the authenticationSuccessHandler. We are declaring a
  ServerAuthenticationSuccessHandler bean. This interface declares
  the Mono<Void> onAuthenticationSuccess(WebFilterExchange
  webFilterExchange, Authentication authentication) method,
  and it gets invoked when the application authenticates successfully. All
  of this is necessary because our UI will point to the /login endpoint,
  and this configuration will generate the necessary SESSION ID that is
  required for subsequent calls. In other words, this call will be the one
  that has the response that the UI needs, in this case the JSON object that
  is expecting to contain all the User info, so it can apply the necessary
  logic in the front end to differentiate from ADMIN and USER roles.

As you can see, we need to add a little more configuration for Reactive web apps, but
the primary purpose of using Spring Security is the same, and Spring Boot helps a lot by
removing the boilerplate that we would be needed to add if we were using only Spring

apps. Before you proceed to the next section, review the code and analyze it at your own pace. Remember that we are including both authentication and authorization by integrating both apps—the Users App to retrieve user information, and the My Retro App to apply the security based on the rules we set in the ServerHttpSecurity configuration.

Next, let's test the My Retro App.

Unit Testing My Retro App Security for Authorization

To test the My Retro App, we are going to do a unit test rather than an integration test. However, we are going to use the @MockBean that we used in Chapter 8 for integration testing to test the authorization that we set for some of the endpoints because we are going to mock authentication.

Open/create the RetroBoardWebFluxTests class. See Listing 9-14.

Listing 9-14. src/test/java/apress/com/myretro/RetroBoardWebFluxTests.java

```
package com.apress.myretro;

import com.apress.myretro.board.Card;
import com.apress.myretro.board.CardType;
import com.apress.myretro.board.RetroBoard;
import com.apress.myretro.client.UserClient;
import com.apress.myretro.security.RetroBoardSecurityConfig;
import com.apress.myretro.service.RetroBoardService;
import com.apress.myretro.web.RetroBoardController;
import org.junit.jupiter.api.Test;
import org.mockito.Mockito;
import org.springframework.beans.factory.annotation.Autowired;
import org.springframework.boot.test.autoconfigure.web.reactive.
WebFluxTest;
import org.springframework.boot.test.mock.mockito.MockBean;
import org.springframework.context.annotation.Import;
import org.springframework.http.HttpHeaders;
import org.springframework.http.MediaType;
import org.springframework.security.test.context.support.WithMockUser;
import org.springframework.test.web.reactive.server.WebTestClient;
import org.springframework.web.reactive.function.BodyInserters;
```

```
import reactor.core.publisher.Flux;
import reactor.core.publisher.Mono;

import java.util.Arrays;
import java.util.UUID;

@Import({RetroBoardSecurityConfig.class})
@WithMockUser
@WebFluxTest(controllers = {RetroBoardController.class})
public class RetroBoardWebFluxTests {

    @MockBean
    RetroBoardService retroBoardService;

    @MockBean
    UserClient userClient;

    @Autowired
    private WebTestClient webClient;

    @Test
    void getAllRetroBoardTest(){

        Mockito.when(retroBoardService.findAll()).thenReturn(Flux.just(
                new RetroBoard(UUID.randomUUID(),"Simple Retro",
                Arrays.asList(
                        new Card(UUID.randomUUID(),"Happy to be here",
                        CardType.HAPPY),
                        new Card(UUID.randomUUID(),"Meetings everywhere",
                        CardType.SAD),
                        new Card(UUID.randomUUID(),"Vacations?",
                        CardType.MEH),
                        new Card(UUID.randomUUID(),"Awesome Discounts",
                        CardType.HAPPY),
                        new Card(UUID.randomUUID(),"Missed my train",
                        CardType.SAD)
                ))
        ));
```

```
    webClient.get()
            .uri("/retros")
            .accept(MediaType.APPLICATION_JSON)
            .exchange()
            .expectStatus().isOk()
            .expectBody().jsonPath("$[0].name").isEqualTo("Simple
            Retro");

    Mockito.verify(retroBoardService,Mockito.times(1)).findAll();
}

@Test
void findRetroBoardByIdTest() {
    UUID uuid = UUID.randomUUID();
    Mockito.when(retroBoardService.findById(uuid)).
    thenReturn(Mono.just(
            new RetroBoard(uuid, "Simple Retro", Arrays.asList(
                    new Card(UUID.randomUUID(), "Happy to be here",
                    CardType.HAPPY),
                    new Card(UUID.randomUUID(), "Meetings everywhere",
                    CardType.SAD),
                    new Card(UUID.randomUUID(), "Vacations?",
                    CardType.MEH),
                    new Card(UUID.randomUUID(), "Awesome Discounts",
                    CardType.HAPPY),
                    new Card(UUID.randomUUID(), "Missed my train",
                    CardType.SAD)
            ))
    ));

    webClient.get()
            .uri("/retros/{uuid}", uuid.toString())
            .header(HttpHeaders.ACCEPT, MediaType.APPLICATION_
            JSON_VALUE)
            .exchange()
            .expectStatus().isOk()
            .expectBody(RetroBoard.class);
```

```
        Mockito.verify(retroBoardService, Mockito.times(1)).findById(uuid);
    }

    @Test
    @WithMockUser(roles = "ADMIN")
    void saveRetroBoardTest(){
        RetroBoard retroBoard = new RetroBoard();
        retroBoard.setName("Simple Retro");

        Mockito.when(retroBoardService.save(retroBoard))
                .thenReturn(Mono.just(retroBoard));

        webClient.post()
                .uri("/retros")
                .contentType(MediaType.APPLICATION_JSON)
                .body(BodyInserters.fromValue(retroBoard))
                .exchange()
                .expectStatus().isOk();

        Mockito.verify(retroBoardService,Mockito.times(1)).
        save(retroBoard);
    }

    @Test
    @WithMockUser(roles = "ADMIN")
    void deleteRetroBoardTest(){
        UUID uuid = UUID.randomUUID();
        Mockito.when(retroBoardService.delete(uuid)).thenReturn(Mono.empty());

        webClient.delete()
                .uri("/retros/{uuid}",uuid.toString())
                .exchange()
                .expectStatus().isOk();

        Mockito.verify(retroBoardService,Mockito.times(1)).delete(uuid);
    }

}
```

The `RetroBoardWebFluxTests` class includes the following annotations:

- `@MockBean`: We are using this annotation in two main classes, `RetroBoardService` and `UserClient`, because we want just to apply the authorization and see if the roles work.

- `@WithMockUser`: We are using this annotation at the class level, but in the `saveRetroBoardTest` and `deleteRetroBoardTest` methods, we are overwriting it by adding the `roles` parameter and defining the `ADMIN` role. If you remove this annotation from these methods, the tests will fail.

If you run the tests, they should pass without any issues. Again, as shown at the end of Listing 9-14, we are doing just unit testing by mocking some services, but testing the authorization is one of the important key elements we added in this project.

Putting Everything Together: UI, Users App, and Retro App

To put everything together, we are going to use the My Retro App UI (front-end app), which you can find in the `09-security/myretro-ui` folder. Please review and analyze it before running it.

Let's first review the important file that includes the security configuration. Open/create in the `myretro-ui` folder the `login.js` file. See Listing 9-15.

Listing 9-15. myretro-ui/assets/js/login.js

```
$(function() {

  $('#btn__login').click(function(e) {
    e.preventDefault();
    let user_email = $('input[name="user_email"]').val();
    let user_password = $('input[name="user_password"]').val();
    $.ajax({
      url: 'http://localhost:8081/login',
      method: 'POST',
        data: {
            username: user_email,
            password: user_password
        },
```

```
    xhrFields: {
        withCredentials: true
    },
  headers:{
    'Access-Control-Allow-Origin': '*'
  }
})
    .done(function( response, textStatus, xhr ) {
      console.log(response)

      if(xhr.status === 200 &&
          response.active === true &&
          response.email === user_email &&
          response.password === user_password ) {

        let isAdmin = response.userRole.includes("ADMIN");

        localStorage.setItem('ls_name', response.name);
        localStorage.setItem('ls_email', response.email);
        localStorage.setItem('ls_gravatar', response.gravatarUrl);
        localStorage.setItem('ls_role', response.userRole);

        document.cookie = xhr.getResponseHeader('x-myretro');

        console.log(xhr.getAllResponseHeaders());

        if (isAdmin){
          window.location.href = "admin-dashboard.html";
        }else {
          window.location.href = "admin-users.html";
        }

        return;
      }
    });
  });
});
```

In the $.ajax call in login.js, we are using the POST HTTP method and pointing to the /login endpoint. We didn't specify this endpoint in the My Retro App, but Spring Security defines it by default; the same happens for the /logout endpoint. Also note that we are using the xhrFields.withCredentials that will allow us to send credentials (cookies, headers, TLS client certificates) for cross-site access control requests (remember that this is an external app, the UI, and we required to do this). After a successful login, it will save the session ID for subsequent calls.

Now, it's time to run the apps and see if works. These are the main step to make it work:

1. Run the Users App, which will run on port 8080.

2. Run the My Retro App, which will run on port 8081.

3. Run the *myretro-ui* using the index.html file. You can use your IDE (such as VS Code or WebStorm) or use Python.
    If you use Python, you need to be in the root of the myretro-ui folder and execute the following command,:

```
python3 -m http.server
```

This will open port 8000. Now you can go to your browser and open it. You should see the Login page shown in Figure 9-7.

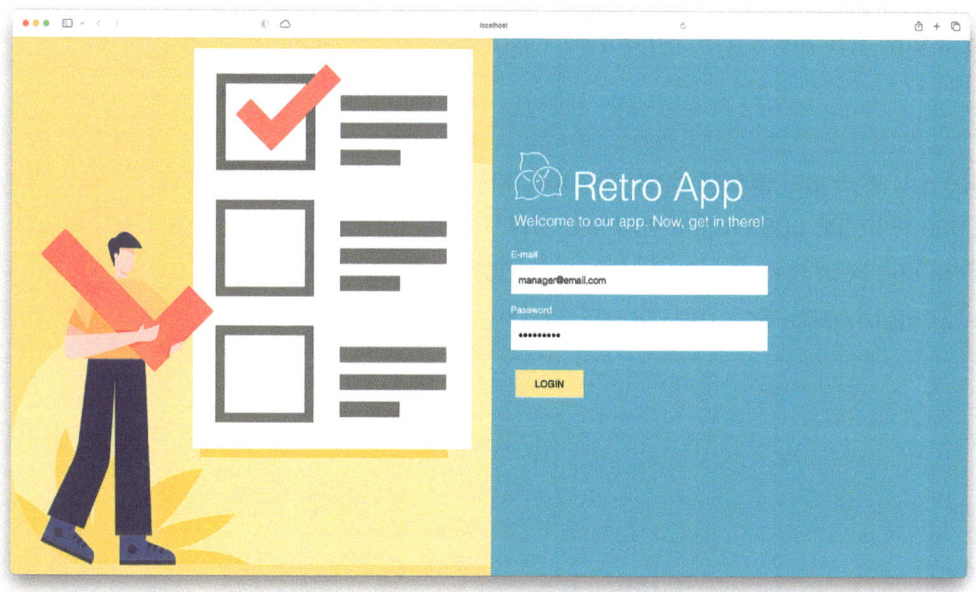

Figure 9-7. *Login page for My Retro App (the same as Users App)*

Enter manager@email.com in the E-mail field and aw2s0meR! in the Password field
and then click the Login button.

You should see the dashboard of the My Retro App, shown in Figure 9-8, where you
can manage Retros or Users.

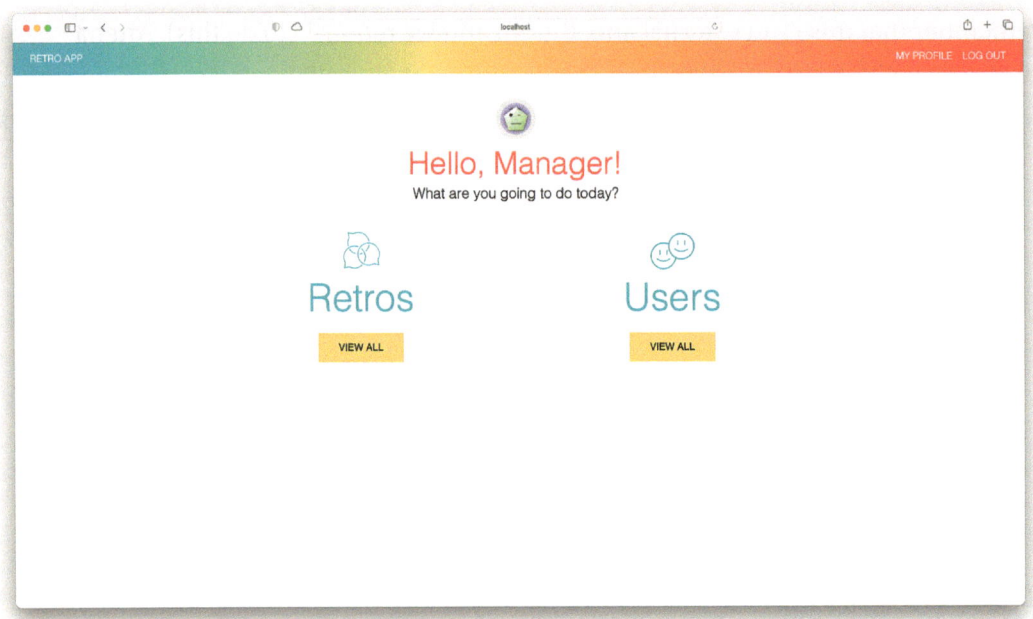

Figure 9-8. *My Retro App after a successful login*

Note   This is just to test Security, and not too much on functionality. Please, play
around with the apps and if you find any UI bug, let me know, or you can help by
submitting your own PR in GitHub with more UI details: https://github.com/
felipeg48/pro-spring-boot-3rd.

Adding Social Login with OAuth2

Nowadays, many applications require users to authenticate and be authorized by third-party entities, and normally the entities are social media platforms such as Facebook, Google, X (formally Twitter), GitHub, and so forth. In this section, we are going to configure the Users App to authenticate through GitHub. I will take you step by step through the configuration, and you will discover that the process is very simple by using Spring Boot, which will help to wire up everything so that you can use any of the social media entities.

By using a Social media platform you are providing to your apps a single sign-on (SSO) capability that allows you to simplify registration and logins for your end users. When doing Social media Login, you get some useful features:

- Email verification

- User profiles

- One-click experience

And if you think about it, because social media is the most popular way to communicate to the world, this capability will bring more registrations and more traffic to your applications. So, let's get started with the Users App.

Social Login to the Users App

You have access to the code, so go ahead and import the Users App in 09-security-social/users to your favorite IDE. Then, open the build.gradle file. See Listing 9-16.

Listing 9-16. build.gradle

```
plugins {
    id 'java'
    id 'org.springframework.boot' version '3.2.3'
    id 'io.spring.dependency-management' version '1.1.4'
}

group = 'com.apress'
version = '0.0.1-SNAPSHOT'
sourceCompatibility = '17'
```

```
repositories {
    mavenCentral()
}

dependencies {
    implementation 'org.springframework.boot:spring-boot-starter-web'
    implementation 'org.springframework.boot:spring-boot-starter-
    validation'
    implementation 'org.springframework.boot:spring-boot-starter-security'
    implementation 'org.springframework.boot:spring-boot-starter-data-jpa'
    implementation 'org.springframework.boot:spring-boot-starter-
    oauth2-client'

    runtimeOnly 'com.h2database:h2'
    runtimeOnly 'org.postgresql:postgresql'

    compileOnly 'org.projectlombok:lombok'
    annotationProcessor 'org.projectlombok:lombok'

    // Web
    implementation 'org.webjars:bootstrap:5.2.3'

    testImplementation 'org.springframework.boot:spring-boot-starter-test'
    testImplementation 'org.springframework.security:spring-security-test'
}

tasks.named('test') {
    useJUnitPlatform()
}

test {
    testLogging {
        events "passed", "skipped", "failed" //, "standardOut",
        "standardError"

        showExceptions true
        exceptionFormat "full"
        showCauses true
        showStackTraces true
```

```
        // Change to `true` for more verbose test output
        showStandardStreams = false
    }
}
```

Listing 9-16 shows that we are adding to the dependencies the familiar `spring-boot-starter-security` and the new dependency `spring-boot-starter-oauth2-client`. Once Spring Boot identifies this new dependency (through auto-configuration), it will wire up everything that your app needs to use Social Login using the OAuth 2 protocol and client.

Next, open/create the `UserSecurityConfig` class. See Listing 9-17.

Listing 9-17. src/main/java/apress/com/users/security/UserSecurityConfig.java

```
package com.apress.users.security;

import org.springframework.context.annotation.Bean;
import org.springframework.context.annotation.Configuration;
import org.springframework.security.config.Customizer;
import org.springframework.security.config.annotation.web.builders.
HttpSecurity;
import org.springframework.security.web.SecurityFilterChain;
import org.springframework.security.web.csrf.CookieCsrfTokenRepository;
import org.springframework.security.web.util.matcher.AntPathRequestMatcher;

@Configuration
public class UserSecurityConfig {

    @Bean
    SecurityFilterChain filterChain(HttpSecurity http) throws Exception {
        http
                .csrf( c -> c.csrfTokenRepository(CookieCsrfTokenReposito
                ry.withHttpOnlyFalse()))
                .authorizeHttpRequests( auth ->
                        auth
                                .requestMatchers(new
                                AntPathRequestMatcher("/index.html"),
```

```
                                    new AntPathRequestMatcher("/
                                    webjars/**"),
                                    new AntPathRequestMatcher("/
                                    error")).permitAll()
                               .anyRequest().authenticated()
                    )
                .logout( l -> l.logoutSuccessUrl("/index.html").
                permitAll() )
                .oauth2Login(Customizer.withDefaults());

        return http.build();
    }
}
```

Let's review UserSecurityConfig class. First, we are defining our security filter by providing the SecurityFilterChain bean and configuring the HttpSecurity class. We are configuring the CSRF filter and setting the token repository withHttpOnlyFalse, which is necessary in the Spring Security OAuth 2 client because it allows the CSRF token to be read by JavaScript, and because we are going to use GitHub, the pages of which need this feature. Also, we are permitting access to the index.html page, the /error endpoint, and the assets that live in the /webjars/** path; everything else should be authenticated. We had declared the /logout and redirecting to the home (/index.html) with no security. Finally, the aouth2Login(Customizer.withDefaults()) configuration will use the social login mechanism to go to the Github page, let you know that you are about to use the Users App and that you need to give the right permissions, then, after accepting the access, you are allowed to use the Users App endpoints. Behind the scenes, Spring Boot already configured everything that is necessary to start the OAuth 2 protocol with GitHub. See Figure 9-9.

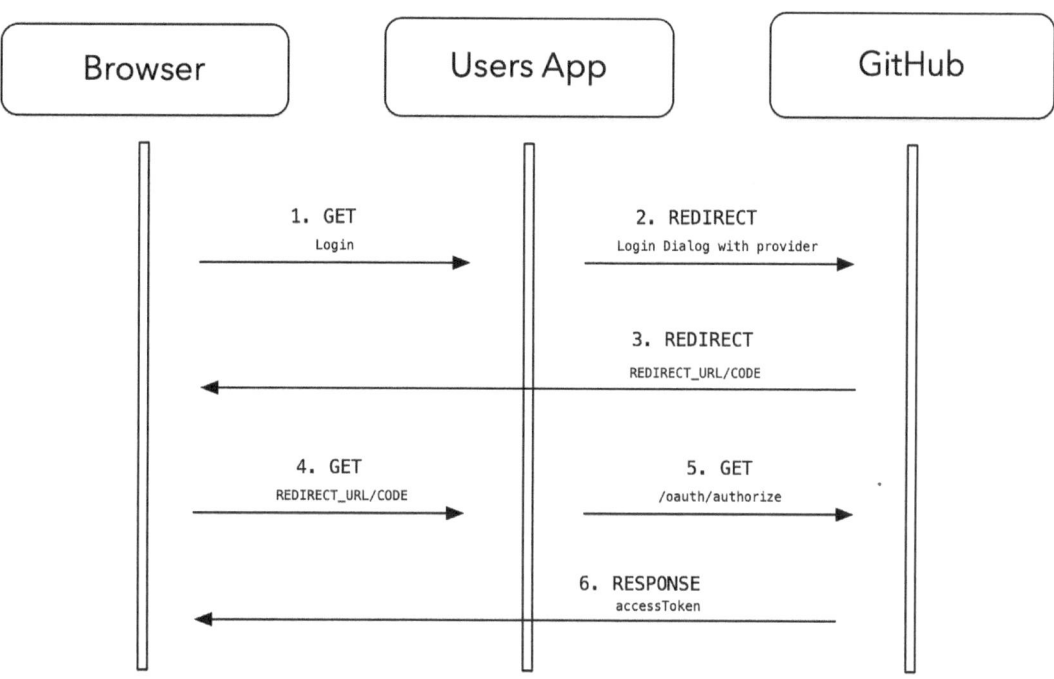

Figure 9-9. *Sequence diagram for Social Login*

Next, open/create the `application.yaml` file. See Listing 9-18.

Listing 9-18. src/main/resources/application.yaml

```
spring:
  h2:
    console:
      enabled: true
  datasource:
    generate-unique-name: false
    name: test-db
  jpa:
    show-sql: true

  security:
      oauth2:
        client:
          registration:
            github:
```

```
        client-id: ${GITHUB_CLIENT_ID}
        client-secret: ${GITHUB_CLIENT_SECRET}
        scope:
          - user:email
          - https://github.com/login/oauth/authorize

logging:
  level:
    org:
      springframework:
        security: DEBUG
```

Listing 9-18 shows the application.yaml properties. The important part is the security.oauth2.client.* properties. We are using the registration.github.* and we need to provide the client-id and the client-secret keys. In this case, Spring Boot reads this file and looks for such keys using the environment variables GITHUB_CLIENT_ ID and GITHUB_CLIENT_SECRET. Keeping your secret and password keys as environment variables or secrets is a best practice if you are using Kubernetes infrastructure for deployment.

The following is the process to get these keys:

1. Go to https://github.com/settings/developers and sign in. If you don't have an account, click the link to create an account (it's free), create your account, and then sign in. Make sure you are authenticated.

2. Click the New OAuth App button on the right and fill out the following information in the registration form (see Figure 9-10):

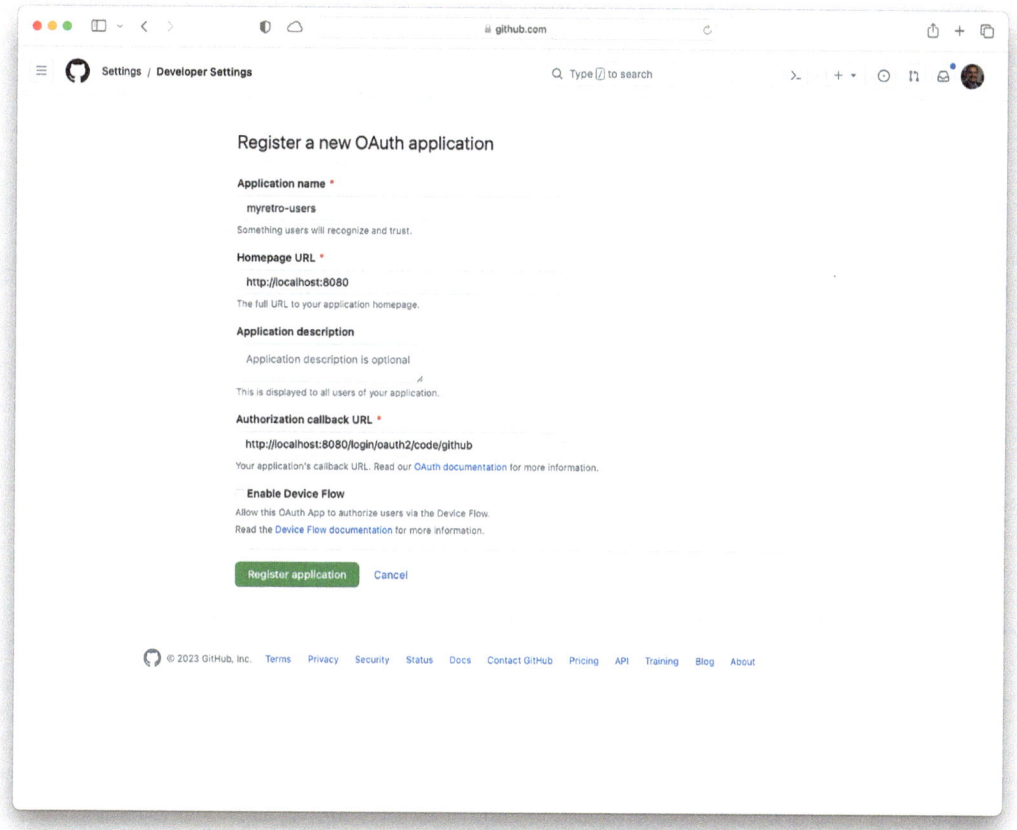

Figure 9-10. *Registering a new OAuth app on GitHub*

- Application name: `myretro-users`

- Homepage URL: `http://localhost:8080`

- Authorization callback URL: `http://localhost:8080/login/oauth2/code/github`

3. Click the Register Application button. You will see the page shown in Figure 9-11.

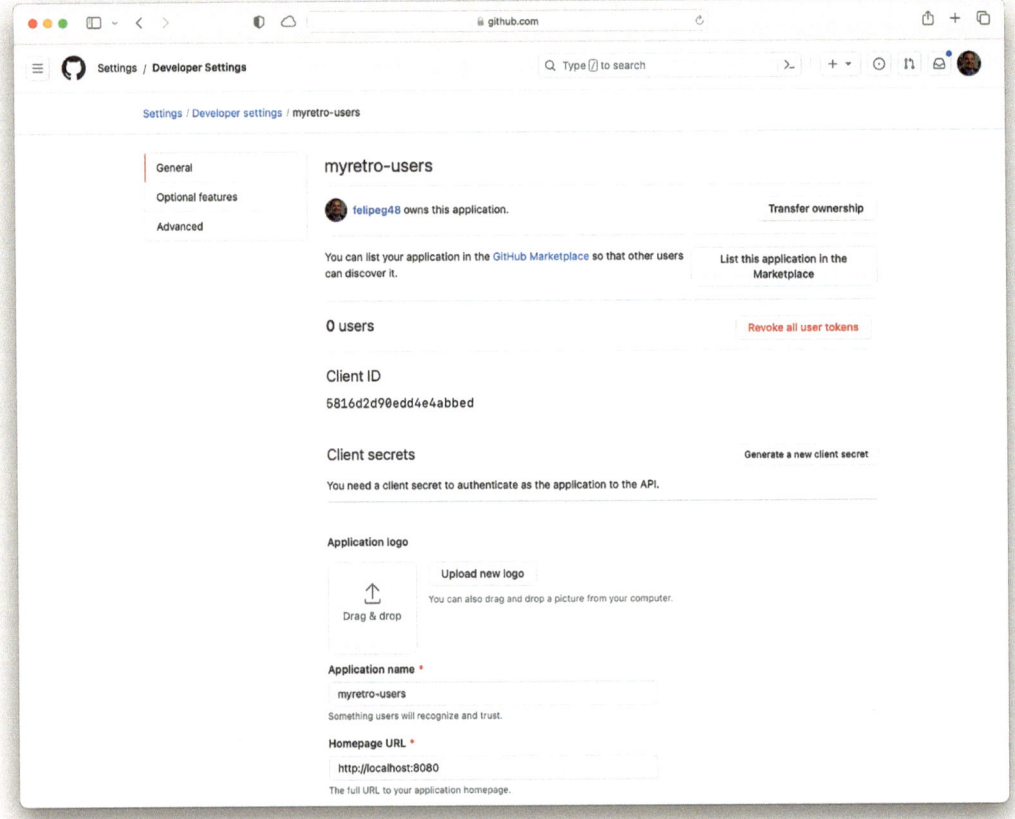

Figure 9-11. *New OAuth app*

4. In the Client Secrets section, click the Generate a New Client
    Secret button. Copy the key that is displayed in the pane. See
    Figure 9-12.

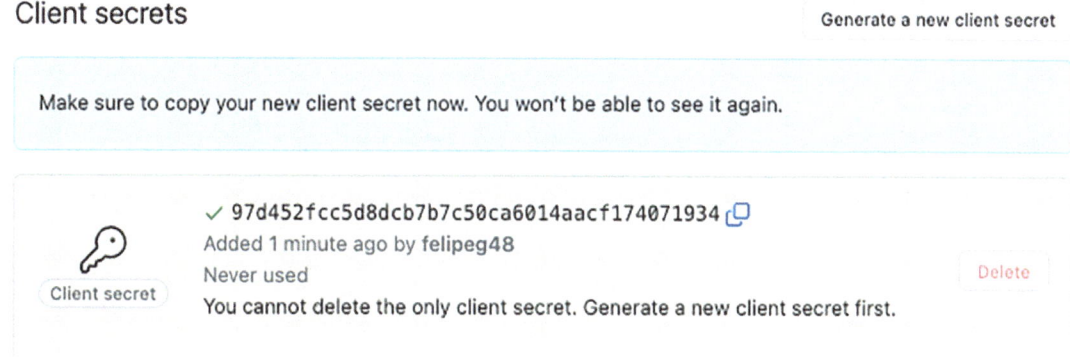

Figure 9-12. *Copying the client secret*

So, copy both keys, the Client ID and your new Client Secrets keys (these keys will be in the environment variable we saw earlier). Then, click the Update Application button, and that's it. The Users App will update to include GitHub as the social login provider.

Running the Users App with Social Login

Now, it's time to run the User App and see the results of the new login functionality. Before you run the Users App, make sure you set the GITHUB_CLIENT_ID and GITHUB_CLIENT_SECRET environment variables with the Client ID and Client secret key from the GitHub pages. You can use you IDE to set those variables (consult your IDE's documentation for help on how to set up environment variables).

If you are working from the command line, you can use something like this:

```
GITHUB_CLIENT_ID=xxxx GITHUB_CLIENT_SECRET=xxxx ./gradlew clean bootRUN
```

For example, I'm using IntelliJ, so I can configure the environment variables in Run/Debug Configuration dialog box, as shown in Figure 9-13.

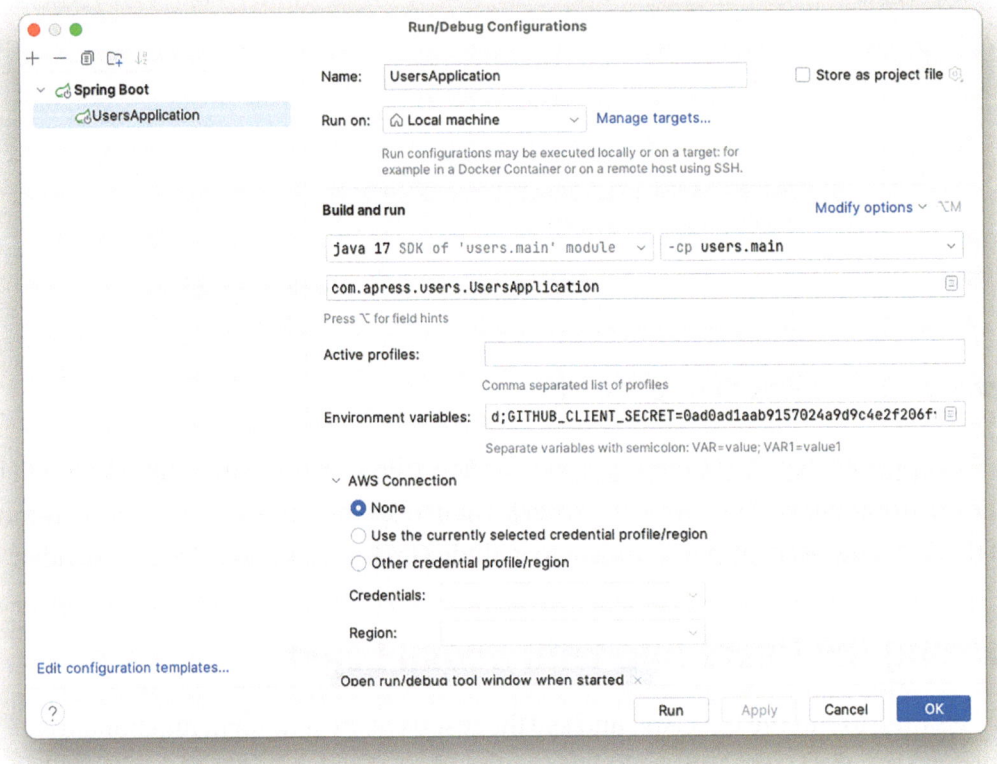

Figure 9-13. *IntelliJ Run/Debug Configurations dialog box*

Once you have the Users App up and running, open your browser and go to: `http://localhost:8080`; you will see the home page shown in Figure 9-14.

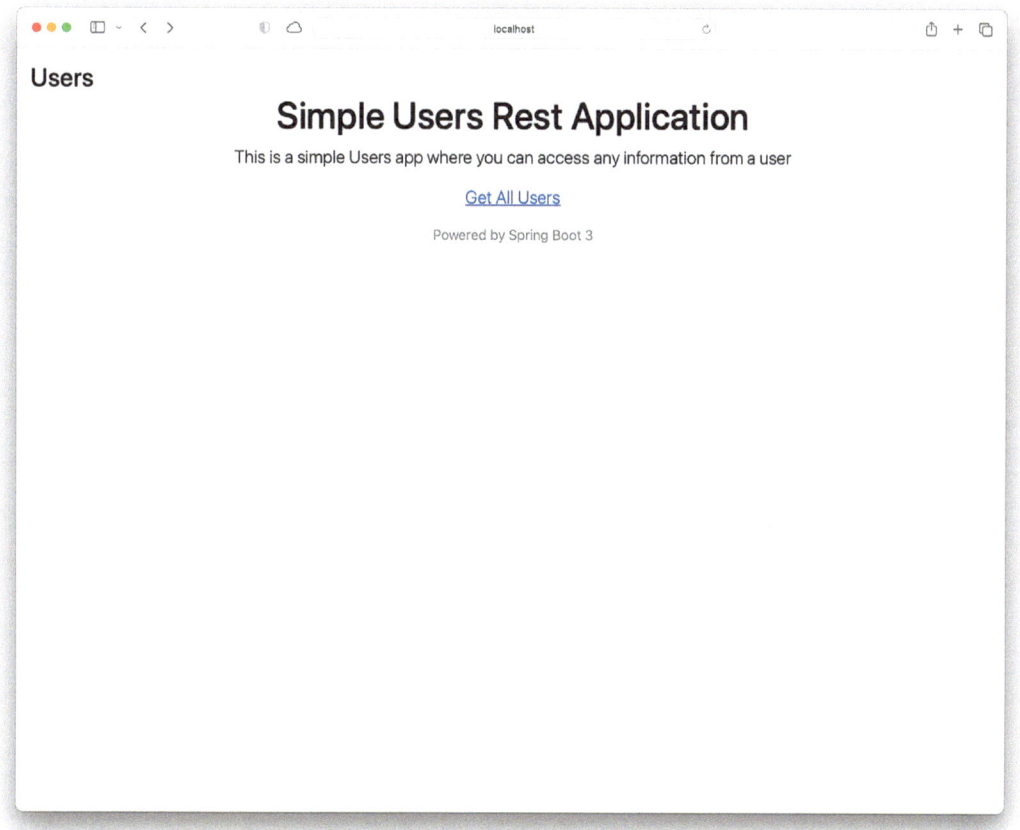

Figure 9-14. *Home page at* `http://localhost:8080`

Click the Get All Users link, and you will be redirected to a GitHub page similar to the one shown in Figure 9-15.

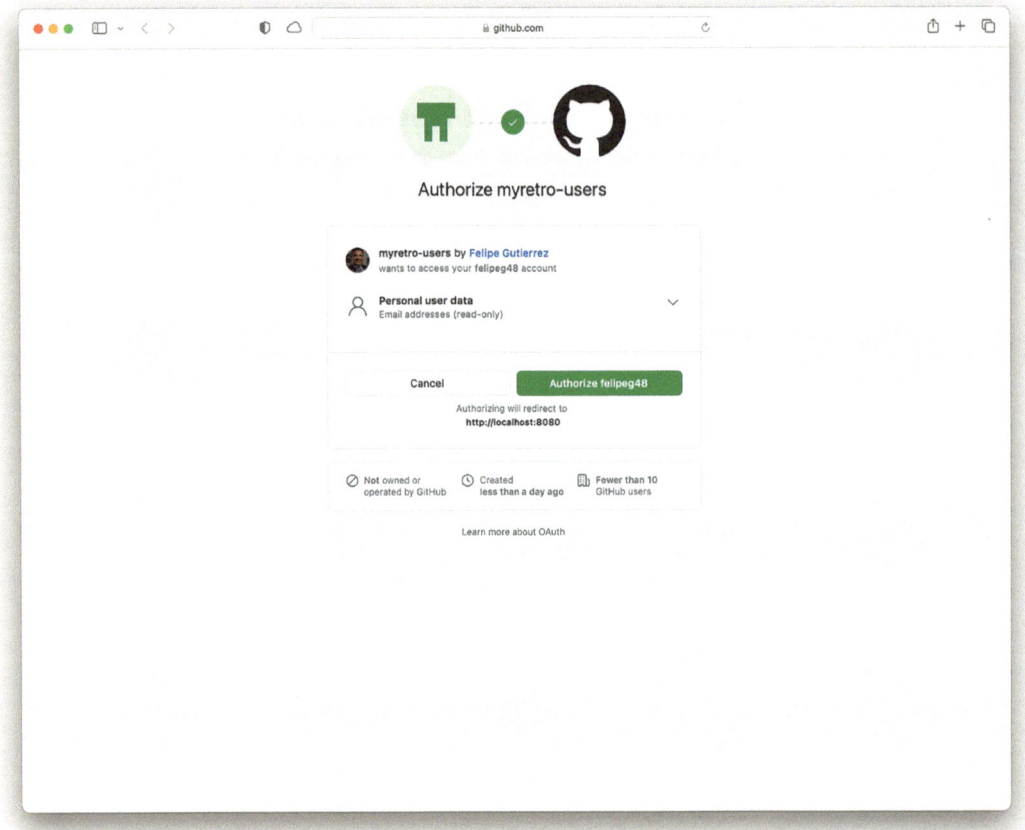

Figure 9-15. *GitHub prompt for authorization*

Where either you need to log in and then authorize that app myretro-users to use the email for read only. After you accept the authorization, you will see the JSON response with all the users.

That's it! Now you have a Social Login for the Users App. If you review it, it's very simple. And remember that behind the scenes, Spring Boot is doing most of the heavy lifting with the Spring Security framework.

Spring Authorization Server

The Spring Security team recently released the new *Spring Authorization Server*, which provides the implementation of the OAuth 2.1 and OpenID Connect 1.0 specifications. Of course, it is built on top of the Spring Security Framework. You can get more insight here: `https://spring.io/projects/spring-authorization-server`.

Getting Started

To use Spring Authorization Server, you need to add the dependency `spring-boot-starter-oauth2-authorization-server` and perform a bit of configuration. If you are following along, open the project `09-oauth2/oauth2-authentication-server` directory.

If you want to start the project from scratch, go to the Spring Initializr (`https://start.spring.io`), click Add Dependencies, and select the dependencies OAuth2 Authorization Server and Spring Web. Set the Group field to `apress.security` and the Artifact and Name fields to `oauth2`. Accept the defaults for the rest of the options and click Generate.

Next, open the `build.gradle` file; it should look like Listing 9-19.

Listing 9-19. build.gradle

```
plugins {
    id 'java'
    id 'org.springframework.boot' version '3.2.3'
    id 'io.spring.dependency-management' version '1.1.4'
}

group = 'com.apress.security'
version = '0.0.1-SNAPSHOT'

java {
    sourceCompatibility = '17'
}

repositories {
    mavenCentral()
}

dependencies {
    implementation 'org.springframework.boot:spring-boot-starter-web'
    implementation 'org.springframework.boot:spring-boot-starter-security'
    implementation 'org.springframework.boot:spring-boot-starter-oauth2-
    authorization-server'
    testImplementation 'org.springframework.boot:spring-boot-starter-test'
}
```

```
tasks.named('test') {
    useJUnitPlatform()
}
```

Next, we are going to add some users that will be authenticated in Spring Authorization Server. Open/create the Oauth2Config class. See Listing 9-20.

Listing 9-20. src/main/java/apress/security/oauth2/Oauth2Config.java

```java
package com.apress.security.oauth2;

import org.springframework.context.annotation.Bean;
import org.springframework.context.annotation.Configuration;
import org.springframework.security.core.userdetails.User;
import org.springframework.security.core.userdetails.UserDetails;
import org.springframework.security.crypto.bcrypt.BCryptPasswordEncoder;
import org.springframework.security.crypto.password.PasswordEncoder;
import org.springframework.security.provisioning.
InMemoryUserDetailsManager;

@Configuration
public class Oauth2Config {

    @Bean
    InMemoryUserDetailsManager inMemoryUserDetailsManager(PasswordEncoder
    passwordEncoder){
        UserDetails admin = User
                .builder()
                .username("admin")
                .password(passwordEncoder.encode("admin"))
                .authorities("users.read","users.write")
                .build();

        UserDetails manager = User
                .builder()
                .username("manager@email.com")
                .password(passwordEncoder.encode("aw2sOmeR!"))
                .authorities("users.read","users.write")
                .build();
```

```
        UserDetails user = User
                .builder()
                .username("user@email.com")
                .password(passwordEncoder.encode("aw2sOmeR!"))
                .authorities("users.read")
                .build();

        return new InMemoryUserDetailsManager(manager,user,admin);
    }

    @Bean
    PasswordEncoder passwordEncoder(){
        return new BCryptPasswordEncoder();
    }
}
```

Note in Listing 9-20 that we are using the InMemoryUserDetailsManager for simplicity, but you already know how to use the JDBC or any custom access to user details.

Finally, we need to add the configuration. Open/create the application.yaml file and use the content shown in Listing 9-21.

Listing 9-21. src/main/resources/application.yaml

```
## Server
server:
  port: ${PORT:9000}

## Logging
logging:
  level:
    org.springframework.security: trace

## Bcrypt - Cost Factor 10
## https://bcrypt.online/
## Spring Security
spring:
  security:
    oauth2:
```

```yaml
  authorizationserver:
    client:
      users-client:
        registration:
          client-name: "Users' Client"
          client-id: "users-client"
          client-secret: "$2y$10$4Da3ibamZ5Jo34gs7HUhLuNmbPdEhxqWrR9
v/Z.qosEmWbYYVHCZe"
          client-authentication-methods:
            - "client_secret_basic"
          authorization-grant-types:
            - "authorization_code"
            - "refresh_token"
            - "client_credentials"
          redirect-uris:
            - "http://127.0.0.1:8181/login/oauth2/code/users-client-
authorization-code"
            - "http://127.0.0.1:8181/users"
          post-logout-redirect-uris:
            - "http://127.0.0.1:8181/"
          scopes:
            - "openid"
            - "profile"
            - "users.read"
            - "users.write"
        require-authorization-consent: true
      myretro-client:
        registration:
          client-name: "MyRetro Client"
          client-id: "myretro-client"
          client-secret: "$2y$10$KzJqWNyybMyX8.OEJNAGI.YRI4M/
FuOZcizXQboZ4YDfQfzG9ZmrK"
          client-authentication-methods:
            - "client_secret_basic"
          authorization-grant-types:
```

```
    - "authorization_code"
    - "refresh_token"
    - "client_credentials"
  redirect-uris:
    - "http://127.0.0.1:8080/login/oauth2/code/myretro-client-
    authorization-code"
  post-logout-redirect-uris:
    - "http://127.0.0.1:8080/"
  scopes:
    - "openid"
    - "profile"
    - "retros:read"
    - "retros:write"
  require-authorization-consent: true
```

In the `application.yaml` file, we are defining the `oauth2` configuration needed for our two projects. The idea is that both projects can use Spring Authorization Server as the authentication and authorization mechanism instead of having that logic locally. Review the configuration.

It's important to notice that after the `spring.security.oauth2.authorizationserver.client.*` properties, you specify your custom registration, in this case the `users-client.*` (for the Users App) and `myretro-client.*` (for the My Retro App).

Look at the `client-secret` property. It's already hashed using BCrypt as the algorithm, with a cost factor of 10, the actual word is `secret`, and its BCrypt hash is `$2y$10$7jJ/Gil7n5tlQJOuSbMiMOcrxH7m9SmSwtBwdFS4XEzBHUCJkYeQG`. You can use the Bcrypt Hash Generator at `https://bcrypt.online/` to hash any other secret.

Also, look at the `redirect-uris` (some of them must be used in the apps that are going to be authenticated against the server) and the `scopes` that are required to consent to the access to the page. These will be displayed once you log in.

That's it! The process is similar to the process described in the previous section for creating the access in the GitHub Developer page, where the `client-id` and `client-secret` were generated by GitHub.

Now, you can run your Spring Authorization Server app. This will start listening on the port 9000.

Using Spring Authentication Server with the Users App

To use Spring Authorization Server with the Users App, you need to add two dependencies: `spring-boot-starter-oauth2-authorization-server` and `spring-boot-starter-oauth2-client`. You can open the `09-oauth2/users` project and import it in your favorite IDE.

Open the `build.gradle` file. It should look like Listing 9-22.

Listing 9-22. build.gradle

```
plugins {
    id 'java'
    id 'org.springframework.boot' version '3.2.3'
    id 'io.spring.dependency-management' version '1.1.4'
}

group = 'com.apress'
version = '0.0.1-SNAPSHOT'
sourceCompatibility = '17'

repositories {
    mavenCentral()
}

dependencies {
    implementation 'org.springframework.boot:spring-boot-starter-web'
    implementation 'org.springframework.boot:spring-boot-starter-webflux'
    implementation 'org.springframework.boot:spring-boot-starter-
    validation'
    implementation 'org.springframework.boot:spring-boot-starter-security'
    implementation 'org.springframework.boot:spring-boot-starter-oauth2-
    authorization-server'
    implementation "org.springframework.boot:spring-boot-starter-
    oauth2-client"
    implementation 'org.springframework.boot:spring-boot-starter-data-jpa'

    runtimeOnly 'com.h2database:h2'
    runtimeOnly 'org.postgresql:postgresql'
```

```
    compileOnly 'org.projectlombok:lombok'
    annotationProcessor 'org.projectlombok:lombok'

    // Web
    implementation 'org.webjars:bootstrap:5.2.3'
    implementation 'org.webjars:webjars-locator-core'

    testImplementation 'org.springframework.boot:spring-boot-starter-test'
    testImplementation 'org.springframework.security:spring-security-test'
}

tasks.named('test') {
    useJUnitPlatform()
}

test {
    testLogging {
        events "passed", "skipped", "failed" //, "standardOut",
        "standardError"

        showExceptions true
        exceptionFormat "full"
        showCauses true
        showStackTraces true

        // Change to `true` for more verbose test output
        showStandardStreams = false
    }
}
```

Next, open the UserSecurityConfig class. See Listing 9-23.

Listing 9-23. src/main/java/apress/com/users/security/UserSecurityConfig.java

```
import org.springframework.context.annotation.Bean;
import org.springframework.context.annotation.Configuration;
import org.springframework.security.config.Customizer;
import org.springframework.security.config.annotation.web.builders.
HttpSecurity;
```

```
import org.springframework.security.config.annotation.web.configuration.
WebSecurityCustomizer;
import org.springframework.security.web.SecurityFilterChain;
import org.springframework.web.cors.CorsConfiguration;
import org.springframework.web.cors.CorsConfigurationSource;
import org.springframework.web.cors.UrlBasedCorsConfigurationSource;

import java.util.Arrays;

@Configuration
public class UserSecurityConfig {

    @Bean
    public WebSecurityCustomizer webSecurityCustomizer() {
        return (web) -> web.ignoring().requestMatchers("/webjars/**");
    }

    @Bean
    SecurityFilterChain filterChain(HttpSecurity http,
    CorsConfigurationSource corsConfigurationSource) throws Exception {
        http
                .csrf(csrf -> csrf.disable())
                .cors(cors -> cors.configurationSource(corsConfiguration
                Source))
                .authorizeHttpRequests( auth ->
                        auth

                                .requestMatchers("/users/**").
                                hasAnyAuthority("SCOPE_users.read",
                                "SCOPE_users.write")
                                .requestMatchers("/logout").permitAll())

                .oauth2Login(login -> login
                        .loginPage("/oauth2/authorization/users-client-
                        authorization-code")
                        .defaultSuccessUrl("/users")
                        .permitAll())
```

CHAPTER 9   SPRING BOOT SECURITY

```
        .oauth2ResourceServer(config -> config.jwt(Customizer.
        withDefaults()));

    return http.build();
}

@Bean
CorsConfigurationSource corsConfigurationSource() {
    CorsConfiguration configuration = new CorsConfiguration();

    configuration.setAllowedOrigins(Arrays.asList("*"));
    configuration.setAllowedMethods(Arrays.asList("*"));
    configuration.setAllowedHeaders(Arrays.asList("*"));

    UrlBasedCorsConfigurationSource source = new
    UrlBasedCorsConfigurationSource();
    source.registerCorsConfiguration("/**", configuration);
    return source;
}

}
```

Note that the requestMatchers() method now includes SCOPE_users.read and SCOPE_users.write. By convention, the SCOPE_ is used when you are using OAuth 2 (and ROLE_ is used when you are using database or other authentication and authorization mechanisms). When Spring Authorization Server provides the scopes (based on the configuration shown in Listing 9-21), notice that they are not with the SCOPE_. Spring Security will add them and must match with the ones you are providing here.

Also notice in Listing 9-23 that now we have the oauth2Login that we are configuring with .loginPage("/oauth2/authorization/users-client-authorization-code"). These values must match the ones declared in Listing 9-21. And there is no special configuration for the resource server (with oauth2ResourceServer), so the default will be sufficient.

Next, to see the configuration needed, open the application.properties file. See Listing 9-24.

Listing 9-24. src/main/resources/application.properties

```
# Server
server.port=${PORT:8181}
oauth2.server.port=${OAUTH2_PORT:9000}

# H2 Console
spring.h2.console.enabled=false

# Datasource
spring.datasource.generate-unique-name=false
spring.datasource.name=test-db

# JPA
spring.jpa.show-sql=true
```

Authorization Server
spring.security.oauth2.resourceserver.jwt.issuer-uri=http://localhost:9000

Client Registration
spring.security.oauth2.client.registration.users-client-authorization-code.
provider=spring
spring.security.oauth2.client.registration.users-client-authorization-code.
client-id=users-client
spring.security.oauth2.client.registration.users-client-authorization-code.
client-secret=secret
spring.security.oauth2.client.registration.users-client-authorization-code.
authorization-grant-type=authorization_code
spring.security.oauth2.client.registration.users-client-authorization-code.
redirect-uri=http://127.0.0.1:8181/login/oauth2/code/{registrationId}

spring.security.oauth2.client.registration.users-client-authorization-code.
scope=openid,profile,users.read,users.write
spring.security.oauth2.client.registration.users-client-authorization-code.
client-name=users-client-authorization-code

 # Provider
spring.security.oauth2.client.provider.spring.issuer-uri=http://
localhost:9000

```
# Logging
logging.level.org.springframework.security=TRACE
```

One of the most important pieces in Listing 9-24 is the `*.client.registration.*` properties. The names of these properties must be exactly the same as the names of the `*.users-client-authorization-code.*` properties when we declared the `redirect-uris` in Listing 9-21.

So, in Spring Authorization Server, we declared `http://127.0.0.1:8181/login/oauth2/code/users-client-authorization-code`, the last segment of which (`users-client-authorization-code`) must be part of the registration properties.

Also, we are declaring that the provider is Spring Authorization Server, `http://localhost:9000`. Again, this configuration is the same as we did in the "Social Login to the Users App" section with GitHub.

Running the Users App

Before running the Users App, make sure Spring Authorization Server is up and running. Then, run the Users App, and you should see some activity in the log console for Spring Authorization Server.

Once the Users App is running, open your browser and go to `http://localhost:8181/users`. Automatically it will redirect you to a Login page (`http://localhost:9000/login`). See Figure 9-16.

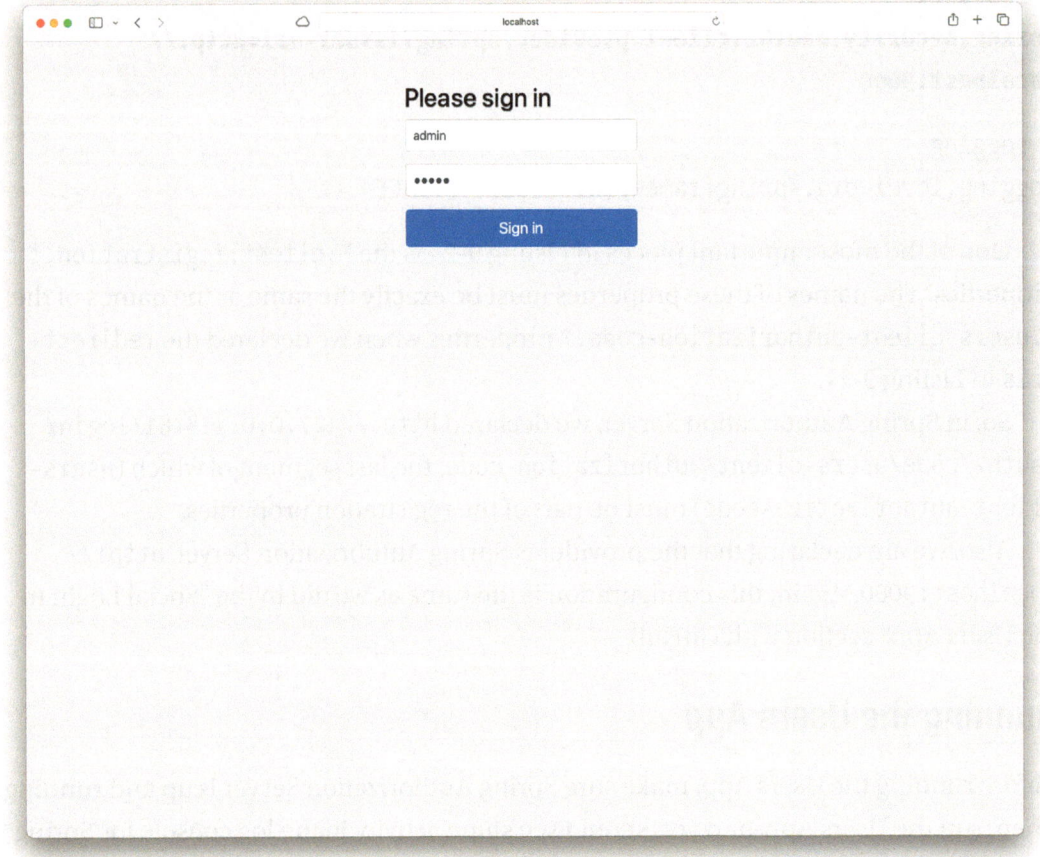

Figure 9-16. *http://localhost:8181/users automatically redirected to http://*
localhost:9000/login

In the fields, enter the credentials admin and admin. These are set in Spring
Authorization Server (see the configuration in Listing 9-20).

Click the Sign In button, and you will be redirected to the Consent Required page
shown in Figure 9-17.

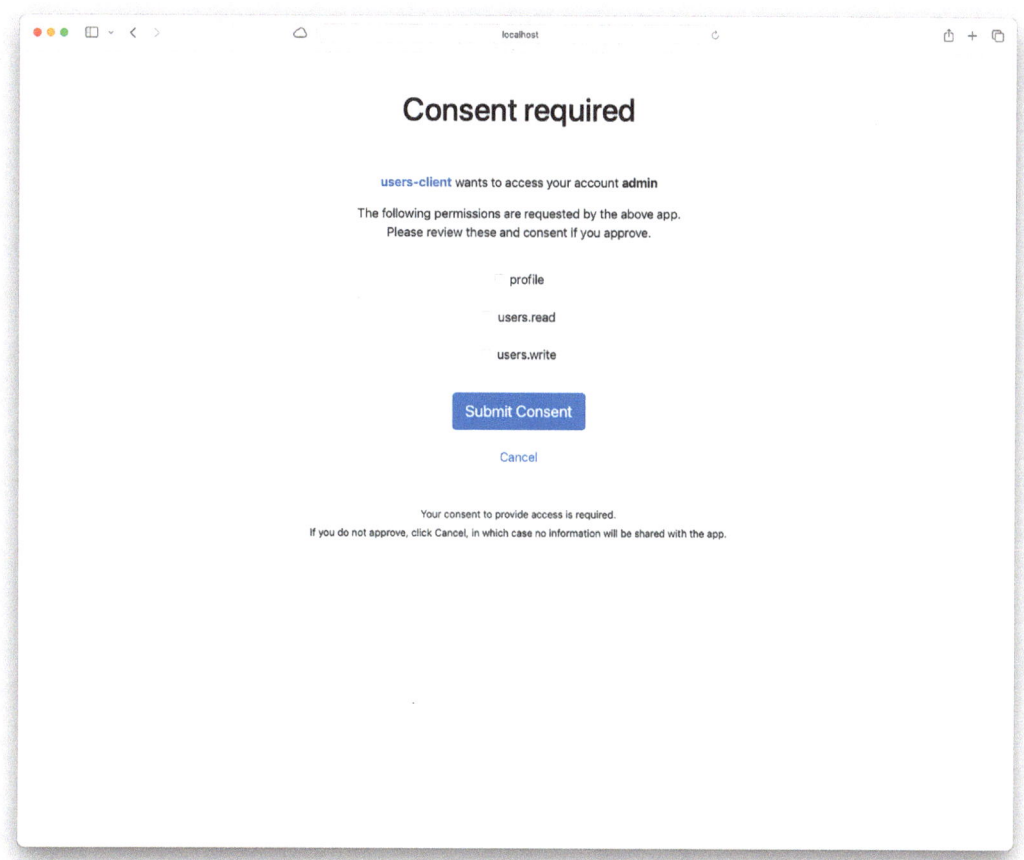

Figure 9-17. *Consent page*

On the Consent Required page, you need to select what you want to share as part of the authorization process. Of course, the /users endpoint requires either users.read or users.write, which will be transformed to SCOPE_users.read and SCOPE_users.write, respectively.

Once you click Submit Consent, you will be presented with the original request, http://localhost:8181/users. See Figure 9-18.

475

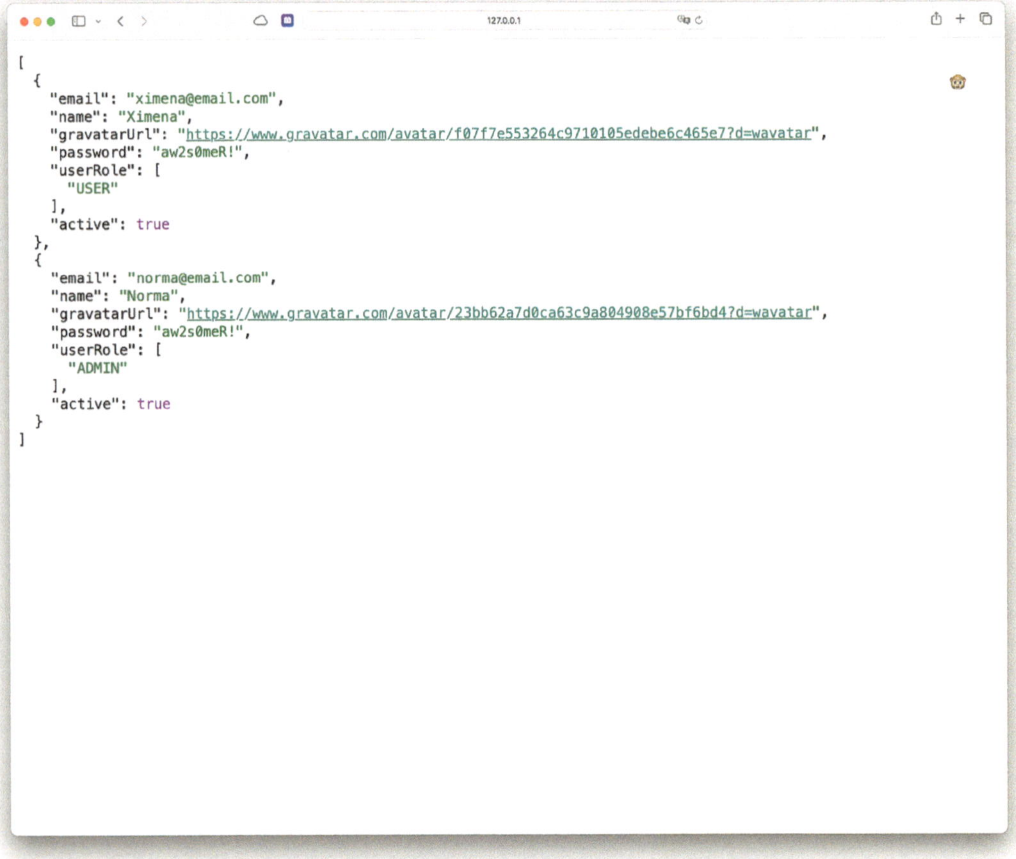

```json
[
  {
    "email": "ximena@email.com",
    "name": "Ximena",
    "gravatarUrl": "https://www.gravatar.com/avatar/f07f7e553264c9710105edebe6c465e7?d=wavatar",
    "password": "aw2s0meR!",
    "userRole": [
      "USER"
    ],
    "active": true
  },
  {
    "email": "norma@email.com",
    "name": "Norma",
    "gravatarUrl": "https://www.gravatar.com/avatar/23bb62a7d0ca63c9a804908e57bf6bd4?d=wavatar",
    "password": "aw2s0meR!",
    "userRole": [
      "ADMIN"
    ],
    "active": true
  }
]
```

Figure 9-18. */users endpoint access after acceptance*

And that's it! You now have another option if your company wants to use Spring Authorization Server as a mechanism for authentication and authorization.

Wait... and My Retro App?

The My Retro App follows the same pattern described for the Users App in the previous section. So, you are encouraged to open it and review it (located in the `09-oauth2/myretro` directory). You will find a small change in the scopes: whereas they are declared as `retros:read` and `retros:write` in the Users App, in the My Retro App they are declared as `users.read` and `users.write` (i.e., with a period instead of a colon). And this difference is to show you that there is no rule of thumb on how you label your security scopes—that particular requirement depends on the business rules of the company.

Summary

In this chapter we covered a lot of Spring Security and how Spring Boot helps to auto-configure most of the beans required for its functionality. You learned different ways to secure your web apps.

You learned how to configure Spring Security for Servlet and Reactive web applications. You learned that you can provide any kind of security simply by using UserDetails, from in-memory, to JDBC, and even custom security.

You learned that Spring Security works with the chain of responsibility pattern and gives you the facility to provide your own filter, in this case just by providing the SecurityFilterChain (for Servlet web apps) or SecurityWebFilterChain (for Reactive web apps) beans and their configuration to secure your requests using roles or scopes.

You learned how to include Social Login in your apps. And how to connect them together. And finally, you learned about the new Spring Authorization Server and how it works and how to configure it.

In Chapter 10, we'll explore the details of messaging with Spring Boot.

Messaging with Spring Boot

This chapter covers the most commonly used messaging brokers for event-driven solutions and how you can use them with Spring Boot, from legacy systems (and still used) such as Java Message Service (JMS) with ActiveMQ, Advanced Message Queuing Protocol (AMQP) with RabbitMQ, pub/sub (topics) solutions with Redis and Kafka, to Simple (or Streaming) Text-Oriented Message Protocol (STOMP) using WebSockets.

Messaging As a Concept

Messaging is a way for different parts of a computer system, or even different systems, to communicate with each other. Messaging is used in many ways, from instant messaging or email, to enterprise applications.

Messaging can be *synchronous*, meaning that the sender and receiver must both be available at the same time, or *asynchronous*, meaning that the sender and receiver can communicate at different times. Messaging can also be *publish-and-subscribe (pub/sub)*, meaning that messages are published to a topic and any interested subscribers can receive them, or *peer-to-peer*, meaning that messages are sent directly from one sender to one receiver.

Messaging is a very powerful way to build distributed systems because it allows different parts of a system to communicate with each other without having to be tightly coupled. This means that changes to one part of the system do not necessarily affect the other parts.

Using Spring Boot to create messaging systems is easy due to the abstractions that simplify how these systems are built. Spring Boot provides several features that make it easy to develop and manage messaging applications, including support for popular messaging brokers such as RabbitMQ, Apache Kafka, Apache Pulsar, and Apache ActiveMQ.

© Felipe Gutierrez 2024
F. Gutierrez, *Pro Spring Boot 3*, https://doi.org/10.1007/978-1-4842-9294-5_10

The examples in this chapter will show you how to use Spring Boot to build messaging applications using different technologies and messaging brokers.

Messaging with Spring Boot

The Spring Framework contains a dedicated technology for messaging, which you can find as a dependency (`spring-messaging`). Around 2013/2014, all the core messaging was part of the Spring Integration technology, but the Spring Integration team decided to move it at the core-level package, so it's easy for future messaging technologies to use such components; and thanks to the Spring Integration team, we now have an amazing technology that can help us integrate different systems not only internally or locally but also remotely.

The Spring Messaging technology provides a simple model for implementing complex enterprise integration solutions with ease. It also facilitates asynchronous and message-driven behavior by promoting components that are loosely coupled for modularity and testability and making easy the separation of concerns between business logic and integration logic; all this promotes reusability and portability.

These are some of the main components for the `spring-messaging` technology:

- `Message`: This is an interface that belongs to the `org.springframework.messaging` package and defines a generic wrapper for any Java object. It has a payload and headers. Consider the following code snippet:

  ```
  package org.springframework.messaging;

  public interface Message<T> {
      T getPayload();
      MessageHeaders getHeaders();
  }
  ```

  The `getHeaders()` method returns a `MessageHeaders` class that implements the `Map<String, Object>` and `Serializable` interfaces. This class includes several useful properties, such as `ID`, `timestamp`, `correlation ID`, and `return address`, among others. Thanks to these properties, serialization and deserialization can happen. More about this later in the chapter.

- MessageChannel: This is a functional interface that defines an overload method send (and it uses the Message interface as a parameter) and its uses as pipe (in a pipe-filter architecture). This interface can be used in point-to-point and publish-and-subscribe scenarios. The following snippet is its definition:

```
package org.springframework.messaging;

@FunctionalInterface
public interface MessageChannel {

    long INDEFINITE_TIMEOUT = -1;

    default boolean send(Message<?> message) {
        return send(message, INDEFINITE_TIMEOUT);
    }

    boolean send(Message<?> message, long timeout);
}
```

- *Message endpoint*: As the name suggests, a *message endpoint* is not an interface or class but rather an endpoint through which *inbound* and *outbound* messages pass and can be enhanced, filtered, and more. You can see a message endpoint as a filter. Spring Messaging and the other messaging technologies that we are going to discuss in this chapter use this concept. And just to mention some of them: *message transformer, message filter, message router, splitter, aggregator, service activator, channel adapter,* and *endpoint bean names.*

Note   If you want more insight into this, check out my book *Spring Boot Messaging* (Apress, 2017; `https://link.springer.com/book/10.1007/978-1-4842-1224-0`), which will help you to understand not only this concept about Messaging but also other messaging technologies.

Using spring-messaging with Spring Boot is even easier, because all the manual configuration that you normally do in a regular Spring app is performed by Spring Boot auto-configuration.

This intro to `spring-messaging` will help you to understand how the different technologies such as JMS, AMQP, Kafka, and Pulsar by Spring are very similar in context and to know what is really happening behind the scenes.

Events with Spring Boot

In my opinion, *events* are an essential way to implement messaging across modules or components within your applications, and normally they are used internally, in your business logic. That said, before we delve into the specific message broker technologies, let's start with events and see how easy it is to use them with Spring Boot.

One of the main features of the Spring Framework is its event handling. The event handling in the Spring `ApplicationContext` is provided through the `ApplicationEvent` class and the `ApplicationListener` interface.

In fact, every time your Spring Boot app starts, there are many events happing in the Spring `ApplicationContext`, and you can listen to such events. For example, the Spring Framework provides some built-in events such as `ContextRefreshedEvent`, `ContextStartedEvent`, `ContextStoppedEvent`, and `ContextClosedEvent`, among others. And, of course, Spring Boot provides its own built-in events, such as `ApplicationReadyEvent`, which happens when your app is ready; the `CommandLineRunner` or `ApplicationRunner` interface implementations run if they are found in your code, and these provide such event. (Does this sound familiar? **TIP**: Check your configuration classes.) Other built-in events include `ApplicationStartingEvent`, `ApplicationFailedEvent`, and many more (some of these events are listeners of the Spring events, that bubble up—sending all events—anything that your application needs).

The steps for using events in Spring are as follows:

1. Create your event by extending the `ApplicationEvent` class.

2. Implement the `ApplicationEventPublishAware` interface and use the `ApplicationEventPublisher` class to publish your `ApplicationEvent` (from the previous step).

3. Implement the `ApplicationListener` based on your event.

4. Implement the `onApplicationEvent` method.

But, is there any other simpler way? Yes, Spring also provides annotation-based event listeners with @EventListener annotation and asynchronous listeners with the combination of the @Async annotation that help you to create with ease your own custom events.

Adding Events to the Users App

Let's add two events to our Users App:

- UserActivatedEvent: Sent when a User changes its status of active or inactive

- UserRemovedEvent: Sent when a User is removed from the database

You have access to the code at the Apress website or here: https://github.com/felipeg48/pro-spring-boot-3rd. You can use the code from the ch10-messaging-events/users folder. Or you can use any other Users App project.

We are going to create a new package named events. Open/create the UserActivatedEvent and UserRemovedEvent classes with the code shown in Listings 10-1 and 10-2, respectively.

Listing 10-1. src/main/java/apress/com/users/events/UserActivatedEvent.java

```
package com.apress.users.events;

import lombok.AllArgsConstructor;
import lombok.Data;
import lombok.NoArgsConstructor;

@AllArgsConstructor
@NoArgsConstructor
@Data
public class UserActivatedEvent {

    private String email;

    private boolean active;
}
```

As you can see in Listing 10-1, the UserActivatedEvent class is a very simple POJO (plain old Java object). The UserRemovedEvent in Listing 10-2 is also very simple.

Listing 10-2. src/main/java/apress/com/users/events/UserRemovedEvent.java

```
package com.apress.users.events;

import lombok.AllArgsConstructor;
import lombok.Data;
import lombok.NoArgsConstructor;

import java.time.LocalDateTime;

@AllArgsConstructor
@NoArgsConstructor
@Data
public class UserRemovedEvent {

    private String email;

    private LocalDateTime removed;
}
```

Next, create the UserLogs class that will be the listener of such events. See Listing 10-3.

Listing 10-3. src/main/java/apress/com/users/events/UserLogs.java

```
package com.apress.users.events;

import lombok.extern.slf4j.Slf4j;
import org.springframework.context.event.EventListener;
import org.springframework.scheduling.annotation.Async;
import org.springframework.stereotype.Component;

@Slf4j
@Component
public class UserLogs {
    @Async
    @EventListener
```

```java
  void userActiveStatusEventHandler(UserActivatedEvent event){
      log.info("User {} active status: {}",event.getEmail(),
      event.isActive());
  }

  @Async
  @EventListener
  void userDeletedEventHandler(UserRemovedEvent event){
      log.info("User {} DELETED at {}",event.getEmail(),
      event.getRemoved());
  }

}
```

The UserLogs class is marked with the @Component annotation, so the other two annotations, @Async and @EventListener, are registered in the app context. This will enable the methods to listen to such events. Right now, these events are log into the console, but in reality these events could be saved or send a message to an external broker.

Next, open/create the UserService class. See Listing 10-4.

Listing 10-4. src/main/java/apress/com/users/service/UserService.java

```java
package com.apress.users.service;

import com.apress.users.events.UserActivatedEvent;
import com.apress.users.events.UserRemovedEvent;
import com.apress.users.model.User;
import com.apress.users.repository.UserRepository;
import lombok.AllArgsConstructor;
import org.springframework.context.ApplicationEventPublisher;
import org.springframework.stereotype.Service;

import java.time.LocalDateTime;
import java.util.Optional;

@AllArgsConstructor
@Service
```

```
public class UserService {

    private UserRepository userRepository;
    private ApplicationEventPublisher publisher;

    public Iterable<User> getAllUsers(){
        return this.userRepository.findAll();
    }

    public Optional<User> findUserByEmail(String email){
        return this.userRepository.findById(email);
    }

    public User saveUpdateUser(User user){
        User userResult = this.userRepository.save(user);
        this.publisher.publishEvent(new UserActivatedEvent(userResult.
        getEmail(),userResult.isActive()));
        return userResult;
    }

    public void removeUserByEmail(String email){
        this.userRepository.deleteById(email);
        this.publisher.publishEvent(new UserRemovedEvent(email,
        LocalDateTime.now()));
    }
}
```

Listing 10-4 shows that the UserService class is marked with the @Service annotation. Also, we are using the ApplicationEventPublisher class. This class will help to publish the events with the publishEvent method.

That's it. Very easy, right? If you run the UsersHttpRequestTests tests, you will see in the console logs something like this:

```
...
2023-10-11T14:29:31.535-04:00  INFO 47670 --- [    Test worker] com.apress.
users.events.UserLogs         : User dummy@email.com active status: false
2023-10-11T14:29:31.542-04:00  INFO 47670 --- [    Test worker] com.
apress.users.events.UserLogs            : User dummy@email.com DELETED at
2023-10-11T14:29:31.541750
```

The configuration for the My Retro App is similar. You can look at the `10-messaging-events/myretro` folder and the `events` and `service` packages. Very simple.

JMS with Spring Boot

This section shows how you can integrate Java Message Service with Spring Boot apps, using our two main projects. JMS has been around since the late 1980s and is still being used by quite a few big companies, primarily financial institutions. JMS was developed by Sun Microsystems, and then it went to the Java Community Process (JCP) with Oracle for development of version 2.0. JMS continues to evolve and currently is part of the *Jakarta EE ecosystem* and named *Jakarta Messaging*. JMS 3.0 is still under development.

It's important to know that JMS supports both the *point-to-point* and *publish-and-subscribe* messaging models. In a point-to-point model, the *Queue* is the hub of communication, and in the publish-and-subscribe model, the *Topic* is the hub. See Figure 10-1.

point-to-point

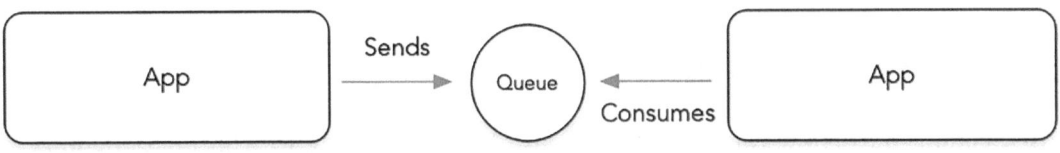

publish-and-subscribe

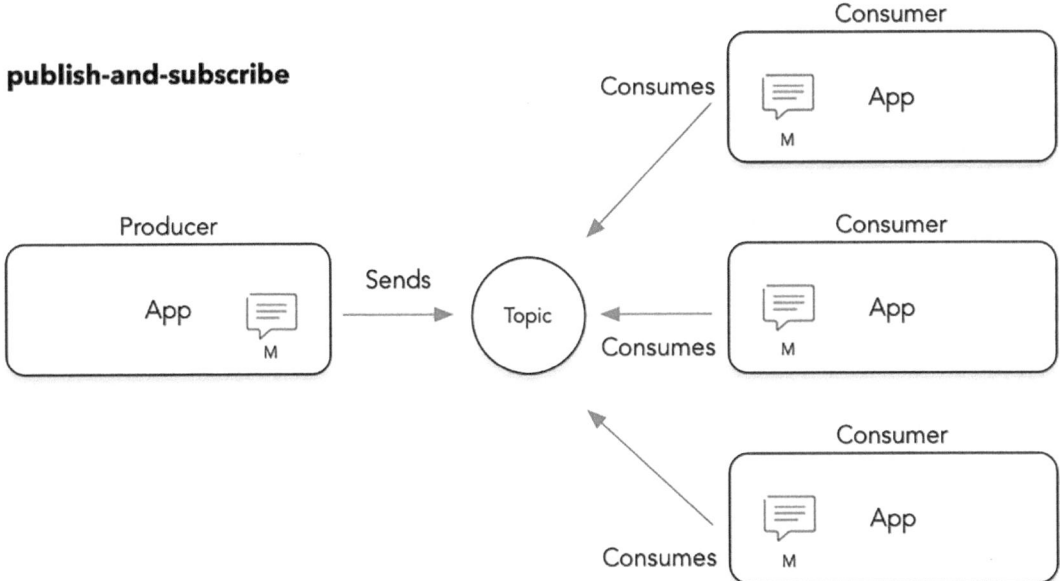

Figure 10-1. *Point-to-point and publish-and-subscribe messaging models*

The Spring Framework supports JMS with several useful classes that can provide the message publishing and consuming. The `JmsTemplate` class wraps up all the boilerplate of connection, session, multithreading, reconnection, and much more. The `JdbcTemplate` class will be practically the same as in Chapters 4 and 5, but for messaging of course. Also, Spring provides several annotations that can help to build consumers with ease, such as the `@JmsListener` annotation and, if you want to do manual message handling, the `SimpleMessageListenerContainer` annotation, which gives you more control over your messaging solution.

Which Broker Are We Using?

Currently, in the IT market there are many JMS broker implementations available, such as Amazon SQS, Apache ActiveMQ Classic and Artemis, IBM MQ, and JBoss Messaging Open Message Queue, among others. Each has some extra features, such as handling multiple protocols besides JMS.

For purposes of this demonstration, we are going to use the Apache ActiveMQ Artemis broker. This broker includes the latest JMS implementation, plus it comes with new protocol implementations such as AMQP, STOMP, MQTT, and OpenWire and additional features. If you want to know more about ActiveMQ Artemis, go to `https://activemq.apache.org/components/artemis/`. The code we are going to use will be the same for any other broker you choose. The important broker-specific information required will be the broker URL and credentials.

Users App with JMS

Let's start with the Users App and see how easy it's to add JMS to this project. You can get the code from the Apress website or from the GitHub repo: `https://github.com/felipeg48/pro-spring-boot-3rd`. You can find the code in the `10-messaging-jms/users` folder.

If you are starting from scratch with the Spring Initializr (`https://start.spring.io`), click Add Dependencies, type JMS in the search field, and choose the Spring for Apache ActiveMQ Artemis dependency. Also add the Web, Validation, JPA, H2, PostgreSQL, Lombok, and Docker Compose Support dependencies. Set the Group ID field to `com.apress` and the Artifact and Name fields to `users`. Then, click Generate, download the project, unzip it, and import it into your favorite IDE.

Open the `build.gradle` file. See Listing 10-5.

Listing 10-5. build.gradle

```
plugins {
    id 'java'
    id 'org.springframework.boot' version '3.2.3'
    id 'io.spring.dependency-management' version '1.1.4'
}
```

```
group = 'com.apress'
version = '0.0.1-SNAPSHOT'
sourceCompatibility = '17'

repositories {
    mavenCentral()
}

dependencies {
    implementation 'org.springframework.boot:spring-boot-starter-web'
    implementation 'org.springframework.boot:spring-boot-starter-
    validation'

    implementation 'org.springframework.boot:spring-boot-starter-artemis'
    implementation 'com.fasterxml.jackson.datatype:jackson-datatype-jsr310'

    developmentOnly 'org.springframework.boot:spring-boot-docker-compose'

    implementation 'org.springframework.boot:spring-boot-starter-data-jpa'
    runtimeOnly 'com.h2database:h2'
    runtimeOnly 'org.postgresql:postgresql'

    compileOnly 'org.projectlombok:lombok'
    annotationProcessor 'org.projectlombok:lombok'

    // Web
    implementation 'org.webjars:bootstrap:5.2.3'

    testImplementation 'org.springframework.boot:spring-boot-starter-test'
}

tasks.named('test') {
    useJUnitPlatform()
}
```

Because we have the `spring-boot-starter-artemis` dependency in the build.
gradle file, Spring Boot will auto-configure the `ConnectionFactory` (with the default
`broker-url`, mode, and other properties), the `JmsTemplate`, the `MessageConverter`
(default for classes that implement `Serializable`, are `Map` or `String` type based), and

other important classes so that they are ready when you use them. The jackson-datatype-jsr310 dependency will help with the event serialization that has the LocalDateTime as a property/field. Also, we are using the spring-boot-docker-compose dependency, so we can run our app and it will look for the docker-compose.yaml file and start the services we need.

We are going to use the Events from the previous section, and we need to add some features that will help us to add the JSON serialization. Next, open/create the UserRemovedEvent class. See Listing 10-6.

Listing 10-6. src/main/java/apress/com/users/events/UserRemovedEvent.java

```
package com.apress.users.events;

import com.fasterxml.jackson.annotation.JsonFormat;
import com.fasterxml.jackson.databind.annotation.JsonSerialize;
import com.fasterxml.jackson.datatype.jsr310.ser.LocalDateTimeSerializer;
import lombok.AllArgsConstructor;
import lombok.Data;
import lombok.NoArgsConstructor;

import java.time.LocalDateTime;

@AllArgsConstructor
@NoArgsConstructor
@Data
public class UserRemovedEvent {

    private String email;

    @JsonFormat(pattern = "yyyy-MM-dd HH:mm:ss")
            @JsonSerialize(using = LocalDateTimeSerializer.class)
    private LocalDateTime removed;
}
```

We added the @JsonFormat and @JsonSerialize annotations to the UserRemovedEvent class. These annotations help to serialize the LocalDateTime with the format yyy-MM-dd HH:mm:ss. If you were to remove it, you would get a long way (unnecessary information, in my opinion) to express something simple.

Next, the UserActivatedEvent class is the same code as shown previously in Listing 10-1, so there's no need to present it again.

Next, open/create the UserLogs class. See Listing 10-7.

Listing 10-7.   src/main/java/apress/com/users/events/UserLogs.java

```java
package com.apress.users.events;

import lombok.AllArgsConstructor;
import lombok.extern.slf4j.Slf4j;
import org.springframework.context.event.EventListener;
import org.springframework.jms.core.JmsTemplate;
import org.springframework.scheduling.annotation.Async;
import org.springframework.stereotype.Component;

import static com.apress.users.jms.UserJmsConfig.DESTINATION_ACTIVATED;
import static com.apress.users.jms.UserJmsConfig.DESTINATION_REMOVED;

@AllArgsConstructor
@Slf4j
@Component
public class UserLogs {

    private JmsTemplate jmsTemplate;

    @Async
    @EventListener
    void userActiveStatusEventHandler(UserActivatedEvent event){
        this.jmsTemplate.convertAndSend(DESTINATION_ACTIVATED,event);
        log.info("User {} active status: {}",event.getEmail(),
        event.isActive());
    }

    @Async
    @EventListener
    void userDeletedEventHandler(UserRemovedEvent event){
        this.jmsTemplate.convertAndSend(DESTINATION_REMOVED,event);
        log.info("User {} DELETED at {}",event.getEmail(),
        event.getRemoved());
    }

}
```

Listing 10-7 shows that now we not only are listening for the events but also sending the events as messages by using the JmsTemplate and calling the convertAndSend method. This method is overloaded to accept different values, and here we are using the destination (this can be the name of the *Queue*, because this is a point-to-point model; in the publish-and-subscribe model, it would be the name of the *Topic*), which will be the name of the queue (DESTINATION_ACTIVATED=activated-users and DESTINATION_REMOVED=removed-users) that will receive the message and the actual message (event, either UserActivatedEvent or UserRemovedEvent) as a second value. The JmsTemplate uses the pattern fire-and-forget unless indicate that you required an immediate response (sendAndReceive), and it will convert the message using a message converter, by default will try to do execute the implementation of the MessageConverter interface, and it will attempt to discover if the message is a String, Map<?,?>, Serializable, or byte[] type. If it can't discover the type, it will throw a MessageConversionException.

We need to send these events as JSON. To do that, we can tell the JmsTemplate to use a JSON message converter. Next, open/create the UserJmsConfig class. See Listing 10-8.

Listing 10-8. src/main/java/apress/com/users/jms/UserJmsConfig.java

```
package com.apress.users.jms;

import org.springframework.context.annotation.Bean;
import org.springframework.context.annotation.Configuration;
import org.springframework.jms.support.converter.MappingJackson2Messa
geConverter;
import org.springframework.jms.support.converter.MessageConverter;
import org.springframework.jms.support.converter.MessageType;

@Configuration
public class UserJmsConfig {

    public final static String DESTINATION_ACTIVATED = "activated-users";
        public final static String DESTINATION_REMOVED =
        "removed-users";

    @Bean
    public MessageConverter messageConverter(){
MappingJackson2MessageConverter converter = new
                                    MappingJackson2Message
                                    Converter();
```

```
        converter.setTypeIdPropertyName("_type");
        converter.setTargetType(MessageType.TEXT);
        return converter;
    }
}
```

The UserJmsConfig class is marked using the @Configuration annotation, which will be picked up when the application starts and help Spring to look for any @Bean definition and configure it accordingly. In this class we are defining two constants, DESTINATION_ACTIVATED=activated-users and DESTINATION_REMOVED=removed-users, which will be the names of the queues (or the *Topics*, depending on the model used). By default, Spring Boot will create these Queues when the JmsTemplate executes the convertAndSend method, so there is no need to create the Queues manually or programmatically. Of course, you can override the default if you want, but in this case, we are using the easier option.

We are defining the MessageConverter; we are overriding it because, by default, Spring Boot will configure this bean. We are creating a MappingJackson2MessageConverter; we are setting the target type as TEXT, meaning that our event object will be sent as text in JSON format, and we are defining a TypeId that will help the converter to identify which class type it needs to do the serialization and deserialization. The TypeId can be any text you like; this is just a hint for the converter (it could be _class_, _id_, or custom, for example), and in this case we are using _type.

Next, open/create the docker-compose.yaml file. See Listing 10-9.

Listing 10-9. docker-compose.yaml

```
version: "3"
services:
  artemis:
    container_name: artemis
    hostname: artemis
    image: apache/activemq-artemis:latest-alpine
    platform: linux/amd64
    restart: always
    environment:
```

```
    EXTRA_ARGS: "--nio --relax-jolokia --http-host 0.0.0.0"
  ports:
    - "61616:61616"
    - "8161:8161"
```

As you can see, the docker-compose.yaml file is very straightforward. When the app starts and sees the spring-boot-docker-compose dependency, it will look for this file and start the services.

By default, the Apache ActiveMQ Artemis credentials are artemis/artemis, so it's necessary to provide such properties spring.artermis.* properties in the application.properties file. See Listing 10-10.

Listing 10-10. src/main/resources/application.properties

```
spring.h2.console.enabled=true
spring.datasource.generate-unique-name=false
spring.datasource.name=test-db

# JMS Remote
spring.artemis.user=artemis
spring.artemis.password=artemis

#spring.artemis.broker-url=tcp://localhost:61616
#spring.artemis.mode=native
```

Now we are ready to publish our events as JSON messages to the JMS broker.

Running the Users App

Go ahead and run the Users App. It will start by starting the services from the docker-compose.yaml file. And you should see the logs from the events:

```
...
User ximena@email.com active status: true
User norma@email.com active status: false
User dummy@email.com active status: false
User dummy@email.com DELETED at 2023-10-12T15:43:10.798841
...
```

Remember that this is happening because in the UserConfiguration class we have the CommandLineRunner init(UserService userService) bean declaration that will be executed once the app is ready and the UserService has the event publisher, then the UserLogs class will log the appropriate events.

Now, let's see the ActiveMQ Artemis broker. Open your browser and go to http://localhost:8161 to access the ActiveMQ Artemis Management Console. See Figure 10-2.

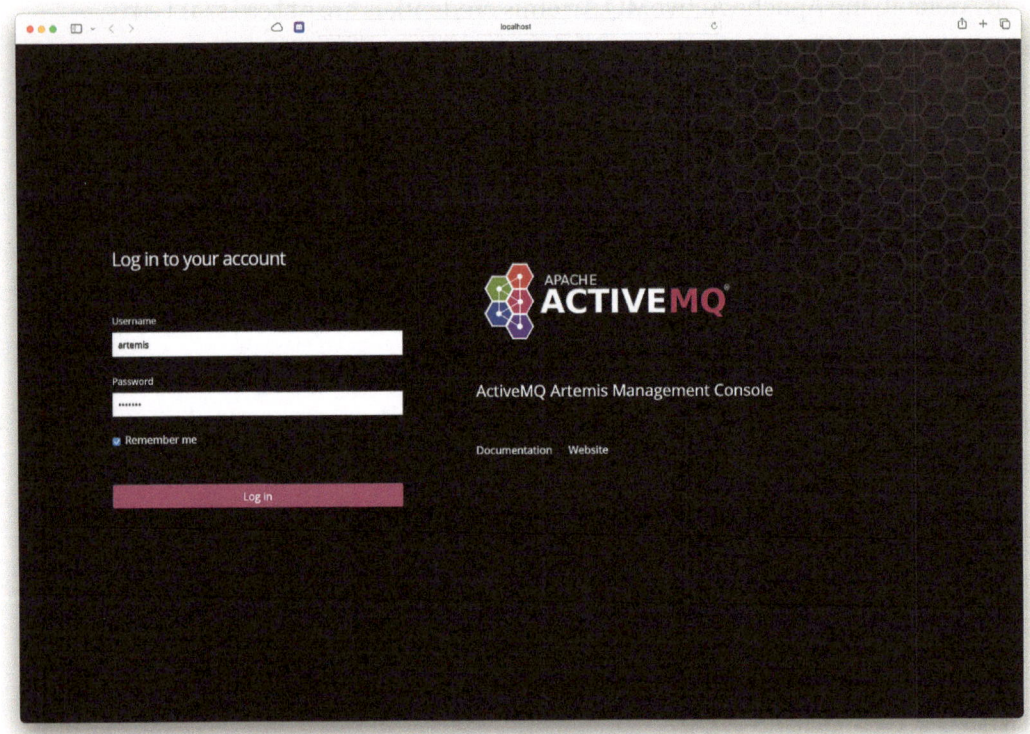

Figure 10-2. *ActiveMQ Artemis Management Console (http://localhost:8161)*

As previously indicated, the username and password both are `artemis`. After you enter them and click Log In, expand the navigation pane, select addresses, and click the Queues tab, as shown in Figure 10-3. You will see the `activated-users` and `removed-users` queues with some messages (indicated in the Message Count column).

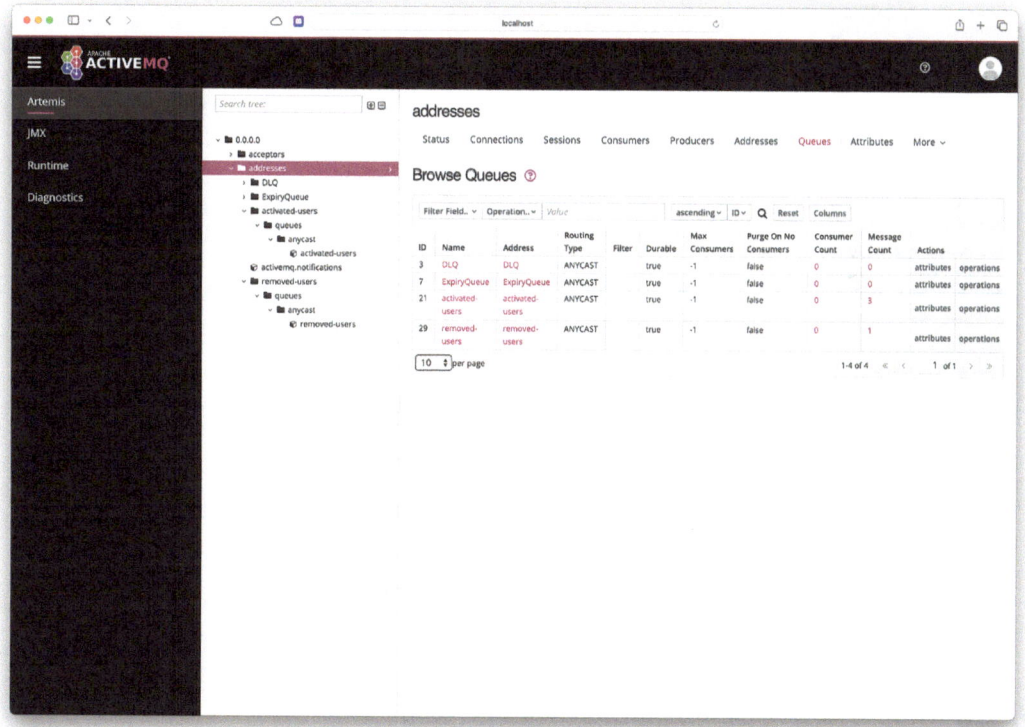

Figure 10-3. *ActiveMQ Artemis Management Console ActiveMQ Console: Queues tab*

Select the `activated-users` queue in the navigation pane to see the messages that are in the queue at that moment. See Figure 10-4.

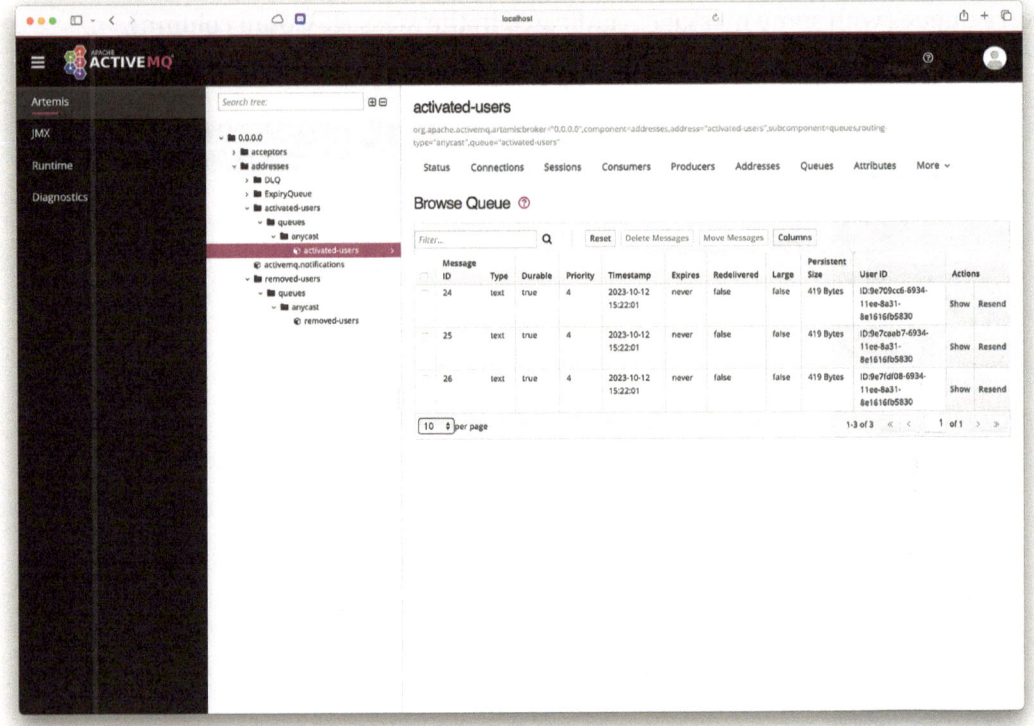

Figure 10-4. *Message list for activated-users queue*

Now, you can select one of the messages and see its contents. Figure 10-5 shows an example of an open message.

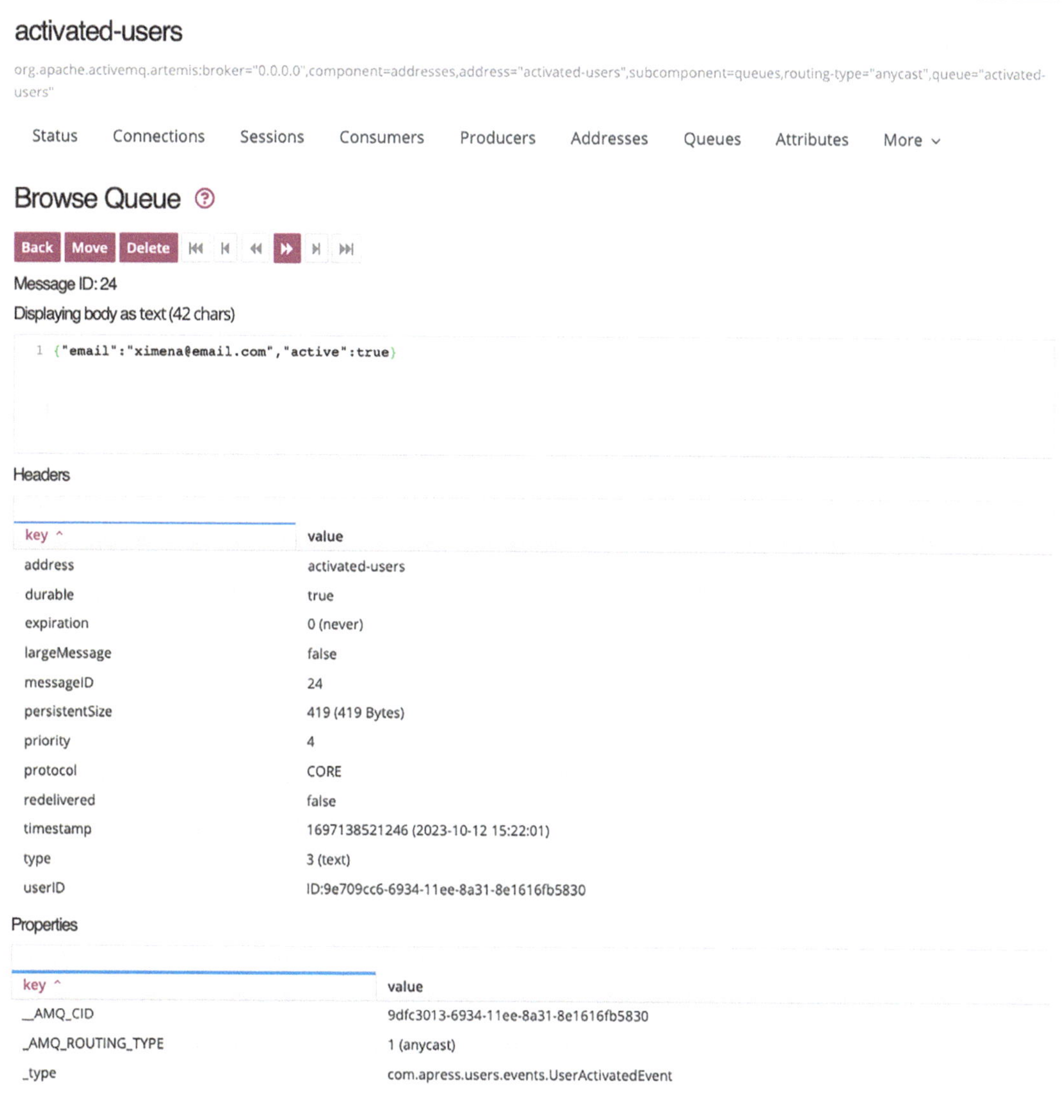

activated-users

org.apache.activemq.artemis:broker="0.0.0.0",component=addresses,address="activated-users",subcomponent=queues,routing-type="anycast",queue="activated-users"

| Status | Connections | Sessions | Consumers | Producers | Addresses | Queues | Attributes | More ˅ |

Browse Queue ⑦

Back Move Delete |◀◀ |◀ ◀◀ ▶▶ ▶| ▶▶|

Message ID: 24

Displaying body as text (42 chars)

```
1 {"email":"ximena@email.com","active":true}
```

Headers

key ^	value
address	activated-users
durable	true
expiration	0 (never)
largeMessage	false
messageID	24
persistentSize	419 (419 Bytes)
priority	4
protocol	CORE
redelivered	false
timestamp	1697138521246 (2023-10-12 15:22:01)
type	3 (text)
userID	ID:9e709cc6-6934-11ee-8a31-8e1616fb5830

Properties

key ^	value
__AMQ_CID	9dfc3013-6934-11ee-8a31-8e1616fb5830
_AMQ_ROUTING_TYPE	1 (anycast)
_type	com.apress.users.events.UserActivatedEvent

Figure 10-5. *activated-users message example*

The message is in JSON format. Take a look at the Headers and Properties sections. The Properties section shows our key _type and the value com.apress.users.events. UserActivatedEvent. Remember that this property is useful for giving a hint to the MessageConverter for deserialization (and serialization) when using a listener.

Next, select the removed-users queue in the navigation pane and open a message to see something like the example shown in Figure 10-6.

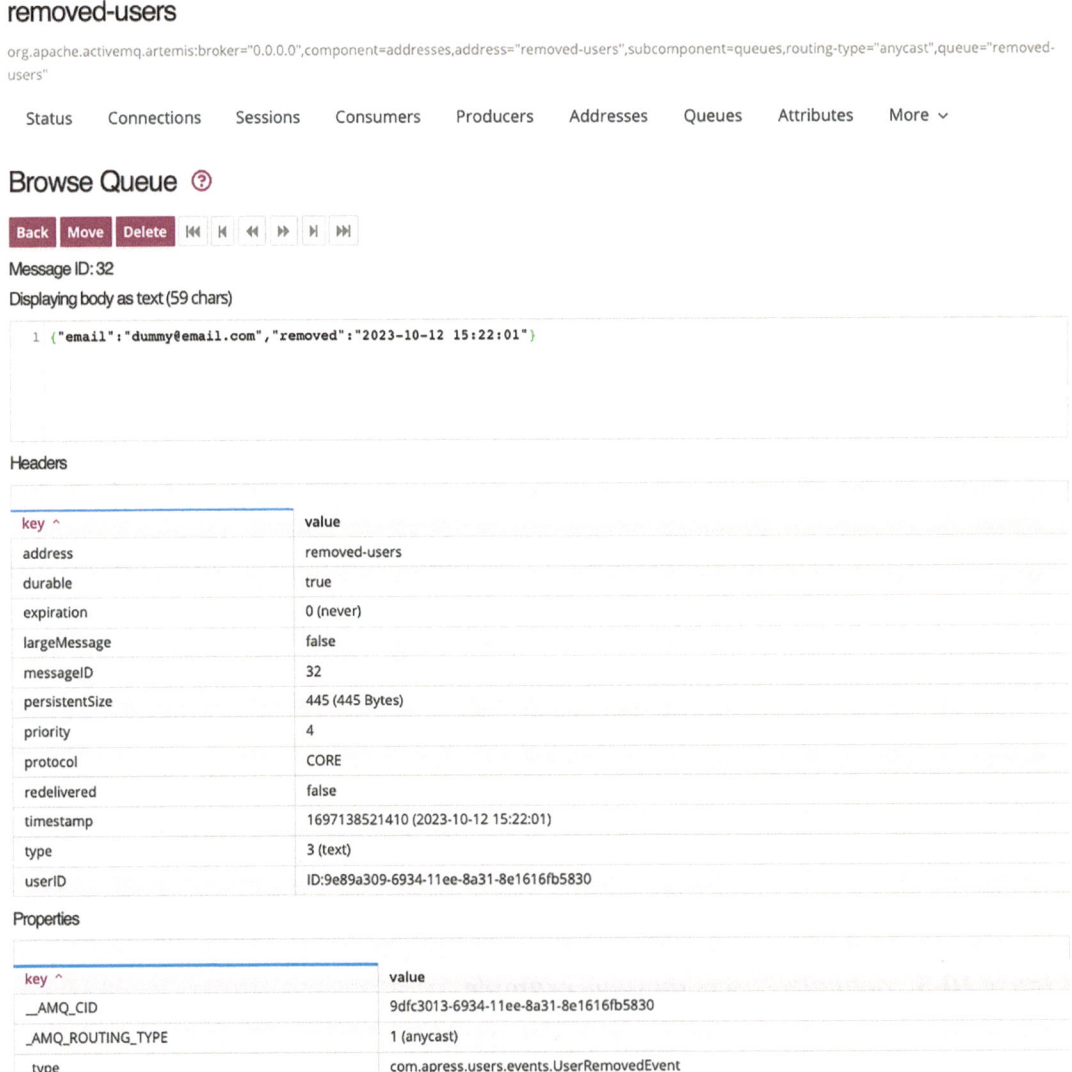

Figure 10-6. *removed-users message example*

Note in Figure 10-6 that the event message uses the JSON format for the LocalDateTime: yyyy-MM-dd HH:mm:ss. This is in the UserRemovedEvent and the @JsonFormat(pattern = "yyyy-MM-dd HH:mm:ss") annotation.

Listen for Incoming User Events with JMS

Next, create/open the UserEventListeners class. See Listing 10-11.

Listing 10-11. src/main/java/apress/com/users/jms/UserEventListeners.java

```java
package com.apress.users.jms;

import com.apress.users.events.UserActivatedEvent;
import jakarta.jms.JMSException;
import lombok.extern.slf4j.Slf4j;
import org.springframework.jms.annotation.JmsListener;
import org.springframework.stereotype.Component;

@Slf4j
@Component
public class UserEventListeners {

    @JmsListener(destination = UserJmsConfig.DESTINATION_ACTIVATED)
        public void onActivatedUserEvent(UserActivatedEvent event){
          log.info("JMS User {}",event);
        }

        @JmsListener(destination = UserJmsConfig.DESTINATION_REMOVED)
        public void onRemovedUserEvent(Object event) throws
        JMSException {
          log.info("JMS User DELETED message: {} ", event);
        }
}
```

Let's review the UserEventListeners class:

- @Component: This annotation is necessary for Spring to wire up every listener found.

- @JmsListener: This annotation needs the destination name (Queue or Topic); it has more parameters, but this one will be enough.

- onActivatedUserEvent: This method accepts a UserActivatedEvent as a parameter value, which will be passed once the message is received and converted back (deserialization) from JSON to object.

- onRemovedUserEvent: This method accepts an Object (this is on purpose) because I want to show you what type of object is being received.

Rerunning the Users App

If you run the Users App again with the new changes thus far, you will have the following output:

```
...
JMS User UserActivatedEvent(email=ximena@email.com, active=true)
JMS User UserActivatedEvent(email=norma@email.com, active=false)
JMS User UserActivatedEvent(email=dummy@email.com, active=false)

JMS User DELETED message: ActiveMQMessage[ID:93288cbb-6937-11ee-8a2a-8
e1616fb5830]:PERSISTENT/ClientMessageImpl[messageID=98, durable=true,
address=removed-users,userID=93288cbb-6937-11ee-8a2a-8e1616fb5830,
properties=TypedProperties[__AMQ_CID=92b75453-6937-11ee-8a2a-8e1616fb5830,
_type=com.apress.users.events.UserRemovedEvent,_AMQ_ROUTING_TYPE=1]]
...
```

The onRemovedUserEvent(Object) gets back an ActiveMQMessage (this class implements the jakarta.jms.Message interface). Review this message and notice that it comes with a lot of information that can be useful for other business logic. You probably are wondering how I can use the actual UserRemovedEvent object instead of the generic, right? The answer, next.

If you change the signature as in the following snippet:

```
@JmsListener(destination = UserJmsConfig.DESTINATION_REMOVED)
public void onRemovedUserEvent(UserRemovedEvent event) throws
JMSException {
    log.info("JMS User DELETED message: {} ", event);
}
```

You will get a MessageConversionException saying that it cannot deserialize the LocalDateTime object. Although we configured a MessageConverter, we need to tell (by creating a custom `MessageConverter`) how to deserialize it. By default, the `com.fasterxml.jackson.databind.ObjectMapper` is configured with its default (for a serialization is easy, but not for the deserialization), but we need to add a way to handle the `LocalDateTime` type.

So, in the `UserJmsConfig` class, replace the `MessageConverter` declaration as shown in the following snippet:

```
@Bean
public MessageConverter messageConverter(){
    ObjectMapper mapper = new ObjectMapper();
    mapper.registerModule(new JavaTimeModule());

    MappingJackson2MessageConverter converter = new
    MappingJackson2MessageConverter();
    converter.setTypeIdPropertyName("_type");
    converter.setTargetType(MessageType.TEXT);
    converter.setObjectMapper(mapper);
    return converter;
}
```

And that's it! Now you can use the `onRemovedUserEvent(UserRemovedEvent event)` signature and everything will work.

Remember that normally a JMS listener is another app that consumes messages asynchronously and does some process with the messages. Here we have the producer and consumer in the same base code. A use case for our project will be to notify the My Retro App when a user has been created or deleted and, for example, do a server push to refresh the web page.

So, now you know how to use the JMS point-to-point model with ActiveMQ Artemis. Next, we'll explore how to use the JMS publish-and-subscribe model (refer to Figure 10-1).

Using JMS Topics with My Retro App

Using JMS with My Retro App to listen for events that occur in the Users App will provide a good example of the publish-and-subcribe model. A user is either activated or not, and if a user is removed from the system, the app must send the events act accordingly. JMS implements the pub/sub model by using *Topics*. In other words, we need to subscribe to any of those events.

You can find the source code in the 10-messaging-jms/myretro folder. Or, if you are starting from scratch using the Spring Initializr (https://start.spring.io), add the same dependencies from the "Users App with JMS" section.

Open/create the build.gradle file as shown in Listing 10-12.

Listing 10-12. build.gradle

```
plugins {
    id 'java'
    id 'org.springframework.boot' version '3.2.3'
    id 'io.spring.dependency-management' version '1.1.4'
}

group = 'com.apress'
version = '0.0.1-SNAPSHOT'
sourceCompatibility = '17'

repositories {
    mavenCentral()
}

dependencies {
    implementation 'org.springframework.boot:spring-boot-starter-web'
    implementation 'org.springframework.boot:spring-boot-starter-validation'

    implementation 'org.springframework.boot:spring-boot-starter-artemis'
    implementation 'com.fasterxml.jackson.datatype:jackson-datatype-jsr310'

    implementation 'org.springframework.boot:spring-boot-starter-data-jpa'

    annotationProcessor 'org.springframework.boot:spring-boot-
    configuration-processor'

    runtimeOnly 'com.h2database:h2'
    runtimeOnly 'org.postgresql:postgresql'

    compileOnly 'org.projectlombok:lombok'
    annotationProcessor 'org.projectlombok:lombok'

    testImplementation 'org.springframework.boot:spring-boot-starter-test'
}
```

```
tasks.named('test') {
    useJUnitPlatform()
}
```

Listing 10-12 shows that we are using the `spring-boot-starter-artemis` and `jackson-datatype-jsr310` dependencies. We don't need `spring-boot-docker-compose` here because we are going to connect to the one the Users App is using.

Note   The code base used here (My Retro App project) is using the `spring-boot-starter-data-jpa` (no reactive in this case).

Next, open/create the UserEvent class. See Listing 10-13.

Listing 10-13.   src/main/java/apress/com/myretro/jms/UserEvent.java

```
package com.apress.myretro.jms;

import com.fasterxml.jackson.annotation.JsonFormat;
import com.fasterxml.jackson.annotation.JsonIgnoreProperties;
import com.fasterxml.jackson.databind.annotation.JsonSerialize;
import com.fasterxml.jackson.datatype.jsr310.ser.LocalDateTimeSerializer;
import lombok.AllArgsConstructor;
import lombok.Data;
import lombok.NoArgsConstructor;

import java.time.LocalDateTime;

@AllArgsConstructor
@NoArgsConstructor
@Data
@JsonIgnoreProperties(ignoreUnknown = true)
public class UserEvent {

    private String email;
    private String action;
    private boolean active;
```

```
@JsonFormat(pattern = "yyyy-MM-dd HH:mm:ss")
@JsonSerialize(using = LocalDateTimeSerializer.class)
private LocalDateTime removed;
```

}

You can think of the UserEvent class as a merger of the information from the UserActivatedEvent and UserRemovedEvent classes from the Users App. Note also that we have a new field, String action, which will hold the event action description (Activated or Removed); this can be useful for new features. Also, we are using the @JsonIgnoreProperties (ignoreUnknown = true) annotation, which can be useful when listening for JSON data that might not have the same fields.

Next, open/create the UserEventListeners class. See Listing 10-14.

Listing 10-14. src/main/java/apress/com/myretro/jms/UserEventListeners.java

```
package com.apress.myretro.jms;

import lombok.extern.slf4j.Slf4j;
import org.springframework.jms.annotation.JmsListener;
import org.springframework.jms.annotation.JmsListeners;
import org.springframework.stereotype.Component;

@Slf4j
@Component
public class UserEventListeners {

    @JmsListeners({
            @JmsListener(destination = "${jms.user-events.queue.1}"),
            @JmsListener(destination = "${jms.user-events.queue.2}")
    })
    public void onUserEvent(UserEvent userEvent) {
        log.info("UserEventListeners.onUserEvent: {}", userEvent);
    }
}
```

The UserEventListeners class includes the following annotations (a few of which should be familiar to you) and method:

- @Component: This annotation marks the class as a Spring bean, and it is necessary for the other annotations to get registered as well.

- @JmsListeners: This annotation accepts multiple @JmsListener annotations. This is useful in our case because we have only one message to receive, the UserEvent object.

- @JmsListener: This annotation is useful to connect to the topics (or queues, in a point-to-point model), and in this case we are using some properties that are external and will be found in the application.properties file.

- onUserEvent: This method is marked as a listener and it will receive the UserEvent messages. But how can we do that if we have a JSON message, one for the *activated* users and another for the *removed* users?

Next, create/open the UserEventConfig class. See Listing 10-15.

Listing 10-15. src/main/java/apress/com/myretro/jms/UserEventConfig.java

```
package com.apress.myretro.jms;

import com.fasterxml.jackson.core.JsonProcessingException;
import com.fasterxml.jackson.databind.ObjectMapper;
import com.fasterxml.jackson.datatype.jsr310.JavaTimeModule;
import jakarta.jms.JMSException;
import jakarta.jms.Message;
import jakarta.jms.Session;
import org.springframework.context.annotation.Bean;
import org.springframework.context.annotation.Configuration;
import org.springframework.jms.support.converter.
MessageConversionException;
import org.springframework.jms.support.converter.MessageConverter;
```

```java
@Configuration
public class UserEventConfig {

    @Bean
    MessageConverter messageConverter() {
        return new MessageConverter() {

            @Override
            public Message toMessage(Object object,
            Session session) throws JMSException,
            MessageConversionException {
            throw new UnsupportedOperationException("Not
            supported yet.");
                }

            @Override
            public Object fromMessage(Message message) throws
            JMSException,
MessageConversionException {

ObjectMapper mapper = new ObjectMapper();
            mapper.registerModule(new JavaTimeModule());
            try {
                String _type = message.getStringProperty("_type");
                UserEvent event =  mapper.readValue(message.
                getBody(String.class), UserEvent.class);
                event.setAction("Activated");
                if (_type.contains("Removed"))
                    event.setAction("Removed");

                return event;
            } catch (JsonProcessingException e) {
                throw new RuntimeException(e);
            }
        }
            };
    }
}
```

The UserEventConfig class includes the following:

- @Configuration: As you know already, this annotation helps Spring to look for any @Bean definition and configure it accordingly.

- MessageConverter: This bean is necessary; we don't need the default because we are dealing with JSON objects, and some have, we need to do the deserialization. We are returning a new implementation of the MessageConverter. This interface provides two methods to implement, toMessage (which we don't need because we are not sending any message; we are just receiving) and fromMessage. We are using the jackson library to use the ObjectMapper, both because we need to use the JavaTimeModule (due the LocalDateTime type we are using) and because we need to fill out the action field with either Activated or Removed based on the event.

Take a moment to review the code in Listing 10-15 carefully.

Next, open the application.properties file. See Listing 10-16.

Listing 10-16. src/main/resources/application.properties

```
server.port=${PORT:8181}

## Data
spring.h2.console.enabled=true
spring.datasource.generate-unique-name=false
spring.datasource.name=test-db
# spring.jpa.show-sql=true

## JMS Remote
#spring.artemis.broker-url=tcp://localhost:61616
#spring.artemis.mode=native
spring.artemis.user=artemis
spring.artemis.password=artemis

## JMS Topic
spring.jms.pub-sub-domain=true

## User Event Queues
jms.user-events.queue.1=activated-users
jms.user-events.queue.2=removed-users
```

At the end of the file, note that two queues are listed. The values are injected in the @JmsListener annotations and these are the Topic's names. So, how can we change into a *Topics* (pub/sub) instead of *Queues* (point-to-point), as easy as set the spring.jms.pub-sub-domain=true property, and that's it. Of course, it's important that the sender/publisher of the message also sets this value. This means that the Users App needs to use this property as well.

Running My Retro App

To run the My Retro App, you need to do the following:

1. Go to the Users App and add the spring.jms.pub-sub-domain=true property to the application.properties file and start the application. This runs the docker-compose.yaml file and launches the ActiveMQ Artemis service. If you are curious, look at the Queues in the ActiveMQ Artemis Console (http://localhost:8161, credentials artemis/artemis); they will be the same, but you will find some duplicates with a Router type of *MULTICAST* (this feature is more related to ActiveMQ Artemis). This will allow the Users App to subscribe multiple consumers so they can receive the messages published to these queues or topics in this case.

2. Run the My Retro App.

3. Add a user to the Users API. You can use the user.http file with the HTTP client from VS Code or IntelliJ to test the API, or in a terminal, you can execute

    ```
    curl -i -s -d '{"name":"Dummy","email":"dummy@email.
    com","password":
    "aw2sOmeR!","userRole":["INFO"],"active":true}' \
    -H "Content-Type: application/json" \
    http://localhost:8080/users
    ```

    ```
    HTTP/1.1 201
    Location: http://localhost:8080/users/dummy@email.com
    Content-Type: application/json
    Transfer-Encoding: chunked
    ```

```
Date: Fri, 13 Oct 2023 19:31:30 GMT

{"email":"dummy@email.com","name":"Dummy","gravatarUrl":null,"
password":
"aw2s0meR!","userRole":["INFO"],"active":true}
```

If you check the logs on both apps, you will see in the My Retro App the following:

```
UserEventListeners.onUserEvent: UserEvent(email=dummy@email.com,
action=Activated, active=true, removed=null)
```

Now, you can delete the users you have just created with

```
curl -i -s -XDELETE http://localhost:8080/users/dummy@email.com
HTTP/1.1 204
Date: Fri, 13 Oct 2023 19:35:05 GMT
```

Then in the logs, you will see

```
UserEventListeners.onUserEvent: UserEvent(email=dummy@email.com,
action=Removed, active=false, removed=2023-10-13T15:35:05)
```

On each log, we have set the action, either `Activated` or `Removed`.

Now you know how to enable Topics (pub/sub) using JMS. Next we'll look at another messaging implementation, AMQP with Spring Boot.

AMQP with Spring Boot

Early messaging protocols, such as those developed by Sun Microsystems, Oracle, IBM, and Microsoft, were proprietary, making it difficult to mix technologies or programming languages.

In response to this challenge, a team at JPMorgan Chase created the Advanced Message Queuing Protocol (AMQP). AMQP is an open standard application layer protocol for message-oriented middleware (MOM). AMQP is a wire-level protocol, which means that applications can use any technology or programming language to communicate with AMQP-compliant messaging brokers.

There are many different AMQP messaging brokers available, but RabbitMQ is one of the most popular. RabbitMQ is easy to use and scale and is very fast, so we are going to use it in this section.

Installing RabbitMQ

Before we discuss RabbitMQ, you can install it (this is optional), even though we are going to use Docker Compose to play with it. If you are using macOS or Linux, you can use the `brew` command:

```
$ brew upgrade
$ brew install rabbitmq
```

If you are using another UNIX system or a Windows system, you can go to the RabbitMQ website and use the installers (`https://www.rabbitmq.com/docs/download`). RabbitMQ is written in Erlang, so its major dependency is to install the Erlang runtime in your system. Nowadays, all the RabbitMQ installers come with all the Erlang dependencies. Make sure the executables are in your `PATH` variable (for Windows and Linux, depending on what OS you are using). If you are using `brew`, you don't need to worry about setting the `PATH` variable.

RabbitMQ/AMQP: Exchanges, Bindings, and Queues

AMQP defines three concepts that are a little different from the JMS world, but very easy to understand. AMQP defines *exchanges*, which are entities to which the messages are sent. Every exchange takes a message and routes it to zero or more *queues*. This routing involves an algorithm that is based on the exchange type and rules, called *bindings*.

AMPQ defines five exchange types (in addition to the default type): *Direct, Fanout, Topic, Consistent Hash,* and *Headers*. Figure 10-7 shows these different exchange types.

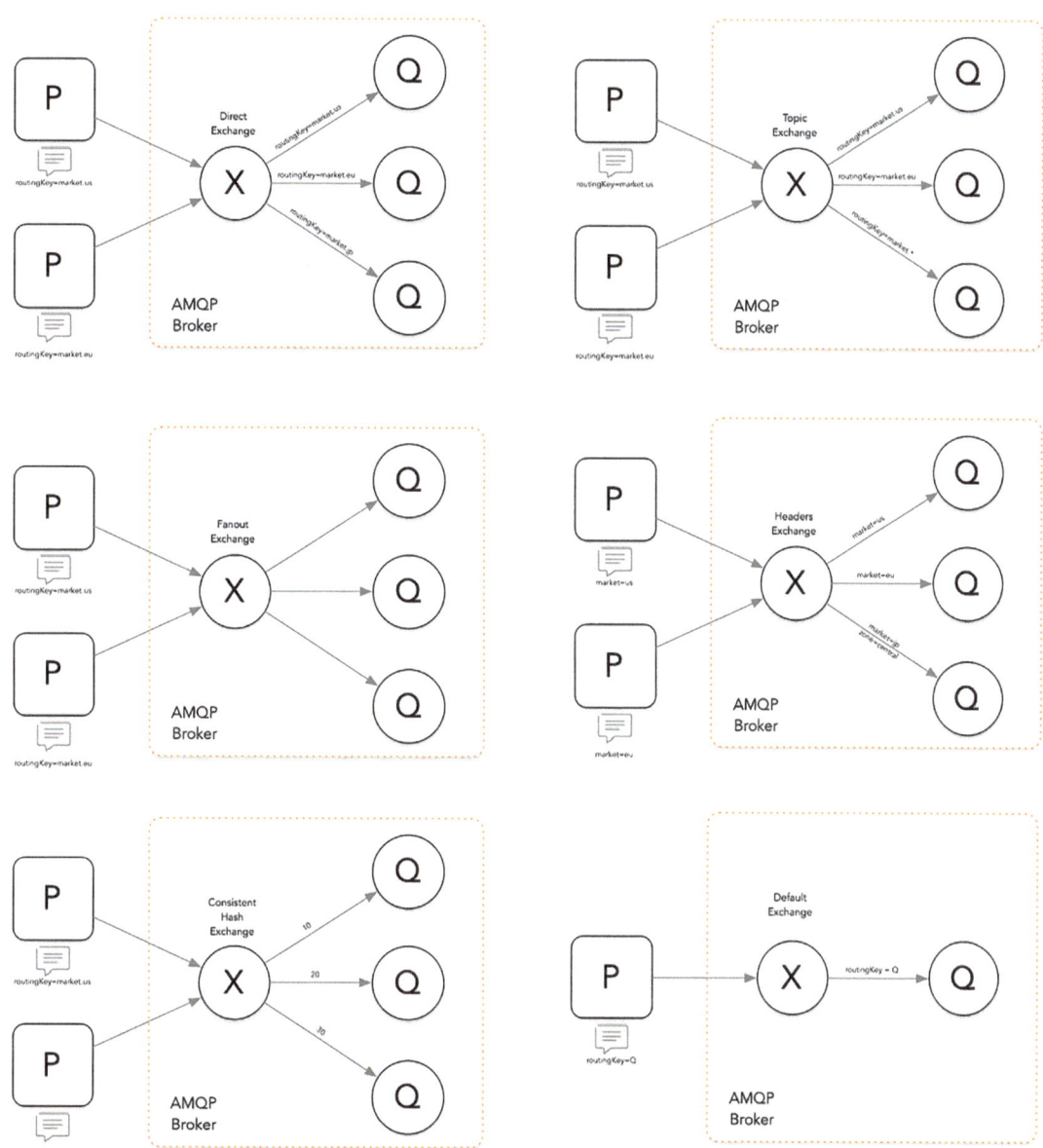

Figure 10-7. *AMQP exchanges, bindings, and queues*

So, the main idea of AMPQ is to send a message to an exchange, including a routing key, and then the exchange, based on its type, will deliver the message to the queue (or it won't if the routing key doesn't match).

The *default exchange* is bound automatically to every queue created. The *direct exchange* is bound to a queue by a *routing key*; you can see this exchange type as one-to-one binding. The *topic exchange* is similar to the *direct exchange*; the only difference is that in its *binding* you can add a wildcard into its *routing key* (You can use the * -asterisk-that matches exactly one word in a routing key, or you can use the # -hash- that matches zero or more words in a routing key, or any other combination). The *headers exchange* is like the topic exchange; the only difference is that the binding is based on the message headers (this is a very powerful exchange, and you can do all and any expressions for its headers). The *fanout exchange* copies the message to all the bound queues; you can see this exchange as a message broadcast. The Consistent Hash Exchange is a specialized exchange type that distributes messages across multiple queues using a consistent hashing algorithm. This ensures that messages with the same routing key are consistently sent to the same queue, even when queues are added or removed. This makes it ideal for scaling out consumers while maintaining message ordering within specific groups (defined by the routing key).

You can get more information about these concepts and exchange types at `https://rabbitmq.com/tutorials/amqp-concepts.html`.

Using AMQP with RabbitMQ in the Users App

Now it's time to use the power of RabbitMQ in the Users App. You can download the source (from the `10-messaging-rmq/users` folder), or if you are using the Spring Initializr (`https://start.spring.io`), make sure to select Spring for RabbitMQ, Web, JPA, Validation, Docker Compose, Lombok, H2, and PostgreSQL as dependencies. Set the Group field to `com.apress` and the Artifact and Name fields to `users`, generate the project, download it, unzip it, and import it into your favorite IDE.

Open the `build.gradle` file, which should look like Listing 10-17.

Listing 10-17. build.gradle

```
plugins {
    id 'java'
    id 'org.springframework.boot' version '3.2.3'
    id 'io.spring.dependency-management' version '1.1.4'
}
```

```
group = 'com.apress'
version = '0.0.1-SNAPSHOT'
sourceCompatibility = '17'

repositories {
    mavenCentral()
}

dependencies {
    implementation 'org.springframework.boot:spring-boot-starter-web'
    implementation 'org.springframework.boot:spring-boot-starter-
    validation'

    implementation 'org.springframework.boot:spring-boot-starter-amqp'
    implementation 'com.fasterxml.jackson.datatype:jackson-datatype-jsr310'

    developmentOnly 'org.springframework.boot:spring-boot-docker-compose'

    implementation 'org.springframework.boot:spring-boot-starter-data-jpa'
    runtimeOnly 'com.h2database:h2'
    runtimeOnly 'org.postgresql:postgresql'

    compileOnly 'org.projectlombok:lombok'
    annotationProcessor 'org.projectlombok:lombok'

    // Web
    implementation 'org.webjars:bootstrap:5.2.3'

    testImplementation 'org.springframework.boot:spring-boot-starter-test'
}

tasks.named('test') {
    useJUnitPlatform()
}
```

The important dependencies here are spring-boot-amqp and jackson-datatype-jsr310, which will help us to deal with the LocalDateTime serialization and deserialization.

When Spring Boot starts, it will identify that you have the AMQP dependency and auto-configure all the necessary default Spring beans that will help to connect to the RabbitMQ broker.

Next, open/create the events package and the UserActivatedEvent and UserRemovedEvent classes that we are going to send to RabbitMQ. They are similar to the same classes in the previous sections, but I want to show you more alternatives; the choice of how to configure these classes will depend on your business logic. See Listings 10-18 and 10-19.

Listing 10-18. src/main/java/apress/com/users/events/UserActivatedEvent.java

```java
package com.apress.users.events;

import lombok.AllArgsConstructor;
import lombok.Data;
import lombok.NoArgsConstructor;

@AllArgsConstructor
@NoArgsConstructor
@Data
public class UserActivatedEvent {

    private final String action = "ACTIVATION_STATUS";

    private String email;

    private boolean active;
}
```

Listing 10-19. src/main/java/apress/com/users/events/UserRemovedEvent.java

```java
package com.apress.users.events;

import com.fasterxml.jackson.annotation.JsonFormat;
import lombok.AllArgsConstructor;
import lombok.Data;
import lombok.NoArgsConstructor;

import java.time.LocalDateTime;

@AllArgsConstructor
@NoArgsConstructor
@Data
public class UserRemovedEvent {
```

```java
    private final String action = "REMOVED";

    private String email;

    @JsonFormat(shape=JsonFormat.Shape.STRING,pattern = "yyyy-MM-dd
    HH:mm:ss")
    private LocalDateTime removed;
}
```

As you can see, we are now adding the actions as a constant in both classes, an easy way to cheat instead of discovering the action based on the name, as we did previously.

Publishing Messages to RabbitMQ

Let's see how we can publish messages to RabbitMQ. Remember that a publisher requires the following steps: open a Connection (to the RabbitMQ broker), create a Channel, and send the Message to an Exchange. Fortunately, these steps are simplified with spring-amqp and Spring Boot.

Next, open/create the UserLogs class. See Listing 10-20.

Listing 10-20. src/main/java/apress/com/users/events/UserLogs.java

```java
package com.apress.users.events;

import lombok.AllArgsConstructor;
import lombok.extern.slf4j.Slf4j;
import org.springframework.amqp.rabbit.core.RabbitTemplate;
import org.springframework.context.event.EventListener;
import org.springframework.scheduling.annotation.Async;
import org.springframework.stereotype.Component;

import static com.apress.users.amqp.UserRabbitConfiguration.USERS_
ACTIVATED;
import static com.apress.users.amqp.UserRabbitConfiguration.USERS_REMOVED;

@Slf4j
@AllArgsConstructor
@Component
public class UserLogs {
```

```
private RabbitTemplate rabbitTemplate;

@Async
@EventListener
void userActiveStatusEventHandler(UserActivatedEvent event){
    this.rabbitTemplate.convertAndSend(USERS_ACTIVATED,event);
    log.info("User {} active status: {}",event.getEmail(),event.
    isActive());
}

@Async
@EventListener
void userDeletedEventHandler(UserRemovedEvent event){
    this.rabbitTemplate.convertAndSend(USERS_REMOVED,event);
    log.info("User {} DELETED at {}",event.getEmail(),event.
    getRemoved());
}
}
```

Listing 10-20 shows that the UserLogs class is similar to its configuration in the previous section (see Listing 10-7), but in this case we are using the RabbitTemplate class. This class implements the Template design pattern, removing the boilerplate of connecting to the broker, session management, message conversion (based on the Spring Boot auto-configuration use of the SimpleMessageConverter), reconnecting when there is an error, retries, connection pool, channels, and much more.

In this example, the RabbitTemplate is using the convertAndSend method that accepts the *routing key* as the first parameter (in this case, either users.activated or users.removed) and the event as the second parameter. This method has many overloads, so you can choose what fits for your business logic.

Note   I previously noted that a publisher always connects to an Exchange, and in this class an Exchange is specified, but it will be declared in the RestTemplate configuration, so that using the convertAndSend method is easier.

Next, open/create the UserRabbitConfiguration class. See Listing 10-21.

Listing 10-21. src/main/java/apress/com/users/amqp/
UserRabbitConfiguration.java

```
package com.apress.users.amqp;

import com.fasterxml.jackson.databind.ObjectMapper;
import com.fasterxml.jackson.datatype.jsr310.JavaTimeModule;
import org.springframework.amqp.core.Binding;
import org.springframework.amqp.core.Queue;
import org.springframework.amqp.core.TopicExchange;
import org.springframework.amqp.rabbit.connection.ConnectionFactory;
import org.springframework.amqp.rabbit.core.RabbitTemplate;
import org.springframework.amqp.support.converter.
Jackson2JsonMessageConverter;
import org.springframework.context.annotation.Bean;
import org.springframework.context.annotation.Configuration;

@Configuration
public class UserRabbitConfiguration {

    public static final String USERS_EXCHANGE = "USERS";
    public static final String USERS_STATUS_QUEUE = "USER_STATUS";
    public static final String USERS_REMOVED_QUEUE = "USER_REMOVED";
    public static final String USERS_ACTIVATED = "users.activated";
    public static final String USERS_REMOVED = "users.removed";

    @Bean
    RabbitTemplate rabbitTemplate(ConnectionFactory connectionFactory){
        RabbitTemplate rabbitTemplate = new RabbitTemplate(connectio
        nFactory);
        rabbitTemplate.setExchange("USERS");
        rabbitTemplate.setMessageConverter(new Jackson2JsonMessageConverter
        (new ObjectMapper().registerModule(new JavaTimeModule())));
        return rabbitTemplate;
    }
```

```
@Bean
TopicExchange exchange(){
    return new TopicExchange(USERS_EXCHANGE);
}

@Bean
Queue userStatusQueue(){
    return new Queue(USERS_STATUS_QUEUE,true,false,false);
}

@Bean
Queue userRemovedQueue(){
    return new Queue(USERS_REMOVED_QUEUE,true,false,false);
}

@Bean
Binding userStatusBinding(){
    return new Binding(USERS_STATUS_QUEUE,Binding.DestinationType.
    QUEUE,USERS_EXCHANGE,USERS_ACTIVATED,null);
}

@Bean
Binding userRemovedBinding(){
    return new Binding(USERS_REMOVED_QUEUE,Binding.DestinationType.
    QUEUE,USERS_EXCHANGE,USERS_REMOVED,null);
}
}
```

Let's review the UserRabbitConfiguration class:

- @Configuration: Again, this annotation is a marker for the class that prompts Spring Boot to configure any @Bean declared in this class.

- ConnectionFactory: This interface gathers all the necessary values from the auto-configuration, such as host (where the RabbitMQ broker is running), username, password, virtualhost, and any listeners. This interface implementation is required for the RabbitTemplate class.

- RabbitTemplate: This class implements all the necessary logic to open a connection (a pool of connections), channels (a pool of channels), retries (in case of connection lost to the broker), and it can be used to override and set new values, and in this case we are setting the *Exchange*, in this case a Topic Exchange named USERS. Also, we are overriding the message converter using the Jackson2JsonMessageConverter class.

- Jackson2JsonMessageConverter: This class has a hierarchy that extends some abstract classes and implements the MessageConverter interface. Here we are using it to convert our event into JSON format and we are registering the JavaTimeModule class module, which will deal with the LocalDateTime format.

- TopicExchange: This class creates a durable *Topic Exchange*, which will allow us to create some keys that may contain some regular expressions and use the routing capabilities that RabbitMQ offers by default with this type of *Exchange*.

- Queue: We are declaring two queues, USERS_STATUS and USERS_ REMOVED. These queues have the properties of durable set to true (which means that in case of a crash or restart, the queue is still there), exclusive set to false (which means that we can use multiple consumers for this queue), and autodelete set to false (which means that even though a consumer gets disconnected from the queue, the queue will be there).

- Binding: As previously depicted in the diagram in Figure 10-7, every *Exchange* is connected to a *Queue* through a *Binding* that can have or not (depending on the *Exchange*) a *Routing Key*. In this case, we are creating a *Binding* per *Queue* and assigning a *Routing Key* (users. activated and users.removed).

It's important to notice that we are creating the Exchange, Queues, and Bindings programmatically, but you can create them manually in the RabbitMQ web console. Next, add/open the docker-compose.yaml file. See Listing 10-22.

Listing 10-22. docker-compose.yaml

```
version: "3"
services:
  rabbitmq:
    container_name: rabbitmq
    hostname: rabbitmq
    image: rabbitmq:management-alpine
    restart: always
    ports:
      - "15672:15672"
      - "5672:5672"
```

Running the Users App with RabbitMQ

To run the Users App with RabbitMQ, you can use the command line or your IDE. When you run it, you will see the following in the console:

```
...
Created new connection: rabbitConnectionFactory#7fff419d:0/
SimpleConnection@6f5f892c [delegate=amqp://guest@127.0.0.1:5672/,
localPort=63724]
UserLogs: User ximena@email.com active status: true
UserLogs: User norma@email.com active status: false
UserLogs: User dummy@email.com active status: false
UserLogs: User dummy@email.com DELETED at 2023-10-19T15:57:46.049020
...
```

Let's review what just happened. Open your browser and point it to http://localhost:15672, which is the RabbitMQ web console. See Figure 10-8.

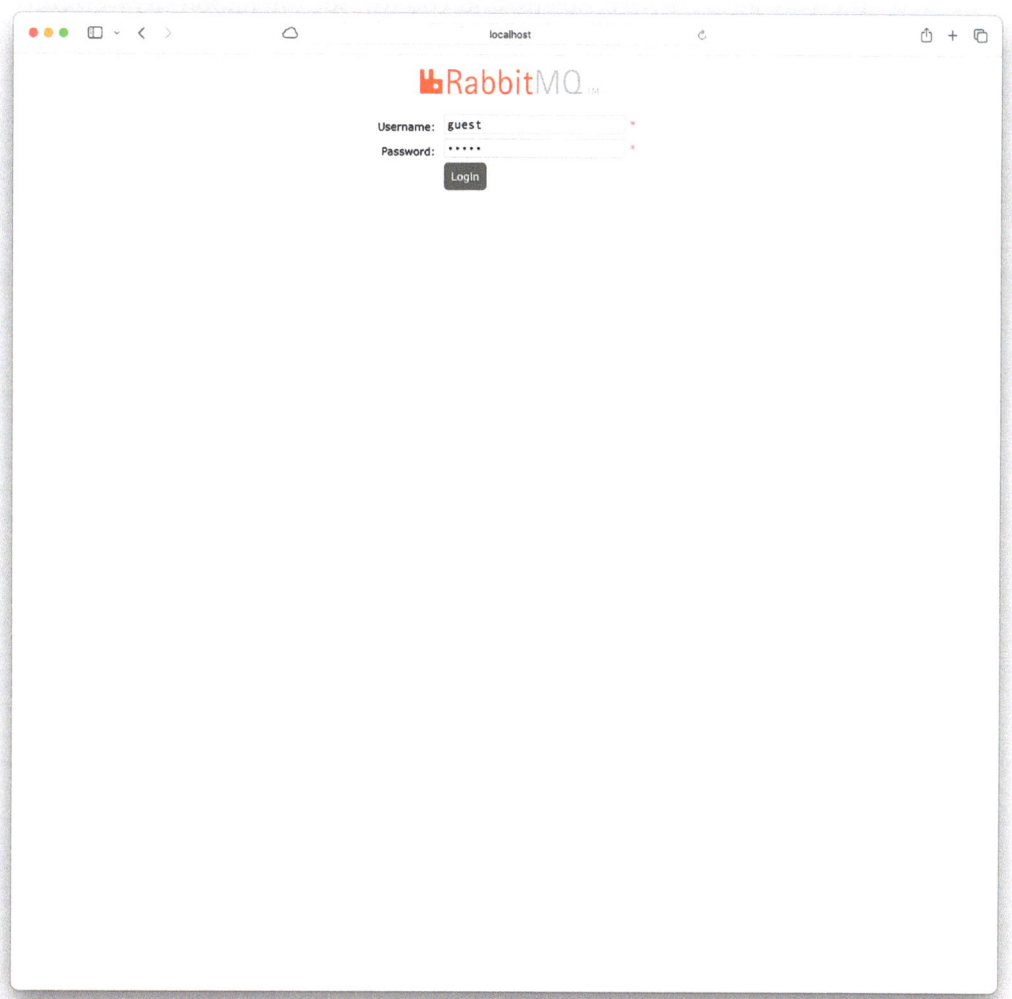

Figure 10-8. *RabbitMQ web console (`http://localhost:15672`) login page*

The credentials are guest/guest. Log in, and you will see the RabbitMQ web console screen shown in Figure 10-9.

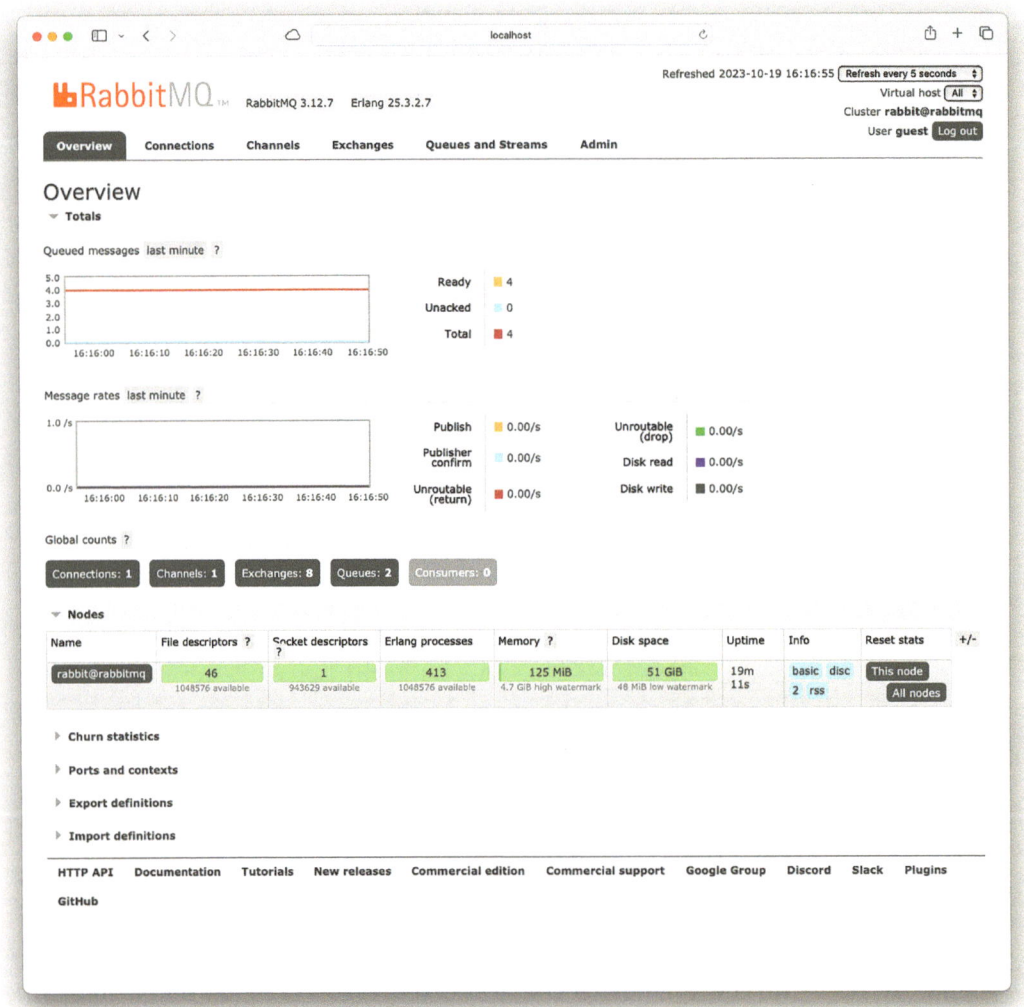

Figure 10-9. *RabbitMQ web console main page*

Here you can see everything Statistics in real time, all the messages that are passing through, get consumption information or any other important properties, such as memory, disk space, and much more. Next, click the Exchanges tab. See Figure 10-10.

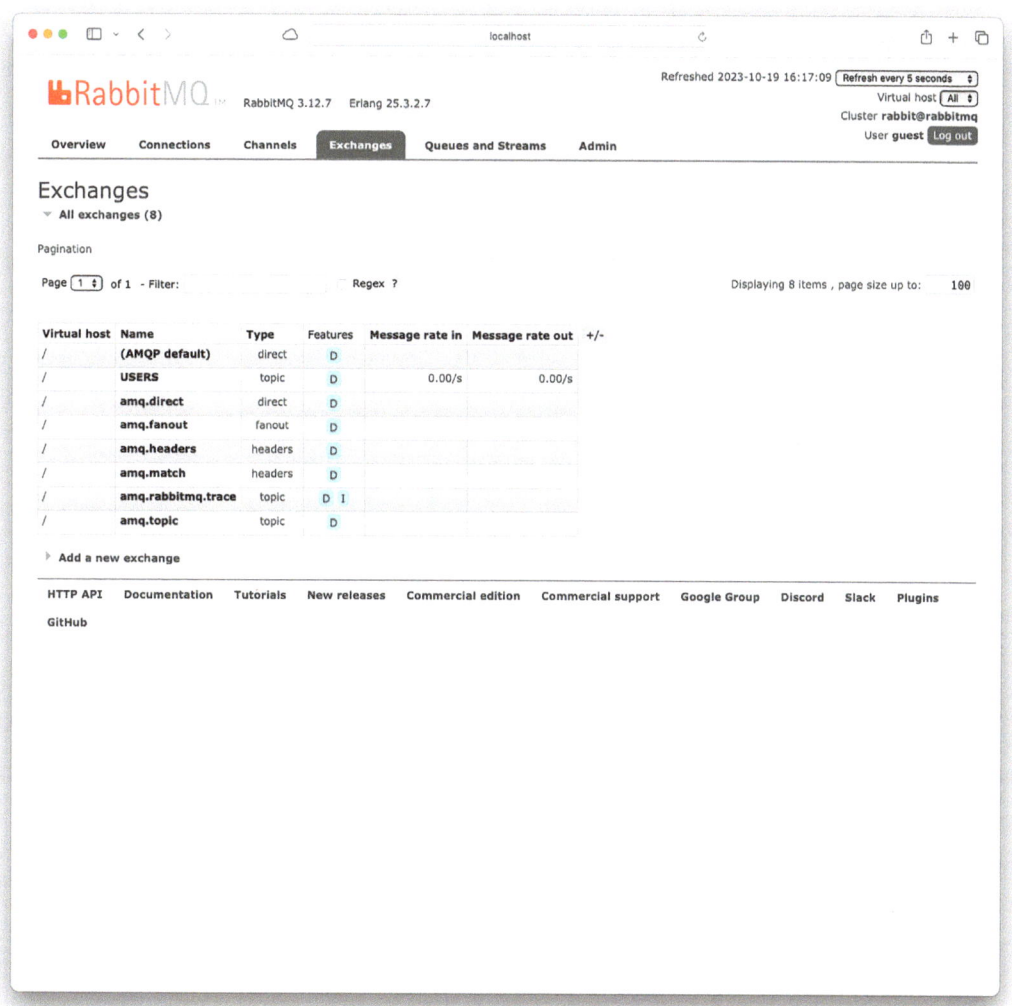

Figure 10-10. *RabbitMQ web console Exchanges tab*

By default, several exchanges are already predefined. And note that the USERS *Exchange* (type Topic) is declared.

Next, click the Queues and Streams tab. See Figure 10-11.

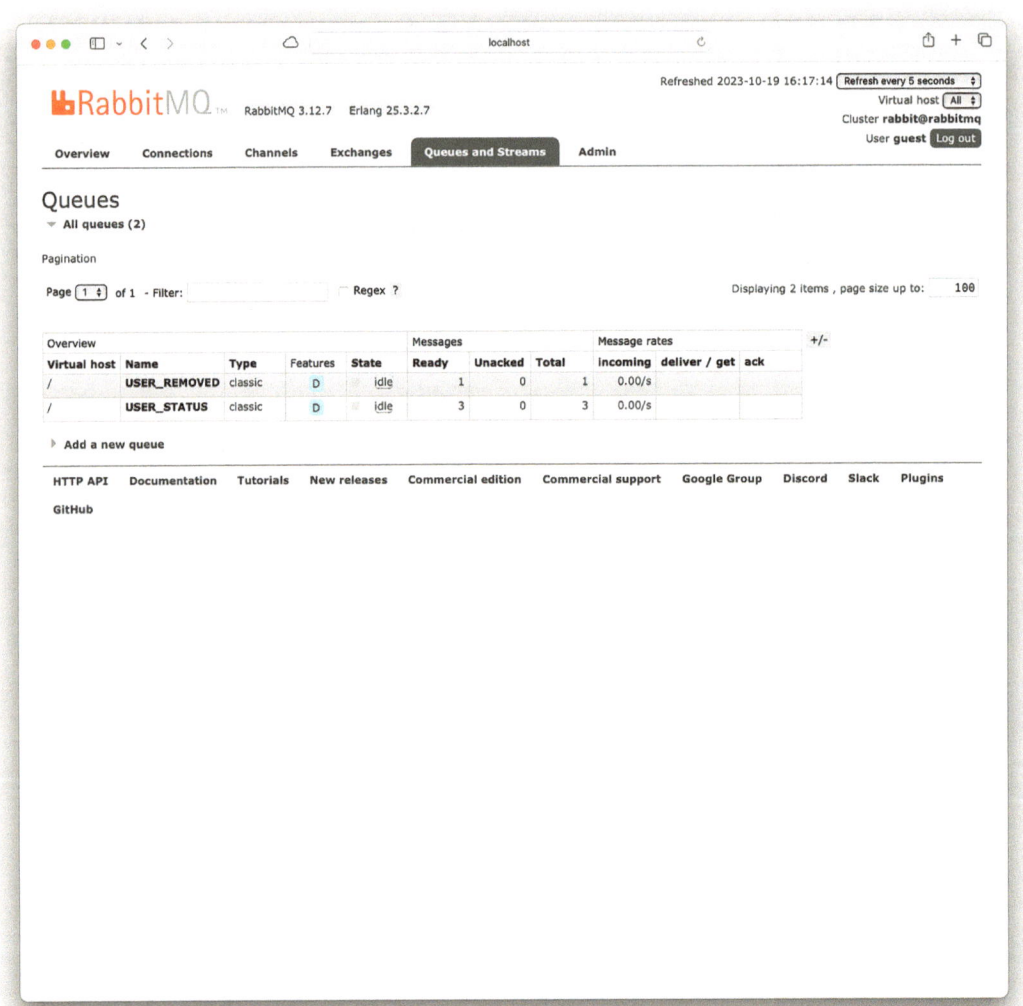

Figure 10-11. *RabbitMQ web console Queues and Streams tab*

The USER_STATUS and USER_REMOVED Queues both are defined as *Classic Queue*, which is just fine. There is also a *Quorum Queue*, which can be used when you need more high availability and are doing clustering; and in our case, we don't need this (for now).

Next, click the USER_STATUS *Queue* to see details, as shown in Figure 10-12.

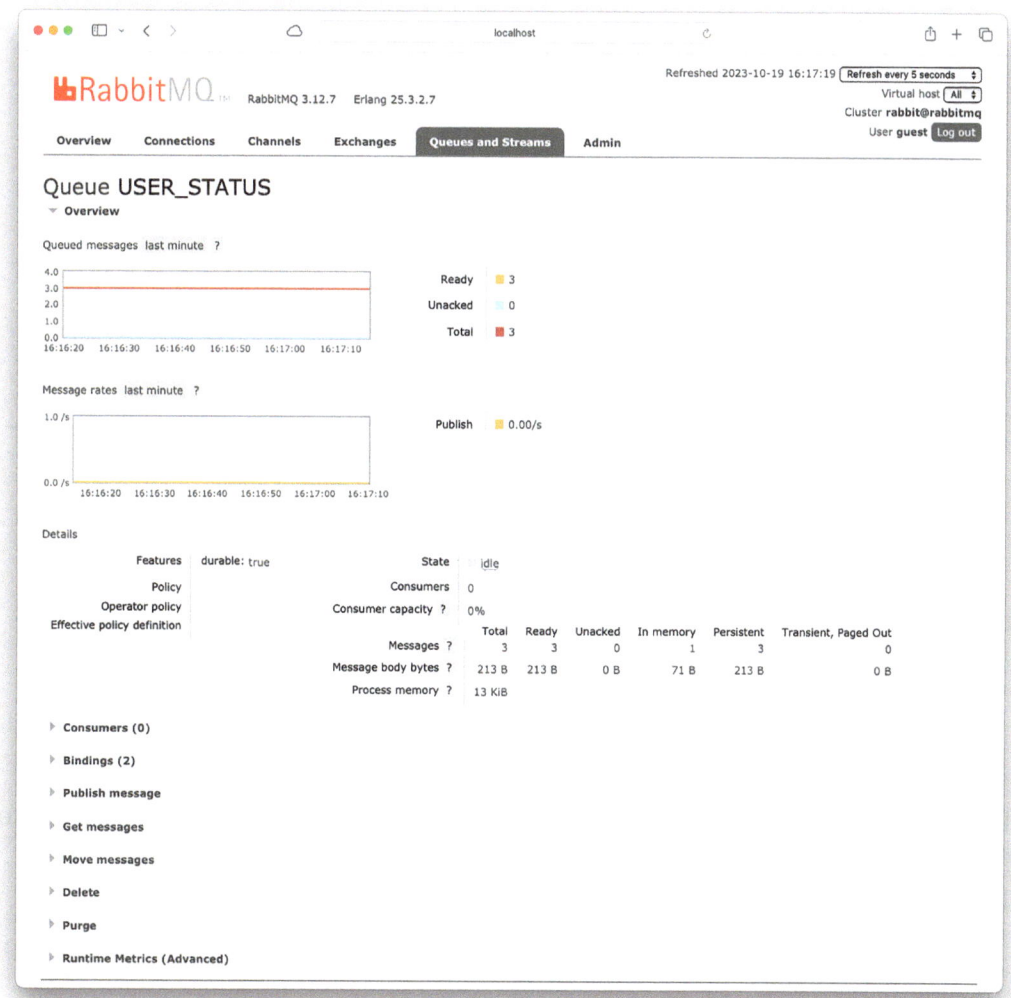

Figure 10-12. *RabbitMQ web console with USER_STATUS Queue displayed*

You should have some messages already there. Scroll down the page and expand the Get Message section. Click the Get Message(s) button, and you should see something like Figure 10-13.

Figure 10-13. *RabbitMQ web console USER_STATUS Queue with Get Messages section expanded*

The Payload section at the bottom shows the event message converted into JSON format. In the Properties section, note that in the headers field there is a key of __TypeId__ and its value is com.apress.users.event.UserActivatedEvent; this is like TypeId in JMS, but here it was not necessary to explicitly declare it, by default Spring AMQP will generate it.

If you review the message from the USER_REMOVED Queue, you should see something like Figure 10-14.

Figure 10-14. *RabbitMQ web console USER_REMOVED Queue with Get Messages expanded*

The headers field in the Properties section and the message in the Payload section indicate we have what we need.

Now you can stop your Users App.

Consuming Messages from RabbitMQ

To consume messages from RabbitMQ, a consumer must follow these steps: open a Connection (to the RabbitMQ broker), create a Channel, connect to a Queue, and consume the Message (with auto-acknowledge or manual acknowledge).

We are consuming events in JSON format, so we need the same configuration as in the JMS configuration: we need to override the default listener and register the JavaTimeModule to be used for the deserialization.

So, let's create the final version of the UserRabbitConfiguration class. See Listing 10-23.

Listing 10-23. src/main/java/apress/com/users/amqp/ UserRabbitConfiguration.java

```
package com.apress.users.amqp;

import com.fasterxml.jackson.databind.ObjectMapper;
import com.fasterxml.jackson.datatype.jsr310.JavaTimeModule;
import org.springframework.amqp.rabbit.config.SimpleRabbitListener
ContainerFactory;
```

```
import org.springframework.amqp.rabbit.connection.ConnectionFactory;
import org.springframework.amqp.rabbit.core.RabbitTemplate;
import org.springframework.amqp.support.converter.
Jackson2JsonMessageConverter;
import org.springframework.context.annotation.Bean;
import org.springframework.context.annotation.Configuration;

@Configuration
public class UserRabbitConfiguration {

    public static final String USERS_EXCHANGE = "USERS";
    public static final String USERS_STATUS_QUEUE = "USER_STATUS";
    public static final String USERS_REMOVED_QUEUE = "USER_REMOVED";
    public static final String USERS_ACTIVATED = "users.activated";
    public static final String USERS_REMOVED = "users.removed";

    @Bean
    RabbitTemplate rabbitTemplate(ConnectionFactory connectionFactory){
        RabbitTemplate rabbitTemplate = new RabbitTemplate(connectio
        nFactory);
        rabbitTemplate.setExchange("USERS");
        rabbitTemplate.setMessageConverter(new Jackson2JsonMessageConverter
        (new ObjectMapper().registerModule(new JavaTimeModule())));
        return rabbitTemplate;
    }

    @Bean
    public SimpleRabbitListenerContainerFactory rabbitListenerContainer
    Factory(ConnectionFactory connectionFactory) {
        SimpleRabbitListenerContainerFactory factory = new
        SimpleRabbitListenerContainerFactory();
        factory.setConnectionFactory(connectionFactory);
        factory.setMessageConverter(new Jackson2JsonMessageConverter(new
        ObjectMapper().registerModule(new JavaTimeModule())));
        return factory;
    }

}
```

Listing 10-23 shows that we removed the *Exchange, Queue,* and *Binding* beans; I will show you another way to create them. Next, we are declaring the SimpleRabbitListernerContainerFactory (very similar to the JMS version) and we are using the well-known ConnectionFactory and registering the JavaTimeModule.

Next, open/create the UserListeners class. See Listing 10-24.

Listing 10-24. src/main/java/apress/com/users/amqp/UserListeners.java

```java
package com.apress.users.amqp;

import com.apress.users.events.UserActivatedEvent;
import com.apress.users.events.UserRemovedEvent;
import lombok.extern.slf4j.Slf4j;
import org.springframework.amqp.rabbit.annotation.Exchange;
import org.springframework.amqp.rabbit.annotation.Queue;
import org.springframework.amqp.rabbit.annotation.QueueBinding;
import org.springframework.amqp.rabbit.annotation.RabbitListener;
import org.springframework.stereotype.Component;

import static com.apress.users.amqp.UserRabbitConfiguration.USERS_
ACTIVATED;
import static com.apress.users.amqp.UserRabbitConfiguration.USERS_EXCHANGE;
import static com.apress.users.amqp.UserRabbitConfiguration.USERS_REMOVED;
import static com.apress.users.amqp.UserRabbitConfiguration.USERS_
REMOVED_QUEUE;
import static com.apress.users.amqp.UserRabbitConfiguration.USERS_
STATUS_QUEUE;

@Component
@Slf4j
public class UserListeners {

    @RabbitListener(
            bindings = @QueueBinding(value = @Queue(name = USERS_STATUS_
            QUEUE, durable = "true", autoDelete = "false")
                    ,exchange = @Exchange(name=USERS_EXCHANGE,type =
                    "topic"),key=USERS_ACTIVATED))
```

```
public void userStatusEventProcessing(UserActivatedEvent
activatedEvent){
    log.info("[AMQP - Event] Activated Event Received: {}",
    activatedEvent);
}

@RabbitListener(
        bindings = @QueueBinding(value = @Queue(name = USERS_REMOVED_
        QUEUE, durable = "true", autoDelete = "false")
                ,exchange = @Exchange(name=USERS_EXCHANGE,
                type="topic"),key=USERS_REMOVED))
public void userRemovedEventProcessing(UserRemovedEvent removedEvent){
    log.info("[AMQP - Event] Activated Event Received: {}",
    removedEvent);
}
}
```

The UserListeners class includes the following annotations:

- @RabbitMQListener: This annotation creates a RabbitListener that will take the default implementation unless overridden; and in this case we are creating the SimpleRabbitListenerContainerFactory that will bring our own MessageConverter configuration (in this case the Jackson2JsonMessageConverter). With this annotation you can define *Queues, Exchanges,* and *Bindings* (through other annotations), making this simpler and more readable code because you know where the listener is consuming from.

- @QueueBinding: This annotation allows you to create a binding from the Exchange to the Queue. One of the benefits here is that if the *Queue, Exchange,* and *Bindings* are not present, this annotation will attempt to create them.

- @Queue: This annotation allows you to declare a *Queue* in the broker with some properties, such as durable and autodelete among others. If the *Queue* is not in the broker, it will attempt to create it.

- @Exchange: This annotation declares an *Exchange* in the broker with some properties. If the *Exchange* is not present in the broker, then it will attempt to create it.

Note   If for some reason you change a property (any property) of a *Queue* (for example, set `durable` to `true` the first time and then to `false` the next time), the queue will fail. So, carefully design the types of *Exchanges*, *Routing Keys*, and *Queue* types and properties you will use to avoid any errors.

Running the Users App with RabbitMQ to Consume Messages

Running the Users App with RabbitMQ, you will have a publisher and consumers, and you will get the following output:

```
...
User ximena@email.com active status: true
User norma@email.com active status: false
User dummy@email.com active status: false
[AMQP - Event] Activated Event Received: UserActivatedEvent(action=
ACTIVATION_STATUS, email=ximena@email.com, active=true)
[AMQP - Event] Activated Event Received: UserActivatedEvent(action=
ACTIVATION_STATUS, email=norma@email.com, active=false)
[AMQP - Event] Activated Event Received: UserActivatedEvent(action=
ACTIVATION_STATUS, email=dummy@email.com, active=false)
User dummy@email.com DELETED at 2023-10-19T17:27:26.442656
[AMQP - Event] Activated Event Received: UserRemovedEvent(action=REMOVED,
email=dummy@email.com, removed=2023-10-19T17:27:26)
...
```

Awesome! Now, you know how to use RabbitMQ for messaging. Of course, this is just a tiny bit of what you can do with it. Continue reading to discover other features.

Using RabbitMQ in My Retro App to Consume User Events

Now that we've established how to consume messages from RabbitMQ (based on the Users App and its User Events), we can replicate the same pattern in the My Retro App. We need to declare the SimpleRabbitListenerContainerFactory, add the @RabbitListener annotation to a method that will receive the two events, and create the event classes—basically just copy and paste from the Users App, right? But we haven't explored one of the best features of RabbitMQ, the routing capabilities. If our intention is to capture every event from the users, we don't need to create a class per event or a listener per event; instead, we can use the *Topic Exchange* capability to route messages based on the routing key, and because this is a *Topic* type, we can use wildcards for the keys. See Figure 10-15.

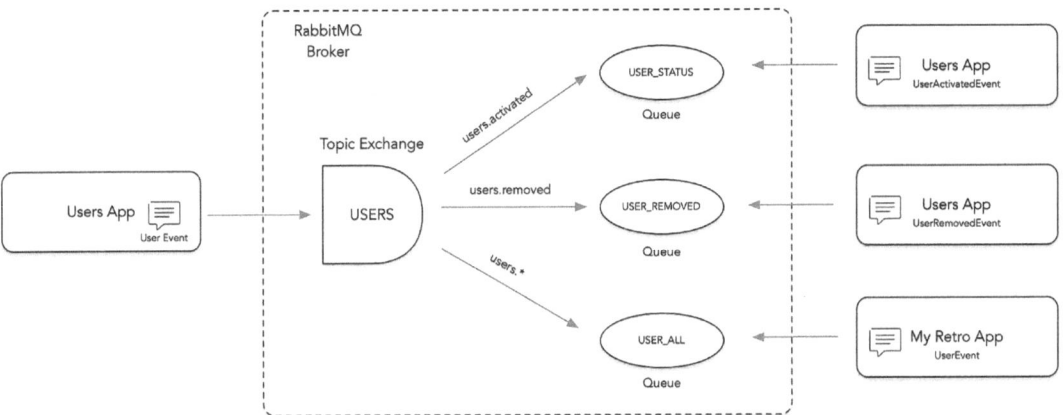

Figure 10-15. *Users App interaction with My Retro App via Topic Exchange*

Figure 10-15 shows the interaction that we'll have with the Users App. Recall that we've just created consumers that are listening for the UserActivadedEvent (based on the routing key users.activated that is bound to the USERS_STATUS *Queue*) and the UserRemovedEvent (based on the routing key users.removed that is bound to the USER_REMOVED *Queue*). We are going to create another *Queue* (named USER_ALL) that will receive the copy of the two events (either UserActivatedEvent or UserRemovedEvent), and this is based on the routing key users.*; this is a wildcard that works only with *Topic Exchanges* allowing to have more flexibility for any type of routing. Remember that if we use a * (asterisk) will match exactly one word in a routing keyand if we use the # (hash) will match zero or more words in a routing key.

So, you can open (or create the My Retro App project based on the JPA). You can find the code in the 10-messageing-amqp/myretro folder.

Open/create the UserEvent class. See Listing 10-25.

Listing 10-25. src/main/java/apress/com/myretro/amqp/UserEvent.java

```
package com.apress.myretro.amqp;

import com.fasterxml.jackson.annotation.JsonFormat;
import com.fasterxml.jackson.annotation.JsonIgnore;
import com.fasterxml.jackson.annotation.JsonIgnoreProperties;
import lombok.Data;

import java.time.LocalDateTime;

@Data
@JsonIgnoreProperties(ignoreUnknown = true)
public class UserEvent {

    private String action;

    private String email;

    private boolean active;

    @JsonFormat(shape=JsonFormat.Shape.STRING,pattern = "yyyy-MM-dd
    HH:mm:ss")
    private LocalDateTime removed;
}
```

The UserEvent class will be used for both events; as you can see, it's kind of a merge of the UserActivatedEvent and UserRemovedEvent fields.

Next, open/create the UserListener class. See Listing 10-26.

Listing 10-26. src/main/java/apress/com/myretro/amqp/UserListener.java

```
package com.apress.myretro.amqp;

import lombok.extern.slf4j.Slf4j;
import org.springframework.amqp.rabbit.annotation.Exchange;
import org.springframework.amqp.rabbit.annotation.Queue;
import org.springframework.amqp.rabbit.annotation.QueueBinding;
import org.springframework.amqp.rabbit.annotation.RabbitListener;
import org.springframework.stereotype.Component;
```

```
@Component
@Slf4j
public class UserListener {

    private static final String USERS_ALL = "users.*";
    private static final String USERS_ALL_QUEUE = "USER_ALL";
    private static final String USERS_EXCHANGE = "USERS";

    @RabbitListener(
            bindings = @QueueBinding(value = @Queue(name = USERS_ALL_QUEUE,
            durable = "true", autoDelete = "false")
                    ,exchange = @Exchange(name=USERS_EXCHANGE,type =
                    "topic"),key=USERS_ALL))
    public void userStatusEventProcessing(UserEvent userEvent){
        log.info("[AMQP - Event] Received: {}", userEvent);
    }
}
```

Listing 10-26 shows that the UserListener class is almost the same as in the Users App, but here we are declaring the new USER_ALL *Queue*, the *Binding* between the *Queue* and the USERS *Exchange* using the *Routing Key* users.* that allows to receive a copy of every event message, and the users.activated and users.removed messages. To have an effective way to use the *Topic Exchange* and its Routing capabilities, it is important to have a good naming convention for your routing keys, which is why we named them with the prefix users.; this enables us to have in the future users.status.activated, users. status.removed, users.status.deactivated, users.admin.activated, users.admin. deactivated, and even some wildcards such as users.*.activated (this will include the users.status.activated and users.admin.activated), etc. The *Topic Exchange* also accepts the # symbol, which will match zero or more words, and the * symbol, which will math one word.

Next, open/create the UserRabbitConfiguration class. See Listing 10-27.

Listing 10-27. src/main/java/apress/com/myretro/amqp/
UserRabbitConfiguration.java

```
package com.apress.myretro.amqp;

import com.fasterxml.jackson.databind.ObjectMapper;
import com.fasterxml.jackson.datatype.jsr310.JavaTimeModule;
import org.springframework.amqp.rabbit.config.
SimpleRabbitListenerContainerFactory;
import org.springframework.amqp.rabbit.connection.ConnectionFactory;
import org.springframework.amqp.support.converter.
Jackson2JsonMessageConverter;
import org.springframework.context.annotation.Bean;
import org.springframework.context.annotation.Configuration;

@Configuration
public class UserRabbitConfiguration {

    @Bean
    public SimpleRabbitListenerContainerFactory rabbitListenerContainerFact
    ory(ConnectionFactory connectionFactory) {
        SimpleRabbitListenerContainerFactory factory = new
        SimpleRabbitListenerContainerFactory();
        factory.setConnectionFactory(connectionFactory);
        factory.setMessageConverter(new Jackson2JsonMessageConverter(new
        ObjectMapper()
                .registerModule(new JavaTimeModule())));
        return factory;
    }

}
```

As we did earlier for the UserRabbitConfiguration class, we are declaring the
SimpleRabbitListenerContainerFactory bean and configuring the MessageConverter
with the Jackson2JsonMessageConverter to register the JavaTimeModule that will help
with the deserialization of the LocalDateTime type.

Running the My Retro App to Listen for User Events from RabbitMQ

Make sure you have the Users App up and running; the Users App has the docker-compose.yaml file that will be used by Spring Boot to start the RabbitMQ broker.

Next, run the My Retro App. You can check out the RabbitMQ web console and see that now we have three Queues, as shown in Figure 10-16.

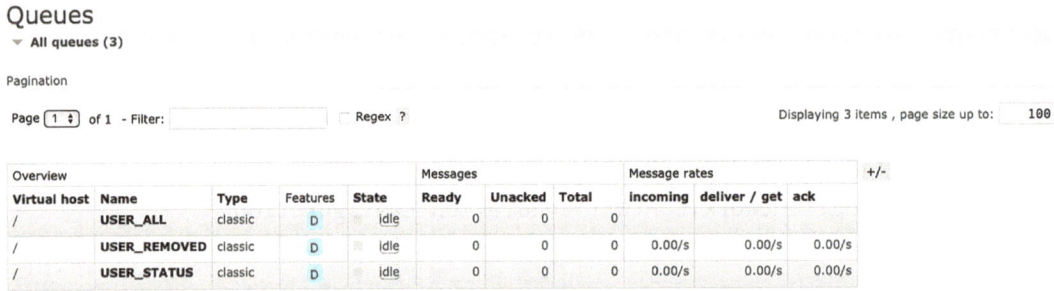

Figure 10-16.   *RabbitMQ web console with three Queues*

Next, you can send a new user in the terminal with the following command:

```
curl -i -s -d '{"name":"Dummy","email":"dummy@email.com","password":
"aw2s0meR!","userRole":["INFO"],"active":true}' \
-H "Content-Type: application/json" \
http://localhost:8080/users
```

```
HTTP/1.1 201
Location: http://localhost:8080/users/dummy@email.com
Content-Type: application/json
Transfer-Encoding: chunked
Date: Sat, 21 Oct 2023 17:32:34 GMT
```

```
{"email":"dummy@email.com","name":"Dummy","gravatarUrl":null,"password":
"aw2s0meR!","userRole":["INFO"],"active":true}
```

And if you check out the My Retro Console you should have the event:

```
[AMQP - Event] Received: UserEvent(action=ACTIVATION_STATUS,
email=dummy@email.com, active=true, removed=null)
```

You can experiment also with deleting a user with the following command:

```
curl -i -s -XDELETE http://localhost:8080/users/dummy@email.com
```

```
HTTP/1.1 204
Date: Sat, 21 Oct 2023 17:34:12 GMT
```

And the My Retro App console will have the following:

```
[AMQP - Event] Received: UserEvent(action=REMOVED, email=dummy@email.com,
active=false, removed=2023-10-21T13:34:12)
```

Nice, now you've integrated both apps/services that use events that are sent to RabbitMQ as your message broker.

WebSockets with Spring Boot

WebSockets is one of my favorite technologies because it allows near real-time event processing. WebSockets is a communication protocol that provides full-duplex channels over TCP. The WebSocket protocol defines two types of messages, text and binary, and normally is used with the definition of a subprotocol that allows to send and receive messages.

One of the most interesting parts about this protocol is that there is a handshake between the client and the server, which starts with the client sending an HTTP request with special headers that allow a direct connection through TCP using a unique hash key between the client and server, then the communication can start by sending and receiving WebSocket data frames. This is how it looks behind the scenes:

Client request:

```
GET /chat HTTP/1.1
Upgrade: websocket
Connection: Upgrade
Sec-WebSocket-Key: dGhlIHNhbXBsZSBub25jZSB3aGljaCBrZXk=
```

Server response:

```
HTTP/1.1 101 Switching Protocols
Upgrade: websocket
Connection: Upgrade
Sec-WebSocket-Accept: s3pPLMBiTxaQ9kYGzzh5fWusIqFw=
```

In this section we'll use Spring's STOMP (Simple Text-Oriented Messaging Protocol) support (as the subprotocol) and see how Spring WebSocket acts as the STOMP broker to clients using Spring Boot.

Adding WebSockets to the Users App

To add WebSockets to the Users App, simply add the spring-boot-starter-websocket dependency, which will be enough for Spring Boot to auto-configure all the necessary defaults to provide a WebSocket communication framework that enables you to create event-driven and real-time applications with ease.

You can use the code in the 10-messaging-websockets/users folder or you can start from scratch with the Spring Initializr (https://start.spring.io). For the latter, add the dependencies WebSocket, Validation, JPA, Lombok, H2, and Postgresql, set the Group field to com.apress and the Artifact and Name fields to users, generate and download the project, unzip it, and import it into your favorite IDE. The main codebase will be from the Users Web/JPA.

We need to add some extra dependencies, so open the build.gradle file. See Listing 10-28.

Listing 10-28. build.gradle

```
plugins {
    id 'java'
    id 'org.springframework.boot' version '3.2.3'
    id 'io.spring.dependency-management' version '1.1.4'
}

group = 'com.apress'
version = '0.0.1-SNAPSHOT'
sourceCompatibility = '17'

repositories {
    mavenCentral()
}

dependencies {
    implementation 'org.springframework.boot:spring-boot-starter-websocket'
    implementation 'org.springframework.boot:spring-boot-starter-validation'
```

```
implementation 'org.webjars:webjars-locator-core'
implementation 'org.webjars:sockjs-client:1.5.1'
implementation 'org.webjars:stomp-websocket:2.3.4'

implementation 'com.fasterxml.jackson.datatype:jackson-datatype-jsr310'

implementation 'org.springframework.boot:spring-boot-starter-data-jpa'
runtimeOnly 'com.h2database:h2'
runtimeOnly 'org.postgresql:postgresql'

compileOnly 'org.projectlombok:lombok'
annotationProcessor 'org.projectlombok:lombok'

// Web
implementation 'org.webjars:bootstrap:5.2.3'
implementation 'org.webjars:jquery:3.7.1'

testImplementation 'org.springframework.boot:spring-boot-starter-test'
}

tasks.named('test') {
    useJUnitPlatform()
}
```

Listing 10-28 shows that we are only using the spring-boot-starter-websocket; all the web dependencies come with the websocket dependency, so no need to declare them. Also, we are using a sockjs-client and stomp-websocket that will help on the UI (client) side. We are going to send JSON messages, so we need the jackson-datatype-jsr310 dependency as well.

Next, open/create the UserSocket class. See Listing 10-29.

Listing 10-29. src/main/java/apress/com/users/web/socket/UserSocket.java

```
package com.apress.users.web.socket;

import lombok.AllArgsConstructor;
import org.springframework.messaging.core.MessageSendingOperations;
import org.springframework.stereotype.Component;

import java.time.LocalDateTime;
import java.time.format.DateTimeFormatter;
```

```java
import java.util.HashMap;
import java.util.Map;

@AllArgsConstructor
@Component
public class UserSocket {

    private MessageSendingOperations<String> messageSendingOperations;

    public void userLogSocket(Map<String,Object> event){
        Map<String, Object> map = new HashMap<>(){{
            put("event",event);
            put("version","1.0");
            put("time",LocalDateTime.now().format(DateTimeFormatter.
            ofPattern("yyyy-MM-dd HH:mm:ss")));
        }};
        this.messageSendingOperations.convertAndSend("/topic/user-
        logs",map);
    }
}
```

The important element in the UserSocket class is the MessageSendingOperations interface, which belongs to the org.springframework.messaging.core package from the spring-message dependency. This interface is a common way to send messages regardless of the technology used; you can think of it as a low-level communication. You can use any protocol you need—in fact, the Spring JMS and Spring AMQP use this interface for specialization of the JmsMessageOperations and RabbitMessageOperations interfaces, respectively, for sending messages.

You can see that we are sending just a Map<String, Object> as a message or event; you can send anything really, but I wanted to show you that you can do basic classes, such as a Map.

Next, open/create the UserSocketConfiguration class. As you can imagine, we need to do some configuration. See Listing 10-30.

Listing 10-30. src/main/java/apress/com/users/web/socket/
UserSocketConfiguration.java

```
package com.apress.users.web.socket;

import org.springframework.context.annotation.Configuration;
import org.springframework.messaging.simp.config.MessageBrokerRegistry;
import org.springframework.web.socket.config.annotation.
EnableWebSocketMessageBroker;
import org.springframework.web.socket.config.annotation.
StompEndpointRegistry;
import org.springframework.web.socket.config.annotation.
WebSocketMessageBrokerConfigurer;

@Configuration
@EnableWebSocketMessageBroker
public class UserSocketConfiguration implements
WebSocketMessageBrokerConfigurer {
    @Override
    public void configureMessageBroker(MessageBrokerRegistry config) {
        config.enableSimpleBroker("/topic");
        config.setApplicationDestinationPrefixes("/app");
    }

    @Override
    public void registerStompEndpoints(StompEndpointRegistry registry) {
        registry.addEndpoint("/logs").withSockJS();
    }
}
```

The UserSocketConfiguration class includes the following:

- @Configuration: We need this class to be part of the configuration when the app starts, so we need to mark it with this annotation.

- @EnableWebSocketMessageBroker: We need this annotation because we are going to do some *broker-backed* messaging over WebSocket using a higher-level messaging subprotocol; in other words, we are going to be the broker that handles the STOMP protocol over WebSocket.

- `WebSocketMessageBrokerConfigurer`: This interface is useful because it defines methods for configuring message handling through simple protocols, such as STOMP. With this implementation we define our message converters, custom return value handlers, STOMP endpoints, the usage of the `MessageChannel` from the core of `spring-messaging`, and much more.

- `MessageBrokerRegistry`: This is one of the parameters of the `ConfigureMessageBroker` method implementation, and it helps to configure the message broker options. In this case we are configuring the broker destination prefix with `/topic` and the application destination prefix with `/app`.

- `StompEndpointRegistry`: The `StompEndpointRegistry` parameter is a contract for registering STOMP over WebSocket endpoints and in this case the `/logs` endpoint.

This completes the back end. Yes, that's it! Next, let's look from the UI side.

Using the WebSocket Client in the Users App to Consume Events

Next, open/create the `app.js` file that will contain the client side. See Listing 10-31.

Listing 10-31. src/main/resources/static/js/app.js

```
let stompClient = null;

function connect() {
    let socket = new SockJS('/logs');
    stompClient = Stomp.over(socket);
    stompClient.connect({}, function (frame) {
            console.log('Connected: ' + frame);
            stompClient.subscribe('/topic/user-logs', function
            (response) {
                    console.log(response);
                    showLogs(response.body);
            });
    });
}
```

```
function showLogs(message) {
    $("#logs").append("<tr><td>" + message + "</td></tr>");
}

$(function () {
    connect();
});
```

The app.js file is a plain JavaScript file in which we are using the jQuery library. We are using the SockJS class that comes from the socks-client JavaScript dependency. The SockJS class is using the /logs endpoint we defined in the UserSocketConfiguration class (Listing 10-30). We are also creating the Stomp client that will allow us to connect and then subscribe for any incoming message into the /topic/user-logs destination; this was configured in the UserSocketConfiguration class and in the UserSocket class (Listing 10-29) when we used the convertAndSend method call. Once we get the user event, we use the showLogs method and append it to the #logs HTML element (this is a <div/> element).

Next, open/create the index.html file. See Listing 10-32.

Listing 10-32. src/main/resources/static/index.html

```
<!DOCTYPE html>
<html lang="en">
<head>
    <meta charset="UTF-8">
    <link rel="stylesheet" type="text/css"
          href="webjars/bootstrap/5.2.3/css/bootstrap.min.css">
    <script src="/webjars/jquery/3.7.1/jquery.min.js"></script>
    <script src="/webjars/sockjs-client/1.5.1/sockjs.min.js"></script>
    <script src="/webjars/stomp-websocket/2.3.4/stomp.min.js"></script>
    <script src="/js/app.js"></script>
    <title>Welcome - Users App</title>
</head>
<body class="d-flex h-100 text-center">
```

```html
<div class="cover-container d-flex w-100 h-100 p-3 mx-auto flex-column">
    <header class="mb-auto">
        <div>
            <h3 class="float-md-start mb-0">Users</h3>
        </div>
    </header>

    <main class="px-3">
        <h1>Simple Users Rest Application</h1>
        <p class="lead">This is a simple Users app where you can access any
        information from a user</p>
        <p class="lead">
            <a href="/users">Get All Users</a>
        </p>
    </main>

    <footer class="mt-auto text-black-50">
        <p>Powered by Spring Boot 3</p>
        <table>
            <div id="logs">

            </div>
        </table>
    </footer>
</div>

</body>
</html>
```

Listing 10-32 shows that we are using the jQuery, sockjs, and the stomp libraries that are embedded in the org.webjars dependencies. And we are use a <div/> with id=logs that will be the element where we are appending the user event message we will receive from the server.

Running the Users App with WebSockets

Now, we are ready to see in action the Users App with WebSockets. So, run the Users App either from the command line or using your IDE, and then open your browser and point it to `http://localhost:8080`. Next, open the web developer console. All browsers have them.

The web developer console should show something similar to Figure 10-17.

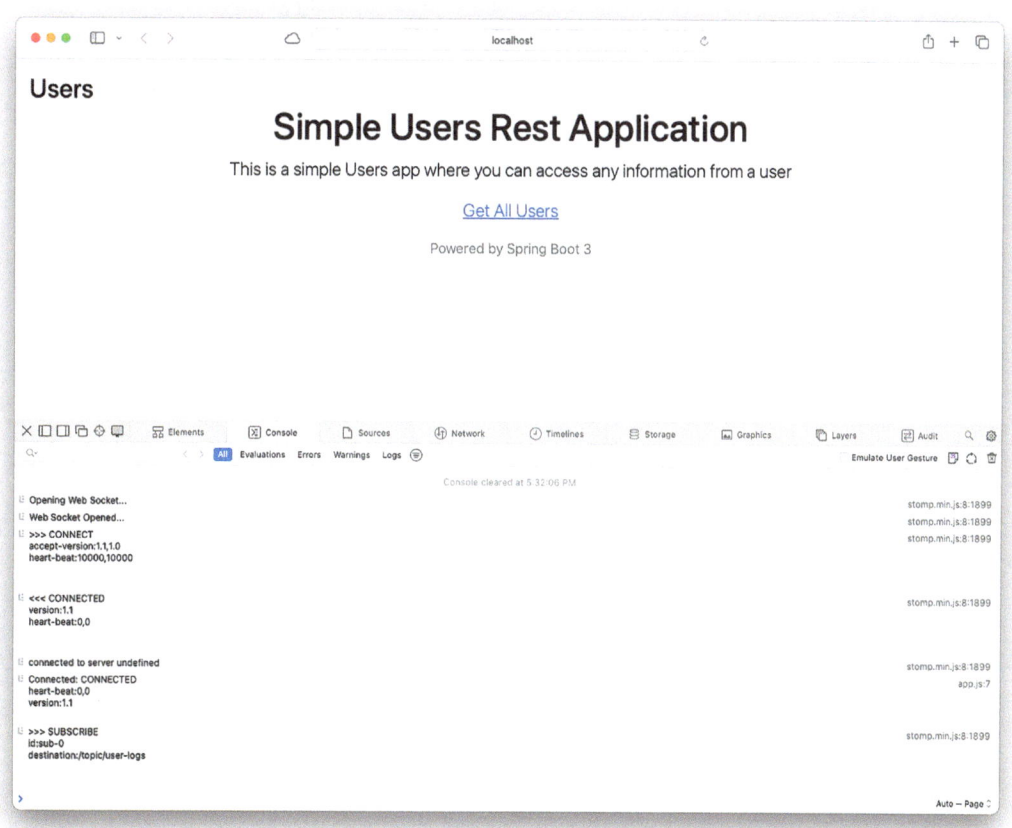

Figure 10-17. *Web developer console (`http://localhost:8080`)*

As shown in Figure 10-17, you should see that you are connected to the server and the destination `/topic/user-logs`. Now, if you add and remove a user using the command line:

```
curl -i -s -d '{"name":"Dummy","email":"dummy@email.com","password":"aw2sOm
eR!","userRole":["INFO"],"active":true}' \
-H "Content-Type: application/json" \
http://localhost:8080/users
```

```
curl -i -s -XDELETE http://localhost:8080/users/dummy@email.com
```

you should see in the web developer console something similar to Figure 10-18.

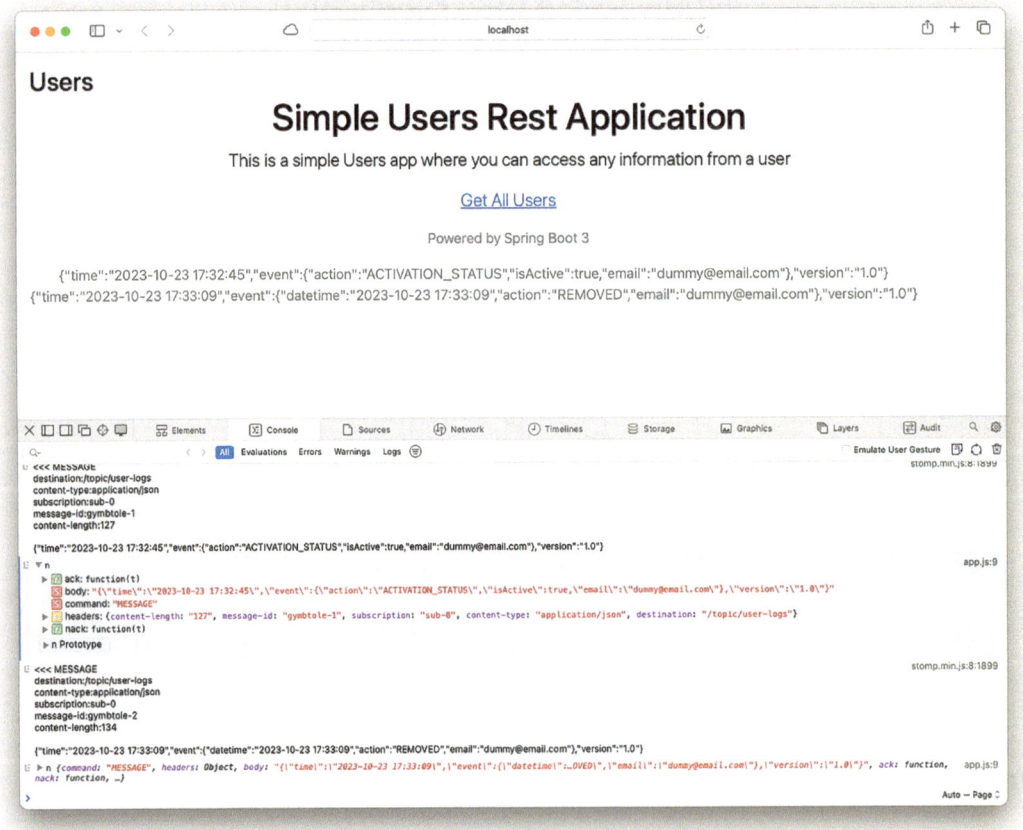

Figure 10-18. *User events in web developer console*

Figure 10-18 shows the result of adding and deleting a user. Notice that the message comes in JSON format.

Using WebSockets in My Retro App to Consume Events

One of the most common questions from my students is that if we can actually consume from another app, can we use WebSockets from My Retro App and receive the User events? Of course we can, so let's do it.

If you are starting from scratch, you can follow the same process from previous sections. Be sure to add the spring-boot-starter-websocket dependency and, of course, the others we are using in this project. You can use the code from the 10-messaging-websocket/myretro folder.

Open the build.gradle file. See Listing 10-33.

Listing 10-33. build.gradle

```
plugins {
    id 'java'
    id 'org.springframework.boot' version '3.2.3'
    id 'io.spring.dependency-management' version '1.1.4'
}

group = 'com.apress'
version = '0.0.1-SNAPSHOT'
sourceCompatibility = '17'

repositories {
    mavenCentral()
}

dependencies {

    implementation 'org.springframework.boot:spring-boot-starter-websocket'
    implementation 'org.springframework.boot:spring-boot-starter-
    validation'

    implementation 'com.fasterxml.jackson.datatype:jackson-datatype-jsr310'

    implementation 'org.springframework.boot:spring-boot-starter-data-jpa'

    annotationProcessor 'org.springframework.boot:spring-boot-
    configuration-processor'
```

```
    runtimeOnly 'com.h2database:h2'
    runtimeOnly 'org.postgresql:postgresql'

    compileOnly 'org.projectlombok:lombok'
    annotationProcessor 'org.projectlombok:lombok'

    testImplementation 'org.springframework.boot:spring-boot-starter-test'
}

tasks.named('test') {
    useJUnitPlatform()
}
```

Next, open/create the Event and UserEvent classes. See Listings 10-34 and 10-35, respectively.

Listing 10-34. src/main/java/com/apress/myretro/web/socket/Event.java

```
package com.apress.myretro.web.socket;

import lombok.Data;
import java.time.LocalDateTime;
import com.fasterxml.jackson.annotation.JsonFormat;

@Data
public class Event {
    private String version;
@JsonFormat(shape=JsonFormat.Shape.STRING,pattern = "yyyy-MM-dd HH:mm:ss")
    private LocalDateTime time;
    private UserEvent event;
}
```

Listing 10-35. src/main/java/com/apress/myretro/web/socket/UserEvent.java

```
package com.apress.myretro.web.socket;

import com.fasterxml.jackson.annotation.JsonFormat;
import com.fasterxml.jackson.annotation.JsonIgnoreProperties;
import lombok.Data;

import java.time.LocalDateTime;
```

```
@JsonIgnoreProperties(ignoreUnknown = true)
@Data
public class UserEvent {
    private String email;
    private boolean active;
    private String action;
    @JsonFormat(shape=JsonFormat.Shape.STRING,pattern = "yyyy-MM-dd
    HH:mm:ss")
    private LocalDateTime datetime;
}
```

We need to create the Event and UserEvent classes because we are sending the composite event message. Also, we need to use the @JsonFormat annotation to have a standard data-time pattern that comes from the User Service.

Next, open/create the consumer, the UserSocketClient class. See Listing 10-36.

Listing 10-36. src/main/java/com/apress/myretro/web/socket/ UserScoketClient.java

```
package com.apress.myretro.web.socket;

import lombok.extern.slf4j.Slf4j;
import org.springframework.messaging.simp.stomp.StompCommand;
import org.springframework.messaging.simp.stomp.StompHeaders;
import org.springframework.messaging.simp.stomp.StompSession;
import org.springframework.messaging.simp.stomp.StompSessionHandlerAdapter;
import org.springframework.stereotype.Component;

@Slf4j
@Component
public class UserSocketClient extends StompSessionHandlerAdapter {

    @Override
    public void afterConnected(StompSession session, StompHeaders
    connectedHeaders) {
        log.info("Client connected: headers {}", connectedHeaders);
        session.subscribe("/topic/user-logs", this);
    }
```

```
@Override
public void handleFrame(StompHeaders headers, Object payload) {
    log.info("Client received: payload {}, headers {}", payload,
    headers);
}

@Override
public void handleException(StompSession session, StompCommand command,
                            StompHeaders headers, byte[] payload,
                            Throwable exception) {
    log.error("Client error: exception {}, command {}, payload {},
    headers {}",
            exception.getMessage(), command, payload, headers);
}

@Override
public void handleTransportError(StompSession session, Throwable
exception) {
    log.error("Client transport error: error {}", exception.
    getMessage());
}
}
```

The UserSocketClient class is important because we need to handle the session, manage any exception, and receive the message (*payload*); you can see this class as a low-level way to get the insight of the WebSocket frames. Let's review it:

- @Component: We need this class as a Spring bean, so we need to mark it using the @Component annotation.

- StompSessionHandlerAdapter: This is an abstract adapter that doesn't have any default implementation because, normally, you must deal with the session and the socket frame itself. This abstract class implements the StompSessionHandler interface, so you need to implement how you want to manage the exceptions and errors that come from the frame or the session, and what else needs to be done when the connection is successful (afterConnected method)

- StompSession: This interface will be implemented behind the scenes and thanks to the configuration provided. This interface is used for sending or receiving messages and creating subscriptions. In this case the afterConnected method uses the StompSession to subscribe to the /topic/users-logs destination.

- StompHeaders: This is the same as the Headers message we saw in previous sections. This class will implement a MultiValueMap interface. This contains useful information such as content-type, content-length, receipt, host, and more. These properties are necessary for the protocol.

Next, open/create the UserSocketMessageConverter class. See Listing 10-37.

Listing 10-37. src/main/java/com/apress/myretro/web/socket/
UserSocketMessageConverter.java

```
package com.apress.myretro.web.socket;

import com.fasterxml.jackson.databind.ObjectMapper;
import com.fasterxml.jackson.datatype.jsr310.JavaTimeModule;
import lombok.extern.slf4j.Slf4j;
import org.springframework.messaging.Message;
import org.springframework.messaging.MessageHeaders;
import org.springframework.messaging.converter.MessageConverter;
import org.springframework.stereotype.Component;

@Slf4j
@Component
public class UserSocketMessageConverter implements MessageConverter {

    public Object fromMessage(Message<?> message, Class<?> targetClass) {
        ObjectMapper mapper = new ObjectMapper().registerModule(new
        JavaTimeModule());
        Event userEvent = null;
        try {
            String m = new String((byte[]) message.getPayload());
            userEvent = mapper.readValue(m, Event.class);
```

```
    }catch(Exception ex){
        log.error("Cannot Deserialize - {}",ex.getMessage());
    }
    return userEvent;
}

@Override
public Message<?> toMessage(Object payload, MessageHeaders headers) {
    throw new UnsupportedOperationException();
}
}
```

The UserSocketMessageConverter class includes the following:

- @Component: We need this custom MessageConverter as a Spring bean because we are going to register later, so we need to mark it using the @Component annotation.

- MessageConverter: As you already know, this is how we can create our custom MessageConverter. We only need to implement the fromMessage method, where we are registering the JavaTimeModule that will deal with the LocalDateTime type when deserializing.

Next, open/create the UserSocketConfiguration class. See Listing 10-38.

Listing 10-38. src/main/java/com/apress/myretro/web/socket/ UserSocketConfiguration.java

```
package com.apress.myretro.web.socket;

import com.apress.myretro.config.RetroBoardProperties;
import lombok.AllArgsConstructor;
import org.springframework.context.annotation.Bean;
import org.springframework.context.annotation.Configuration;
import org.springframework.messaging.simp.stomp.StompSessionHandler;
import org.springframework.web.socket.client.WebSocketClient;
import org.springframework.web.socket.client.standard.
StandardWebSocketClient;
import org.springframework.web.socket.messaging.WebSocketStompClient;
```

```
import org.springframework.web.socket.sockjs.client.
RestTemplateXhrTransport;
import org.springframework.web.socket.sockjs.client.SockJsClient;
import org.springframework.web.socket.sockjs.client.Transport;
import org.springframework.web.socket.sockjs.client.WebSocketTransport;

import java.util.ArrayList;
import java.util.List;

@AllArgsConstructor
@Configuration
public class UserSocketConfiguration {

    RetroBoardProperties retroBoardProperties;

    @Bean
    public WebSocketStompClient webSocketStompClient(WebSocketClient
    webSocketClient, UserSocketMessageConverter userSocketMessageConverter,
                                            StompSessionHandler
                                            userSocketClient) {
        WebSocketStompClient webSocketStompClient = new WebSocketStomp
        Client(webSocketClient);
        webSocketStompClient.setMessageConverter(userSocketMessageConverter);
        webSocketStompClient.connect(retroBoardProperties.
        getUsersService().getHostname() + retroBoardProperties.
        getUsersService().getBasePath(), userSocketClient);
        return webSocketStompClient;
    }

    @Bean
    public WebSocketClient webSocketClient() {
        List<Transport> transports = new ArrayList<>();
        transports.add(new WebSocketTransport(new
        StandardWebSocketClient()));
        transports.add(new RestTemplateXhrTransport());
        return new SockJsClient(transports);
    }
}
```

The UserSocketConfiguration class will help us to stablish the connection to the Users App. We require a client, in this case a WebSocket/STOMP client, so let's review the class:

- RetroBoardProperties: This class binds the external properties in which we can hold information of the remote server (hostname) and its path (basePath). In this case, we need to declare the properties either in the application.properties file or as environment variables (this is your choice).

- StompSessionHandler: This class defines the custom StompSessionHandler, in this case the UserSocketClient class (see Listing 10-34); this class is needed as a part of the constructor for the WebSocketStompClient.

- WebSocketClient: This interface is essential because it initiates the WebSocket request. It requires a list of transports that will be responsible for managing the protocol—in this case, the *SockJS* based on the RestTemplate that we already covered.

- WebSocketStompClient: This class uses the STOMP over WebSocket protocol and gets connected via WebSocketClient. This the main core of the protocol handler, where the life cycle starts. It requires the session handler (see Listing 10-34) as part of its constructor and this class is the one that connects to the WebSocket server by calling the connect method and passing the URI (the remote server based on the RetroBoardProperties values), and how it will be connected (WebSocketClient).

In the UserSocketConfiguration, we used the RetroBoardProperties class, which is used to bind the properties declared in the application.properties file. So, open/create the application.properties file. See Listing 10-39.

Listing 10-39. src/main/resources/application.properties

```
# Port
server.port=${PORT:8081}

# Data
spring.h2.console.enabled=true
```

```
spring.datasource.generate-unique-name=false
spring.datasource.name=test-db
#spring.jpa.show-sql=true

# My Retro Properties
myretro.users-service.hostname=http://localhost:8080
myretro.users-service.base-path=/logs
```

Listing 10-39 shows the users-service.* properties that are necessary to connect to the WebSockets/STOMP server that lives in the Users app.

Running My Retro App Using WebSockets/Stomp

Now it's time to run both apps and see how they interact. So, make sure the Users App is up and running; it should run in port 8080. Now, run the My Retro App. Once it is up and running, take a look at the Users App console; the last output should be something similar to this:

```
WebSocketSession[1 current WS(1)-HttpStream(0)-HttpPoll(0), 1 total, 0
closed abnormally (0 connect failure, 0 send limit, 0 transport error)],
stompSubProtocol[processed CONNECT(1)-CONNECTED(1)-DISCONNECT(0)],
stompBrokerRelay[null], inboundChannel[pool size = 0, active threads = 0,
queued tasks = 0, completed tasks = 6], outboundChannel[pool size = 0,
active threads = 0, queued tasks = 0, completed tasks = 1], sockJs
Scheduler[pool size = 8, active threads = 1, queued tasks = 2, completed
tasks = 428]
WebSocketSession[1 current WS(1)-HttpStream(0)-HttpPoll(0), 1 total, 0
closed abnormally (0 connect failure, 0 send limit, 0 transport error)],
stompSubProtocol[processed CONNECT(1)-CONNECTED(1)-DISCONNECT(0)],
stompBrokerRelay[null], inboundChannel[pool size = 0, active threads = 0,
queued tasks = 0, completed tasks = 6], outboundChannel[pool size = 0,
active threads = 0, queued tasks = 0, completed tasks = 1], sockJs
Scheduler[pool size = 8, active threads = 1, queued tasks = 2, completed
tasks = 861]
```

This means that there is a connection established between the Users App and the My Retro App. Now, use the previous command to send a new User:

```
curl -i -s -d '{"name":"Dummy","email":"dummy@email.com","password":
"aw2sOmeR!","userRole":["INFO"],"active":true}' \
-H "Content-Type: application/json" \
http://localhost:8080/users
```

With this command, you will see in the My Retro App console the following output:

```
2023-10-27T17:01:01.811-04:00  INFO 53760 --- [lient-AsyncIO-2]
c.a.myretro.web.socket.UserSocketClient  : Client received: payload
Event(version=1.0, time=2023-10-27T17:00:53, event=UserEvent(
email=dummy@email.com, active=false, action=ACTIVATION_STATUS,
datetime=null)), headers {destination=[/topic/user-logs], content-
type=[application/json], subscription=[0], message-id=[c0396b652a9f42c388
4d9f4666b8115d-4], content-length=[127]}
```

You are getting the *payload* that in this case is the Event type. And if you remove the user:

```
curl -I -s -XDELETE http://localhost:8080/users/dummy@email.com
```

you will get something like this:

```
2023-10-27T17:01:55.261-04:00  INFO 53760 --- [lient-AsyncIO-5]
c.a.myretro.web.socket.UserSocketClient  : Client received: payload
Event(version=1.0, time=2023-10-27T17:01:55, event=UserEvent(
email=dummy@email.com, active=false, action=REMOVED, datetime=
2023-10-27T17:01:55)), headers {destination=[/topic/user-logs], content-
type=[application/json], subscription=[0], message-id=[c0396b652a9f42c388
4d9f4666b8115d-5], content-length=[134]}
```

Nice! Now you know how to use WebSockets between apps.

RSocket with Spring Boot

RSocket is one of the newest technologies being used for messaging. This technology is an application protocol for multiplexed, duplex communication over TCP, WebSocket, and stream transports. RSocket has different interaction models:

- *Request/response*: You send a message, and you get a message back.

- *Request/stream*: You send a message, and you get back a stream of messages.

- *Fire-and-forget*: You send a message, and you get no response.

- *Channel*: A stream of messages is sent in both directions.

Some of the key features of the RSocket protocol are Reactive Streams semantics, request throttling, session resumption, fragmentation and reassembly of large messages, and keepalive frames (heartbeats). When using RSocket with Spring, the `spring-messaging` module supplies the `RSocketRequester` and annotated responders such as `@MessageMapping` annotation.

In this section we will use only the request/response interaction model to demonstrate another way to establish interaction with our two apps. So, let's get started and see how easy it to add RSocket technologies to the Users App and My Retro App.

Using RSocket with the Users App with Spring Boot

You can find all the source code in the `10-messaging-rsocket/users` folder. We are going to use WebFlux code from previous chapters. If you are starting from scratch with the Spring Initializr (`https://start.spring.io`), set the Group field to `com.apress` and the Artifact and Name fields to `users` and add the dependencies RSocket, WebFlux, Validation, R2DBC, H2, and Lombok. Generate and download the project, unzip it, and import it into your favorite IDE.

Let's start with the `build.gradle` file. It should look like Listing 10-40.

Listing 10-40. build.gradle

```
plugins {
    id 'java'
    id 'org.springframework.boot' version '3.2.3'
```

```gradle
    id 'io.spring.dependency-management' version '1.1.4'
}

group = 'com.apress'
version = '0.0.1-SNAPSHOT'
sourceCompatibility = '17'

repositories {
    mavenCentral()
}

dependencies {
    implementation 'org.springframework.boot:spring-boot-starter-webflux'
    implementation 'org.springframework.boot:spring-boot-starter-rsocket'
    implementation 'org.springframework.boot:spring-boot-starter-validation'

    implementation 'com.fasterxml.jackson.datatype:jackson-datatype-jsr310'

    implementation 'org.springframework.boot:spring-boot-starter-
    data-r2dbc'
    runtimeOnly 'io.r2dbc:r2dbc-h2'
    runtimeOnly 'org.postgresql:r2dbc-postgresql'

    compileOnly 'org.projectlombok:lombok'
    annotationProcessor 'org.projectlombok:lombok'

    // Web
    implementation 'org.webjars:bootstrap:5.2.3'

    testImplementation 'org.springframework.boot:spring-boot-starter-test'
    testImplementation 'io.projectreactor:reactor-test'
}

tasks.named('test') {
    useJUnitPlatform()
}
```

In the build.gradle file, we are adding the spring-boot-starter-webflux, spring-boot-starter-rsocket, and r2dbc-h2 (database) dependencies.

Next, open/create the UserRSocket class. See Listing 10-41.

Listing 10-41. src/main/java/com/apress/users/rsocket/UserRSocket.java

```java
package com.apress.users.rsocket;

import com.apress.users.model.User;
import com.apress.users.service.UserService;
import lombok.AllArgsConstructor;
import org.springframework.messaging.handler.annotation.MessageMapping;
import org.springframework.stereotype.Controller;
import reactor.core.publisher.Flux;
import reactor.core.publisher.Mono;

@AllArgsConstructor
@Controller
public class UserRSocket {

    UserService userService;

    @MessageMapping("new-user")
    Mono<User> newUser(final User user){
        return this.userService.saveUpdateUser(user);
    }

    @MessageMapping("all-users")
    Flux<User> getAllUsers(){
        return this.userService.getAllUsers();
    }

    @MessageMapping("user-by-email")
    Mono<User> findUserByEmail(final String email){
        return this.userService.findUserByEmail(email);
    }

    @MessageMapping("remove-user-by-email")
    Mono<Void> removeUserByEmail(final String email){
        this.userService.removeUserByEmail(email);
        return Mono.empty();
    }

}
```

The important feature in the UserRSocket class is the @MessageMapping annotation, which is a responder in an RSocket app; it's like the annotations you are used to, such as @GetMapping, @PostMapping, etc. With this annotation, you need to define a *destination*, and it can accept more parameters such as java.security.Principal, @Header, @Payload, etc. The annotated method can return a STOMP over WebSocket value, or a custom definition with @SendTo or @SendToUser annotations. In this case we are using Mono and Flux types.

Next, open the application.properties file, the contents of which should be similar to Listing 10-42.

Listing 10-42. src/main/resources/application.properties

```
# RSocket
spring.rsocket.server.port=9898

# R2DBC
spring.r2dbc.properties.initialization-mode=always
spring.r2dbc.generate-unique-name=false
spring.r2dbc.name=retro-db

# Logging
logging.level.org.springframework.r2dbc=DEBUG
```

In the application.properties file, the important property is spring.rsocket.server.port, which maps to port 9898.

And that's it! You don't need anything else. Basically, when Spring Boot finds the spring-boot-starter-rsocket dependency, it will auto-configure the RSocketServer, the handler, the message converter, and so forth with the default values, and in this case, it will register all the destinations based on the @MessageMapping annotation responders.

Running the Users App with RSocket and Spring Boot

Now, you can run the Users App. You don't need to do anything special here. The most important information is in the logs. So, once you run the app (either from the command line with ./gradlew bootRun or from your IDE), take a look at the logs. You should see something similar to the following:

```
...
--- [            main] o.s.b.web.embedded.netty.NettyWebServer   : Netty
started on port 8080
--- [            main] o.s.b.rsocket.netty.NettyRSocketServer    : Netty
RSocket started on port(s): 9898
--- [            main] com.apress.users.UsersApplication         : Started
UsersApplication in 2.576 seconds
```

Remember that this is a WebFlux app, so the Netty server is used and configured by default by Spring Boot. Also, a NettyRSockerServer is being started, and it's listening on port 9898.

Add some users using the command line:

```
curl -i -s -d '{"name":"Dummy","email":"dummy@email.com","password":
"aw2s0meR!","userRole":["INFO"],"active":true}' \
-H "Content-Type: application/json" \
http://localhost:8080/users

curl -i -s -d '{"name":"Ximena","email":"ximena@email.com","password":
"aw2s0meR!","userRole":["INFO","ADMIN"],"active":true}' \
-H "Content-Type: application/json" \
http://localhost:8080/users
```

Now you are ready to request RSocket messages.

Requesting RSocket Messages in My Retro App

Now, let's see what we need to do to incorporate the RSocket technology into the My Retro App. If you have downloaded the code, take a look at the 10-messagin-rsocket/ myretro folder. If you are starting from scratch with the Spring Initializr (https:// start.spring.io), set the Group field to com.apress and the Artifact and Name fields to myretro, and add the dependencies RSocket, WebFlux, Validation, Mongo Reactive, Docker Compose, and Lombok. Generate and download the project, unzip it, and import it into your favorite IDE.

Let's start with the build.gradle file. See Listing 10-43.

Listing 10-43. build.gradle

```
plugins {
    id 'java'
    id 'org.springframework.boot' version '3.1.5'
    id 'io.spring.dependency-management' version '1.1.3'
}

group = 'com.apress'
version = '0.0.1-SNAPSHOT'
sourceCompatibility = '17'

repositories {
    mavenCentral()
}

dependencies {

    implementation 'org.springframework.boot:spring-boot-starter-webflux'
    implementation 'org.springframework.boot:spring-boot-starter-rsocket'
    implementation 'org.springframework.boot:spring-boot-starter-
    validation'

    implementation 'com.fasterxml.jackson.datatype:jackson-datatype-jsr310'

    implementation 'org.springframework.boot:spring-boot-starter-data-
    mongodb-reactive'

    annotationProcessor 'org.springframework.boot:spring-boot-
    configuration-processor'

    developmentOnly 'org.springframework.boot:spring-boot-docker-compose'

    compileOnly 'org.projectlombok:lombok'
    annotationProcessor 'org.projectlombok:lombok'

    testImplementation 'org.springframework.boot:spring-boot-starter-test'
}

tasks.named('test') {
    useJUnitPlatform()
}
```

Listing 10-43 shows that we are using the `spring-boot-starter-rsocket` dependency.

Next, open/create the User and UserRole classes, shown in Listings 10-44 and 10-45, respectively.

Listing 10-44. src/main/java/com/apress/myretro/rsocket/User.java

```java
package com.apress.myretro.rsocket;

import com.fasterxml.jackson.annotation.JsonIgnoreProperties;
import lombok.AllArgsConstructor;
import lombok.Data;
import lombok.NoArgsConstructor;

import java.util.Collection;
import java.util.Collections;
import java.util.UUID;

@NoArgsConstructor
@AllArgsConstructor
@Data
@JsonIgnoreProperties(ignoreUnknown = true)
public class User {

    UUID id;

    private String email;

    private String name;

    private String gravatarUrl;

    private Collection<UserRole> userRole = Collections.emptyList();

    private boolean active;
}
```

Listing 10-45. src/main/java/com/apress/myretro/rsocket/UserRole.java

```
package com.apress.myretro.rsocket;

public enum UserRole {
    USER, ADMIN, INFO
}
```

As you can see, the User and UserRole classes are the same as in the Users App. Next, open/create the UserClient interface. See Listing 10-46.

Listing 10-46. src/main/java/com/apress/myretro/rsocket/UserClient.java

```
package com.apress.myretro.rsocket;

import org.springframework.messaging.rsocket.service.RSocketExchange;
import org.springframework.stereotype.Component;
import reactor.core.publisher.Flux;

@Component
public interface UserClient {

    @RSocketExchange("all-users")
    Flux<User> getAllUsers();

}
```

The UserClient interface is marking a method with the @RSocketExchange annotation, which defines the RSocket endpoint all-users. Behind the scenes and thanks to Spring Boot, this endppoint will be able to reach out to the RSocket server and look for that destination (responder), in this case the Users App. For now, we are going to reach out to only one destination; you can do the same for any other destination (@ MessageMapping).

Next, open/create the Config class. See Listing 10-47.

Listing 10-47. src/main/java/com/apress/myretro/rsocket/Config.java

```
package com.apress.myretro.rsocket;

import lombok.extern.slf4j.Slf4j;
import org.springframework.beans.factory.annotation.Value;
```

```java
import org.springframework.boot.CommandLineRunner;
import org.springframework.context.annotation.Bean;
import org.springframework.context.annotation.Configuration;
import org.springframework.messaging.rsocket.RSocketRequester;
import org.springframework.messaging.rsocket.service.
RSocketServiceProxyFactory;

@Slf4j
@Configuration
public class Config {

    @Bean
    public RSocketServiceProxyFactory getRSocketServiceProxyFactory(
    RSocketRequester.Builder requestBuilder, @Value("${myretro.users-
    service.host:localhost}")String host, @Value("${myretro.users-service.
    port:9898}")int port) {
        RSocketRequester requester = requestBuilder.tcp(host, port);
        return RSocketServiceProxyFactory.builder(requester)
                .build();
    }

    @Bean
    public UserClient getClient(RSocketServiceProxyFactory factory) {
        return factory.createClient(UserClient.class);
    }

    @Bean
    CommandLineRunner commandLineRunner(UserClient userClient) {
        return args -> {
            userClient.getAllUsers().doOnNext(
                    user -> log.info("User: {}", user)
            ).subscribe();
        };
    }
}
```

Let's review the Config class:

- RSocketRequester.Builder: This interface is configured by Spring
  Boot auto-configuration. This is an RSocket wrapper that can send
  and return objects. It has the capacity to route and prepare other
  useful metadata. In this case we are using the values of the myretro.
  user-service.* properties (to get the host and port where the
  RSocket server is running) to create an RSocketRequester object.
  This object can create the factory we need.

- RSocketServiceProxyFactory: This is a factory for creating a client
  proxy based on the RSocket service interface that is used with the @
  RSocketExchange methods (UserClient interface).

- UserClient: This interface is an RSocket service. Behind the scenes,
  there is an implementation that does all the work of connecting,
  sending, receiving, etc. to the RSocket server.

Note that we are using a CommandLineRunner, which means that when the app is
ready, it will execute the code, and in this case it will use the UserClient#getAllUsers
method to make a request to the all-users destination that lives in the Users App.

Next, open the application.properties file, shown in Listing 10-48.

Listing 10-48. src/main/resources/application.properties

```
# MongoDB
#spring.data.mongodb.uri=mongodb://retroadmin:aw2sOme@127.0.0.1:27017/
retrodb?directConnection=true&serverSelectionTimeoutMS=2000&authSource=
admin&appName=mongosh+1.7.1
#spring.data.mongodb.database=retrodb

# App
server.port=8081

# Users Service
myretro.users-service.host=localhost
myretro.users-service.port=9898
```

In the `application.properties` file, we define the `users-service` properties to indicate where the RSocket server is running, in this case in the same machine (localhost) and port 9898.

Finally, we need to add the `docker-compose.yaml` file. Remember that we use the `spring-boot-docker-compose` dependency, so it can start up the MongoDB service. See Listing 10-49.

Listing 10-49. docker-compose.yaml

```
version: "3.1"
services:
  mongo:
    image: mongo
    restart: always
    environment:
      MONGO_INITDB_DATABASE: retrodb
    ports:
      - "27017:27017"
```

Running the RSocket Requester in My Retro App with Spring Boot

Now, it's time to run the RSocket requester in the My Retro App. You can do it from the command line or in your IDE. After you run it, you should see the following output:

```
...
--- User: User(id=7d8f43c1-4911-4945-8a5d-301b365e82ce, email=dummy@email.
com, name=Dummy, gravatarUrl=https://www.gravatar.com/avatar/fb651279f471
2e209991e05610dfb03a?d=wavatar, userRole=[INFO], active=true)
--- User: User(id=2074203e-add0-40ca-95c2-82b2a23e1f13, email=ximena@email.
com, name=Ximena, gravatarUrl=https://www.gravatar.com/avatar/f07f7e55326
4c9710105edebe6c465e7?d=wavatar, userRole=[INFO, ADMIN], active=true)
...
```

As you can see, we are requesting to the `all-users` destination, using RSocket!

Other Messaging Frameworks

In the Apress GitHub repository for this book (`https://www.apress.com/gp/services/source-code`) or in my personal GitHub repository (`https://github.com/felipeg48/pro-spring-boot-3rd`), you can find additional examples of messaging frameworks such as Kafka, Redis, and more. Covering them all would require another entire book. So, please stay tuned for more code and features in the repo. I think the most important element of developing this kind of applications, is that it reflects the Spring way to create apps, because it allows you to use Spring/Spring Boot with ease to build any messaging solutions using this programming style.

Summary

In this chapter you learned about different messaging technologies that can help you to create synchronous and asynchronous messaging systems. You discovered that the main core of Spring messaging is in the `spring-messaging` module, and when you use it with Spring Boot, the auto-configuration will set all the defaults, including the connection, session, and reconnection handling, message conversion, and much more.

Also, you learned that the `spring-messaging` module includes interfaces and implementations such as the Template design pattern that is implemented in every technology and brings concrete class implementations such as `JmsTemplate`, `RabbitTemplate`, `RedisTemplate`, and `PulsarTemplate` (and many more) that help you to take care of the business logic and not spend time on the underlying technology, making development much easier.

CHAPTER 11

Spring Boot Actuator

Nowadays in software development and system management, visibility and observability play crucial roles in ensuring the smooth operation, performance, and reliability of applications and infrastructure. While these terms are often used interchangeably, there's a subtle yet significant distinction between the two:

- *Visibility*: Refers to the ability to gather and monitor data from a system or application. This data can include metrics, logs, and events that provide insights into the system's current state and behavior. Visibility tools help organizations understand what is happening within their systems, enabling them to identify potential issues and take corrective actions before they escalate into major problems.

- *Observability*: Goes beyond mere visibility and encompasses the ability to understand the internal workings of a system and the relationships between its components. Observability tools provide deeper contextual insights into system behavior, allowing developers and operators to diagnose issues more effectively, predict potential failures, and optimize performance.

The benefits of visibility and observability extend far beyond simply identifying and resolving problems. They also provide these benefits:

- *Empower organizations to respond to hidden demands*: By understanding how users are interacting with their systems, organizations can proactively identify and address potential bottlenecks and capacity constraints before they impact user experience.

- *Enable developers to act quickly*: Observability tools provide developers with real-time insights into the impact of their code changes, allowing them to quickly identify and fix bugs and improve performance.

© Felipe Gutierrez 2024
F. Gutierrez, *Pro Spring Boot 3*, https://doi.org/10.1007/978-1-4842-9294-5_11

- *Speed up development and deployment*: By reducing the time spent
  on troubleshooting and debugging, visibility and observability
  tools help teams move faster through the development and
  deployment cycle.

In today's dynamic and complex IT environments, visibility and observability are essential tools for ensuring the health, performance, and reliability of systems. By providing deep insights into system behavior, these tools empower organizations to respond effectively to hidden demands, enable developers to act quickly, and ultimately speed up development and deployment.

What Is Spring Boot Actuator?

Spring Boot Actuator is a powerful tool that can help developers in various ways, including:

- *Monitoring application health*: Actuator provides a set of endpoints
  that expose key metrics about the health of a Spring Boot
  application, such as CPU usage, memory consumption, and thread
  pool utilization. This information can help developers identify
  potential performance issues and take corrective actions before they
  impact users.

- *Debugging and troubleshooting*: Actuator provides endpoints that
  allow developers to dump thread stacks, inspect environment
  variables, and view logs. This information can be invaluable when
  debugging and troubleshooting issues in a running application.

- *Enabling remote management*: Actuator can be configured to expose
  endpoints over HTTPS, allowing developers to manage and monitor
  their applications remotely. This can be particularly useful for
  applications deployed in production environments.

- *Integrating with external monitoring tools*: Actuator can be configured
  to export metrics to external monitoring tools, such as Prometheus or
  Grafana. This allows developers to centralize their monitoring data
  and gain deeper insights into application performance. All this is
  thanks to the Micrometer technology (`https://micrometer.io/`).

- *Customizing actuator endpoints*: Actuator provides a flexible API that allows developers to create custom endpoints to expose additional information or functionality from their applications.

As developers, Spring Boot Actuator can help us with these tasks:

- *Identify performance bottlenecks*: By monitoring CPU, memory, and thread pool usage, developers can identify performance bottlenecks in their applications.

- *Detect memory leaks*: Actuator's dump endpoints allow developers to inspect the heap and identify memory leaks.

- *Troubleshoot database connectivity issues*: Actuator's connection pool metrics can help developers identify database connectivity issues.

- *Track application startup and shutdown*: Actuator's lifecycle events endpoint provides insights into application startup and shutdown behavior.

- *Monitor application security*: Actuator's security endpoints provide information about the security configuration of the application.

Overall, Spring Boot Actuator is a valuable tool for developers that can help them monitor, debug, troubleshoot, and manage their Spring Boot applications. Let's start adding Spring Boot Actuator to our main apps.

Users App with Spring Boot Actuator

Let's start with the Users App. You have access to the source code in the 11-actuator/ users folder. If you want to start from scratch, we are going to use the Users App with JPA. In the Spring Initializr (https://start.spring.io), set the Group field to com. apress and the Artifact and Name fields to users, and add the dependencies Web, Validation, Data JPA, Actuator, H2, PostgreSQL, and Lombok. Generate and download the project, unzip it, and import it into your favorite IDE.

Let's check the build.gradle file. See Listing 11-1.

Listing 11-1. build.gradle

```
plugins {
    id 'java'
    id 'org.springframework.boot' version '3.2.3'
    id 'io.spring.dependency-management' version '1.1.4'
}

group = 'com.apress'
version = '0.0.1-SNAPSHOT'
sourceCompatibility = '17'

configurations {
    compileOnly {
        extendsFrom annotationProcessor
    }
}

repositories {
    mavenCentral()
}

dependencies {
    implementation 'org.springframework.boot:spring-boot-starter-web'
    implementation 'org.springframework.boot:spring-boot-starter-
    validation'

    implementation 'org.springframework.boot:spring-boot-starter-data-jpa'

    implementation 'org.springframework.boot:spring-boot-starter-actuator'

    runtimeOnly 'com.h2database:h2'
    runtimeOnly 'org.postgresql:postgresql'

    compileOnly 'org.projectlombok:lombok'
    annotationProcessor 'org.projectlombok:lombok'
    annotationProcessor 'org.springframework.boot:spring-boot-
    configuration-processor'
```

```
    // Web
    implementation 'org.webjars:bootstrap:5.2.3'

    testImplementation 'org.springframework.boot:spring-boot-starter-test'
}

tasks.named('test') {
    useJUnitPlatform()
}
```

The only new dependency is the `spring-boot-starter-actuator`. Just by adding this, the Spring Boot auto-configuration will set up all the Actuator defaults. This means that you instantly have *production-ready features*! Spring Boot Actuator defines several endpoints that allow you to have these out-of-the-box, production-ready features, such as metrics, environment vartiables, events, sessions, thread dumps, scheduled task, loggers, health indicators, and much more.

/actuator

By default, Spring Boot Actuator defines the `/actuator` endpoint as a prefix for other actuator endpoints. The only actuator endpoint enabled is mapped to the `/actuator/health` endpoint, responsible for showing the application health information.

Note   If you started from scratch, you could copy the version of the Users App that incorporates the JPA technology.

Let's run the application. No modification is required. Adding the `spring-boot-starter-actuator` dependency is sufficient. You can run the Users App either by using your IDE or by using the following command line:

```
./gradlew bootRun
...

...

INFO 11418 --- [main] o.s.b.a.e.web.EndpointLinksResolver    : Exposing 1
endpoint(s) beneath base path '/actuator'

...

...
```

Notice the log about exposing the /actuator endpoint.

Now, open your browser and go to http://localhost:8080/actuator, and you should see the /actuator endpoint (prefix) response, shown in Figure 11-1.

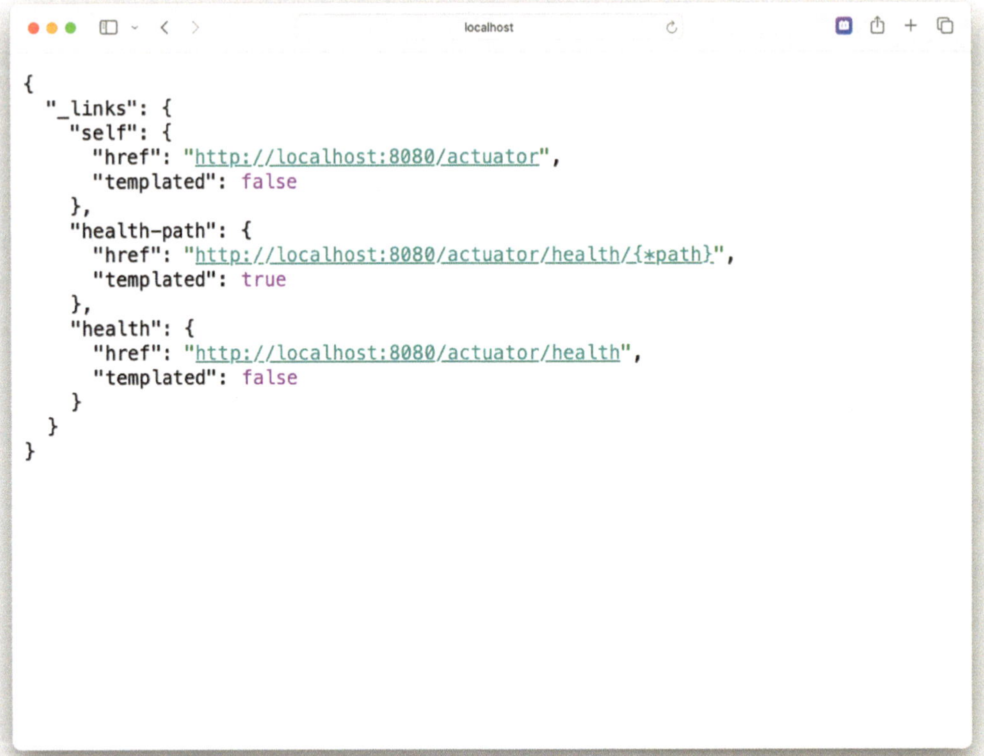

```
{
  "_links": {
    "self": {
      "href": "http://localhost:8080/actuator",
      "templated": false
    },
    "health-path": {
      "href": "http://localhost:8080/actuator/health/{*path}",
      "templated": true
    },
    "health": {
      "href": "http://localhost:8080/actuator/health",
      "templated": false
    }
  }
}
```

Figure 11-1. *Default prefix /actuator (http://localhost:8080/actuator)*

The /actuator endpoint (prefix) response is a HATEOAS (Hypermedia As The Engine Of Application State) response. One of the main features is that includes resource links that can be accessed directly in the browser and can be used for more specialized jobs such as web scraping to find out how the app defines its API, and much more. So, click the last reference, http://localhost:8080/actuator/health, and you will see the following response:

```
{
  "status": "UP"
}
```

The /actuator/health endpoint provides detailed information about the health of the application. I will show you more details about this endpoint in later sections. For now, this endpoint is the only one that is enabled by default.

Configuring /Actuator Endpoints

You can configure, override, enable, and do much more to the actuator endpoints. As you already saw, the prefix of the actuator is /actuator and you can easily override it with the following property:

management.endpoints.web.base-path

You can add that property to the application.properties file and assign the value /users-system-management. See Listing 11-2.

Listing 11-2. src/main/resources/application.properties

```
# Spring Properties
spring.h2.console.enabled=true
spring.datasource.generate-unique-name=false
spring.datasource.name=test-db

# Actuator Config
management.endpoints.web.base-path=/users-system-management

# Spring Info
spring.application.name=Users App
```

Listing 11-2 shows that we also added a new property, spring.application.name, with the value Users App, which will be useful to test some of the features of Spring Boot Actuator later in the chapter.

After the preceding property change, rerun the Users App and point your browser to http://localhost:8080/users-system-management. See Figure 11-2.

577

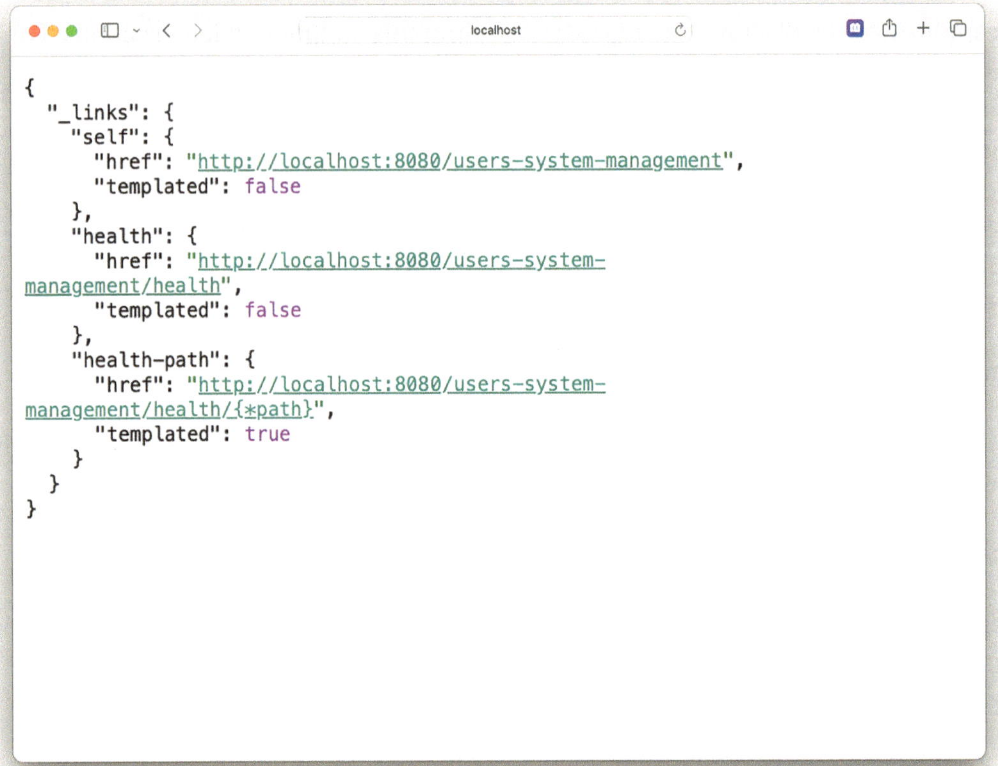

```json
{
  "_links": {
    "self": {
      "href": "http://localhost:8080/users-system-management",
      "templated": false
    },
    "health": {
      "href": "http://localhost:8080/users-system-
management/health",
      "templated": false
    },
    "health-path": {
      "href": "http://localhost:8080/users-system-
management/health/{*path}",
      "templated": true
    }
  }
}
```

Figure 11-2. `http://localhost:8080/users-system-management`

The properties `management.endpoints.web.*` have more ways to override the defaults, and we are going to review them soon.

Next, comment out the property we just added by putting a # before the property declaration, so that we can work just with the `/actuator` (a short prefix).

Using Spring Profiles for Actuator Endpoints

Before continuing with Spring Boot Actuator, let's use the Spring profiles to test the Actuator features. Create a blank file named `application-actuator.properties` in the `src/main/resources` folder, in the same location of the `application.properties` file.

Enabling Actuator Endpoints

So far, we have just worked with the /actuator/health endpoint, the default. Now it's time to review the other endpoints. Spring Boot Actuator has a convention to use the endpoints with the pattern /actuator/{id}, where the {id} is the endpoint you want to access; in other words, each feature has its own *id*, and to use the endpoints we need to enable them with the following property:

management.endpoints.web.exposure.[include|exclude]=<id>[,<id>]|*

So, you can choose to enable them all by using this syntax:

management.endpoints.web.exposure.include=*

Or you can enable an actuator feature individually, separated by commas if you want to enable more than one. For example:

management.endpoints.web.exposure.include=health,info,env,shutdown,bea ns,metrics

By default, these endpoints are also exposed through Java Management Extensions (JMX). You can also include or exclude actuator endpoints for JMS with

management.endpoints.jmx.exposure.[include|exclude]=[*|<id>[,<id>]

For example, to include all actuator endpoints:

management.endpoints.jmx.exposure.include=*

/actuator/info

This actuator endpoint provides general information about your application, and accessing it only requires a GET request to /actuator/info. One of the cool things you can do with this endpoint is get the information about the Git (branch, commit, etc.) and the Build (artifact, version, group), and you can even have your own information about your app. In other words, you can add after the property info.* anything you want that makes sense for your app.

So, in the application-actuator.properties file, add the properties shown in Listing 11-3.

Listing 11-3. src/main/resources/application-actuator.properties

```
management.endpoints.web.exposure.include=health, info

# Actuator Info
management.info.env.enabled=true

info.application.name=${spring.application.name}
info.developer.name=Felipe
info.developer.email=felipe@email.com
info.api.version=1.0
```

Listing 11-3 shows the properties added to the application-actuator.properties file. First, note that we are exposing just the health and info endpoints. Also, we are enabling the info environment (with management.info.env.enabled=true) and adding useful information for our Users App with the info.* properties. We are calling the base (application.properties) with the info.application.name property.

Next, let's run our application, but first make sure to add the profile either as an environment variable (SPRING_PROFILES_ACTIVE=actuator) or as a parameter. If you are using an IDE, check in your IDE documentation how to set a profile. In IntelliJ, in the top menu Run -> Edit Configurations, you can find the Active Profiles field, or you can set the environment variable in your OS. If you are running your app from the command line, you can execute the following:

```
./gradlew bootRun --args='--spring.profiles.active=actuator'
```

Now, if you direct your browser to http://localhost:8080/actuator/info, you will see something similar to the code shown in Figure 11-3.

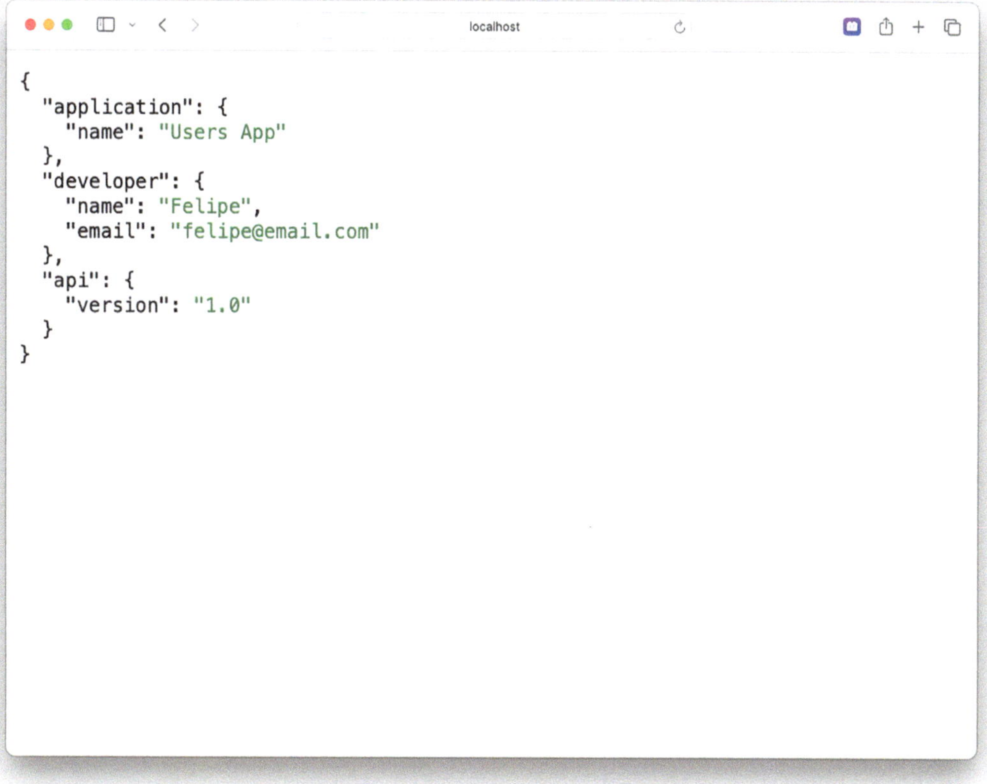

Figure 11-3. */actuator/info endpoint*

To enable Git and Build Info, we need to add a plugin for the Git information (`com.gorylenko.gradle-git-properties`) and a call for the Build Info (`buildInfo()`) to the build.gradle file. See Listing 11-4.

Listing 11-4. build.gradle

```
plugins {
    id 'java'
    id 'org.springframework.boot' version '3.2.3'
    id 'io.spring.dependency-management' version '1.1.4'
    id "com.gorylenko.gradle-git-properties" version "2.4.1"
}

group = 'com.apress'
```

```
version = '0.0.1-SNAPSHOT'
sourceCompatibility = '17'

configurations {
    compileOnly {
        extendsFrom annotationProcessor
    }
}

repositories {
    mavenCentral()
}

springBoot {
    buildInfo()
}

dependencies {
    implementation 'org.springframework.boot:spring-boot-starter-web'
    implementation 'org.springframework.boot:spring-boot-starter-
    validation'

    implementation 'org.springframework.boot:spring-boot-starter-data-jpa'

    implementation 'org.springframework.boot:spring-boot-starter-actuator'

    runtimeOnly 'com.h2database:h2'
    runtimeOnly 'org.postgresql:postgresql'

    compileOnly 'org.projectlombok:lombok'
    annotationProcessor 'org.projectlombok:lombok'
    annotationProcessor 'org.springframework.boot:spring-boot-
    configuration-processor'

    // Web
    implementation 'org.webjars:bootstrap:5.2.3'

    testImplementation 'org.springframework.boot:spring-boot-starter-test'
}
```

```
tasks.named('test') {
    useJUnitPlatform()
}
```

Listing 11-4 shows the build.gradle file where we added the plugin necessary for the Git information and the call for the Build information.

Next, you can rerun the Users App. Don't forget the actuator profile. From now on, we are going to use this profile, so be ready. Once you rerun your app, take a look at http://localhost:8080/actuator/info. You will see something like Figure 11-4.

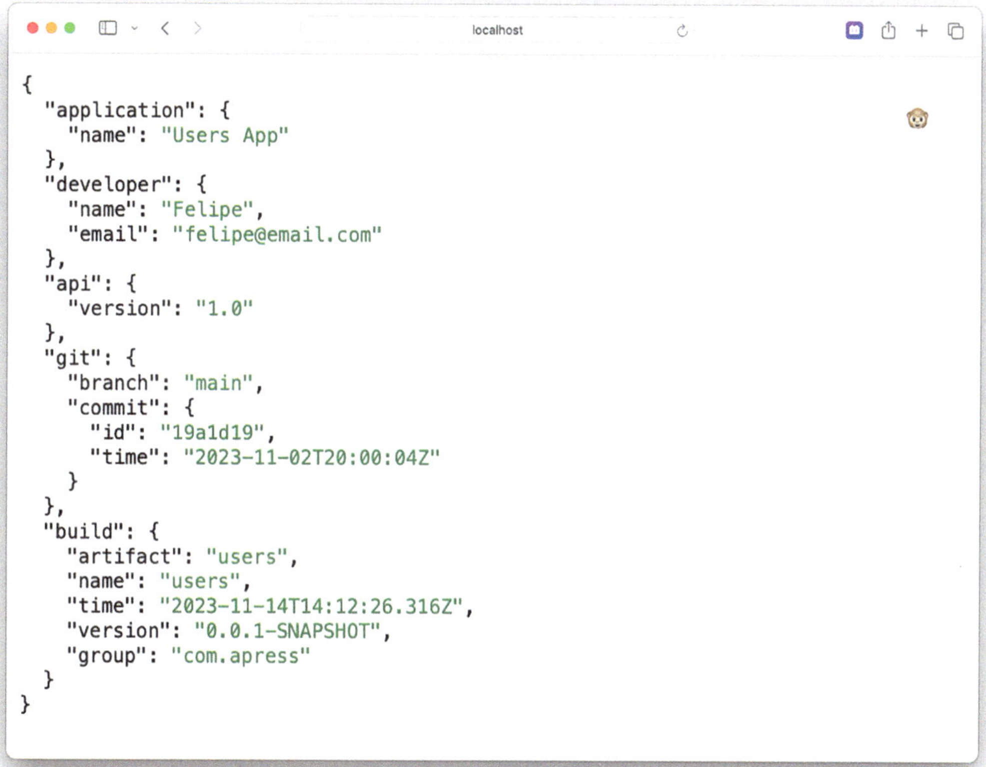

Figure 11-4. */actuator/info with Git and Build information*

Figure 11-4 shows you the /actuator/info endpoint with the Git and Build info. You can get more details from the Git info. You can get the *user*, *messages*, *remote*, *tags*, etc. To enable the full details, you can add the following property:

```
management.info.git.mode=full
```

/actuator/env

This actuator endpoint provides information about the application's environment. This is useful when you want to see what environment variables your app has access to. To test it, you can add it to the list of endpoints:

management.endpoints.web.exposure.include=health,info,**env**

You can access this endpoint with the following GET request to /actuator/env, and you should see similar output to that shown here:

```
curl -s http://localhost:8080/actuator/env | jq .
{
  "activeProfiles": [
    "actuator"
  ],
  "propertySources": [
    {
      "name": "server.ports",
      "properties": {
        "local.server.port": {
          "value": "******"
        }
      }
    },
    {
      "name": "servletContextInitParams",
      "properties": {}
    },
    {
      "name": "systemProperties",
      "properties": {
        "java.specification.version": {
          "value": "******"
        },
```

```
      "sun.jnu.encoding": {
        "value": "******"
      },
...
...
}
```

/actuator/beans

This endpoint provides information about the application's Spring beans. You can get access to this endpoint with the following GET request to /actuator/beans. This is useful when you want to see if your configuration picks up the declared beans or if you have any conditional bean that needs to be present depending on your business conditions.

```
curl -s http://localhost:8080/actuator/beans | jq .
```

```
{
  "contexts": {
    "Users App": {
      "beans": {
        "userLogs": {
          "aliases": [],
          "scope": "singleton",
          "type": "com.apress.users.events.UserLogs",
          "resource": "file [/Users/felipeg/Progs/Books/pro-spring-
          boot-3rd/java/11-actuator/users/build/classes/java/main/com/
          apress/users/events/UserLogs.class]",
          "dependencies": [
            "logEventEndpoint"
          ]
        ...
        ...

        "jdbcTemplate": {
          "aliases": [],
          "scope": "singleton",
          "type": "org.springframework.jdbc.core.JdbcTemplate",
```

```
      "resource": "class path resource [org/springframework/boot/
      autoconfigure/jdbc/JdbcTemplateConfiguration.class]",
      "dependencies": [
        "dataSourceScriptDatabaseInitializer",
        "org.springframework.boot.autoconfigure.jdbc.
        JdbcTemplateConfiguration",
        "dataSource",
        "spring.jdbc-org.springframework.boot.autoconfigure.jdbc.
        JdbcProperties"
      ]
    }
  }
}
}
}
```

/actuator/conditions

This endpoint provides information about the evaluation of conditions in the
configuration and auto-configuration classes. In other words, any classes marked with
the @Configuration annotation and other conditional annotations will be evaluated and
the information provided will be shown here. To access this endpoint, you can do the
following GET request to /actuator/conditions:

```
curl -s http://localhost:8080/actuator/conditions | jq .

{
  "contexts": {
    "Users App": {
      "positiveMatches": {
      ...
      ...
        "BeansEndpointAutoConfiguration": [
          {
            "condition": "OnAvailableEndpointCondition",
```

```
        "message": "@ConditionalOnAvailableEndpoint marked as exposed
        by a 'management.endpoints.jmx.exposure' property"
      }
    ],

  ...
  ...

    "JacksonAutoConfiguration.JacksonObjectMapperConfiguration": [
      {
        "condition": "OnClassCondition",
        "message": "@ConditionalOnClass found required class
        'org.springframework.http.converter.json.
        Jackson2ObjectMapperBuilder'"
      }
    ],
  ...
  ...
  }
}
```

/actuator/configprops

This endpoint provides all the information about the classes that provide configuration properties—in other words, classes marked with the @ConfigurationProperties annotation. You can do a GET request to the /actuator/configprops:

```
curl -s http://localhost:8080/actuator/configprops | jq .

{
  "contexts": {
    "Users App": {
      "beans": {
        "spring.jpa-org.springframework.boot.autoconfigure.orm.jpa.
        JpaProperties": {
          "prefix": "spring.jpa",
          "properties": {
            "mappingResources": ,
            "showSql": "******",
```

```
          "generateDdl": "******",
          "properties": {}
        },
        "inputs": {
          "mappingResources": [],
          "showSql": {},
          "generateDdl": {},
          "properties": {}
        }
      },
      ...

    ...
      "dataSource": {
        "prefix": "spring.datasource.hikari",
        "properties": {
          "error": "******"
        },
        "inputs": {
          "error": {}
        }
      },
      ...
        ....
      }
    }
  }
}
```

/actuator/heapdump

This endpoint provides the heap dump from the running app's JVM. To access this
information, you can do the following GET request to the /actuator/heapdump. It's
important to know that this endpoint has a lot of information and the response is
in binary format. So, you might prefer to use the -O option in the curl command to
generate a file named heapdump:

```
curl -s http://localhost:8080/actuator/heapdump -O
```

There are several software tools available for analyzing Java heap dump files. Here are a few of the most popular options (among others): Eclipse Memory Analyzer (MAT), VisualVM, and Jhat.

/actuator/threaddump

This endpoint provides the thread dump as JSON format from the running app's JVM. To access this information, you can do a GET request to the /actuator/threaddump:

```
curl -s http://localhost:8080/actuator/threaddump | jq .

{
  "threads": [
    {
      "threadName": "Reference Handler",
      "threadId": 2,
      "blockedTime": -1,
      "blockedCount": 7,
      "waitedTime": -1,
      "waitedCount": 0,
      "lockOwnerId": -1,
      "daemon": true,
      "inNative": false,
      "suspended": false,
      "threadState": "RUNNABLE",
      "priority": 10,
      "stackTrace": [
        {
          "moduleName": "java.base",
          "moduleVersion": "17.0.5",
          "methodName": "waitForReferencePendingList",
          "fileName": "Reference.java",
          "lineNumber": -2,
          "nativeMethod": true,
          "className": "java.lang.ref.Reference"
```

```
      },
      {
        "moduleName": "java.base",
        "moduleVersion": "17.0.5",
        "methodName": "processPendingReferences",
        "fileName": "Reference.java",
        "lineNumber": 253,
        "nativeMethod": false,
        "className": "java.lang.ref.Reference"
      },
      {

        "moduleName": "java.base",
        "moduleVersion": "17.0.5",
        "methodName": "run",
        "fileName": "Reference.java",
        "lineNumber": 215,
        "nativeMethod": false,
        "className": "java.lang.ref.Reference$ReferenceHandler"
      }
    ],
    "lockedMonitors": [],
    "lockedSynchronizers": []
  },
  ...
  ...
  ...
  ]
}
```

/actuator/mappings

This endpoint provides all the information about the request mappings, everything that is marked with @RestController or @Controller and manually defined. To access this information, you can do a GET request to the /actuator/mappings:

```
curl -s http://localhost:8080/actuator/mappings | jq .
```

```
{
  "contexts": {
    "Users App": {
      "mappings": {
        "dispatcherServlets": {
          "dispatcherServlet": [
            {
              "handler": "Actuator web endpoint 'configprops'",
              "predicate": "{GET [/actuator/configprops], produces
              [application/vnd.spring-boot.actuator.v3+json || application/
              vnd.spring-boot.actuator.v2+json || application/json]}",
              "details": {
                ...
              }
            },
            ...
            ...

            {
              "handler": "Actuator web endpoint 'health'",
              "predicate": "{GET [/actuator/health], produces [application/
              vnd.spring-boot.actuator.v3+json || application/vnd.spring-
              boot.actuator.v2+json || application/json]}",
              "details": {
                ....
              }
            },

            ...
            ....
            {
              "handler": "com.apress.users.web.UsersController#save
              (String)",
              "predicate": "{DELETE [/users/{email}]}",
              "details": {
                ...
```

```
        }
      },
      {
        "handler": "com.apress.users.web.UsersController#getAll()",
        "predicate": "{GET [/users]}",
        "details": {
            ...
        }
      },
      {
        "handler": "ResourceHttpRequestHandler [classpath [META-INF/
        resources/webjars/]]",
        "predicate": "/webjars/**"
      },
      {
        "handler": "ResourceHttpRequestHandler [classpath [META-INF/
        resources/], classpath [resources/], classpath [static/],
        classpath [public/], ServletContext ",
        "predicate": "/**"
      }
    ]
  },
  "servletFilters": [
      ...
      ...
  ]
}
        }
      }
    }
  }
}
```

/actuator/loggers

This endpoint provides the information about the logging level that was set for the application. To access this information, you can do a GET request to the /actuator/loggers:

```
curl -s http://localhost:8080/actuator/loggers | jq .

{
  "levels": [
    "OFF",
    "ERROR",
    "WARN",
    "INFO",
    "DEBUG",
    "TRACE"
  ],
  "loggers": {
    "ROOT": {
      "configuredLevel": "INFO",
      "effectiveLevel": "INFO"
    },
    ...
    ...
  },
  "groups": {
    "web": {
      "members": [
        "org.springframework.core.codec",
        "org.springframework.http",
        "org.springframework.web",
        "org.springframework.boot.actuate.endpoint.web",
        "org.springframework.boot.web.servlet.
        ServletContextInitializerBeans"
      ]
    },
```

```
    "sql": {
      "members": [
        "org.springframework.jdbc.core",
        "org.hibernate.SQL",
        "org.jooq.tools.LoggerListener"
      ]
    }
  }
}
```

/actuator/metrics

This endpoint provides a list of names of all the metrics that you can access by adding the respective name at the end of the endpoint. These metrics have useful information about your application, the system, and the JVM. Also, if you add the Micrometer dependency with Prometheus and Grafana, you can export these metrics as well. Interesting, right? We are going to see this in action in the next sections. For now, see the examples below. To access this information, you can do a GET request to the /actuator/metrics:

```
curl -s http://localhost:8080/actuator/metrics | jq .
```

```
{
  "names": [
    "application.ready.time",
    "application.started.time",
    "disk.free",
    "disk.total",
    "executor.active",
    "executor.completed",
    "executor.pool.core",
    "executor.pool.max",
    "executor.pool.size",
    "executor.queue.remaining",
    "executor.queued",
    "hikaricp.connections",
```

```
"hikaricp.connections.acquire",
"hikaricp.connections.active",
"hikaricp.connections.creation",
"hikaricp.connections.idle",
"hikaricp.connections.max",
"hikaricp.connections.min",
"hikaricp.connections.pending",
"hikaricp.connections.timeout",
"hikaricp.connections.usage",
"http.server.requests",
"http.server.requests.active",
"jdbc.connections.active",
"jdbc.connections.idle",
"jdbc.connections.max",
"jdbc.connections.min",
"jvm.buffer.count",
"jvm.buffer.memory.used",
"jvm.buffer.total.capacity",
"jvm.classes.loaded",
"jvm.classes.unloaded",
"jvm.compilation.time",
"jvm.gc.live.data.size",
"jvm.gc.max.data.size",
"jvm.gc.memory.allocated",
"jvm.gc.memory.promoted",
"jvm.gc.overhead",
"jvm.gc.pause",
"jvm.info",
"jvm.memory.committed",
"jvm.memory.max",
"jvm.memory.usage.after.gc",
"jvm.memory.used",
"jvm.threads.daemon",
"jvm.threads.live",
"jvm.threads.peak",
```

```
    "jvm.threads.started",
    "jvm.threads.states",
    "logback.events",
    "process.cpu.usage",
    "process.files.max",
    "process.files.open",
    "process.start.time",
    "process.uptime",
    "spring.data.repository.invocations",
    "system.cpu.count",
    "system.cpu.usage",
    "system.load.average.1m",
    "tomcat.sessions.active.current",
    "tomcat.sessions.active.max",
    "tomcat.sessions.alive.max",
    "tomcat.sessions.created",
    "tomcat.sessions.expired",
    "tomcat.sessions.rejected"
  ]
}
```

For example, you can get the metrics info of the JVM with /jvm.info:

```
curl -s http://localhost:8080/actuator/metrics/jvm.info | jq .

{
  "name": "jvm.info",
  "description": "JVM version info",
  "measurements": [
    {
      "statistic": "VALUE",
      "value": 1.0
    }
  ],
  "availableTags": [
    {
```

```
      "tag": "vendor",
      "values": [
        "GraalVM Community"
      ]
    },
    {
      "tag": "runtime",
      "values": [
        "OpenJDK Runtime Environment"
      ]
    },
    {
      "tag": "version",
      "values": [
        "17.0.5+8-jvmci-22.3-b08"
      ]
    }
  ]
}
```

More about this endpoint very soon!

/actuator/shutdown

This endpoint gracefully shuts down the application. To make the /shutdown work,
you not only need to include it in the management.endpoints.web.exposure.include
property but also enable it; for example:

```
management.endpoints.web.exposure.include=health,info,event-
config,env,shutdown
management.endpoint.shutdown.enabled=true
```

After these settings, you can do an HTTP POST request to the /actuator/shutdown:

```
curl -s -XPOST http://localhost:8080/actuator/shutdown

{"message":"Shutting down, bye..."}
```

You need to be very careful with the shutdown setting. The best way will be to secure it. Only admins should be able to do this. And, yes, we can add security to this. I'll show you in the next section.

Note   This section introduced the most commonly used actuator endpoints. There are many more endpoints, but they are enabled when the right dependency is in place and auto-configuration is executed. We are going to talk about several of these endpoints in the following sections.

Adding Security

Any actuator endpoint can have sensitive information that you don't want to share, in which case the best option is to secure your site and the actuator endpoints. So, let's add security to the Users App.

In the build.gradle file, add the following dependency:

```
implementation 'org.springframework.boot:spring-boot-starter-security'
```

If you run the application and try to access the actuator endpoint, you will be prompted for a username and password, the user will be user and the password is the one that was printed out in the console.

```
...
WARN 25992 --- [           main] .s.s.UserDetailsServiceAutoConfiguration :

Using generated security password: a84b06a6-1c99-4fb8-969d-6b969f73580a

This generated password is for development use only. Your security
configuration must be updated before running your application in
production.

INFO 25992 --- [           main] o.s.b.a.e.web.EndpointLinksResolver      :
Exposing 16 endpoint(s) beneath base path '/actuator'
...
...
```

Of course, you already knew this from Chapter 8. But, we can add some security configuration so that we don't have to rely on the password each time we restart the app. Open/create the UserSecurityAuditConfiguration class. See Listing 11-5.

Listing 11-5. src/main/java/com/apress/users/config/
UserSecurityAuditConfiguration.java

```
package com.apress.users.config;

import org.springframework.boot.actuate.audit.AuditEventRepository;
import org.springframework.boot.actuate.audit.InMemoryAuditEventRepository;
import org.springframework.context.annotation.Bean;
import org.springframework.context.annotation.Configuration;
import org.springframework.security.config.Customizer;
import org.springframework.security.config.annotation.web.builders.
HttpSecurity;
import org.springframework.security.core.userdetails.User;
import org.springframework.security.core.userdetails.UserDetails;
import org.springframework.security.crypto.bcrypt.BCryptPasswordEncoder;
import org.springframework.security.crypto.password.PasswordEncoder;
import org.springframework.security.provisioning.
InMemoryUserDetailsManager;
import org.springframework.security.provisioning.UserDetailsManager;
import org.springframework.security.web.SecurityFilterChain;
import org.springframework.security.web.servlet.util.matcher.
MvcRequestMatcher;
import org.springframework.web.servlet.handler.HandlerMappingIntrospector;

@Configuration
public class UserSecurityAuditConfiguration {

    @Bean
    public SecurityFilterChain securityFilterChain(HttpSecurity http,
    HandlerMappingIntrospector introspector) throws Exception {
        MvcRequestMatcher.Builder mvcMatcherBuilder = new
        MvcRequestMatcher.Builder(introspector);
        http
```

```
                .csrf(csrf -> csrf.disable())
                .authorizeHttpRequests( auth -> auth
                        .requestMatchers(mvcMatcherBuilder.pattern("/
                        actuator/**")).hasRole("ACTUATOR")
                        .anyRequest().authenticated())
                .formLogin(Customizer.withDefaults())
                .httpBasic(Customizer.withDefaults());

        return http.build();
    }

    @Bean
    UserDetailsManager userDetailsManager(PasswordEncoder passwordEncoder){
        UserDetails admin = User
                .builder()
                .username("admin")
                .password(passwordEncoder.encode("admin"))
                .roles("ADMIN","USER","ACTUATOR")
                .build();

        UserDetails manager = User
                .builder()
                .username("manager")
                .password(passwordEncoder.encode("manager"))
                .roles("ADMIN","USER")
                .build();

        UserDetails user = User
                .builder()
                .username("user")
                .password(passwordEncoder.encode("user"))
                .roles("USER")
                .build();

        return new InMemoryUserDetailsManager(manager,user,admin);
    }

    @Bean
```

```
PasswordEncoder passwordEncoder(){
    return new BCryptPasswordEncoder();
    }
}
```

In the UserSecurityAuditConfiguration class, we are setting a role, ACTUATOR, that can access only the /actuator/** endpoints, and the user admin is the only one with that role. So, if you rerun the Users App, the /actuator/** endpoints will be secure.

/actuator/auditevents

Spring Boot Actuator provides the /actuator/auditevents endpoint when Spring Security is enabled; this allows you to audit every time there is an authentication *success, failure,* or *access denied* event. So, to use this endpoint, it's necessary to define an AuditEventRepository bean that brings a convenient way to store in memory these events with the InMemoryAuditEventRepository class (an implementation of the AuditEventRepository interface).

So, in the UserSecurityAuditConfiguration class (see Listing 11-5), add the following bean declaration:

```
@Bean
public AuditEventRepository auditEventRepository() {
    return new InMemoryAuditEventRepository();
}
```

That's the only thing you need. So, when the app starts the auto-configuration finding that you have security and the AuditEventRepository bean declared, it will enable the /actuator/auditevents endpoint and it will save every event in memory.

Next, rerun the Users App and try to access it with different users and a wrong password. Then, you can take a look at the events with the following:

```
curl -s -u admin:admin http://localhost:8080/actuator/auditevents | jq .

{
  "events": [
    {
      "timestamp": "2023-11-20T19:04:28.683151Z",
      "principal": "manager",
```

```
      "type": "AUTHENTICATION_SUCCESS",
      "data": {
        "details": {
          "remoteAddress": "0:0:0:0:0:0:0:1"
        }
      }
    },
    {
      "timestamp": "2023-11-20T19:04:34.975087Z",
      "principal": "manager",
      "type": "AUTHORIZATION_FAILURE",
      "data": {
        "details": {
          "remoteAddress": "0:0:0:0:0:0:0:1"
        }
      }
    },
    {
      "timestamp": "2023-11-20T19:04:54.906022Z",
      "principal": "admin",
      "type": "AUTHENTICATION_SUCCESS",
      "data": {
        "details": {
          "remoteAddress": "0:0:0:0:0:0:0:1"
        }
      }
    }
  ]
}
```

Remember that admin/admin is the only one that has the ACTUATOR role. If you want to use your app in a production environment, you need to add your own implementation of the AuditEventRepository class and perhaps save this into a database.

If you want to have more control, you can always listen for these events. An AuditApplicationEvent class is being used for these events, and you can listen to it as well. So, if you want, you can add the following code to the UserLogs class (we are going to use it later as well):

```
@EventListener
public void on(AuditApplicationEvent event) {
    log.info("Audit: {}", event);
}
```

In the source code, you can find the complete class in the 11-actuator/users folder (src/main/java/com/apress/users/events/UserLogs.java).

If you rerun your application after this change and try to log in, you will get something similar to the following output:

```
INFO 26910 --- [nio-8080-exec-1] com.apress.users.events.UserLogs          :
Audit Event: org.springframework.boot.actuate.audit.listener.AuditAppl
icationEvent[source=AuditEvent [timestamp=2023-11-20T20:00:02.559298Z,
principal=admin, type=AUTHENTICATION_SUCCESS, data={details=WebAuthenticati
onDetails [RemoteIpAddress=127.0.0.1, SessionId=null]}]]
```

Now, you know how to add security to the actuator endpoints and take advantage of the audit events.

Implementing Custom Actuator Endpoints

Implementing a custom actuator endpoint is easy! To show you how easy it is, let's add a simple feature to our Users App. We are going to add prefix and postfix characters when there is a log event so that we can identify the event immediately.

First, open/create the LogEventConfig class, which will hold our feature data, in this case the prefix and postfix characters that we are going to use. See Listing 11-6.

Listing 11-6. src/main/java/com/apress/users/actuator/LogEventConfig.java

```
package com.apress.users.actuator;

import lombok.AllArgsConstructor;
import lombok.Data;
import lombok.NoArgsConstructor;
```

```
@AllArgsConstructor
@NoArgsConstructor
@Data
public class LogEventConfig {

    private Boolean enabled = true;

    private String prefix = ">> ";

    private String postfix = " <<";
}
```

Listing 11-6 shows you the LogEventConfig class. The simple LogEventConfig class shows that the respective default prefix and postfix values are >> and <<. As you can see very simple.

Next, open/create the LogEventEndpoint class. See Listing 11-7.

Listing 11-7. src/main/java/com/apress/users/actuator/LogEventEndpoint.java

```
package com.apress.users.actuator;

import org.springframework.boot.actuate.endpoint.annotation.Endpoint;
import org.springframework.boot.actuate.endpoint.annotation.ReadOperation;
import org.springframework.boot.actuate.endpoint.annotation.WriteOperation;
import org.springframework.lang.Nullable;
import org.springframework.stereotype.Component;

@Component
@Endpoint(id="event-config")
public class LogEventEndpoint {

    private LogEventConfig config = new LogEventConfig();

    @ReadOperation
    public LogEventConfig config() {
        return config;
    }

    @WriteOperation
    public void eventConfig(@Nullable Boolean enabled,@Nullable String
    prefix, @Nullable String postfix) {
```

```
        if (enabled != null)
            this.config.setEnabled(enabled);
        if (prefix != null)
            this.config.setPrefix(prefix);
        if (postfix != null)
            this.config.setPostfix(postfix);
    }

    public Boolean isEnable() {
        return config.getEnabled();
    }
}
```

The LogEventEndpoint class includes the following annotations:

- @Endpoint: This annotation allows you to create a custom actuator endpoint. It's necessary to provide an id that doesn't collide with the others, in this case event-config. This annotation creates the /actuator/event-config endpoint, and it will be available via the Web (through HTTP) and JMX. If you need it only for the Web, then you can use the @WebEndpoint annotation; if you need it only for JMX, use @JmxEndpoint. In this case, we want it for both, so we are using @Endpoint.

- @ReadOperation: This annotation marks a method that will return a value with a response status of 200 (OK); if this method doesn't return a value, the response status will be 404 (Not Found). This annotation can be accessed through a GET HTTP method. In this case we are just returning the values of the LogEventConfig object.

- @WriteOperation: This annotation marks a method that accepts simple type parameters; they cannot accept custom objects, and this is because the endpoints should be agnostic and they go through a conversion process. This operation accepts only an HTTP POST method. If this operation returns a value, the response status will be 200 (OK); if it doesn't return a value, it will return a response status

of 204 (No Content). The parameters are marked with @Nullable
annotation (it can be either @javax.annotation.Nullable or @org.
springframework.lang.Nullable) because in the JMX context,
the parameters are required by default, but they can be optional by
adding @Nullable.

This class is marked as @Component, meaning that it will be picked up by Spring Boot
auto-configuration and set as a Spring bean.

Even though we didn't use it, there is a @DeleteOperation annotation that accepts
a DELETE HTTP method. If this returns a value, then it will produce a status of 200(OK),
and if does not return a value, then it will respond with status 204 (No Content).

In our example, in the @ReadOperation annotation we are just returning an object,
so it will produce an application/json Content-Type, but we can return an
org.springframework.core.io.Resource, which will produce an application/octect-
stream Content-Type.

Next, open the UserLogs class. See Listing 11-8.

Listing 11-8.   src/main/java/com/apress/users/events/UserLogs.java

```
package com.apress.users.events;

import com.apress.users.actuator.LogEventEndpoint;
import lombok.AllArgsConstructor;
import lombok.extern.slf4j.Slf4j;
import org.springframework.boot.actuate.audit.listener.
AuditApplicationEvent;
import org.springframework.context.event.EventListener;
import org.springframework.scheduling.annotation.Async;
import org.springframework.stereotype.Component;

@AllArgsConstructor
@Slf4j
@Component
public class UserLogs {

    private LogEventEndpoint logEventEndpoint;
```

```
@Async
@EventListener
void userActiveStatusEventHandler(UserActivatedEvent event){
    if (logEventEndpoint.isEnable())
        log.info("{} User {} active status: {} {}",logEventEndpoint.
        config().getPrefix(),
                event.getEmail(),event.isActive(),logEventEndpoint.
                config().getPostfix());
    else
        log.info("User {} active status: {}",event.getEmail(),event.
        isActive());
}

@Async
@EventListener
void userDeletedEventHandler(UserRemovedEvent event){
    if (logEventEndpoint.isEnable())
        log.info("{} User {} DELETED at {} {}",logEventEndpoint.
        config().getPrefix(),
                event.getEmail(),event.getRemoved(),logEventEndpoint.
                config().getPostfix());
    else
        log.info("{} User {} DELETED at {} {}",event.getEmail(),event.
        getRemoved());
}

@EventListener
public void on(AuditApplicationEvent event) {
    log.info("Audit Event: {}", event);
}
```

}

In the UserLogs class, we are using the LogEventEndpoint class and its values for the prefix (>> default value) and postfix (<< default value). We are verifying that the endpoint is enabled to add these prefix and postfix characters. We have the audit security listener at the end.

Next, enable the `/actuator/event-config` in the `application-actuator.`
`properties` file with

management.endpoints.web.exposure.include=health,info,**event-**
config,env,shutdown

This property is enabling the endpoints by id, but keep in mind that you can use * to
add all endpoints. We are enabling the `event-config` endpoint.

Next, run it. Once it is up and running, you can check out the new custom
endpoint with

```
curl -s -u admin:admin http://localhost:8080/actuator/event-config | jq .
{
  "enabled": true,
  "prefix": ">> ",
  "postfix": " <<"
}
```

Nice! So, if you add a new user,

```
curl -XPOST -H "Content-Type: application/json" -d '{"email":"felipe@email.
com","name":"Felipe","gravatarUrl":"https://www.gravatar.com/avatar/23bb62a
7d0ca63c9a804908e57bf6bd5?d=wavatar","password":"awesome","userRole":["USER
","ADMIN"],"active":true}' http://localhost:8080/users
```

you should see console output something like this:

```
>>  User felipe@email.com active status: true  <<
```

Next, let's change the prefix and postfix characters. We can either use an HTTP
POST request or use JMX. An HTTP POST request is as simple as executing the following
command:

```
curl -i -u admin:admin -H"Content-Type: application/json" -XPOST
-d'{"prefix":"[ ","postfix":"]"}' http://localhost:8080/actuator/
event-config
```

We are changing the characters of the prefix to [and the postfix to]. You can verify
the change with

```
curl -s -u admin:admin http://localhost:8080/actuator/event-config | jq .

{
  "enabled": true,
  "prefix": "[ ",
  "postfix": "]"
}
```

So, if we delete the previous user:

```
curl -s -u admin:admin -XDELETE http://localhost:8080/users/felipe@
email.com
```

we should see this in the console output:

```
[  User felipe@email.com DELETED at 2023-11-21T11:14:54.639662 ]
```

Easy! right? Next, let's see how we can use JMX in this case.

Accessing Custom Endpoints with JMX

Java Management Extensions (JMX) is a technology that lets you manage and monitor Java applications and resources. It provides a standard way to expose information about your running application (like performance metrics or configuration settings) and allows you to change those settings dynamically. Think of it as a control panel for your Java application, accessible remotely, that gives you insights into its health and lets you tweak it while it's running. Spring Boot Actuator leverages JMX to expose operational information about your running application. By default, Actuator automatically registers its endpoints as JMX MBeans (Managed Beans), making them accessible through JMX clients like JConsole. This allows you to monitor and manage your application remotely, including health checks, metrics, configuration details, thread dumps, and more. Essentially, JMX serves as a communication channel for Actuator to provide valuable insights and control over your application's runtime behavior. Now, let's use the JMX console in our app. In your terminal, execute the following command to open the window shown in Figure 11-5:

```
jconsole
```

JConsole is a graphical interface that is compliant with the JMX specification, and it's used for JVM instrumentation to observe the performance and resource consumption of the applications running on the JVM.

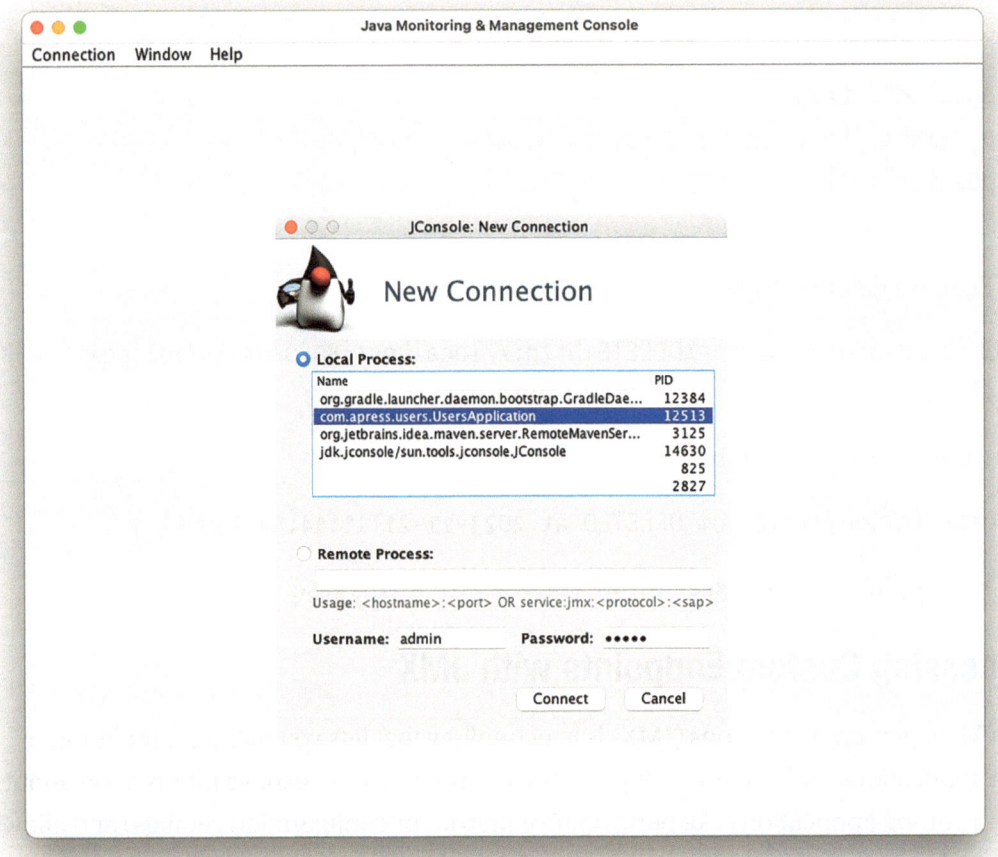

Figure 11-5. *JConsole*

As shown in Figure 11-5, in the Local Process field, select `com.apress.users.`
`UsersApplication`, enter `admin` in both the Username and Password fields, and click
Connect. Then you should see something like Figure 11-6, with the Overview tab
displayed by default.

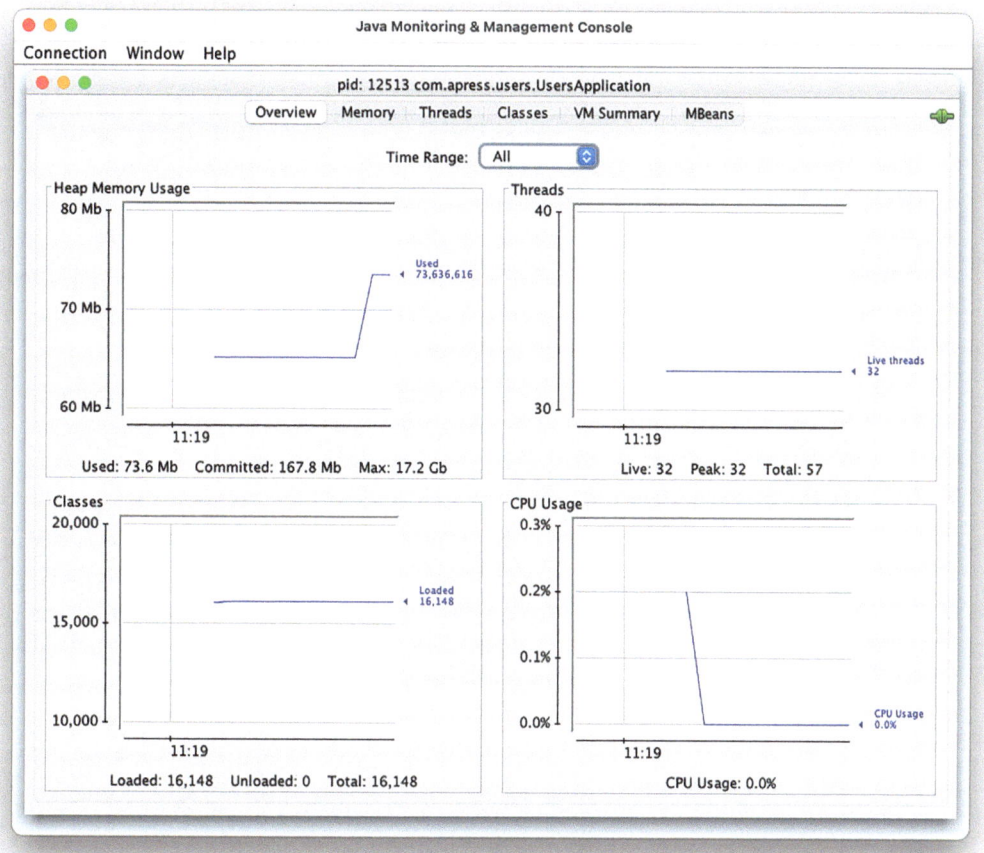

Figure 11-6. *JConsole Overview tab*

Select the last tab, MBeans (Managed Beans). As shown in Figure 11-7, in the navigation pane on the left, expand the node org.springframework.boot ➤ Endpoint ➤ Event-config ➤ Operations and you'll see that we have enabled the /event-config endpoint and the read (config) and write (eventConfig) operations.

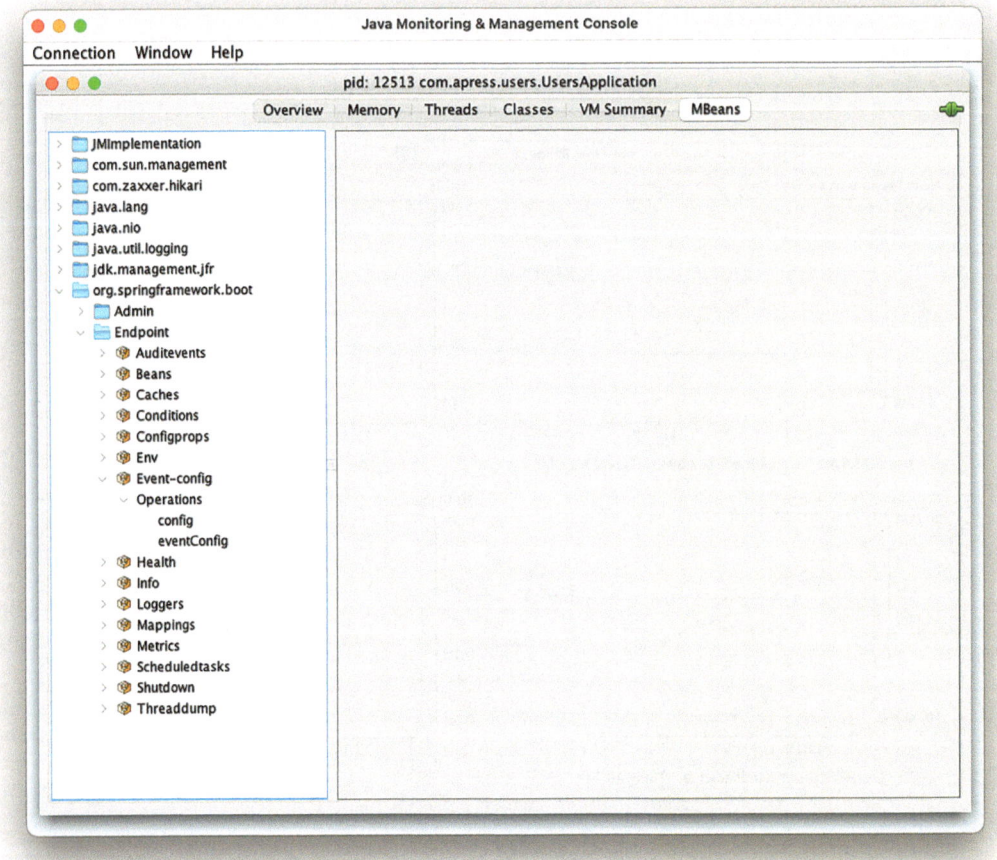

Figure 11-7. *JConsole MBeans tab*

Click the `config` operation and you'll see the information about the read operation, the return type (`Map`), and so forth, as shown in Figure 11-8.

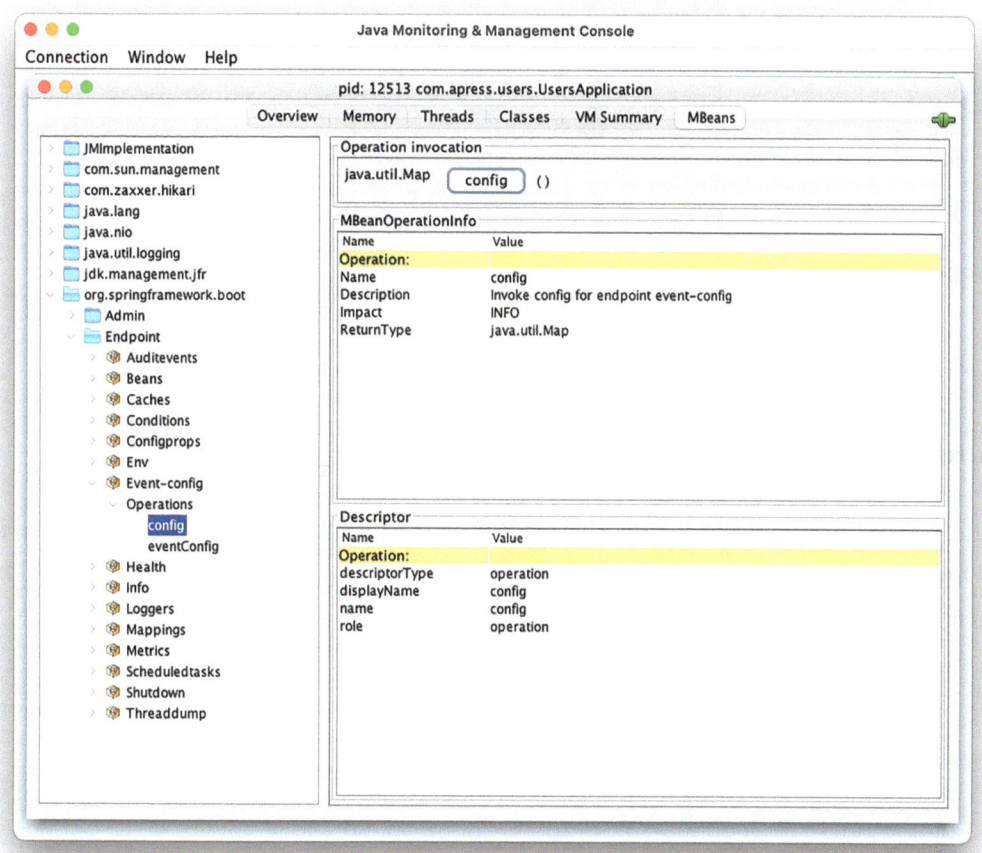

Figure 11-8. *JConsole MBeans tab expanded to Event-config* ➤ *Operations* ➤ *config*

Then, if you click the `config` button in the Operation Invocation section, you will see the pop-up window shown in Figure 11-9.

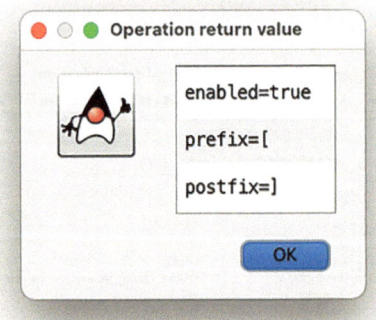

Figure 11-9. *JConsole – MBeans – Event Config – Operations - config*

Return to the navigation pane and select the write operation eventConfig. Then, in the Operation Invocation section, replace **String** with ** in the prefix and postfix fields. See Figure 11-10.

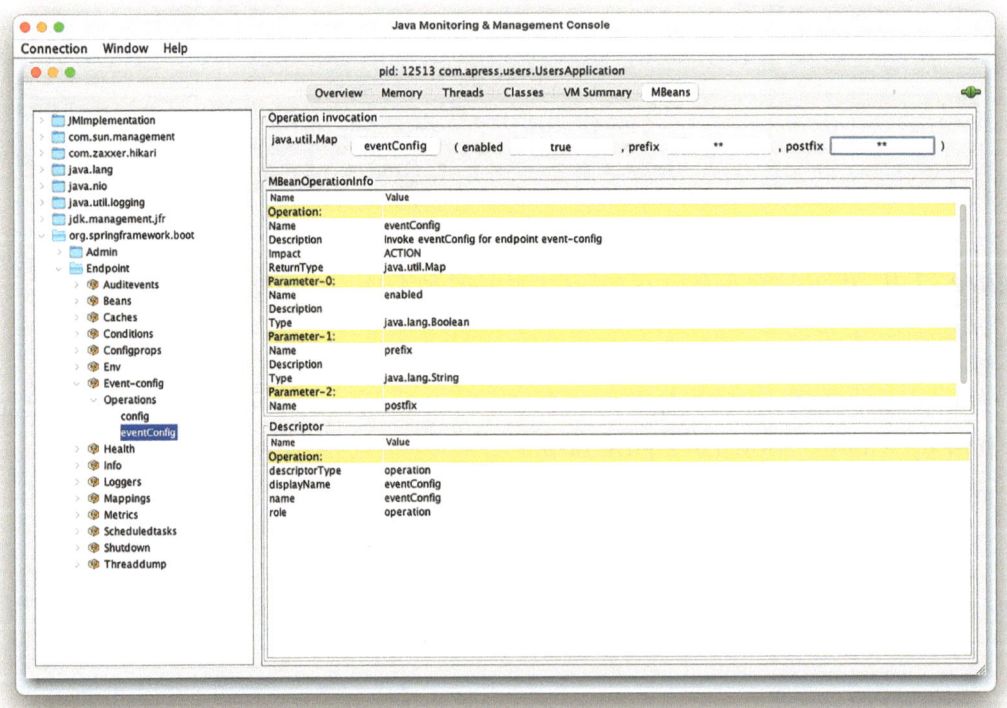

Figure 11-10. *JMX Console - Event Config - Operations*

Next, click the eventConfig button to the left and that will set the new values for the prefix and postfix. You can check them out by selecting config in the navigation pane, or you can add a new user. You can use the same command as before:

```
curl -XPOST -H "Content-Type: application/json" -d '{"email":"felipe@email.
com","name":"Felipe","gravatarUrl":"https://www.gravatar.com/avatar/23bb62a
7d0ca63c9a804908e57bf6bd5?d=wavatar","password":"awesome","userRole":["USER
","ADMIN"],"active":true}' http://localhost:8080/users
```

After executing this command, you should see the following in your console output:

```
** User felipe@email.com active status: true **
```

So, this is how you can interact with JMX and the actuator endpoints. Before you close the JConsole, browse around and view the other endpoints.

More Spring Boot Actuator Configuration

This section introduces several additional configuration options that Spring Boot Actuator offers.

CORS Support

To enable protection against cross-origin resource sharing (CORS), use the following properties syntax:

```
management.endpoints.web.cors.allowed-origins=<*|<site1[,site2]>>
management.endpoints.web.cors.allowed-methods=<*|[http-methods[,]]>
```

For example:

```
management.endpoints.web.cors.allowed-origins=*
management.endpoints.web.cors.allowed-methods=GET,POST,PUT,DELETE,OPTIONS
```

Changing the Server Address, Port, and Base Path

Changing the server port, address, and base path of Spring Boot Actuator could be a good practice for security reasons. Consider the following example:

```
management.server.port=8282
management.server.address=127.0.0.1
management.server.base-path=/management
```

First, we are changing the port to 8282, meaning that Spring Boot Actuator will start serving at that port. Also, we are defining the address as 127.0.01 (localhost), which could be useful if you have multiple network interfaces because you can use a special address only for the Spring Boot Actuatort endpoints that won't be publicly available just for internal use. Finally, we are changing the base path to /management, meaning that access to Spring Boot Actuator will be http://localhost:8282/management/actuator.

If you want to run the Users App with this new configuration, it's important to add the new base path to the security with the following line of code:

```
.requestMatchers(mvcMatcherBuilder.pattern("/management/**")).
hasRole("ACTUATOR")
```

Using SSL with Spring Boot Actuator

You can add an SSL to Spring Boot Actuator with the following properties:

```
management.server.ssl.enabled=true
management.server.ssl.key-store=classpath:keystore.p12
management.server.ssl.key-store-password=changeit
management.server.ssl.key-password=changeit
management.server.ssl.key-store-type=PKCS12
management.server.ssl.key-alias=tomcat
```

Configuring Endpoints

Several actuator endpoints have read operations, such as /actuator/metrics, /actuator/beans, /actuator/env, etc., and these endpoints can cache their values for a period of time. You can control this time period with the following actuator property syntax:

```
management.endpoint.<id>.cache.time-to-live=<period-of-time>
```

The following is an example:

```
management.endpoint.beans.cache.time-to-live=10s
management.endpoint.metrics.cache.time-to-live=10s
management.endpoint.env.cache.time-to-live=10s
```

Overview of /actuator/health

This endpoint is one of the most important features of Spring Boot Actuator because it can help us determine if our app is up or down based on the health of some subsystems or any other business rules. As previously mentioned, this endpoint is activated by default, so when you access it you will see something like this:

```
curl -u admin:admin -s http://localhost:8080/actuator/health | jq .

{
  "status": "UP"
}
```

This is just enough for external systems or clients that make sure their systems are up and running and take the right responses. When you deploy your app into the cloud (described in Chapter 13) to a Kubernetes environment, for example, Spring Boot will auto-configure extra details for the /actuator/health endpoint and it will create the /actuator/health/liveness and /actuator/health/readiness, so it's easy for you to expose this in your YAML deployment files.

One nice feature of the /actuator/health endpoint is that it not only can expose more information about the app but also enables you to create a custom health indicator and build up the necessary status and report it back. So, if you need more information, you can adjust the details with the following syntax:

management.endpoint.health.show-details=<never,when_authorized|always>

If you don't set this property, by default it is never. So, in our case and because we are using Spring Security, we can use the when_authorized value. And not only that, we can have even more granular access with the roles by using the following property syntax:

```
management.endpoint.health.roles=<role-name[,]>
```

So, you can add this to the application-actuator.properties file:

management.endpoint.health.show-details=when_authorized
management.endpoint.health.roles=ACTUATOR

If you rerun the Users App and execute the following command, you will see output similar to what's shown here:

```
curl -u admin:admin -s http://localhost:8080/actuator/health | jq .

{
  "status": "UP",
  "components": {
    "db": {
      "status": "UP",
      "details": {
        "database": "H2",
        "validationQuery": "isValid()"
      }
    },
    "diskSpace": {
      "status": "UP",
      "details": {
        "total": 1000240963584,
        "free": 97917366272,
        "threshold": 10485760,
        "path": "/Users/felipeg/Progs/Books/pro-spring-boot-3rd/java/11-
        actuator/users/.",
        "exists": true
      }
    },
    "ping": {
      "status": "UP"
    }
  }
}
```

Note that this output also shows components such as db, diskSpace, and ping. Those are health indictors, discussed next.

Health Indicators

One of the benefits of using Spring Boot Actuator and the /actuator/health endpoint is that there are several health indicators that you can add out of the box to give you details of your application's health, and in this case the db (where you can see the engine used, in this case H2 and the validation query that is executed to see if the database is alive) and the diskspace (where you can see the details of where this is executed, free space, etc.) are some of them.

These components get all the health information that is collected from a HealthContributorRegistry, by default all the HealthContributor implementations defined in the app context, and in our case in the Users App ApplicationContext.

Other built-in health indicators that you can use are cassandra, couchbase, elasticsearch, hazelcast, influxdb, jms, ldap, mail, mongo, neo4j, ping, rabbit, and redis.

So, let's do an experiment and see our app in action. In the build.gradle file, in the dependencies section, add the following dependency:

```
implementation 'org.springframework.boot:spring-boot-starter-amqp'
```

We are adding AMQP to our project, even though we don't have any code. So, rerun your app and go to the /actuator/health endpoint (you will get some errors about connecting, but disregard them for now):

```
curl -u admin:admin -s http://localhost:8080/actuator/health | jq .

{
  "status": "DOWN",
  "components": {
    "db": {
      "status": "UP",
      "details": {
        "database": "H2",
        "validationQuery": "isValid()"
      }
    },
```

```
    "diskSpace": {
      "status": "UP",
      "details": {
        "total": 1000240963584,
        "free": 97769340928,
        "threshold": 10485760,
        "path": "/Users/felipeg/Progs/Books/pro-spring-boot-3rd/java/11-
        actuator/users/.",
        "exists": true
      }
    },
    "ping": {
      "status": "UP"
    },
    "rabbit": {
      "status": "DOWN",
      "details": {
        "error": "org.springframework.amqp.AmqpConnectException: java.net.
        ConnectException: Connection refused"
      }
    }
  }
}
```

Now, you have the rabbit component (based on the amqp dependency) with the details, and because a RabbitMQ broker is not running, it reports DOWN, and the main health status of the app is also DOWN. In other words, this status bubbles up to the main health details of your app. However, you can change the order in which the status is presented so that the status of the app indicates UP even though the status of the rabbit component is DOWN. Of course, there is a way to indicate the status is UP even though the rabbit component is DOWN (maybe it is not essential that rabbit is up due to some of your business rules). So, to define the order, we can set the following property:

```
management.endpoint.health.status.order=fatal,down,out-of-
service,unknown,up
```

The `management.endpoint.health.status.order` property defines the preceding status order by default, so if you omit the down value like this:

management.endpoint.health.status.order=fatal,out-of-service,unknown,up

and then rerun the app, the main status should be UP, even though the `rabbit` component is reporting DOWN. If you want to see the status of the `rabbit` component as UP, you can start a RabbitMQ through a Docker command and rerun the app:

```
docker run -d --rm --name rabbit -p 15672:15672 -p 5672:5672
rabbitmq:management-alpine
```

```
curl -u admin:admin -s http://localhost:8080/actuator/health | jq .
{
  "status": "UP",
  "components": {
    "db": {
      "status": "UP",
      "details": {
        "database": "H2",
        "validationQuery": "isValid()"
      }
    },
    "diskSpace": {
      "status": "UP",
      "details": {
        "total": 1000240963584,
        "free": 97315823616,
        "threshold": 10485760,
        "path": "/Users/felipeg/Progs/Books/pro-spring-boot-3rd/java/11-
        actuator/users/.",
        "exists": true
      }
    },
    "ping": {
      "status": "UP"
    },
```

```
    "rabbit": {
      "status": "UP",
      "details": {
        "version": "3.12.9"
      }
    }
  }
}
```

Now, the **rabbit** component is reporting the status as UP and the version. If you want, you can stop the RabbitMQ with

```
docker stop rabbit
```

Also, you can comment out the amqp dependency we added in the build.gradle file, because we no longer need it.

If you want to add your own custom health indicator, there are a few options to do it, as discussed in the next section.

A Custom Health Indicator for the Users App

As I previously explained, all the health information is collected from the HealthContributorRegistry, which contains implementations of the HealthContributor interface. HealthContributor is an empty interface, basically a marker that is agnostic of what you will be reporting as health status. Spring Boot Actuator provides a HealthIndicator interface that we can implement to build up a health status.

Open/create the EventsHealthIndicator class. This class will use the LogEventEndpoint and check out if it's enabled or not. See. Listing 11-9.

Listing 11-9. src/main/java/com/apress/users/actuator/ EventsHealthIndicator.java

```
package com.apress.users.actuator;

import lombok.AllArgsConstructor;
import org.springframework.boot.actuate.health.Health;
import org.springframework.boot.actuate.health.HealthIndicator;
```

```
import org.springframework.boot.actuate.health.Status;
import org.springframework.stereotype.Component;

@AllArgsConstructor
@Component
public class EventsHealthIndicator implements HealthIndicator {

    private LogEventEndpoint logEventEndpoint;

    @Override
    public Health health() {
        if (check() )
            return Health.up().build();
        else
            return Health.status(new Status("EVENTS-DOWN","Events are
            turned off!")).build();
    }

    private boolean check(){
        return this.logEventEndpoint.isEnable();
    }
}
```

Listing 11-9 shows that the EventsHealthIndicator class implements the HealthIndicator interface, which is a functional interface that has a default method getHealth(boolean) and the health() method that we are overriding. We are checking whether our endpoint (/actuator/event-config) is enabled of not, and based on that we can build up our status. If is not enabled, we create a new Status with EVENTS-DOWN and a description stating Events are turned off!.

Also notice that the EventsHealthIndicator class is marked as @Component, so it will be picked up by the Spring Boot and Actuator auto-configuration, and this class will be part of the HealthContributorRegistry as an events component, which is a naming convention. So, in other words, your custom health indicator must end with HealthIndicator. The syntax is <Name>HealthIndicator, and because the name is EventsHealthIndicator, it will be registered as events.

Next, in the `application-actuator.properties` file, enable the endpoint and add a new status order:

```
management.endpoints.web.exposure.include=health,info,event-
config,env,shutdown
management.endpoint.health.status.order=events-down,fatal,down,out-of-
service,unknown,up
```

Look that in the order we added the `events-down` that correlate to the new status that we created in Listing 11-9 (EVENTS-DOWN).

Next, run the Users App and go to the `/actuator/health` endpoint:

```
curl -u admin:admin -s http://localhost:8080/actuator/health | jq .
{
  "status": "UP",
  "components": {
    "db": {
      "status": "UP",
      "details": {
        "database": "H2",
        "validationQuery": "isValid()"
      }
    },
    "diskSpace": {
      "status": "UP",
      "details": {
        "total": 1000240963584,
        "free": 98204057600,
        "threshold": 10485760,
        "path": "/Users/felipeg/Progs/Books/pro-spring-boot-3rd/java/11-
        actuator/users/.",
        "exists": true
      }
    },
    "events": {
        "status": "UP"
      },
```

```
    "ping": {
      "status": "UP"
    }
  }
}
```

Now our custom health indicator appears in the /actuator/health endpoint. What happens if we disable this endpoint? You can do it by using JMX with JConsole, or do a POST request to change the value, or set the default field value to false (in the LogEventConfig class).

You will get the following output:

```
curl -u admin:admin -s http://localhost:8080/actuator/health | jq .

{
  "description": "Events are turned off!",
    "status": "EVENTS-DOWN",
  "components": {
    "db": {
      "status": "UP",
      "details": {
        "database": "H2",
        "validationQuery": "isValid()"
      }
    },
    "diskSpace": {
      "status": "UP",
      "details": {
        "total": 1000240963584,
        "free": 98187304960,
        "threshold": 10485760,
        "path": "/Users/felipeg/Progs/Books/pro-spring-boot-3rd/java/
        11-actuator/users/.",
        "exists": true
      }
    },
    "events": {
```

```
        "description": "Events are turned off!",
        "status": "EVENTS-DOWN"
      },
   "ping": {
     "status": "UP"
   }
  }
}
```

As you can see, the status order took precedence and we can say that because the events are down and our app status is down as well. Remember that here we are using a custom Status with value EVENTS-DOWN.

My Retro App with Spring Boot Actuator Observability: Metrics, Logs, and Tracing

In this section we are going discuss observability and how to add this new concept into the My Retro App. Observability is a way to observe internally what is going on in your running system from the outside, normally through logs, metrics, and traces. Spring Boot, with the power of Spring Boot Actuator and Micrometer, helps you to add observability out of the box.

As indicated, you need to use the Spring Boot Actuator and Micrometer framework dependencies. Micrometer provides a vendor-neutral API for collecting and exposing metrics. It supports a variety of monitoring systems, including Prometheus, Datadog, Graphite, InfluxDB, and many more. You can get more info at the main site of Micrometer (https://micrometer.io) and in its documentation (https://micrometer. io/docs).

Let's dig into some of the most common use cases for using observability with Spring Boot and add them to the My Retro App.

Adding Observability into My Retro App

You have access to the source code in the 11-actuator/myretro folder. If you want to start from scratch, we are going to use the Users with JPA. In the Spring Initializr (https://start.spring.io), set the Group field to com.apress and the Artifact and

Name fields to users, and add the dependencies Web, WebFlux, Validation, Data JPA, Actuator, H2, PostgreSQL, Lombok, Prometheus, Distributed Tracing, Zipkin, and Docker Compose Support. Generate and download the project, unzip it, and import it into your favorite IDE.

Let's check the build.gradle file. See Listing 11-10.

Listing 11-10. build.gradle

```
plugins {
    id 'java'
    id 'org.springframework.boot' version '3.2.3
    id 'io.spring.dependency-management' version '1.1.4'
}

group = 'com.apress'
version = '0.0.1-SNAPSHOT'
sourceCompatibility = '17'

configurations {
    compileOnly {
        extendsFrom annotationProcessor
    }
}

repositories {
    mavenCentral()
}

dependencies {
    implementation 'org.springframework.boot:spring-boot-starter-web'
    implementation 'org.springframework.boot:spring-boot-starter-webflux'
    implementation 'org.springframework.boot:spring-boot-starter-
    validation'
    implementation 'org.springframework.boot:spring-boot-starter-data-jpa'
    implementation 'org.springframework.boot:spring-boot-starter-actuator'
    implementation 'org.springframework.boot:spring-boot-starter-aop'

    runtimeOnly     'com.github.loki4j:loki-logback-appender:1.4.1'
    implementation 'io.micrometer:micrometer-tracing-bridge-brave'
```

```
    implementation 'io.zipkin.reporter2:zipkin-reporter-brave'
    implementation  'net.ttddyy.observation:datasource-micrometer-spring-
    boot:1.0.2'

    runtimeOnly 'com.h2database:h2'
    runtimeOnly 'org.postgresql:postgresql'
    runtimeOnly 'io.micrometer:micrometer-registry-jmx'
    runtimeOnly 'io.micrometer:micrometer-registry-prometheus'

    developmentOnly 'org.springframework.boot:spring-boot-docker-compose'

    compileOnly 'org.projectlombok:lombok'
    annotationProcessor 'org.projectlombok:lombok'
    annotationProcessor 'org.springframework.boot:spring-boot-
    configuration-processor'

    testImplementation 'org.springframework.boot:spring-boot-starter-test'
}
tasks.named('test') {
    useJUnitPlatform()
}
```

Listing 11-10 shows that we are using additional dependencies that will help us to bring observability to our application. Important to note here is that we are adding the Web and WebFlux dependencies, so what happens if we try to run the app? What does Spring Boot do? I'll explain why we need this and how to configure Spring Boot to use only a particular type of web application. Also, note that we need the `spring-boot-starter-aop`, `loki-logback-appender`, and `datasource-micrometer-spring-boot` dependencies. The `aop` dependency will be essential for the observability code we are adding, the `loki-logback-appender` dependency will be for the logs, and the `datasource-micrometer-spring-boot` dependency will be for the database observability.

Adding Custom Metrics and Observations

As you already know, Spring Boot Actuator comes with some default metrics that allow us to know more about the memory consumption, CPU, threads, JVM statistics, and much more. However, sometimes these metrics are not sufficient to measure and observe some logic within our application.

The combination of Spring Boot Actuator and Micrometer allows us to add custom metrics and observations to our apps with ease. For example, in our My Retro App, we can add counters that allow us to know how many times a particular endpoint was hit, as well as observe what is happening behind the scenes.

Let's start by opening/creating the `RetroBoardMetricsInterceptor` class. See Listing 11-11.

Listing 11-11. src/main/java/com/apress/myretro/metrics/
RetroBoardMetricsInterceptor.java

```
package com.apress.myretro.metrics;

import io.micrometer.core.instrument.MeterRegistry;
import jakarta.servlet.http.HttpServletRequest;
import jakarta.servlet.http.HttpServletResponse;
import lombok.AllArgsConstructor;
import lombok.extern.slf4j.Slf4j;
import org.springframework.web.servlet.HandlerInterceptor;

@AllArgsConstructor
@Slf4j
public class RetroBoardMetricsInterceptor implements HandlerInterceptor {

    private MeterRegistry meterRegistry;

    @Override
    public void afterCompletion(HttpServletRequest request,
    HttpServletResponse response, Object handler, Exception ex) throws
    Exception {
        String URI = request.getRequestURI();
        String METHOD = request.getMethod();

        if (!URI.contains("prometheus")) {
            log.info("URI: " + URI + " METHOD: " + METHOD);
            meterRegistry.counter("retro_board_api", "URI", URI, "METHOD",
            METHOD).increment();
        }
    }
}
```

In the RetroBoardMetricsInterceptor class, we are implementing a HandlerInterceptor interface. This interface is part of the org.springframework.web.servlet package, and we can use it to intercept every single request. It has some default implementation, and here we are overriding the afterCompletion method. One of the key elements in this class is the MeterRegistry instance (an abstract class that creates and manages your application's set of meters, such as counters, gauges, task timers, and much more), which allows us to use a counter.incremet method that aggregates the value to the defined name (in this case, retro_board_api) and accepts some event tags (key/value pairs) for tracking (in this case, URI=METHOD). In other words, there will be a metric named retro_board_api that will increment its counter every time there is a request to the app endpoint and its value is different from the prometheus string.

Next, open/create the RetroBoardMetrics class. See Listing 11-12.

Listing 11-12. src/main/java/com/apress/myretro/metrics/RetroBoardMetrics.java

```java
package com.apress.myretro.metrics;

import io.micrometer.core.instrument.Counter;
import io.micrometer.core.instrument.MeterRegistry;
import io.micrometer.observation.ObservationRegistry;
import io.micrometer.observation.aop.ObservedAspect;
import org.springframework.context.annotation.Bean;
import org.springframework.context.annotation.Configuration;
import org.springframework.web.servlet.handler.MappedInterceptor;

@Configuration
public class RetroBoardMetrics {

    @Bean
    ObservedAspect observedAspect(ObservationRegistry registry) {
        return new ObservedAspect(registry);
    }

    @Bean
    public Counter retroBoardCounter(MeterRegistry registry){
        return  Counter.builder("retro_boards").description("Number of
        Retro Boards").register(registry);
    }
```

```
@Bean
public MappedInterceptor metricsInterceptor(MeterRegistry registry){
    return new MappedInterceptor(new String[]{"/**"},new RetroBoardMetr
    icsInterceptor(registry));
  }
}
```

Let's review the RetroBoardMetrics class:

- @Configuration: First, it's important to know that we need enable
  these metrics, and a way to do that is to create a configuration
  class. You can add these beans in a main configuration class,
  but in this case, I wanted to demonstrate that you can use the
  @Configuration annotation to create your own project-specific
  class (RetroBoardMetrics in this example) in which to add these
  beans (more on this when we extend our project by creating our own
  spring-boot-starter).

- ObservedAspect/ObservationRegistry: The ObservedAspect class
  is an aspect for intercepting methods that have the @Observerd
  annotation, or all the methods in a class marked with this annotation.
  With this annotation, you will observe the execution of the method
  code, giving you a trace ID and a span (the time interval between
  events). And because we needed to be part of the metrics exposure,
  is necessary to add the ObservationRegistry that will have all the
  context to create the observation we need.

- Counter/MeterRegistry: The Counter bean increases values
  monotonically. This Counter interface extends from the Meter
  interface to provide a fluent API to create a custom metric meter.
  In this case, we need to give it a name (retro_boards) and a
  description, then we need to register it using the MeterRegistry
  abstract class that creates and manages our app's set of meters. In
  this case, we are going to use this counter every time there is a new
  RetroBoard added to the system.

- MappedInterceptor: This bean uses the
  RetroBoardMetricsInterceptor class (see Listing 11-11) that wraps
  a HandlerInterceptor and uses URL patterns to determine whether
  it applies to a given request, and in this case all the /** except for the
  ones that contain the string prometheus.

As you can see, the declaration to add observability and metrics is very
straightforward.

Adding an External Request with the New @HttpExchange

In this section we'll configure our My Retro App to call the Users App by using the
@HttpExchange annotation. This annotation is part of the Spring Framework 6 web core
package that allows us to define declarative HTTP services using Java interfaces. I will
talk more about the @HttpExchange annotation in later chapters, but for now, let's create
a simple interface that will call the Users App endpoint.

Open/create the UserClient interface. See Listing 11-13.

Listing 11-13. src/main/java/com/apress/myretro/client/UserClient.java

```java
package com.apress.myretro.client;

import com.apress.myretro.client.model.User;
import org.springframework.web.bind.annotation.PathVariable;
import org.springframework.web.service.annotation.GetExchange;
import org.springframework.web.service.annotation.HttpExchange;
import reactor.core.publisher.Flux;
import reactor.core.publisher.Mono;

@HttpExchange(url = "/users", accept = "application/json", contentType =
"application/json")
public interface UserClient {

    @GetExchange
    Flux<User> getAllUsers();

        @GetExchange("/{email}")
    Mono<User> getById(@PathVariable String email);
}
```

As you can see, the UserClient interface is very straightforward. You annotate your interface with @HttpExchange; it accepts parameters such as the url, contentType, accept, and method. Then you define method-specific annotation such as @GetExchange, @PostExchange, @PutExchange, @PatchExchange, @DeleteExchange, and more.

In Listing 11-13 we are using the Reactive WebFlux types such as Mono and Flux. Very straightforward.

Next, open/create the User class and the UserRole enum. See Listings 11-14 and 11-15, respectively.

Listing 11-14. src/main/java/com/apress/myretro/client/model/User.java

```
package com.apress.myretro.client.model;

import java.util.Collection;

public record User (String email, String name, String gravatarUrl, String
password, Collection<UserRole> userRole, boolean active)
{}
```

Listing 11-15. src/main/java/com/apress/myretro/client/model/UserRole.java

```
package com.apress.myretro.client.model;

public enum UserRole {
    USER, ADMIN, INFO
}
```

Next, open/create the UserClientConfig class. See Listing 11-16.

Listing 11-16. src/main/java/com/apress/myretro/client/UserClientConfig.java

```
package com.apress.myretro.client;

import org.springframework.beans.factory.annotation.Value;
import org.springframework.context.annotation.Bean;
import org.springframework.context.annotation.Configuration;
import org.springframework.web.reactive.function.client.WebClient;
import org.springframework.web.reactive.function.client.support.
WebClientAdapter;
import org.springframework.web.service.invoker.HttpServiceProxyFactory;
```

```
@Configuration
public class UserClientConfig {

    @Bean
    WebClient webClient(@Value("${users.app.url}") String baseUrl,
                        @Value("${users.app.username}") String username,
                        @Value("${users.app.password}") String password) {
        return WebClient.builder()
                .defaultHeaders(header -> header.setBasicAuth(username,
                 password))
                .baseUrl(baseUrl)
                .build();
    }

    @Bean
    UserClient userClient(WebClient webClient) {
        HttpServiceProxyFactory httpServiceProxyFactory =
                HttpServiceProxyFactory.builderFor(WebClientAdapter.
                create(webClient))
                        .build();
        return httpServiceProxyFactory.createClient(UserClient.class);
    }
}
```

The UserClientConfig class includes the following:

- @Configuration: Here we have declared another configuration class that deals only with the client. This is a good practice when you have different dependencies.

- WebClient: You already know about this class and how we can use it to do the request calls to external services. Note that we are also adding the basic authorization header necessary for our Users service (remember that the Users App already security dependencies included).

- HttpServiceProxyFactory: We are declaring our UserClient interface (see Listing 11-13) that uses the @HttpExchange annotation and declares the @GetExchange annotations for the endpoints. The HttpServiceProxyFactory will create a client proxy from an HTTP service based on the mentioned annotations.

Next, open/create the RetroBoardAndCardService class. See Listing 11-17.

Listing 11-17. src/main/java/com/apress/myretro/service/ RetroBoardAndCardService.java

```
package com.apress.myretro.service;

import com.apress.myretro.board.RetroBoard;
import com.apress.myretro.client.UserClient;
import com.apress.myretro.events.RetroBoardEvent;
import com.apress.myretro.events.RetroBoardEventAction;
import com.apress.myretro.exceptions.RetroBoardNotFoundException;
import com.apress.myretro.persistence.RetroBoardRepository;
import io.micrometer.core.instrument.Counter;
import io.micrometer.observation.annotation.Observed;
import lombok.AllArgsConstructor;
import lombok.extern.slf4j.Slf4j;
import org.springframework.context.ApplicationEventPublisher;
import org.springframework.stereotype.Service;

import java.time.LocalDateTime;
import java.util.UUID;

@AllArgsConstructor
@Slf4j
@Service
@Observed(name = "retro-board-service",contextualName =
"retroBoardAndCardService")
public class RetroBoardAndCardService {

    private RetroBoardRepository retroBoardRepository;

    private ApplicationEventPublisher eventPublisher;
```

```java
private Counter retroBoardCounter;

private UserClient userClient;

public Iterable<RetroBoard> getAllRetroBoards(){
    log.info("Getting all retro boards");
    return this.retroBoardRepository.findAll();
}

public RetroBoard findRetroBoardById(UUID uuid){
    log.info("Getting retro board by id: {}", uuid);
    return this.retroBoardRepository.findById(uuid).orElseThrow(RetroBo
    ardNotFoundException::new);
}

public RetroBoard saveOrUpdateRetroBoard(RetroBoard retroBoard){
    RetroBoard retroBoardResult = this.retroBoardRepository.
    save(retroBoard);
    eventPublisher.publishEvent(new RetroBoardEvent(retroBoardResult.
    getRetroBoardId(), RetroBoardEventAction.CHANGED, LocalDateTime.
    now()));

    this.retroBoardCounter.increment();

    return retroBoardResult;
}

public void deleteRetroBoardById(UUID uuid){
    this.retroBoardRepository.deleteById(uuid);
    eventPublisher.publishEvent(new RetroBoardEvent(uuid,RetroBoard
    EventAction.DELETED,LocalDateTime.now()));
}

public void getAllUsers(){
    log.info("Getting all users");
    userClient.getAllUsers().subscribe(
            user -> log.info("User: {}", user),
            error -> log.error("Error: {}", error),
```

```
            () -> log.info("Completed")
    );
  }
}
```

Listing 11-17 shows that we are marking the RetroBoardAndCardService class using the @Observed annotation, meaning that every method call will have its own observation. Also, we are using the method getAllUsers to have a call of the UserClient.

Next, open/create the RetroBoardController class. It is basically the same as in previous chapters, but here we need to add the call to get all the users from the Users App. See Listing 11-18.

Listing 11-18. src/main/java/com/apress/myretro/web/ RetroBoardController.java

```
package com.apress.myretro.web;

import com.apress.myretro.board.Card;
import com.apress.myretro.board.RetroBoard;
import com.apress.myretro.service.RetroBoardAndCardService;
import jakarta.validation.Valid;
import jakarta.validation.constraints.NotNull;
import lombok.AllArgsConstructor;
import org.springframework.http.HttpStatus;
import org.springframework.http.ResponseEntity;
import org.springframework.web.bind.annotation.*;
import org.springframework.web.util.UriComponentsBuilder;

import java.net.URI;
import java.util.UUID;

@AllArgsConstructor
@RestController
@RequestMapping("/retros")
public class RetroBoardController {

    private RetroBoardAndCardService retroBoardAndCardService;
```

```java
@GetMapping
public ResponseEntity<Iterable<RetroBoard>> getAllRetroBoards(){
    return ResponseEntity.ok(retroBoardAndCardService.
    getAllRetroBoards());
}

@GetMapping("/{uuid}")
public ResponseEntity<RetroBoard> findRetroBoardById(@PathVariable
UUID uuid){
    return ResponseEntity.ok(this.retroBoardAndCardService.
    findRetroBoardById(uuid));
}

@RequestMapping(method = { RequestMethod.POST,RequestMethod.PUT})
public ResponseEntity<RetroBoard>saveRetroBoard(@RequestBody RetroBoard
retroBoard, UriComponentsBuilder componentsBuilder){
    RetroBoard saveOrUpdateRetroBoard = this.retroBoardAndCardService.
    saveOrUpdateRetroBoard(retroBoard);
    URI uri = componentsBuilder.path("/{uuid}").buildAndExpand(saveOrUp
    dateRetroBoard.getRetroBoardId()).toUri();
    return ResponseEntity.created(uri).body(saveOrUpdateRetroBoard);
}

@DeleteMapping("/{uuid}")
@ResponseStatus(HttpStatus.NO_CONTENT)
public ResponseEntity deleteRetroBoardById(@PathVariable UUID uuid){
    this.retroBoardAndCardService.deleteRetroBoardById(uuid);
    return ResponseEntity.noContent().build();
}

// External Call
@GetMapping("/users")
public ResponseEntity<Void> getAllUsers(){
    this.retroBoardAndCardService.getAllUsers();
    return ResponseEntity.ok().build();
}

}
```

Listing 11-18 shows that the only method will be calling with the endpoint /retros/ users, will be the getAllUsers method that calls the services that use the UserClient to call the Users App service.

Adding Logging Using Grafana Loki

The logs are sent to the standard output, but we also can stream them to an external entity for further analysis, and because we added the loki-logback-appender as a dependency, we can use Grafana Loki (https://grafana.com/oss/loki/) to aggregate the logs to store or query them to get insights into the system in a centralized console. To do this, we need to add an XML configuration to the project.

Open/create the logback-spring.xml file in the resources folder. See Listing 11-19.

Listing 11-19. src/main/resources/logback-spring.xml

```xml
<?xml version="1.0" encoding="UTF-8"?>
<configuration>
    <include resource="org/springframework/boot/logging/logback/base.xml"/>
    <springProperty scope="context" name="appName" source="spring.
    application.name"/>

    <appender name="LOKI" class="com.github.loki4j.logback.Loki4jAppender">
        <http>
            <url>http://localhost:3100/loki/api/v1/push</url>
        </http>
        <format>
            <label>
                <pattern>application=${appName},host=${HOSTNAME},level=%
                level</pattern>
            </label>
            <message>
                <pattern>${FILE_LOG_PATTERN}</pattern>
            </message>
            <sortByTime>true</sortByTime>
        </format>
    </appender>
```

```
<root level="INFO">
    <appender-ref ref="LOKI"/>
</root>
</configuration>
```

In the `logback-spring.xml` file, we first need to declare a property that will be an identifier for our logs. We are following the recommended practice and sticking with the name of the application using the `spring.application.name` property from Spring Boot. Next, we are declaring the actual configuration of the log appender with the name `LOKI`, first and the most important part, is where do we need to stream up the logs in this case we are using the *Grafana Loki* service (discussed in more detail a bit later) at the `http://localhost:3100/loki/api/v1/push` endpoint. We are declaring the pattern that the log will have (it's important to establish a good pattern for your apps). For our purposes, just using the application's name, the host, and the level will be sufficient. The level will be the `logging.pattern.correlation` property that we are going to set next.

Next, let's add some properties. Open the `application.properties` file. See Listing 11-20.

Listing 11-20. src/main/resources/application.properties

```
## DataSource
spring.h2.console.enabled=true
spring.datasource.generate-unique-name=false
spring.datasource.name=test-db

spring.jpa.show-sql=true

## Server
server.port=9081

## Docker Compose
spring.docker.compose.readiness.wait=never

## Application
spring.main.web-application-type=servlet
spring.application.name=my-retro-app
logging.pattern.correlation=[${spring.application.name:},%X{traceId:-},%X{spanId:-}]
```

```
## Actuator Info
info.developer.name=Felipe
info.developer.email=felipe@email.com
info.api.version=1.0
management.endpoint.env.enabled=true

## Actuator
management.endpoints.web.exposure.include=health,info,metrics,prometheus

## Actuator Observations
management.observations.key-values.application=${spring.application.name}

## Actuator Metrics
management.metrics.distribution.percentiles-histogram.http.server.
requests=true

## Actuator Tracing
management.tracing.sampling.probability=1.0

## Actuator Prometheus
management.prometheus.metrics.export.enabled=true
management.metrics.use-global-registry=true

## Users App Service
users.app.url=http://localhost:8080
users.app.username=admin
users.app.password=admin
```

Let's review some of the properties in the application.properties file.

Remember that we added the Web and WebFlux dependencies? Well, our app actually needs to be run as a Servlet app (not Reactive), so we added the spring.main. web-application-type=servlet property.

Another important property is spring.docker.compose.readiness.wait=never; because we are going to use multiple services and some of them don't have the readiness, this value must set to never. Another key will be the logging.pattern. correlation that will be injected in the log level, the one we declared in the logback-spring.xml file (see Listing 11-19).

Review the Actuator properties, which are very straightforward to follow. Both prometheus properties are by default, but I wanted to show you that they can be overridden when needed. Also, we are using percentiles, data that we need to expose to create the necessary graphs. The observation will be based on the key application that has the name of the app (my-retro-app). And the sampling that we need is always 1.0 (100%) of the times.

Last, we are declaring the users.app.* properties that are used to retrieve the user information.

Declaring Services in Docker Compose

Next, open/create the docker-compose.yaml file. See Listing 11-21.

Listing 11-21. docker-compose.yaml

```
version: '3'

services:
  tempo:
    image: grafana/tempo
    extra_hosts: [ 'host.docker.internal:host-gateway' ]
    ports:
      - "14268"
      - "9411:9411"
    volumes:
      - ./services/tempo/tempo.yaml:/etc/tempo.yaml:ro
      - ./services/tempo/tempo-data:/tmp/tempo
    command: [ "-config.file=/etc/tempo.yaml" ]

  loki:
    image: grafana/loki
    extra_hosts: [ 'host.docker.internal:host-gateway' ]
    ports:
      - "3100:3100"
    command: [ "-config.file=/etc/loki/local-config.yaml" ]
    environment:
      - JAEGER_AGENT_HOST=tempo
```

```
    - JAEGER_ENDPOINT=http://tempo:14268/api/traces
    - JAEGER_SAMPLER_TYPE=const
    - JAEGER_SAMPLER_PARAM=1

  prometheus:
    image: prom/prometheus
    extra_hosts: ['host.docker.internal:host-gateway']
    volumes:
      - ./services/prometheus/prometheus.yml:/etc/prometheus/
      prometheus.yml:ro
    command:
      - '--config.file=/etc/prometheus/prometheus.yml'
      - '--enable-feature=exemplar-storage'
    ports:
      - "9090:9090"

  grafana:
    image: grafana/grafana
    extra_hosts: ['host.docker.internal:host-gateway']
    environment:
      - GF_AUTH_ANONYMOUS_ENABLED=true
      - GF_AUTH_ANONYMOUS_ORG_ROLE=Admin
      - GF_AUTH_DISABLE_LOGIN_FORM=true
    volumes:
      - ./services/grafana/datasources:/etc/grafana/provisioning/
      datasources:ro
      - ./services/grafana/dashboards:/etc/grafana/provisioning/
      dashboards:ro
    ports:
      - "3000:3000"
```

Take a moment to review the docker-compose.yaml file, which is very straightforward. In this case we required four services, Prometheus (for metrics), Grafana (for graphics), Grafana Loki (for logs), and Grafana Tempo (for tracing).

In Listing 11-21, some of the services have a `volumes` parameter declared and they are pointing to a directory structure `services/` folder. Let's review them:

- `services/prometheus/prometheus.yml`:

```yaml
global:
  scrape_interval: 2s
  evaluation_interval: 2s

scrape_configs:
  - job_name: 'prometheus'
    static_configs:
      - targets: [ 'host.docker.internal:9090' ]
  - job_name: 'retroBoard'
    metrics_path: '/actuator/prometheus'
    static_configs:
      - targets: ['host.docker.internal:9081']
```

This file has the configuration where Prometheus will get its information. We are using the `host.docker.internal:9090` that points to our host machine and the endpoint, in this case `/actuator/prometheus`.

- `services/grafana/datasources/datasources.yaml`:

```yaml
apiVersion: 1

datasources:
  - name: Prometheus
    type: prometheus
    access: proxy
    url: http://prometheus:9090
    editable: false
    jsonData:
      httpMethod: POST
      exemplarTraceIdDestinations:
        - name: trace_id
          datasourceUid: tempo
  - name: Tempo
```

```
    type: tempo
    access: proxy
    orgId: 1
    url: http://tempo:3200
    basicAuth: false
    isDefault: true
    version: 1
    editable: false
    apiVersion: 1
    uid: tempo
    jsonData:
      httpMethod: GET
      tracesToLogs:
        datasourceUid: 'loki'
  - name: Loki
    type: loki
    uid: loki
    access: proxy
    orgId: 1
    url: http://loki:3100
    basicAuth: false
    isDefault: false
    version: 1
    editable: false
    apiVersion: 1
    jsonData:
      derivedFields:
        -   datasourceUid: tempo
            matcherRegex: \[.+,(.+?),
            name: TraceID
            url: $${__value.raw}
```

For Grafana to show graphics, it's necessary to declare some datasources from which it will pick up all the information. We are declaring the three datasources that are going to be useful for our metrics (Prometheus), our logs (Loki), and our tracing (Tempo).

- services/grafana/dashboards/dashboard.yml, log_traces_
  metrics.json, spring_boot_statistics.json: These files declare
  the dashboards that we are going to use for our observability. This
  can be manually created using the Grafana web interface, but I'm
  adding these files here so you don't have to. Also, in the next section
  I will show you how to import some existing dashboards from the
  community.

- services/tempo/tempo.yaml:

```
server:
  http_listen_port: 3200

distributor:
  receivers:
    zipkin:

storage:
  trace:
    backend: local
    local:
      path: /tmp/tempo/blocks
```

This file will set the Grafana Tempo listening port and the storage
for the logs, in this case using the local directory.

Ready to Run!

Now, it's time to run both apps. Start the Users App and make sure it is up and running.
Next, run your My Retro App, which will start the docker compose by running each
service. Wait until you see that the app is ready to be used.

If you take a look at the console, you will see something similar to this:

```
[my-retro-app] [            main] [my-retro-app,656a4219c89834979593
2a1ae70f316b,95932a1ae70f316b]
```

This is the LOG we declared in the logback-spring.xml file (see Listing 11-19): first,
the application name, then the host, then the level (which in this case is the correlation
pattern), the name of the app, the trace id, and the time span for the tracing.

Next, execute the `curl` commands that allow you to generate some logs, metrics, and tracing. For example:

```
curl -s http://localhost:9081/retros | jq .
[
  {
    "retroBoardId": "163c859d-9238-463c-a14b-f245e380ec92",
    "name": "Spring Boot 3 Retro",
    "cards": [
      {
        "cardId": "33f23935-1cbd-46b7-a276-195fa2ebdd3e",
        "comment": "Nice to meet everybody",
        "cardType": "HAPPY",
        "created": "2023-12-01 15:29:13",
        "modified": "2023-12-01 15:29:13"
      },
      {
        "cardId": "e9afa712-6f32-4f85-a4c5-7aa6af62e754",
        "comment": "When are we going to travel?",
        "cardType": "MEH",
        "created": "2023-12-01 15:29:13",
        "modified": "2023-12-01 15:29:13"
      },
      {
        "cardId": "ba87ca9e-164a-45d3-87c5-30f4fdad25a2",
        "comment": "When are we going to travel?",
        "cardType": "SAD",
        "created": "2023-12-01 15:29:13",
        "modified": "2023-12-01 15:29:13"
      }
    ],
    "created": "2023-12-01 15:29:13",
    "modified": "2023-12-01 15:29:13"
  }
]
```

curl -s http://localhost:9081/retros/163c859d-9238-463c-a14b-f245e380ec92 | jq .

```json
{
  "retroBoardId": "163c859d-9238-463c-a14b-f245e380ec92",
  "name": "Spring Boot 3 Retro",
  "cards": [
    {
      "cardId": "33f23935-1cbd-46b7-a276-195fa2ebdd3e",
      "comment": "Nice to meet everybody",
      "cardType": "HAPPY",
      "created": "2023-12-01 15:29:13",
      "modified": "2023-12-01 15:29:13"
    },
    {
      "cardId": "e9afa712-6f32-4f85-a4c5-7aa6af62e754",
      "comment": "When are we going to travel?",
      "cardType": "MEH",
      "created": "2023-12-01 15:29:13",
      "modified": "2023-12-01 15:29:13"
    },
    {
      "cardId": "ba87ca9e-164a-45d3-87c5-30f4fdad25a2",
      "comment": "When are we going to travel?",
      "cardType": "SAD",
      "created": "2023-12-01 15:29:13",
      "modified": "2023-12-01 15:29:13"
    }
  ],
  "created": "2023-12-01 15:29:13",
  "modified": "2023-12-01 15:29:13"
}
```

```
curl -i http://localhost:9081/retros/users
HTTP/1.1 200
Content-Length: 0
Date: Fri, 01 Dec 2023 21:13:46 GMT
```

Execute these commands multiple times. Then we are ready to continue and see what happen with these requests.

Observing with Grafana and Prometheus—Metrics, Logs, and Tracing

Now that the My Retro App is up and running, open a browser and go to Prometheus service: `http://localhost:9090/targets`. See Figure 11-11.

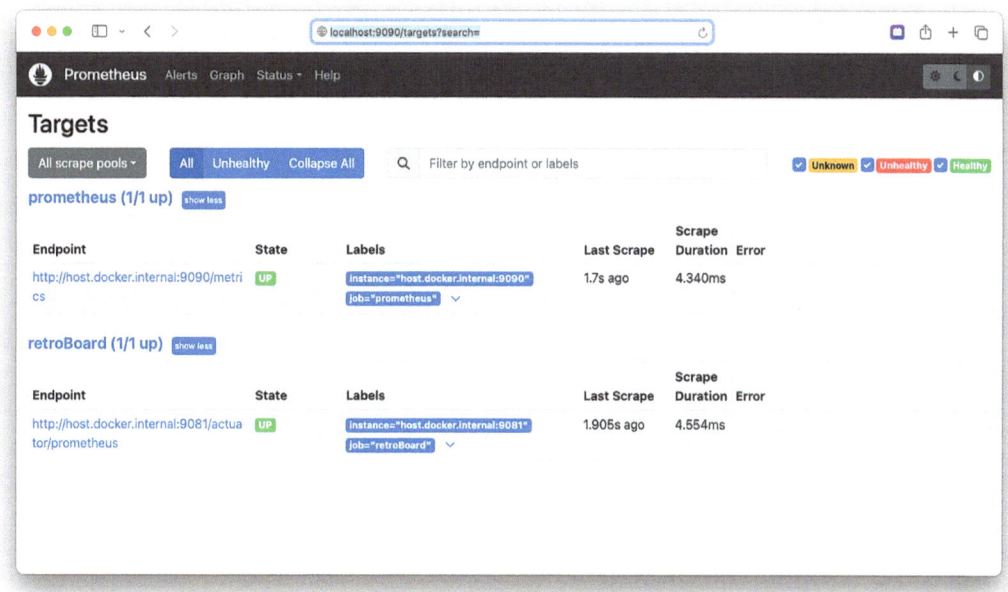

Figure 11-11. *Prometheus (`http://localhost:9090/targets`)*

Figure 11-11 shows the Prometheus service with the two endpoints declared and where are pointing to, in this case our `/actuator/prometheus` endpoint where all the metrics of our My Retro App lives.

Next, open Grafana: `http://localhost:3000`. See Figure 11-12.

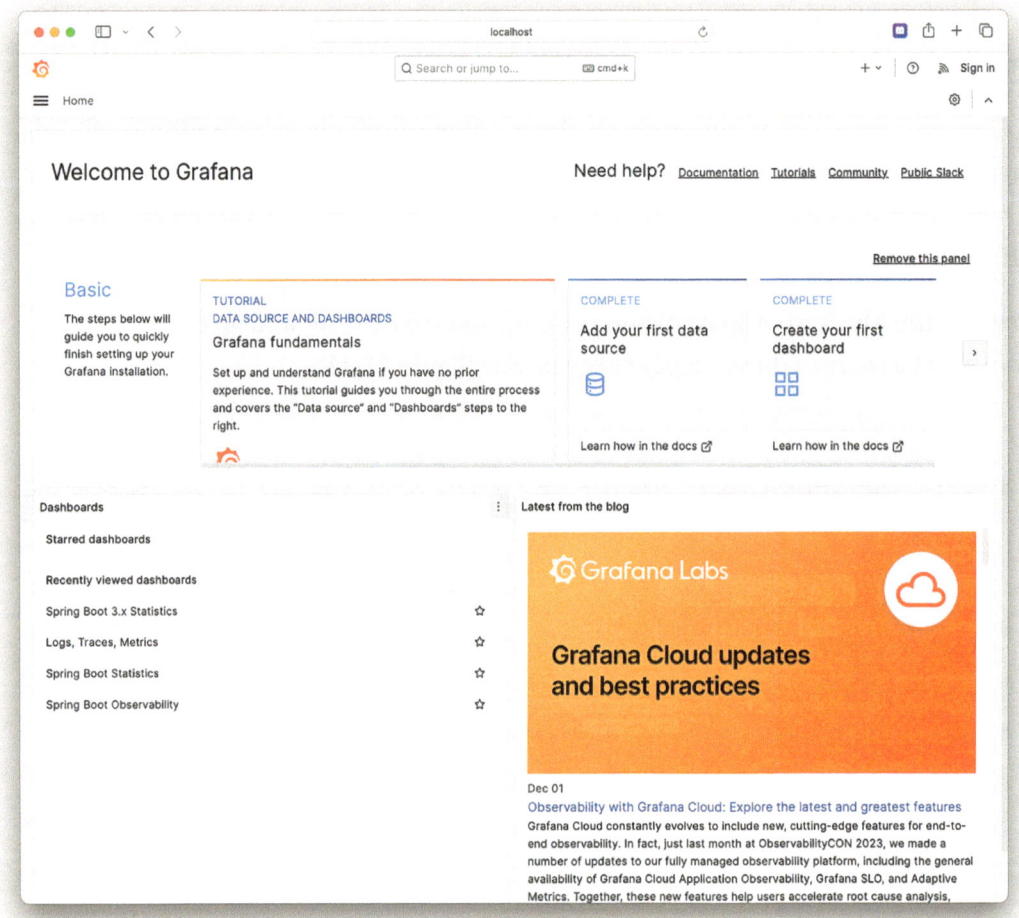

Figure 11-12. *Grafana home page*

Figure 11-12 shows that you are automatically logged in with Admin credentials (based on the environment variables set in the `docker-compose.yaml file`).

Next, click the "hamburger" icon in the top-left corner and select Explore, as shown in Figure 11-13.

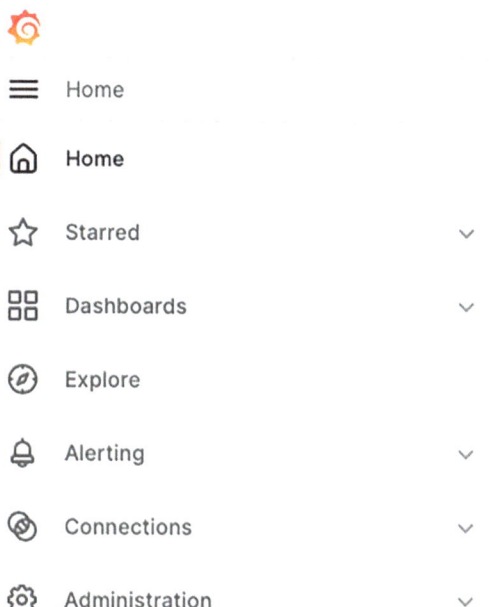

Figure 11-13. *Opening the main menu to select Explore*

After you click Explore, you will be presented with a page that contains all the services that Grafana has access to. They can be queried to get some useful information. In the drop-down menu to the right of Outline, shown in Figure 11-14, select Loki.

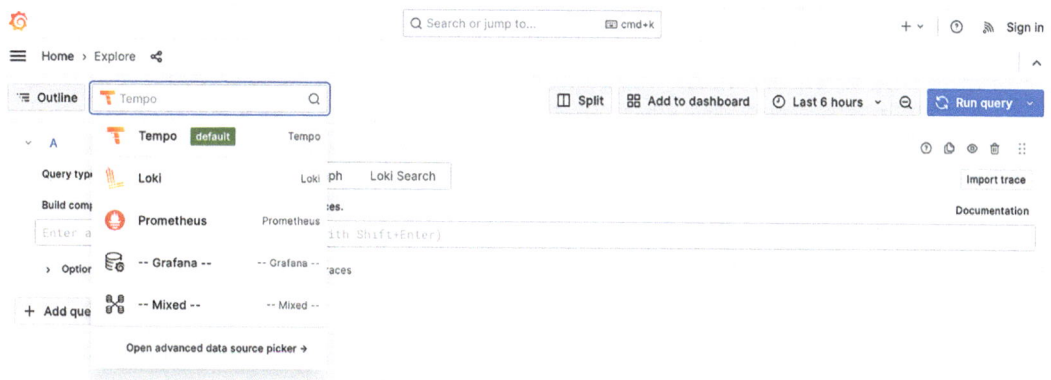

Figure 11-14. *Prometheus - Tempo and Loki Dashboard*

Next, go to the **A** query and select `application` and the `my-retro-app` in the Label Filters drop-down list boxes, as show in Figure 11-15. Then, click the Run Query button in the upper right.

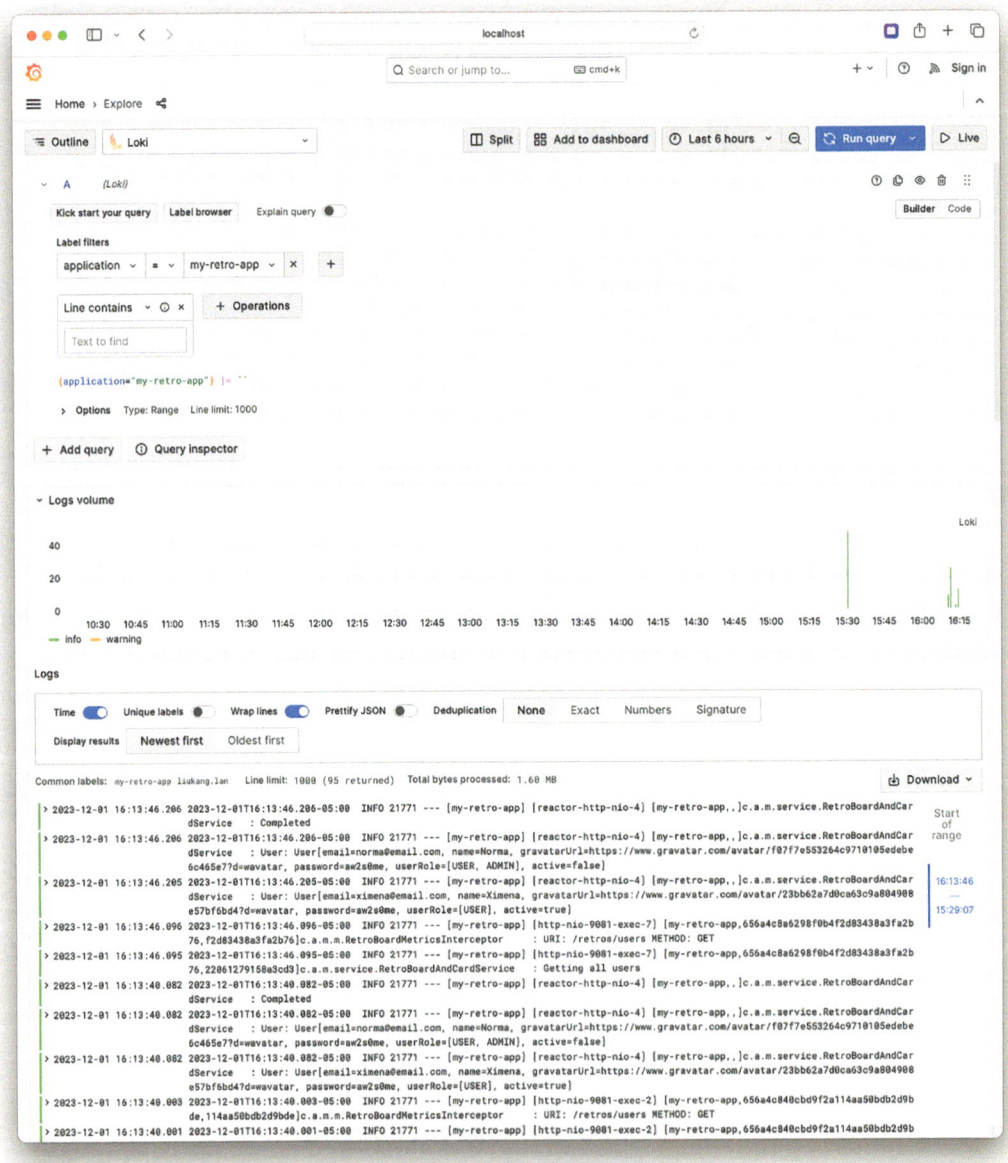

Figure 11-15. *Loki – application/my-retro query*

Figure 11-15 shows the logs of the application, which are sorted by time (newest first). Review them.

In the same query pane under Label Filters, in the Line Contains section, add **Getting all users** as shown in Figure 11-16 and then click the Run Query button again. You will see something like the results shown in Figure 11-16.

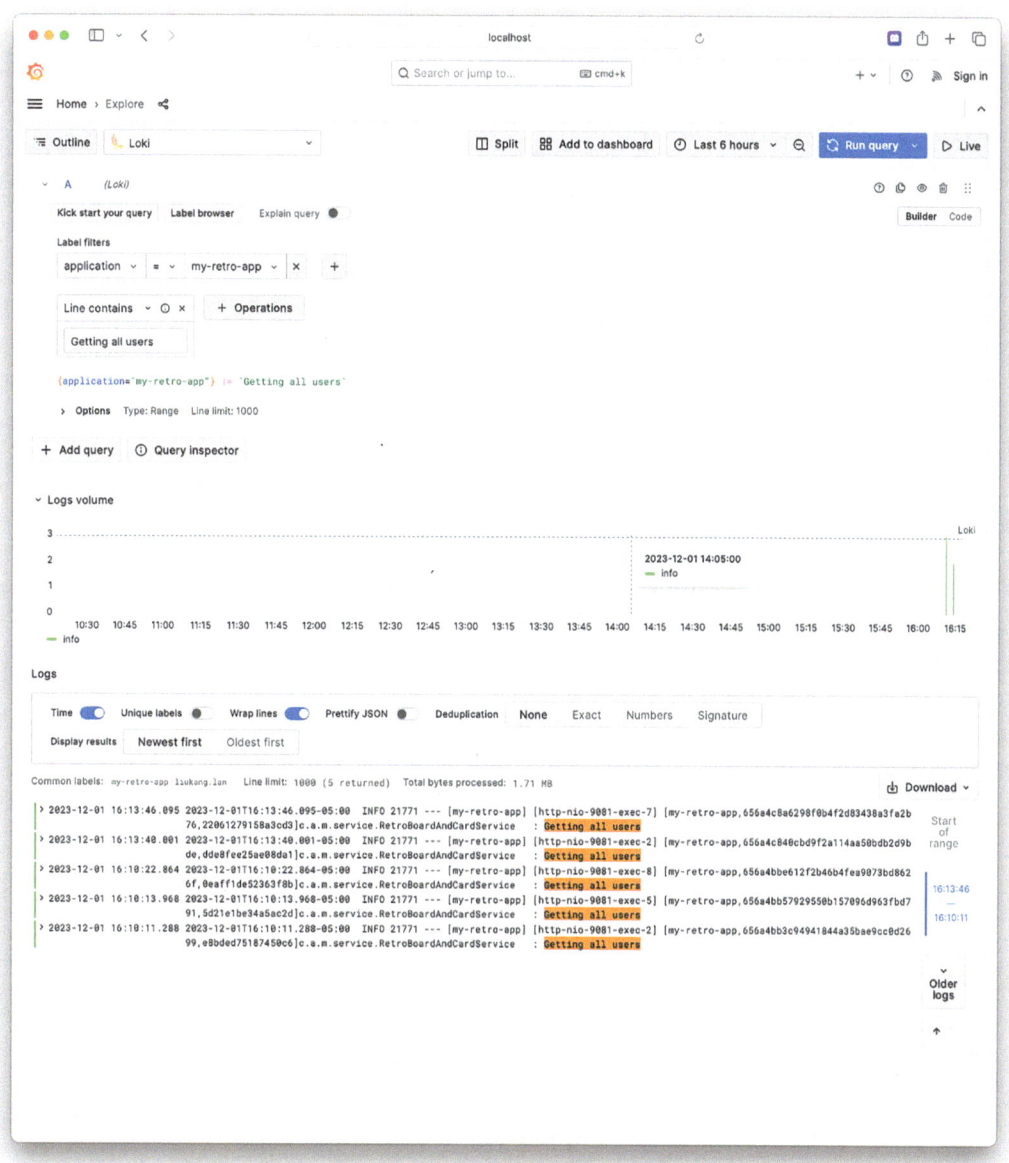

***Figure 11-16.** Loki – application/my-retro-app/Getting all users query*

This new query shows all the requests that we did to the /retros/users endpoint.

Next, if you click one of the Getting all users log, you will see something similar to Figure 11-17.

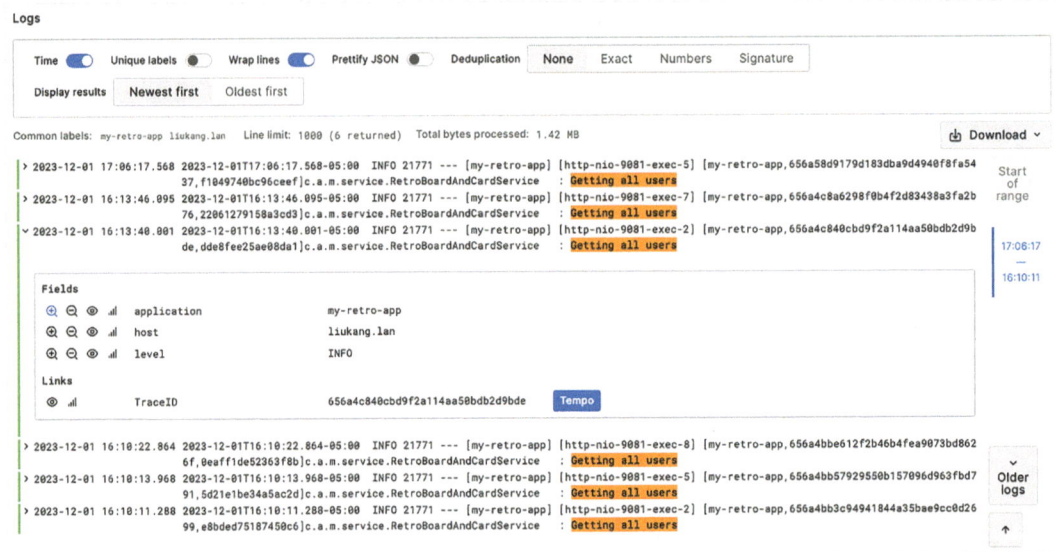

Figure 11-17. *Log details*

Next, click the Tempo button to reveal the trace, which shows what happened when this endpoint was requested. See Figure 11-18.

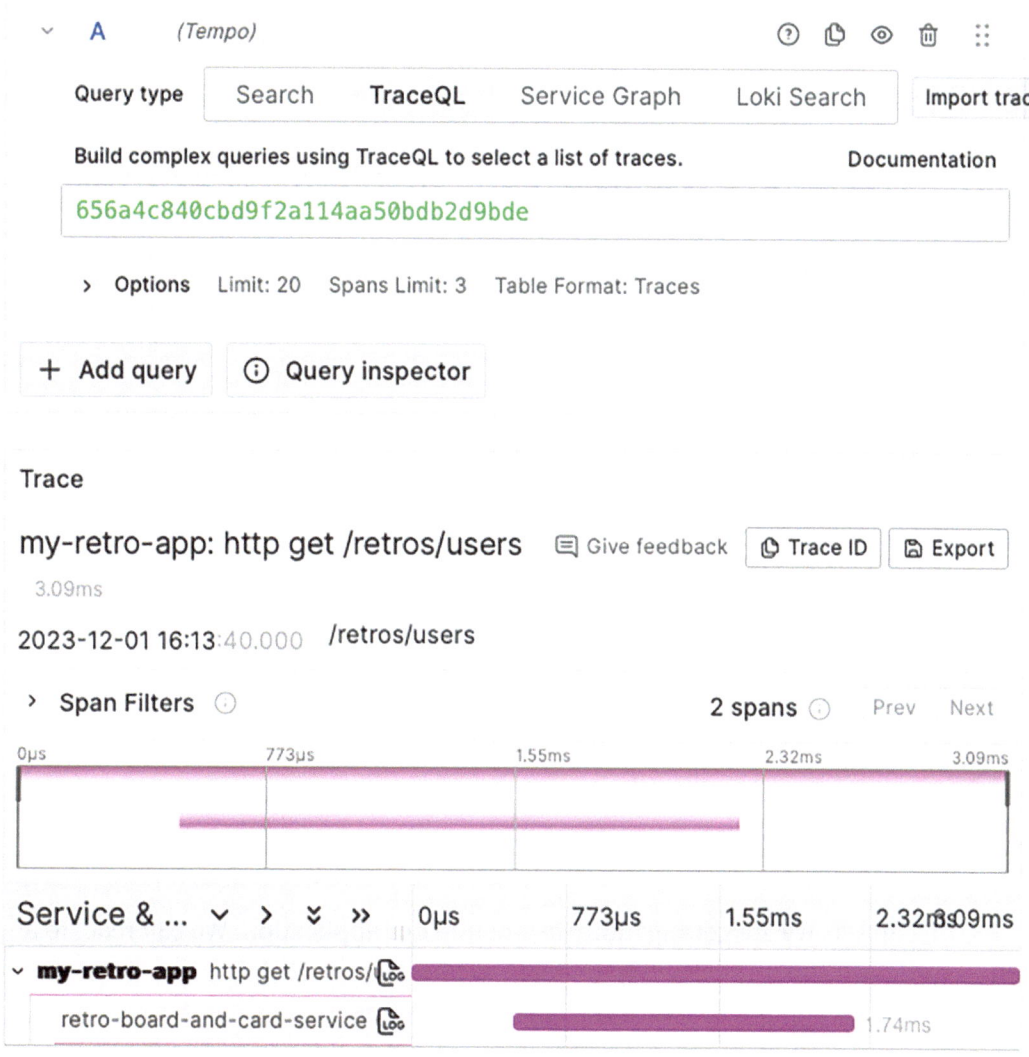

Figure 11-18. *Tempo - Trace ID / Spans – /retros/users*

You can play around and click the requests made and see how much time it took to reach the other service. Now, go back to the Loki pane and change the field Line Contains with the value **Getting all retro boards**. Click one of the logs to see the trace that was made to the database. See Figure 11-19.

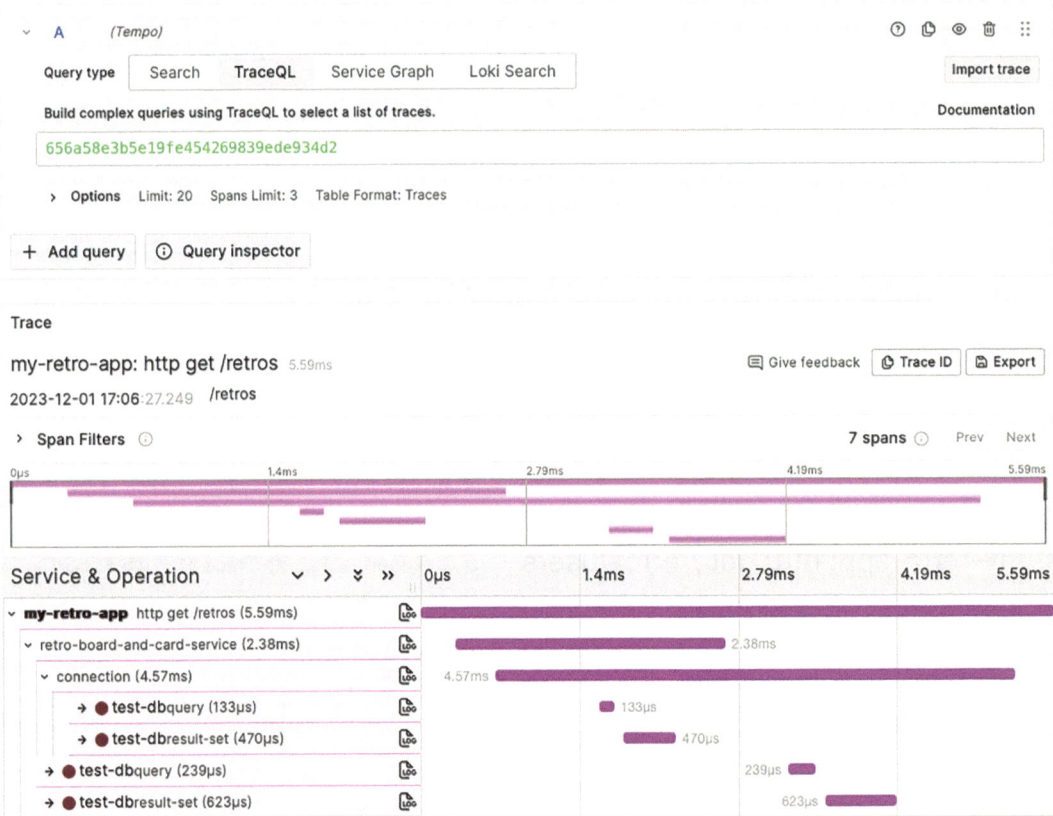

Figure 11-19. *Tempo – Trace ID / Spans - /retros*

As you can see, we are getting more insight into our application. We can react to any bottleneck or any other process that we deem to be taking more time than expected. Take a moment to investigate what else you can do with this views.

What About External Resources: Memory, CPU, Storage?

Glad you asked! So, how can we see the metrics of our app, like the memory, CPU, storage, threads, and more? There are different ways to do this, and normally we need to use Prometheus as the data source and start adding views to a dashboard so that we can have what we need. And even though this is a fun thing to do, I've already provided you with two dashboards that I took from the guy behind Micrometer and Spring Boot Actuator, Jonatan Ivanov! Thanks to Jonatan. Also I found a useful dashboard from Sai Subramanyam, awesome articles from Programming Techie!

You can use one of those dashboard by opening the main menu (the top-left corner, the hamburger icon) and selecting Dashboards. You should see a list of dashboards like the list shown in Figure 11-20.

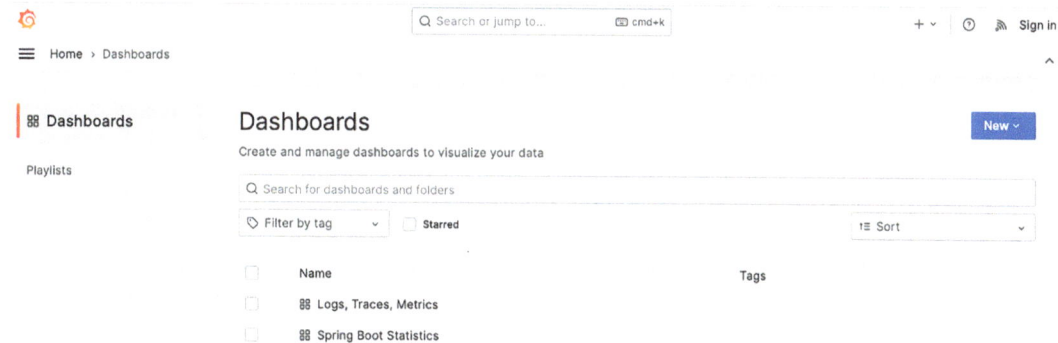

Figure 11-20. *Dashboards*

Choose Spring Boot Statistics and you will see something similar to Figure 11-21.

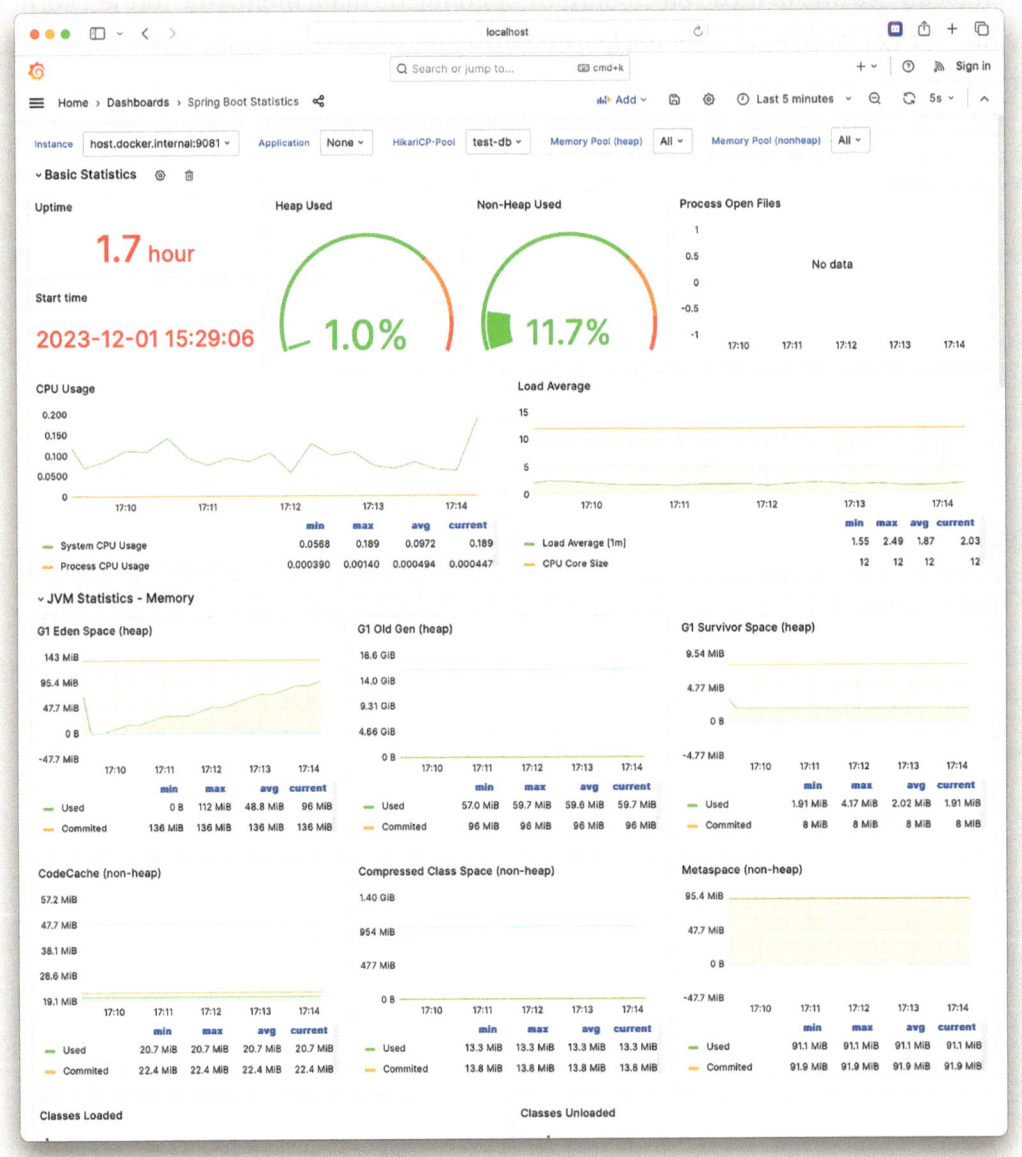

Figure 11-21. *Spring Boot Statistics displaying Prometheus metrics*

You can also import other dashboards from `https://grafana.com/grafana/dashboards/`. You can search for *Spring Boot* (in the search field displaying "Search dashboards") and you will find many! Return to the Dashboards page (see Figure 11-20), click the New button in the top-right corner, and select Import. You will see a page like Figure 11-22.

Import dashboard

Import dashboard from file or Grafana.com

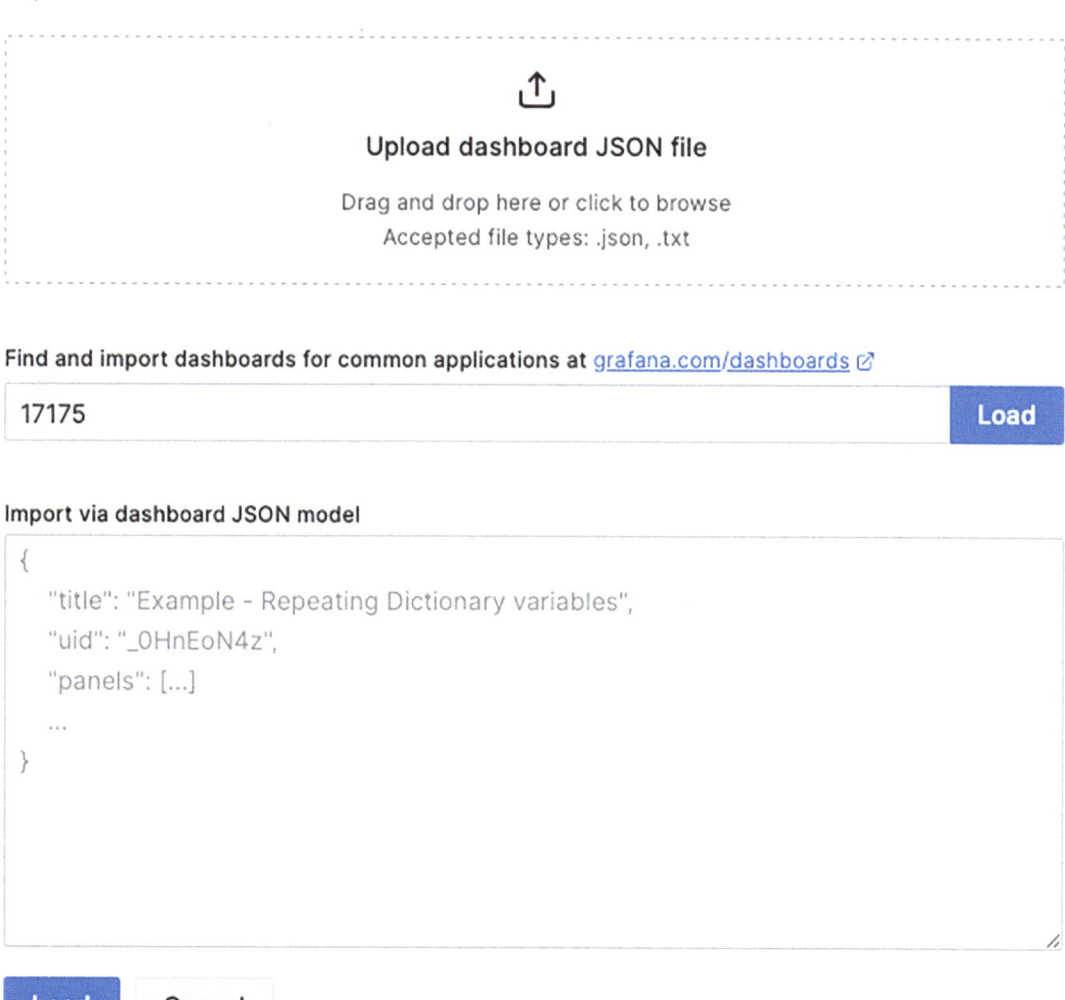

Figure 11-22. *Dashboards - Importing dashboards*

In the "Find and import dashboards..." field, add the ID **17175** (*Spring Boot Observability*, `https://grafana.com/grafana/dashboards/17175-spring-boot-observability/`) and click the Load button. It will take you to the page shown in Figure 11-23 to select the data sources that are already defined, such as Prometheus and Loki. Select them and click Import.

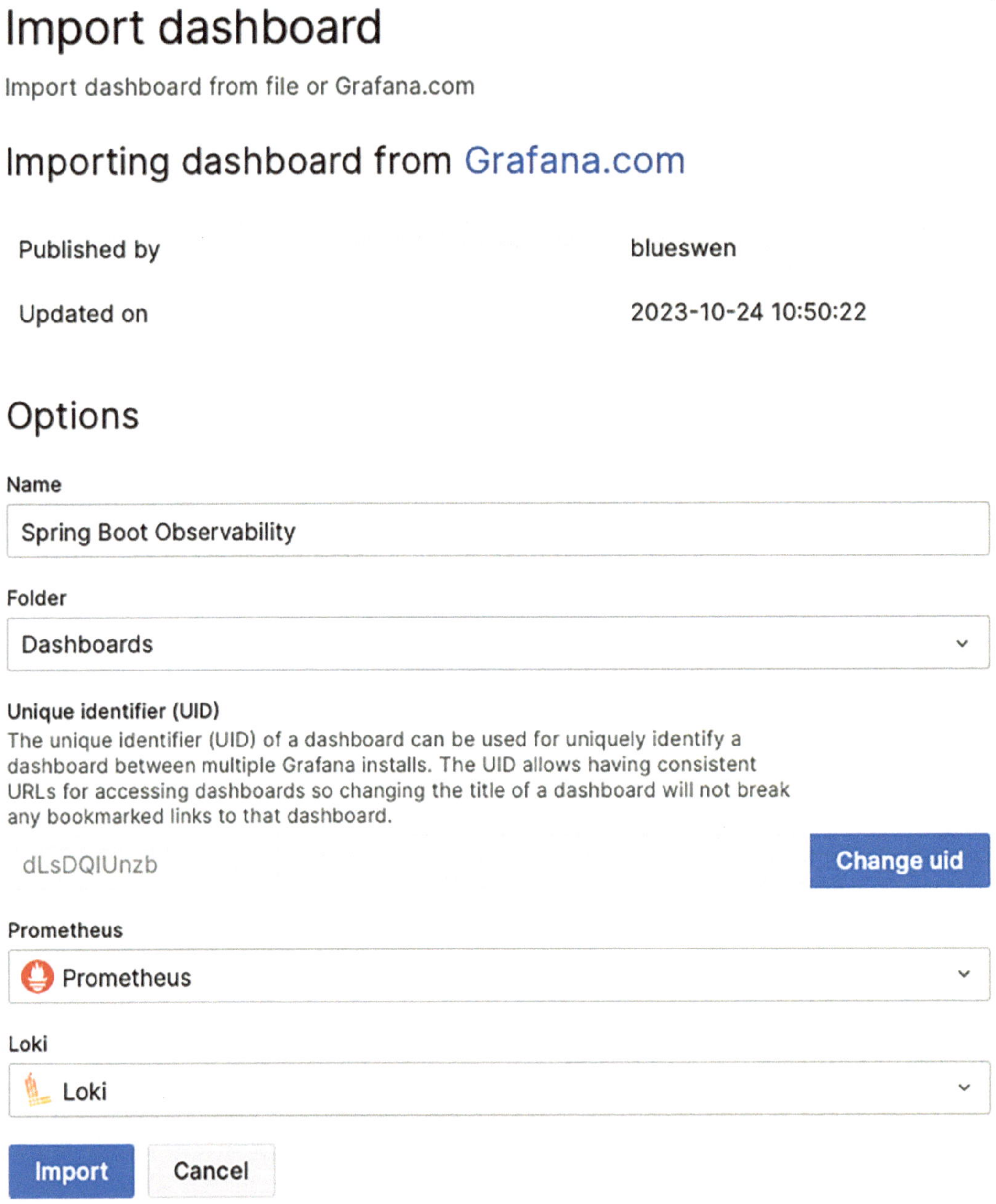

Import dashboard

Import dashboard from file or Grafana.com

Importing dashboard from Grafana.com

Published by	blueswen
Updated on	2023-10-24 10:50:22

Options

Name

Spring Boot Observability

Folder

Dashboards ⌄

Unique identifier (UID)

The unique identifier (UID) of a dashboard can be used for uniquely identify a dashboard between multiple Grafana installs. The UID allows having consistent URLs for accessing dashboards so changing the title of a dashboard will not break any bookmarked links to that dashboard.

dLsDQIUnzb   **Change uid**

Prometheus

🔥 Prometheus ⌄

Loki

Loki ⌄

Import   Cancel

Figure 11-23. *Spring Boot Observability – Dashboard – ID: 17175*

You should see something like the dashboard shown in Figure 11-24.

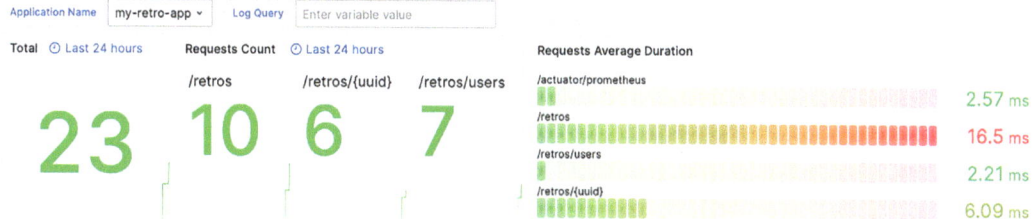

Figure 11-24. *Dashboard – Spring Boot Observability – ID: 17175*

As you can see, you have more options in this dashboard. I like that we have access to the custom metrics we did with the `Counter` class.

Experiment with more dashboards or create new ones and share them. Finally, you can import another one, Spring Boot 3.x Statistics (`https://grafana.com/grafana/dashboards/19004-spring-boot-statistics/`) with ID 19004. Once you have imported it, you should see something like the dashboard shown in Figure 11-25.

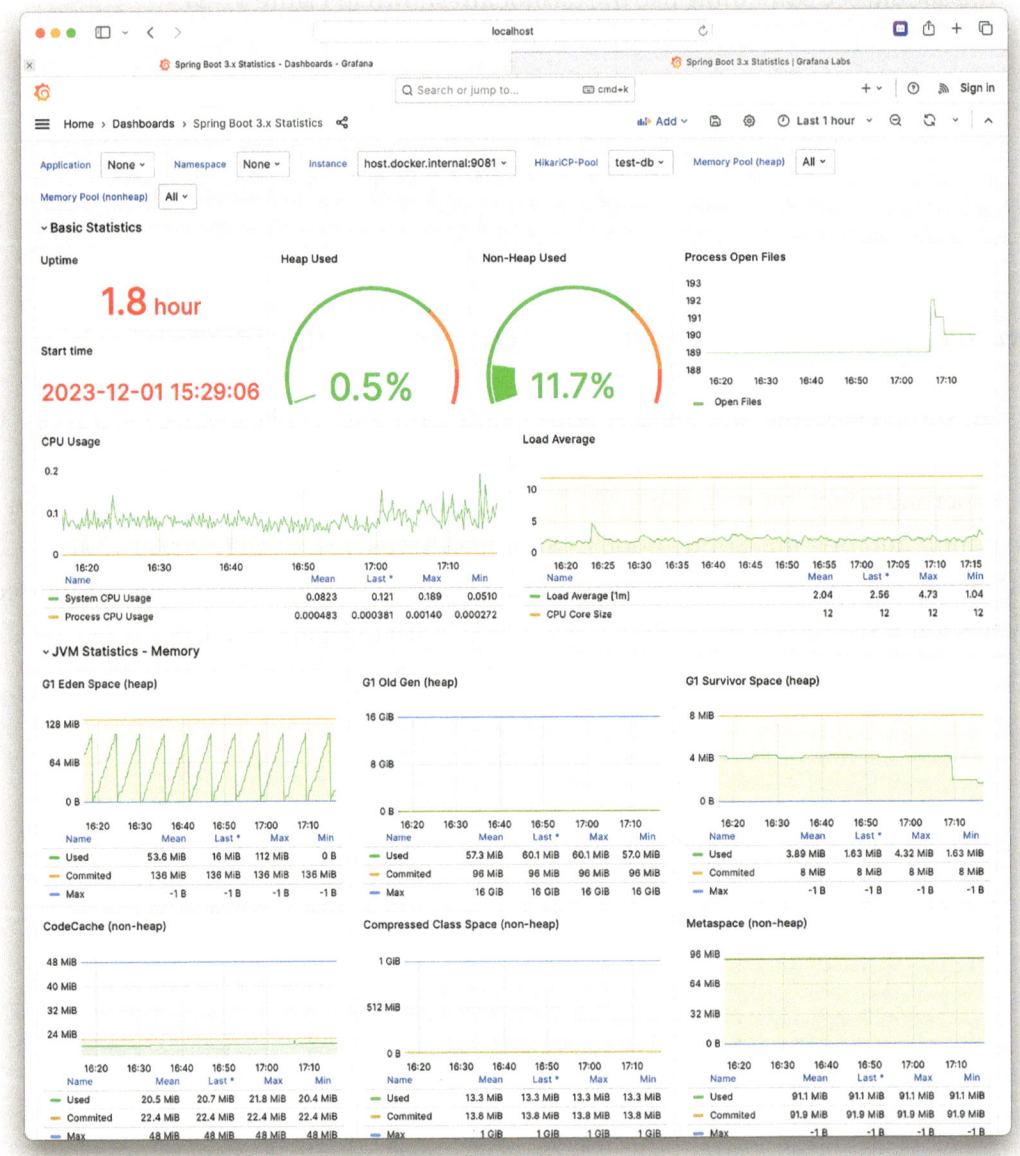

Figure 11-25. *Spring Boot 3.x Statistics Dashboard – ID: 19004*

As demonstrated in this section, you have options to add observability to your projects (logs, metrics, and tracing) all in one console such as Grafana, thanks to the power of Spring Boot Actuator and Micrometer.

Summary

In this chapter you explored the most commonly used actuator endpoints and discovered how easy is to add production-ready features to your application. You learned how to implement your own metrics with the `Counter` and `Request` interceptors, and how to create your own custom health indicators.

You also learned how to add observability to your apps to react when something is wrong (high memory consumption, slow process, etc) by understanding what is happening to your application by checking out the logs, doing tracing, and observing the metrics not only of the app, but how the whole computer behaves.

Finally, you learned how to integrate multiple services such as Prometheus, Grafana, Loki, and Tempo.

CHAPTER 12

Spring Boot Native and AOT

In this chapter I will explain what GraalVM is and how Spring Boot takes advantage of its features by producing native apps that can increase performance, provide faster startup times, improve portability, and much more.

What is Spring AOT?

Spring AOT (Ahead-of-Time) is a new feature introduced in Spring Framework 6 designed to optimize application startup time and reduce runtime overhead, particularly for large and complex applications.

Spring AOT does the following:

- *Early analysis*: AOT processes the application context at build time, performing many of the tasks that traditionally happen during runtime. This includes analyzing bean definitions, resolving dependencies, and verifying configurations.

- *Code generation*: AOT generates optimized code based on the build-time analysis, which gets executed during runtime. This reduces the amount of work done at startup, leading to faster startup times.

- *Reduces reflection*: AOT reduces the reliance on reflection, which can be a bottleneck for startup performance, especially for native images generated by GraalVM.

Here's how Spring AOT is used:

- *Build-time processing*: AOT processing is typically done as part of your project's build process using the Spring Boot build plugins (Maven or Gradle).

665

F. Gutierrez, *Pro Spring Boot 3*, https://doi.org/10.1007/978-1-4842-9294-5_12

- *Runtime execution*: The generated AOT code is then included in your application and used during runtime startup.

- *GraalVM Native Images*: AOT is essential for creating GraalVM native images of Spring applications, as native images require ahead-of-time compilation and have restrictions on reflection usage.

Spring AOT is necessary for the following reasons:

- *Performance optimization*: AOT can significantly improve the startup time of Spring applications, especially for large projects with complex configurations.

- *Resource efficiency*: AOT reduces the memory footprint and CPU usage of Spring applications at runtime.

- *GraalVM Native Image compatibility*: AOT is a prerequisite for creating GraalVM Native Images, which offer even faster startup times and smaller memory footprints.

Use Spring AOT for the following:

- *Large applications*: AOT is most beneficial for large and complex Spring applications where startup time and resource usage are significant concerns.

- *Microservices*: AOT can be useful for microservices architectures where fast startup times are crucial for scaling and responsiveness.

- *GraalVM Native Images*: If you want to build GraalVM Native Images of your Spring application, AOT is mandatory.

Be aware of the following caveats:

- *Reduced flexibility*: AOT imposes some restrictions on dynamic behaviors, such as conditional bean definitions based on runtime conditions.

- *Compatibility*: Not all Spring features are fully compatible with AOT yet, so it's important to check the documentation for limitations.

What Is GraalVM?

GraalVM is a high-performance platform for running applications written in various languages, including Java, JavaScript, Python, Ruby, and more. It offers several key features:

- *High performance*: GraalVM can significantly improve the execution speed of Java applications compared to traditional JVMs. This is achieved through a combination of optimizations, such as ahead-of-time (AOT) compilation and dynamic language support, which we will review in the following sections with Spring Boot.

- *Multilanguage support*: GraalVM can run applications written in various languages, eliminating the need for separate runtimes for each language. This makes it a versatile platform for developing polyglot applications.

- *Native Image*: GraalVM can compile Java applications into stand-alone executables, which can be launched without a JVM. This can further improve performance and reduce memory consumption.

- *Java on Truffle*: GraalVM uses the Truffle framework for dynamic language support. This allows you to run languages like JavaScript and Python alongside Java in the same runtime environment. (Also, as I discovered, you can create your own language easier than ever. If you decide to so so, share it with the world!)

- *Open source*: GraalVM is available as an open source project under the GNU General Public License.

Benefits of Using GraalVM

The following are some of the benefits that GraalVM offers:

- *Faster application startup and execution*: GraalVM can significantly improve the performance of your Java applications, leading to faster startup times and improved responsiveness.

- *Reduced memory footprint*: GraalVM's Native Image technology can reduce the memory footprint of your applications by up to 50 percent.

- *Simplified deployment*: GraalVM's single runtime environment eliminates the need to manage multiple runtimes for different languages.

- *Polyglot application development*: GraalVM allows you to develop applications using various languages, making it easier to build complex and flexible solutions.

- *Improved portability*: GraalVM's Native Image technology allows you to build portable applications that can run on any platform without a JVM.

GraalVM is available for various platforms, including Linux, macOS, Windows, and ARM. You can download the latest version from the GraalVM website: `https://www.graalvm.org/downloads/`. We will cover how to install GraalVM in the section "Creating GraalVM Native Apps" later in the chapter.

Did you know that you can run your Java applications using just GraalVM? Yes, you can run your existing Java applications on GraalVM with no code changes. Simply use the `graalvm` command instead of the standard `java` command. Explore the GraalVM documentation, which provides comprehensive information about the platform, including tutorials, guides, and API references: `https://www.graalvm.org/latest/docs/introduction/`.

Examples of Using GraalVM

Here are some examples of how GraalVM can be used in different scenarios:

- Microservices:

  - GraalVM's fast startup times and low memory footprint make it ideal for building microservices.

  - Its support for multiple languages allows you to use the best language for each service.

  - Native Image allows you to create self-contained microservices that are easy to deploy and manage.

- Serverless computing:

  - GraalVM's small footprint and fast startup times make it well suited for serverless environments.

  - Its support for multiple languages allows you to build serverless functions in the language of your choice.

- Polyglot applications:

  - GraalVM allows you to mix and match different languages within the same application.

  - This can be useful for applications that need to interact with different platforms or technologies.

- High-performance computing:

  - GraalVM's advanced compilers and optimizations can significantly improve the performance of computationally intensive applications.

  - This can be valuable for scientific computing, machine learning, and other high-performance workloads.

- Desktop applications:

  - GraalVM can be used to build native desktop applications with faster startup times and lower memory footprint.

  - This is useful for applications that require high performance and responsiveness.

Here are some specific examples of projects that use GraalVM:

- **Spring Boot**: Spring Boot offers a GraalVM Native Image extension that allows you to build self-contained Spring Boot applications.

- **Apache Kafka**: The open source streaming platform Apache Kafka can be built with GraalVM for improved performance and scalability.

- **GraalVM Enterprise**: Oracle offers a commercial version of GraalVM with additional features and support, including subscription-based access to Oracle JDK and Oracle Cloud Infrastructure integration.

These are just a few examples of how GraalVM can be used. With its unique features and capabilities, GraalVM is a powerful tool that can help developers build faster, more efficient, and more flexible applications.

Spring, Spring Boot, and GraalVM

Now that you know more about GraalVM, it's important to review some key differences between the JVM and GraalVM, see how Spring and Spring Boot support GraalVM, and explore what Spring does with the code to make it easy for GraalVM to generate the native application.

As you already know, GraalVM analyzes the code at compile time to generate the native application based on your OS, so if you are on Windows, GraalVM generates an .exe file, or if you're in a Unix environment, it generates the binary for that Unix OS. For this to work, you must let GraalVM know in advance how your application is built, and if you have dynamic logic in your app (which we do in our project apps, because Spring and Spring Boot have dynamic capabilities at runtime, such as auto-configuration and proxy creation for the Spring beans), you must provide hints in JSON format to let GraalVM know what to do with all these classes.

The following are some of the key differences between the JVM and GraalVM:

- *Static analysis*: GraalVM performs static analysis of the application at build time from the main entry point. This means that if there is code that cannot be reached, then that code won't be part of the native executable.

- *Classpath fixed*: The Classpath must be fixed and cannot be changed at build time.

- *No lazy class loading*: GraalVM doesn't know about dynamic components or any other way to instantiate a class, so we need to tell GraalVM what to do with these resources, serialization, and dynamic proxies. In the executable, all the classes will be loaded into memory at startup.

- *Limitations*: GraalVM still has limitations with some of the Java features. The Spring Boot GitHub site has a Spring Boot with GraalVM page (under the Wiki tab) that provides more details about what is supported and what is not: `https://github.com/spring-projects/spring-boot/wiki/Spring-Boot-with-GraalVM`.

AOT Processing in Spring

As previously mentioned, GraalVM needs help identifying any dynamic component of your application to ensure that the component is part of the final native compiled version. One of the benefits of using Spring with GraalVM is that it has a mechanism that will generate the files that identify an app's dynamic components.

Now, there is a small limitation to this mechanism. Due to the dynamic part of Spring, it is important to notice that any configuration/bean declaration marked with the @Profile annotation has its limitations (@Profile is used in Spring to activate beans conditionally based on active profiles, which are often determined at runtime and the GraalVM's closed-world assumption conflicts with this dynamic behavior. When building a Native Image, GraalVM needs to know upfront which profiles will be active and which beans should be included in the image). To solve this limitation, you need to add extra configuration so that the AOT processing can process any profile you need for your native applications.

If you are using Maven, this is what you need to do if you are using profiles (for example):

```
<profile>
    <id>native</id>
    <build>
        <pluginManagement>
            <plugins>
                <plugin>
                    <groupId>org.springframework.boot</groupId>
                    <artifactId>spring-boot-maven-plugin</artifactId>
                    <executions>
                        <execution>
                            <id>process-aot</id>
                            <configuration>
                                <profiles>profile-cloud,profile-security,
                                profile-staging</profiles>
                            </configuration>
                        </execution>
                    </executions>
```

```
                    </plugin>
                </plugins>
            </pluginManagement>
        </build>
    </profile>
```

If you are using Gradle, then you need to use the following syntax for the profiles:

```
tasks.withType(org.springframework.boot.gradle.tasks.aot.ProcessAot).
configureEach {
    args('--spring.profiles.active=profile-cloud,profile-security,
    profile-staging')
}
```

Also important to note is that if you have properties that change when a bean is created, GraalVM can't identify those changes. This means that if you have any @ Configuration with @ConditionalOnProperty (we haven't covered this annotation yet, but we will when we extend Spring Boot) and any .enable property, won't be supported out of the box. The good news is that Spring can help with that by using its AOT processing!

When the Spring AOT engine is processing your application, it will generate

- The necessary Java source code that GraalVM will understand

- Bytecode for the dynamic proxies

- GraalVM JSON files (called HINT files):

  - Resources (resource-config.json)

  - Reflection (reflect-config.json)

  - Serialization (serialization-config.json)

  - JNI (jni-config.json)

  - Java Proxy (proxy-config.json)

These files will be generated in the target/spring-aot/main/resources folder (if you are using Maven) or build/generated/aotResources folder (if you are using Gradle) and they will be placed in the META-INF/native-image file, where GraalVM will pick them up to generate the Native Image.

Creating GraalVM Native Apps

To create a Native App, we need to prepare our environment/computer by installing the GraalVM tools. My recommendation is to install the *Liberica Native Image Kit (NIK)* from `https://bell-sw.com/pages/downloads/native-image-kit`. There are three NIK versions, corresponding to the JDK 11, 17, and 21 versions, and because we are using the latest version of Spring Boot 3.x, we require at minimum the version JDK 17, so you can choose either NIK 23-JDK 17 or NIK 23-JDK 21.

If you are using macOS or Linux, you can install it using SDKMAN! (`https://sdkman.io/`) by executing the following command (for example):

```
sdk install java 22.3.4.r17-nik
```

Note   You can check out the available *graalvm/Liberica* versions using `sdk list java`; you will see the versions in the last column.

Make sure your terminal/environment is using that version by executing the following command:

```
java -version
openjdk version "17.0.9" 2023-10-17 LTS
OpenJDK Runtime Environment GraalVM 22.3.4 (build 17.0.9+11-LTS)
OpenJDK 64-Bit Server VM GraalVM 22.3.4 (build 17.0.9+11-LTS, mixed mode,
sharing)
```

If you are using Windows, follow the instructions in this blog post that was created by the GraalVM team for Windows users: `https://medium.com/graalvm/using-graalvm-and-native-image-on-windows-10-9954dc071311`. Or, you can follow the instructions from the Windows installer in the Liberica site: `https://bell-sw.com/pages/downloads/native-image-kit`.

Now, you are set with the GraalVM tools.

Creating a Native Users App

Let's see how we can create a Native Users App. If you are following along and want to use the provided source code, go to the 12-native-aot/users folder. If you want to start from scratch with the Spring Initializr (https://start.spring.io), add the dependencies GraalVM Native Support, Web, JPA, Validation, Actuator, H2, PostgreSQL, and Lombok and set the Group field to com.apress and the Artifact and Name fields to users. Click the Generate button, download the project, unzip it, and import it into your favorite IDE.

Open the build.gradle file. See Listing 12-1.

Listing 12-1. build.gradle

```
plugins {
    id 'java'
    id 'org.springframework.boot' version '3.2.3'
    id 'io.spring.dependency-management' version '1.1.4'
    id 'org.hibernate.orm' version '6.4.1.Final'
    id 'org.graalvm.buildtools.native' version '0.9.28'
}

group = 'com.apress'
version = '0.0.1-SNAPSHOT'
sourceCompatibility = '17'

configurations {
    compileOnly {
        extendsFrom annotationProcessor
    }
}

repositories {
    mavenCentral()
}

dependencies {
    implementation 'org.springframework.boot:spring-boot-starter-web'
    implementation 'org.springframework.boot:spring-boot-starter-
    validation'
```

```
    implementation 'org.springframework.boot:spring-boot-starter-data-jpa'

    implementation 'org.springframework.boot:spring-boot-starter-actuator'

    runtimeOnly 'com.h2database:h2'
    runtimeOnly 'org.postgresql:postgresql'

    compileOnly 'org.projectlombok:lombok'
    annotationProcessor 'org.projectlombok:lombok'
    annotationProcessor 'org.springframework.boot:spring-boot-
    configuration-processor'

    // Web
    implementation 'org.webjars:bootstrap:5.2.3'

    testImplementation 'org.springframework.boot:spring-boot-starter-test'
}
tasks.named('test') {
    useJUnitPlatform()
}

hibernate {
    enhancement {
        enableAssociationManagement = true
    }
}
```

Listing 12-1 shows that the plugins section of the build.gradle file now includes the org.graalvm.buildtools.native plugin that will help us to generate the Native application. We are going to use the solution from Chapter 11. If you started from scratch, you can copy and paste the same code. If you are using the 12-native-aot/users folder code, no modifications are necessary. You should have the folder structure shown in Figure 12-1.

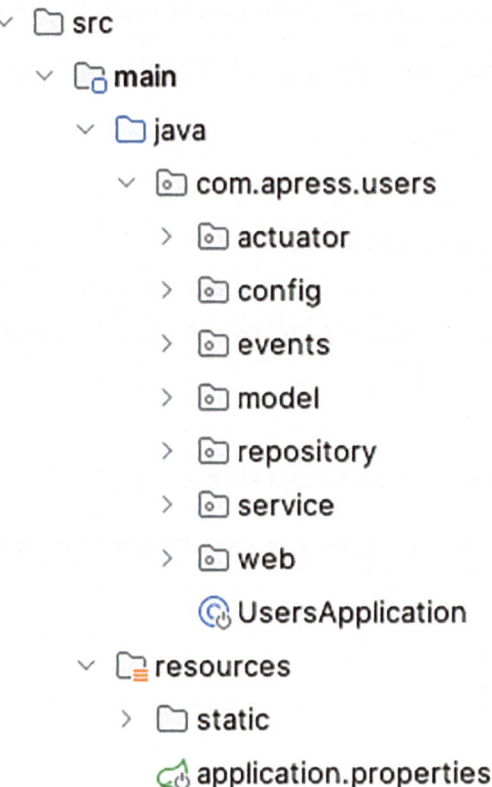

Figure 12-1. *Users App structure*

Before continuing, make sure the code works. You can run it from the command line or from your IDE.

Next, we are going to create the Native Users App. Open a terminal and execute the following command at the root of the `users` project:

```
./gradlew nativeCompile
...
...
BUILD SUCCESSFUL in 1m 45s
9 actionable tasks: 4 executed, 5 up-to-date
```

This can take up to 15 minutes depending on your computer's processor and memory (so, be patient if you have an old computer). I used a Mac with an M3 chip, so it was fast.

If you are using Maven, you can execute the following command:

```
./mvnw -Pnative native:compile
```

The preceding build command (either Gradle or Maven) will generate the executable in the `build/native/nativeCompile/users` folder for Gradle or `users/target/users` folder for Maven.

Let's review what happens before we run the application. There are several stages that happen when we execute the previous command. First, the AOT processing starts and generates all the necessary JSON config files (`proxy-config.json`, `reflect-config.json`, `resource-config.json`, and so on), and then the GraalVM native compilation occurs (remember that the config files are required for GraalVM to know what your app is using to create the native app). At this stage, there are seven internal steps: initializing, building, parsing methods, inlining methods, compiling methods, and creating more classes. You should have a structure similar to Figure 12-2 for Gradle or similar to Figure 12-3 for Maven.

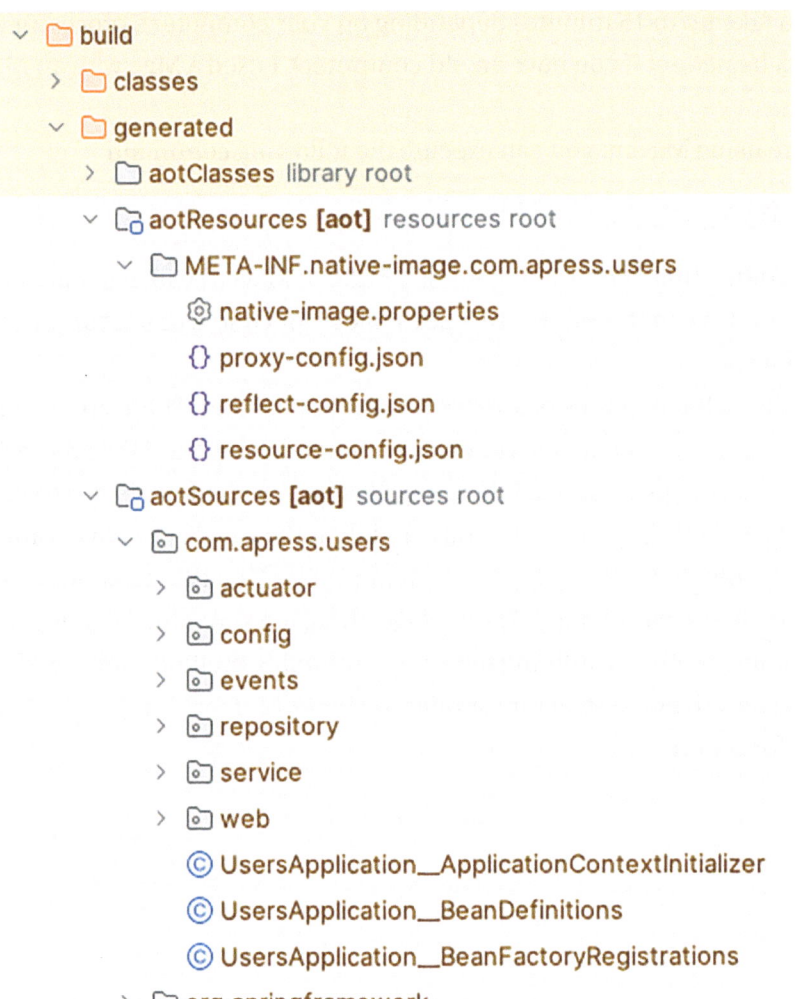

Figure 12-2. *Files and folder structure after using Gradle with GraalVM as native build*

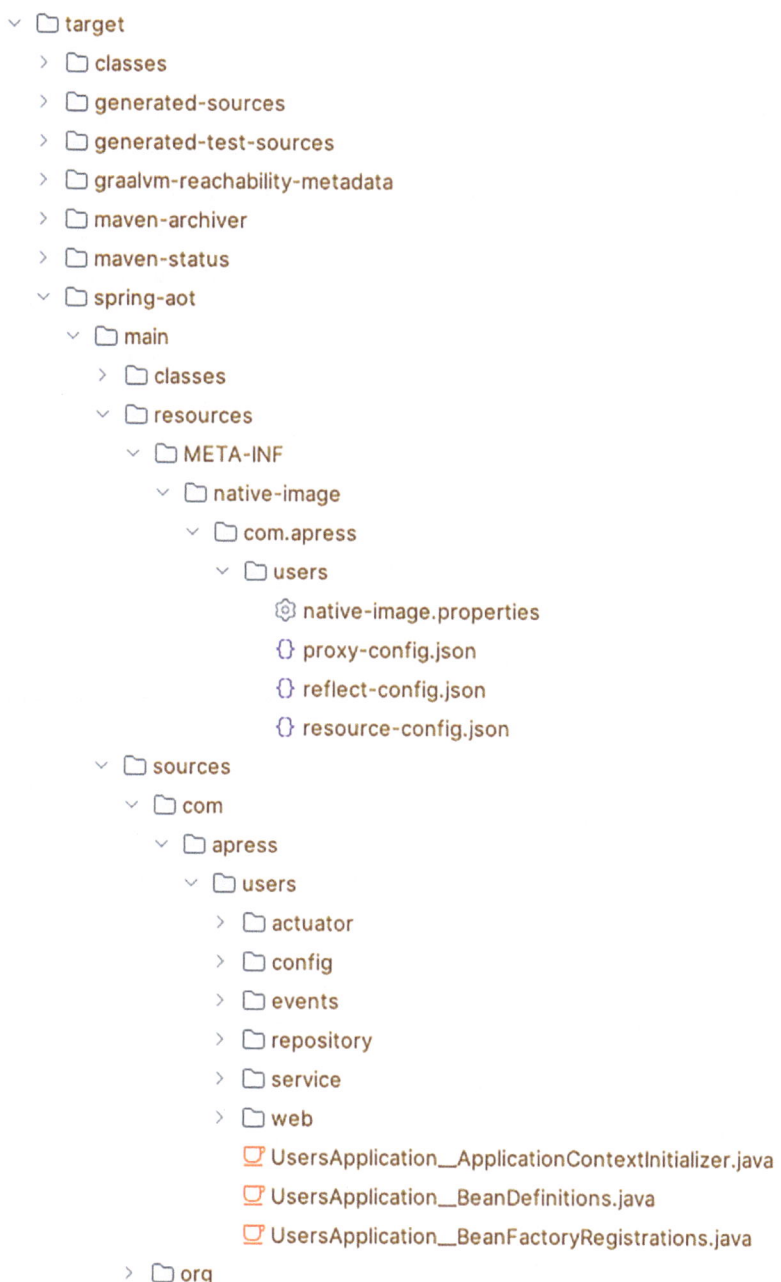

Figure 12-3. *Files and folder structure after using Maven with GraalVM as native build*

Notice that whichever build tool you are using (Gradle or Maven), you will have the same result but in a different folder. Review the UsersApplication_*.java source code. In the UsersApplication__BeanFactoryRegistrations.java file, you will see all the beans that we are using that need to be manually registered to make the application work. The Spring Framework normally registers all the beans at runtime, but in this case we need to notify GraalVM about any dynamic components or proxies.

So, if you are curious about how Spring Framework or Spring Boot works behind the scenes, these classes hold the secret!

Now, you can execute the Users App. Next, in the terminal, execute the following command (if you used Gradle):

```
build/native/nativeCompile/users
```

The app should start faster. My computer took 0.08 seconds to initialize vs. 45 seconds if we do regular JVM. Of course, the Users App is small, and we cannot see the benefits right away, but we can check its memory.

If we take a look at the process ID (PID) of the Users App that is running, we can discover how much memory it is consuming. The following commands identify the total megabytes that the app is using:

```
# We need to get the PID with:

pid=$(ps -fea | grep -E 'nativeCompile.*users' | head -n1 | awk
'{print $2}')

# In my case pid=99008
# Then we need to get the Resident Set Size (RSS) from that PID with:

rss=$(ps -o rss= "$pid" | tail -n1)

# The value was rss=189948
# Then we can calculate by dividing by 1024 with:

mem_usage=$(bc <<< "scale=1; ${rss}/1024")
echo $mem_usage megabytes
186.3 megabytes
```

Alternatively, you can get the same info from a monitoring tool like top/htop for Unix (see Figure 12-4) or Task Manager for Windows.

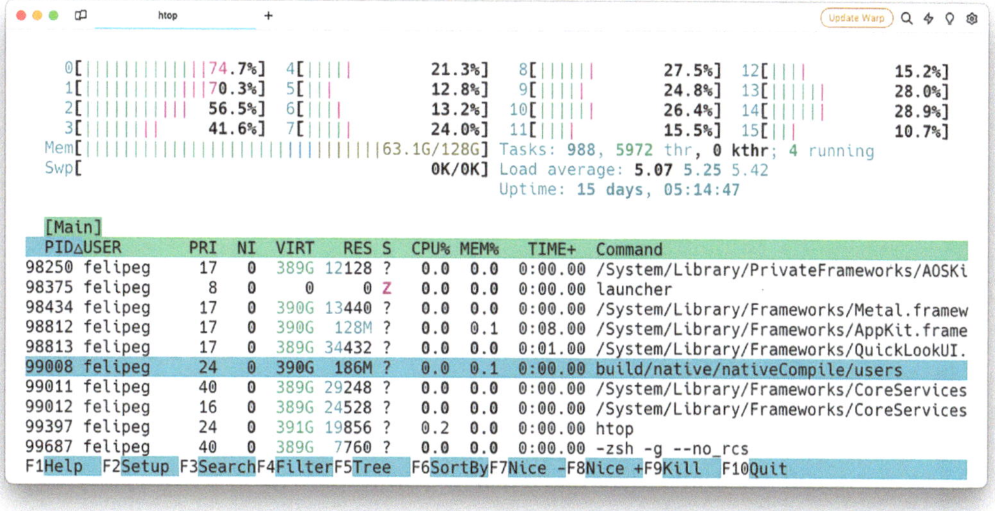

Figure 12-4. *Unix htop showing PID 99008 with 186M*

Because the Users App has the actuator dependency and it's enabled in the application.properties file, you can safely execute the rest call to gracefully turn off the Users App:

```
curl -si -XPOST http://localhost:8080/actuator/shutdown
HTTP/1.1 200
Content-Type: application/vnd.spring-boot.actuator.v3+json
Transfer-Encoding: chunked
Date: Fri, 12 Jan 2024 01:22:44 GMT

{"message":"Shutting down, bye..."}
```

Next, let's do a comparison with the actual JVM. To do this, just build the project to create the JAR file. You can use the following command:

```
./gradlew build
```

And then you can run the app with

```
java -jar build/libs/users-0.0.1-SNAPSHOT.jar
```

First, look that it should take between 4 to 8 seconds to start. Again, this is too low comparing perhaps with a bigger app (more code, more services, etc).

Next, you can repeat the Unix commands to get the memory usage and compare it. It will get around 420M:

```
# Get the PID
pid=$(ps -fea | grep -E 'libs.*users' | tail -n1 | awk '{print $2}')

# Getting the RSS
rss=$(ps -o rss= "$pid" | tail -n1)

# Get the Memory usage
mem_usage=$(bc <<< "scale=1; ${rss}/1024")
echo $mem_usage megabytes
```
416.5 megabytes

As you can see, is the double (memory usage) running in the JVM. You can use your monitor tool of choice. Figure 12-5 shows the results using htop.

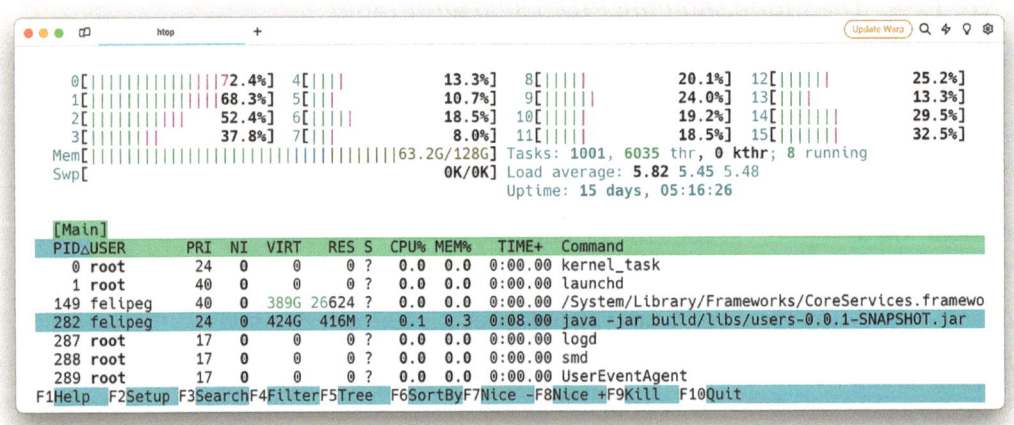

Figure 12-5. *Unix htop showing PID 282 with 416M*

Also, you can use JConsole to see the values, as shown in Figures 12-6 and 12-7.

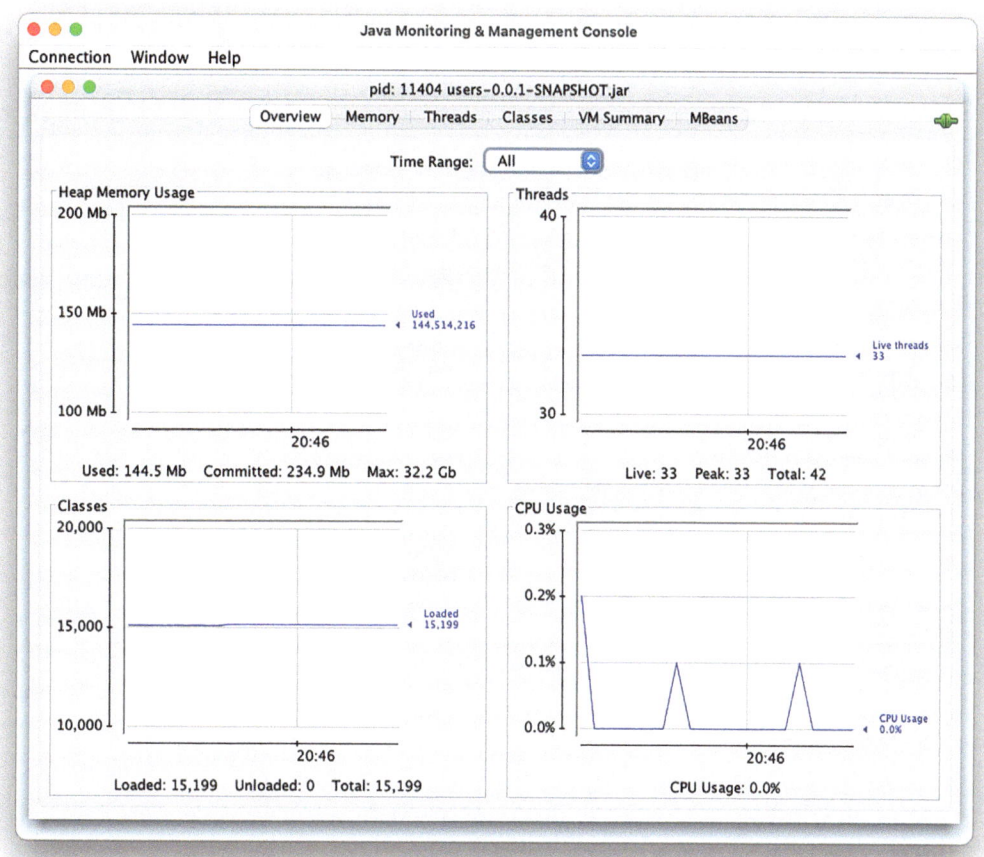

Figure 12-6. *JConsole showing memory usage*

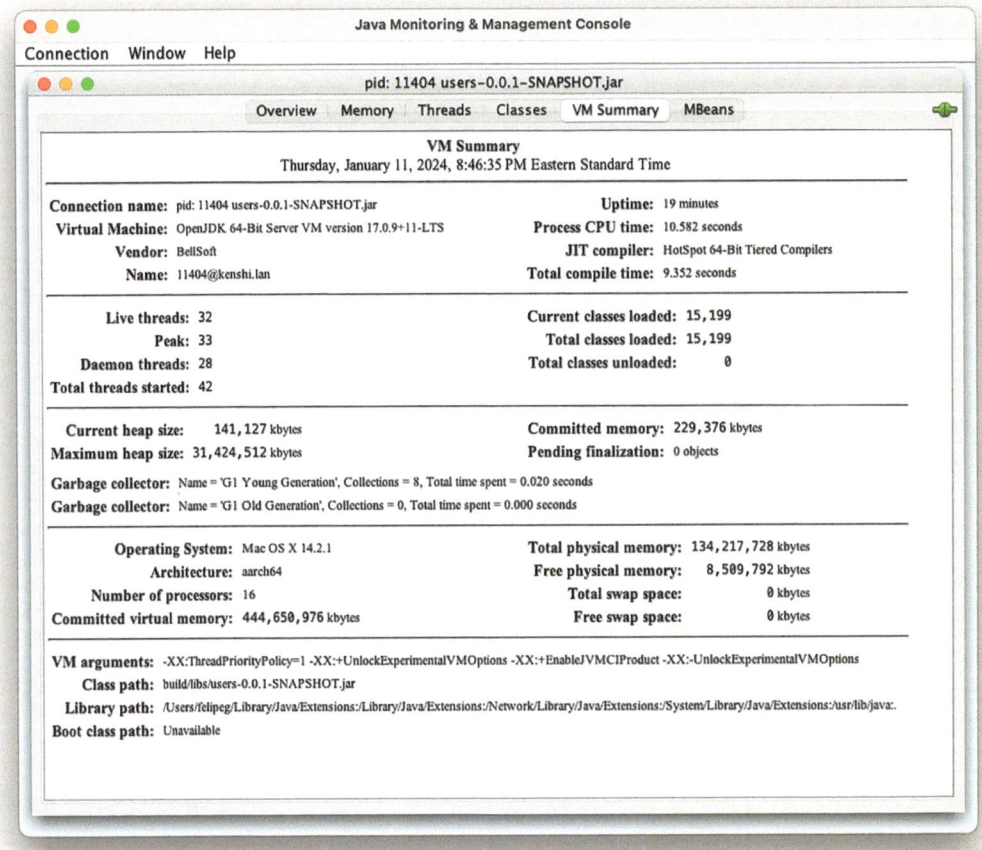

Figure 12-7. *JConsole showing commited virtual memory of 444M*

As you can see, that's a big difference between the JVM and a Native app. Plus, if you start playing around with the API, you will also see some benefits in performance using a native app over a JVM.

You can now shut down the Users App by making a POST request to the `/actuator/shutdown` endpoint.

Creating a Native My Retro App

Now, let's create a native version of the My Retro App. If you have already downloaded the source code, you can find it in the `12-native-aot/myretro` folder. If are starting from scratch with the Spring Initializr (`https://start.spring.io`), add the dependencies

GraalVM Native Support, Web, WebFlux, JPA, Validation, Docker Compose Support, Actuator, H2, PostgreSQL, and Lombok. Set the Group field to com.apress and the Artifact and Name fields to myretro. Click the Generate button, download the project, unzip it, and import it into your favorite IDE.

We are going to use the same code from the previous chapter, so if you are starting from scratch, you can copy and paste the structure. Of course, we need to do a couple of modifications, which I will show and explain to you.

Start by opening the build.gradle file. See Listing 12-2.

Listing 12-2. build.gradle

```
plugins {
    id 'java'
    id 'org.springframework.boot' version '3.2.3'
    id 'io.spring.dependency-management' version '1.1.4'
    id 'org.hibernate.orm' version '6.4.1.Final'
    id 'org.graalvm.buildtools.native' version '0.9.28'
}

group = 'com.apress'
version = '0.0.1-SNAPSHOT'
sourceCompatibility = '17'

configurations {
    compileOnly {
        extendsFrom annotationProcessor
    }
}

repositories {
    mavenCentral()
}

dependencies {
    implementation 'org.springframework.boot:spring-boot-starter-web'
    implementation 'org.springframework.boot:spring-boot-starter-webflux'
    implementation 'org.springframework.boot:spring-boot-starter-
    validation'
```

```
    implementation 'org.springframework.boot:spring-boot-starter-data-jpa'
    implementation 'org.springframework.boot:spring-boot-starter-actuator'
    implementation 'org.springframework.boot:spring-boot-starter-aop'

    runtimeOnly     'com.github.loki4j:loki-logback-appender:1.4.1'
    implementation 'io.micrometer:micrometer-tracing-bridge-brave'
    implementation 'io.zipkin.reporter2:zipkin-reporter-brave'
    implementation   'net.ttddyy.observation:datasource-micrometer-spring-
    boot:1.0.2'

    runtimeOnly 'com.h2database:h2'
    runtimeOnly 'org.postgresql:postgresql'
    runtimeOnly 'io.micrometer:micrometer-registry-jmx'
    runtimeOnly 'io.micrometer:micrometer-registry-prometheus'

    developmentOnly 'org.springframework.boot:spring-boot-docker-compose'

    compileOnly 'org.projectlombok:lombok'
    annotationProcessor 'org.projectlombok:lombok'
    annotationProcessor 'org.springframework.boot:spring-boot-
    configuration-processor'

    testImplementation 'org.springframework.boot:spring-boot-starter-test'
}

tasks.named('test') {
    useJUnitPlatform()
}

hibernate {
    enhancement {
        enableAssociationManagement = true
    }
}
```

Listing 12-2 shows that we are using some of the *actuator* and *micrometer* dependencies. Also, we are using the org.graalvm.buildtools.native plugin that will help us to create the native app.

As mentioned, this project requires some modifications. Remember that the My Retro App communicates with the Users App through the `UserClient` interface. See Listing 12-3.

Listing 12-3. src/main/java/com/apress/myretro/client/UserClient.java

```
package com.apress.myretro.client;

import com.apress.myretro.client.model.User;
import org.springframework.web.bind.annotation.PathVariable;
import org.springframework.web.service.annotation.GetExchange;
import org.springframework.web.service.annotation.HttpExchange;
import reactor.core.publisher.Flux;
import reactor.core.publisher.Mono;

@HttpExchange(url = "/users", accept = "application/json", contentType =
"application/json")
public interface UserClient {

    @GetExchange
        Flux<User> getAllUsers();

        @GetExchange("/{email}")
        Mono<User> getById(@PathVariable String email);
}
```

In the `UserClient` interface, we are using the `@HttpExchange` annotation to declare how we are going to consume the /users endpoint. And, of course, we need to configure the `UserClientConfig` class to use this client in our code. See Listing 12-4.

Listing 12-4. src/main/java/com/apress/myretro/client/UserClientConfig.java

```
package com.apress.myretro.client;

import com.apress.myretro.config.RetroBoardProperties;
import org.springframework.context.annotation.Bean;
import org.springframework.context.annotation.Configuration;
import org.springframework.web.reactive.function.client.WebClient;
import org.springframework.web.reactive.function.client.support.
WebClientAdapter;
```

687

```
import org.springframework.web.service.invoker.HttpServiceProxyFactory;

@Configuration
public class UserClientConfig {

    @Bean
    WebClient webClient(RetroBoardProperties retroBoardProperties) {
        return WebClient.builder()
                .defaultHeaders(header ->
                        header.setBasicAuth(retroBoardProperties.
                        getUsersService().getUsername(),
                                retroBoardProperties.getUsersService().
                                getPassword())))
                .baseUrl(retroBoardProperties.getUsersService().
                getBaseUrl())
                .build();
    }

    @Bean
    UserClient userClient(WebClient webClient) {
        HttpServiceProxyFactory httpServiceProxyFactory =
                HttpServiceProxyFactory.builderFor(WebClientAdapter.
                create(webClient))
                        .build();
        return httpServiceProxyFactory.createClient(UserClient.class);
    }
}
```

Listing 12-4 shows the configuration needed to have access to a WebClient instance. You have seen this in Chapter 11, but this time we are using the RetroBoardProperties to get the username, password, and baseUrl properties identifying where the Users App is running (and although our Users App currently doesn't have security, this is how we would add it to the WebClient).

Next, look at the RetroBoardProperties and UsersService classes, shown in Listings 12-5 and 12-6, respectively. These classes define the configuration properties that we will use for the connection to the Users App.

Listing 12-5. src/main/java/com/apress/myretro/config/
RetroBoardProperties.java

```
package com.apress.myretro.config;
```

```
import lombok.Data;
import org.springframework.boot.context.properties.ConfigurationProperties;
import org.springframework.boot.context.properties.
NestedConfigurationProperty;

@Data
@ConfigurationProperties(prefix = "myretro")
public class RetroBoardProperties {
    @NestedConfigurationProperty
    private UsersService usersService;
}
```

Listing 12-6. src/main/java/com/apress/myretro/config/UsersService.java

```
package com.apress.myretro.config;
```

```
import lombok.Data;

@Data
public class UsersService {
    private String baseUrl;
    private String basePath;
    private String username;
    private String password;
}
```

The RetroBoardProperties and UsersService classes are something that you already know, but RetroBoardProperties includes a new annotation, @NestedConfigurationProperty. This annotation is a *must* if you have nested properties (like in our case) and want to create a native app. Although Spring helps us with the AOT processing, nested properties in this form (RetroBoardProperties and UsersService) won't be detectable and won't be bindable, which is why this annotation is essential for this case.

You can test this app later for a Native app; you can comment out the @NestedConfigurationProperty and compile it (it will compile but it won't run). The RetroBoardProperties class is a dependency for the WebClient to be created, so it must be initialized, and if you don't have this marker (@NestedConfigurationProperty annotation), the GraalVM will compile and create the native app, but then it will fail when it runs because there are no values bindable to the properties.

The application.properties file is shown in Listing 12-7.

Listing 12-7. src/main/resources/application.properties

```
## DataSource
spring.h2.console.enabled=true
spring.datasource.generate-unique-name=false
spring.datasource.name=test-db

spring.jpa.show-sql=true

## Server
server.port=9081

## Docker Compose
spring.docker.compose.readiness.wait=never

## Application
spring.main.web-application-type=servlet
spring.application.name=my-retro-app
logging.pattern.correlation=[${spring.application.name:},%X{traceId:-},%X{spanId:-}]

## Actuator Info
info.developer.name=Felipe
info.developer.email=felipe@email.com
info.api.version=1.0
management.endpoint.env.enabled=true

## Actuator
management.endpoints.web.exposure.include=health,info,metrics,prometheus,shutdown,configprops,env,trace
```

```
## Enable shutdown endpoint
management.endpoint.shutdown.enabled=true

## Actuator Observations
management.observations.key-values.application=${spring.application.name}

## Actuator Metrics
management.metrics.distribution.percentiles-histogram.http.server.
requests=true

## Actuator Tracing
management.tracing.sampling.probability=1.0

## Actuator Prometheus
management.prometheus.metrics.export.enabled=true
management.metrics.use-global-registry=true

## Users App Service
myretro.users-service.base-url=http://localhost:8080
myretro.users-service.base-path=/users
myretro.users-service.username=admin
myretro.users-service.password=admin
```

The new additions to the application.properties file are the final four lines, which define the nested properties to connect to the Users App.

Next, it's time to convert the My Retro App into a Native App. To do this, we first must start up all the service dependencies we have; remember that this app connects to Grafana/Loki. You should have in the root directory (12-native-aot/myretro) the docker-compose.yaml file with all the necessary services declarations.

So, in a terminal window, you can start the services with

```
docker compose up -d
```

Once the services start, execute the following:

```
./gradlew nativeCompile
...
```

```
...
BUILD SUCCESSFUL in 1m 11s
10 actionable tasks: 10 executed
```

When I compiled the app, I used GraalVM 21 (instead of 17), which includes an extra step (performing analysis), but that didn't take too much extra time, and I image that it did some good stuff internally optimizing everything. And as you can see in the preceding output, it took around 30 seconds less, even though the My Retro App is bigger than the Users App.

Before running it, make sure the Users App is up and running, then you can execute:

```
build/native/nativeCompile/myretro
...
...
Completed initialization in 1 ms
```

Next, you can try it by accessing some of the resources. Also check that the Users App access work by going to the /retros/users endpoint. You should see in the console all the users that are being consumed from the Users App.

Let's check out how much memory was consumed:

```
# Get the PID
pid=$(ps -fea | grep -E 'Compile.*myretro' | tail -n1 | awk '{print $2}')

# Getting the RSS
rss=$(ps -o rss= "$pid" | tail -n1)

# Get the Memory usage
mem_usage=$(bc <<< "scale=1; ${rss}/1024")
echo $mem_usage megabytes
```
238.4 megabytes

Only 238.4M for the My Retro App. Let's compare it with a JVM as well. But first, shut down the app with

```
curl -s -XPOST http://localhost:9081/actuator/shutdown
{"message":"Shutting down, bye..."}
```

Next, build it with

```
./gradlew clean build
```

And run it with

```
java -jar build/libs/myretro-0.0.1-SNAPSHOT.jar
```

Now, once started, check the memory usage with

```
# Get the PID
pid=$(ps -fea | grep -E 'libs.*myretro | tail -n1 | awk '{print $2}')

# Getting the RSS
rss=$(ps -o rss= "$pid" | tail -n1)

# Get the Memory usage
mem_usage=$(bc <<< "scale=1; ${rss}/1024")
echo $mem_usage megabytes
```
511.2 megabytes

Yes, the double again (memory usage). Now, you can gracefully shut down the My Retro App with

```
curl -s -XPOST http://localhost:9081/actuator/shutdown
{"message":"Shutting down, bye..."}
```

Note   Getting the PID (Process ID) can be tricky. It depends on the OS and how it shows you the info. Sometimes when executing the command (for example $ ps -aux) goes to the head or sometimes to the tail of the output, just be careful choosing the right PID of your app.

Remember that you also can use Grafana (http://localhost:3000) and use the Spring Boot Statistics Dashboard so that you can compare the performance of the native and the JVM.

GraalVM Native Images, wait... what?

We have just seen the power of GraalVM to create a native application that will increase performance, provide faster starts, and improve memory usage. All of this can be very useful when we deploy our app in the cloud, where we need to have the best performance, less memory consumption so that we can add more simultaneous instances, and easy recovery with faster application restarts when the application shuts down or when there is a new version, where we can apply a deployment solution such as *blue-green deployment* so there is no service interruption for our clients.

GraalVM also can help with these deployments, faster restarts, and much more by combining the power of Cloud Native Buildpacks (CNBs; see `https://buildpacks.io/`) that Spring Boot provides to create Docker images with ease.

Docker image creation has been part of the Spring Boot Plugin (Gradle or Maven build tools) since version 2.3.0. By default, it creates an Open Container Initiative (OCI; see `https://opencontainers.org/`) image using CNBs. In fact, simply using the Spring Boot Plugin is sufficient to create a Docker image. If you don't have the GraalVM dependency, the Spring Boot Plugin by default will create a JAR and use that artifact to create the Docker image, meaning that the process behind the scenes will install the JRE so your JAR can work.

Now, if you include the GraalVM dependency plugin `org.graalvm.buildtools.native`, it will start the AOT process and will create the Docker image with your Native Application, making this a Native Image. So, let's see what we need to do to create a Native Image for our projects.

Creating a Users App Native Image

Creating a Native Image is as simple as executing the following command:

```
./gradlew bootBuildImage
```

That's it. It will start by executing the AOT process and creating the Native Application. Then it will use the Cloud Native Buildpacks to create the Docker image. By default, it will generate the tag `docker.io/library/users:0.0.1-SNAPSHOT`, which means that you can run your app with the following command:

```
docker run --rm -p 8080:8080 --platform linux/amd64 docker.io/library/
users:0.0.1-SNAPSHOT
```

In this command, `--rm` removes the image when we stop it and `-p` exports the port 8080 to the local 8080 (syntax: `HOST-POST:CONTAINER-PORT`). I'm including the `--platform` parameter because I'm using a Mac Silicon (M3-chip) and the

Buildpacks are using Linux AMD 64, and I need to emulate it, which is the function of this parameter. Finally, the long name is the default image name.

If you prefer to add your own tag or name convention, you can add the `imageName` parameter in the command line; for example, my Docker ID is felipeg48, so I can generate the image like this:

```
./gradlew bootBuildImage --imageName=felipeg48/users:v0.0.1
```

This will create my image so that I can run it like this (a bit shorter than the original command):

```
docker run --rm -p 8080:8080 --platform linux/amd64 felipeg48/users:v0.0.1
```

Also, you can publish your image by adding the `publishImage` parameter. You need to be authenticated to publish the image:

```
./gradlew bootBuildImage --imageName=felipeg48/users:v0.0.1 --publishImage
```

Alternatively, you can add configuration directly to the `build.gradle` file:

```
tasks.named("bootBuildImage") {
    imageName.set("felipeg48/users:v0.0.1")
    publish = true
    docker {
        publishRegistry {
            username = "felipe48"
            password = "myAwesome$ecret"
        }
    }
}
```

Note   If you are using Maven, you can create your Native Image with `./mvnw -Pnative spring-boot:build-image`.

Inspecting the Users Native Image

Let's take peek inside the Docker image we have just created. To do so, you can download a tool named dive from https://github.com/wagoodman/dive. You can execute it like this:

```
dive felipeg48/users:v0.0.1
```

You will see something similar to Figure 12-8.

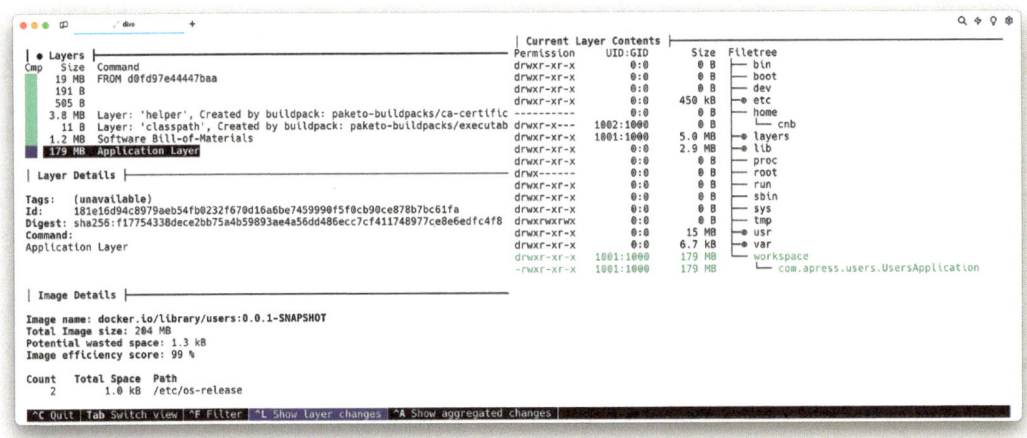

Figure 12-8. *Using the dive tool*

The dive tool provides a visualization of the layers inside the Docker image. You can navigate with the Tab key to the different sections. For example, as shown in Figure 12-8, in the Layers section you can select the Application Layer (with the up/down arrow keyboard keys) to see in the Current Layer Contents pane the path of the application, which in this case is the /workspace/com.apress.users.UsersApplication executable.

Feel free to inspect every layer. In the Current Layer Contents pane you can collapse the folders with the spacebar.

Creating a My Retro App Native Image

Creating this Native Image is the same process as previously described, but remember that you need to start the docker compose up in your terminal.

You can go to the My Retro source code and execute in a terminal the following command:

```
./gradlew bootBuildImage
```

As before, this will generate the `docker.io/library/myretro:0.0.1-SNAPSHOT` image. However, running the image in this context will be different because we need access to the same network where all the services are.

When you executed the `docker compose up`, by default it created a bridged network, which is easy to see if you execute the following command:

```
docker network list
```

```
NETWORK ID      NAME                 DRIVER      SCOPE
f97e915ba54a    bridge               bridge      local
cf3efd768256    host                 host        local
2c3b202a4106    myretro_default      bridge      local
7540438ac56c    none                 null        local
```

It uses the name of the folder in which you executed `docker compose up` and appends `_default` to the name. So, to connect to that network, we are going to use the `--network=<name>` parameter in the `docker run` command. To run our My Retro Native image, execute the following:

```
docker run --rm -p 9081:9081 --network myretro_default --platform linux/
amd64 docker.io/library/myretro:0.0.1-SNAPSHOT
```

That's it! Now, if you try to execute the `/retros/users` endpoint, it will produce an error telling you that it's not reachable. Do you know why? Well, this is because the Users container image needs to run in the same network bridge as the My Retro App. So, you can stop the Users container and run it like this:

```
docker run --rm -p 8080:8080 --network myretro_default --platform linux/
amd64 felipeg48/users:v0.0.1
```

And that's it.

Inspecting the My Retro Native Image

Again, you can take a peek at the My Retro Native Image with the `dive` command:

```
dive docker.io/library/myretro:0.0.1-SNAPSHOT
```

Figure 12-9 shows the result.

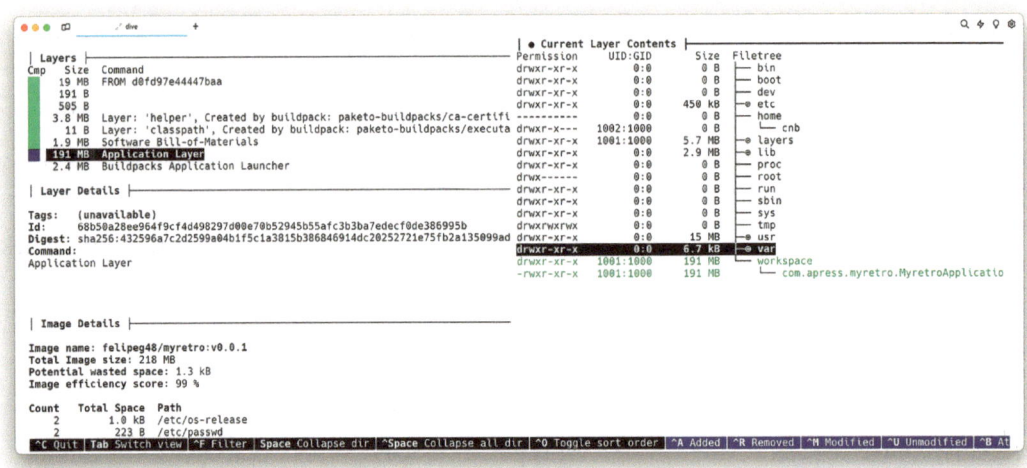

Figure 12-9. *The dive tool with My Retro Native Image*

Now, you can stop the images by making a POST request to the /actuator/ shutdown endpoint and executing the docker compose down command.

Testing Native Apps and Images

If you want to test your apps when creating the Native App/Image, you can do so with the following commands.

If you are using Maven:

```
./mvnw -PnativeTest test
```

If you are using Gradle (as in our case):

```
./ gradle nativeTest
```

AOT Processing with Custom Hints

When you have your own custom AOP (Aspect Oriented Programming) or any other dynamic code (e.g. using reflection), some other resources (configuration files, etc), proxies or serialization process in your app, it all might be skipped by the AOT processing if you don't let the AOT know how your app works. Spring AOT engine gives you a few different ways to provide *custom hints*. You can implement custom hints from the RuntimeHintsRegistrar functional interface via the registerHints method. Or, if your app relies on binding, you can use the @RegisterReflectionForBinding annotation over your JSON serialization bean, for example, or WebClient, RestClient, or RestTemplate directly.

```java
class MyCustomRuntimeHintsRegistrar implements RuntimeHintsRegistrar {

    @Override
    public void registerHints(RuntimeHints hints, ClassLoader
    classLoader) {
        hints.resources().registerPattern("my-data.csv");
                           hints.serialization().registerType
                           (MySerializableConverter.class);
    }
}

@Configuration
class RetroBoardCustomHintsConfiguration {

    @Bean
    @RegisterReflectionForBinding(RetroBoard.class)
    @ImportRuntimeHints(MyCustomRuntimeHintsRegistrar.class)
    ApplicationListener<ApplicationReadyEvent> peopleListener(ObjectMapper
    objectMapper, @Value("classpath:/my-ata.csv") Resource csv) {
        return new ApplicationListener<ApplicationReadyEvent>() {
            @SneakyThrows
            @Override
            public void onApplicationEvent(ApplicationReadyEvent event) {
                try (var in = new InputStreamReader(csv.
                getInputStream())) {
                    var csvData = FileCopyUtils.copyToString(in);
                    Stream.of(csvData.split(System.lineSeparator())).
                    map(line -> line.split(","))
                            .map(row -> new RetroBoard(row[0], row[1])).
                            map(retroBoard -> objectMapper.writeValueAs
                            String(retroBoard)).forEach(System.out::println);
                }
```

```
            }
        };
    }
}
```

As you can see, this is very straightforward, but before moving forward, let's analyze it:

- RuntimeHintsRegistrar: You need to implement from this functional interface by adding code to the registerHints method. Normally, you will add any reflection, serialization, resources (in our example, the my-data.csv resource) and proxies.

- @RegisterReflectionForBinding: If you have a class that will be binding values with some serialization is important to use the @RegisterReflectionForBinding annotation so it can be picked up by the AOT processing.

- @ImportRuntimeHints: This annotation tells the AOT processing what classes are ready for any custom hint with the RuntimeHintsRegistrar.

Again, if we don't add this annotation or register these custom hints, they won't be picked up by the AOT processing.

Summary

In this chapter we covered how to create GraalVM Native applications and Native Images to use with containers such as Docker. You learned that with the native apps you can improve performance, reduce memory consumption, and provide faster starts.

Remember that you need to have GraalVM installed in your system, and only by using the org.graalvm.buildtools.native plugin can you create either a Native app or a Native Image.

In Chapter 13 we launch into the cloud with Spring Cloud.

CHAPTER 13

Spring Cloud with Spring Boot

Spring Cloud is a framework within the Spring ecosystem that provides tools and libraries to simplify the development of distributed systems, particularly for building and deploying microservices-based architectures. It aims to address challenges associated with building, deploying, and managing distributed systems by offering solutions for common patterns and concerns. This chapter introduces some of the most commonly used technologies under the Spring Cloud umbrella to give you a more comprehensive understanding of which tech to use and when.

Challenges Addressed by Spring Cloud

Spring Cloud addresses various challenges in the development of microservices and distributed systems, including but not limited to the following:

- *Service discovery and registration*: Simplifies the registration and discovery of microservices within a distributed system.

- *Load balancing*: Provides tools for load balancing requests across multiple instances of microservices to ensure efficient resource utilization.

- *Configuration management*: Offers solutions for managing configuration settings across microservices in a centralized and dynamic manner.

- *Circuit breakers*: Implements patterns such as the Circuit Breaker pattern (discussed later in the chapter) to handle faults and failures gracefully, preventing cascading failures in distributed systems.

© Felipe Gutierrez 2024
F. Gutierrez, *Pro Spring Boot 3*, https://doi.org/10.1007/978-1-4842-9294-5_13

- *Routing and API gateway*: Facilitates the implementation of routing and API gateway patterns for efficient communication between microservices.

- *Distributed tracing*: Supports distributed tracing to monitor and analyze requests as they traverse through various microservices.

- *Fault tolerance*: Introduces mechanisms for handling faults and failures in a resilient manner, enhancing the overall robustness of distributed systems.

- *Security*: Provides tools for securing communication between microservices and managing authentication and authorization.

Spring Cloud leverages various technologies and components to address the challenges mentioned. The following are some of the key technologies:

- *Netflix OSS components*: Integrates with several Netflix Open Source Software (OSS) components, such as Eureka for service discovery, Ribbon for client-side load balancing, and Hystrix for circuit breakers.

- *Spring Cloud Config*: Enables centralized configuration management for microservices, allowing dynamic updates without requiring application restarts.

- *Spring Cloud Sleuth*: Integrates with distributed tracing systems, providing insights into the flow of requests across microservices. Although this project is still available as a stand-alone project at `https://spring.io/projects/spring-cloud`, the core has been moved to the Micrometer project. So, if you want to learn more about Spring Cloud Sleuth, and use the latest versions of Spring Boot, then the recommendation will be to use *Micrometer Tracing* instead (`https://docs.micrometer.io/tracing/reference/index.html`).

- *Spring Cloud Stream*: Simplifies the development of event-driven microservices by providing abstractions for message-driven communication.

- *Spring Cloud Security*: Offers tools for securing microservices through authentication, authorization, and other security mechanisms.

- *Spring Cloud Bus*: Facilitates the propagation of configuration changes across microservices in a distributed system.

- *Spring Cloud Contract*: Supports consumer-driven contract testing to ensure compatibility between microservices.

- *Spring Cloud Kubernetes*: An extension of the Spring Cloud framework that streamlines the development of microservices for Kubernetes environments. It provides abstractions and integrations that simplify common tasks such as service discovery, configuration management, load balancing, and more, ensuring seamless compatibility with Kubernetes-native features.

- *Spring Cloud Function*: Provides a convenient and flexible framework for building serverless applications and functions. It leverages the strengths of the Spring ecosystem, promotes a consistent development model, and supports multiple programming languages and function as a service (FaaS) providers, allowing developers to focus on writing business logic without being tightly coupled to the underlying infrastructure.

- *Spring Cloud Gateway*: Offers a versatile and extensible gateway solution that facilitates the development of microservices architectures. It addresses the challenges of routing, filtering, load balancing, and other cross-cutting concerns, providing a central point for managing and controlling external access to microservices.

Covering all the projects under the Spring Cloud umbrella is beyond the scope of this book. If you'd like to know about other projects, take a look at the Spring Cloud documentation: `https://spring.io/projects/spring-cloud/`.

In this chapter we'll dig deeper into some of the most commonly used Spring Cloud technologies and use them in our two main projects, the Users App and My Retro App.

Microservices

Microservices refers to an architectural style for developing software applications as a collection of small, independent, and loosely coupled services. Each microservice represents a specific business capability and runs as a separate process, communicating with other microservices through well-defined APIs. Microservices architecture aims to enhance scalability, maintainability, and agility by breaking down complex applications into smaller, independently deployable and scalable services. However, it also introduces challenges like service discovery, load balancing, and distributed tracing. Spring Cloud provides a comprehensive set of tools and libraries that address these challenges, making it easier to build and manage microservices-based applications on the Java platform.

Twelve-Factor App Practices and Spring Boot/Spring Cloud Relationship

The Twelve-Factor App (`https://12factor.net/`) is a set of best practices and principles for building modern, scalable, and maintainable web applications. Developers at Heroku, a cloud platform as a service (PaaS) provider, formulated these practices and they have become widely adopted in the software development industry. While initially designed for monolithic applications, many of the practices and principles align well with microservices architecture.

Here's a quick summary of how microservices relate to the Twelve-Factor App practices and how Spring Boot and Spring Cloud implement them:

1. *Codebase*: Each microservice has its own codebase, managed independently.

    - **Spring Boot**: Supports the development of stand-alone, executable JARs and WARs, making it easy to manage a single codebase.

    - **Spring Cloud**: Enhances the development of microservices architectures, allowing developers to manage multiple codebases for independent microservices.

2. *Dependencies*: Each microservice manages its dependencies, minimizing shared dependencies.

    - **Spring Boot**: Leverages a dependency management system, making it explicit about project dependencies. Developers can easily manage and isolate dependencies.

    - **Spring Cloud**: Integrates with Spring Boot, providing additional features for building distributed systems and managing dependencies between microservices.

3. *Config*: Externalize configuration, allowing dynamic changes without code modifications.

- **Spring Boot**: Encourages externalized configuration, allowing developers to use `application.properties` or `application.yaml` files for configuration. The configuration can be easily overridden using environment variables.

- **Spring Cloud**: Extends Spring Boot's configuration capabilities by providing tools for centralized and dynamic configuration management across microservices.

4. *Backing services*: Microservices interact with databases, queues, and other services as separate entities.

- **Spring Boot**: Easily integrates with various backing services such as databases, message brokers, and caches. It supports configuration properties for connecting to external services.

- **Spring Cloud**: Facilitates the interaction with and discovery of backing services using components like service discovery and client-side load balancing.

5. *Build, release, run*: Microservices are independently built, released, and run.

- **Spring Boot**: Provides an embedded web server, making it easy to package applications as standalone JAR files. The build and run processes are streamlined, simplifying deployment.

- **Spring Cloud**: Builds on Spring Boot's capabilities to enhance the deployment and scaling of microservices, supporting dynamic routing and load balancing.

6. *Processes*: Microservices are designed to be stateless, allowing for easy scalability and resilience.

- **Spring Boot**: Supports the creation of stateless applications. State is typically managed by external services, and the framework facilitates the development of RESTful stateless APIs.

- **Spring Cloud**: Complements Spring Boot in developing stateless microservices, adhering to the distributed nature of cloud-native applications.

7. *Port binding*: Microservices expose APIs and communicate over well-defined ports.

    - **Spring Boot**: Applications can be configured to listen on specific ports, and the framework provides an embedded web server for easy port binding.

    - **Spring Cloud**: Integrates with Spring Boot to manage ports and facilitate communication between microservices using well-defined APIs.

8. *Concurrency*: Microservices scale independently, enabling efficient resource utilization.

    - **Spring Boot**: Supports the creation of concurrent, stateless components. The application can be easily scaled horizontally by deploying multiple instances.

    - **Spring Cloud**: Works seamlessly with Spring Boot to scale microservices independently, providing tools for service discovery and load balancing.

9. *Disposability*: Microservices are designed to be disposable, allowing for quick deployment and scaling.

    - **Spring Boot**: Enables fast startup and graceful shutdown, aligning with the disposability principle. It's designed to be suitable for cloud-native environments.

    - **Spring Cloud**: Aligns with Spring Boot's disposability features, enabling the rapid scaling and deployment of microservices.

10. *Dev/prod parity*: Aim for consistency between development, testing, and production environments.

    - **Spring Boot**: Provides a consistent development model across different environments, minimizing the disparity between development, testing, and production environments.

- **Spring Cloud**: Ensures consistency in development, testing, and production environments, promoting a unified development and deployment model.

11. *Logs*: Microservices generate logs that are often aggregated into centralized systems for monitoring.

    - **Spring Boot**: Integrates with logging frameworks and provides flexible logging configurations, allowing logs to be treated as event streams. Centralized logging can be easily implemented.

    - **Spring Cloud**: Integrates with Spring Boot to support centralized logging, ensuring that logs can be treated as event streams in a distributed system.

12. *Admin processes*: Administrative tasks can be executed independently for each microservice.

    - **Spring Boot**: Supports the implementation of admin or management tasks as one-off processes, separate from the main application logic.

    - **Spring Cloud**: Provides additional tools for managing and monitoring distributed systems, supporting administrative processes and tasks.

Spring Boot and Spring Cloud complement each other to implement the Twelve-Factor App principles effectively, particularly in the context of building cloud-native applications and microservices architectures. They provide a cohesive and comprehensive framework for developing, deploying, and managing modern, scalable applications.

Cloud Development

So far with our two projects, Users App and My Retro App, we have been developing microservices that complement each other to create a complete solution, and both projects work independently of each other, meaning that we can access the Users App directly or through the My Retro App. If you review the Twelve-Factor App practices again with our projects in mind, you'll discover that we cover all of them, providing a perfect scenario for cloud development—but what does "cloud development" actually mean?

Cloud development refers to the entire process of building, testing, deploying, and running software applications in the cloud instead of on physical, onsite servers. It leverages the resources and services provided by cloud computing platforms like Amazon Web Services (AWS), Microsoft Azure, and Google Cloud Platform (GCP).

Some key characteristics of cloud development:

- *Location independence*: Development and deployment happen within the cloud platform, accessible from anywhere with an Internet connection.

- *Scalability*: Resources like storage, computing power, and memory can be easily scaled up or down based on demand.

- *Flexibility*: Cloud platforms offer a wide range of services and tools that can be integrated into the development process.

- *Collaboration*: Teams can work together on projects in real time regardless of their physical location.

- *Cost-effectiveness*: Cloud eliminates the need for upfront hardware investments and provides pay-as-you-go pricing.

There are two main approaches to cloud development:

- *Cloud-based development*: This involves using traditional development tools and methodologies but deploying the application to the cloud.

- *Cloud-native development*: This involves designing and building applications specifically for the cloud, taking advantage of its unique characteristics. And, yes, we've already seen some of this in the previous chapter with GraalVM.

Cloud development offers several advantages over traditional on-premises development, including

- *Reduced costs*: Eliminates hardware and maintenance costs

- *Increased agility*: Faster development and deployment cycles

- *Improved scalability*: Easier to handle fluctuating demand

- *Enhanced collaboration*: Enables teams to work together more effectively

- *Greater innovation*: Access to a wider range of tools and services

However, it's important to consider the potential challenges of cloud development, such as:

- *Vendor lock-in*: Dependence on a specific cloud provider

- *Security concerns*: Reliant on implementation of robust security measures

- *Network reliability*: Reliant on Internet connectivity

Overall, cloud development is a growing trend that offers many benefits for organizations of all sizes. If you're considering moving to the cloud, carefully evaluate your needs and choose a platform that meets your requirements.

Using Spring Cloud Technologies

In this section, we are going to explore the following Spring Cloud technologies that can be used on any local (for development purposes) or in any cloud platform:

- *Spring Cloud Consul* for service discovery and external configuration management. HashiCorp Consul, a widely-adopted service networking solution, excels at service discovery, health checking, and configuration management in distributed systems. Recognizing its power, Spring Cloud Consul seamlessly integrates Consul into the Spring ecosystem. This enables Spring Boot applications to leverage Consul for service discovery and external configuration management, providing a robust foundation for building resilient and scalable microservices architectures.

- *Spring Cloud Vault* for database connections and secrets. HashiCorp Vault is a powerful secrets management tool that provides secure storage and access control for sensitive data like passwords, API keys, and certificates. Spring Cloud Vault, a module within the Spring Cloud project, simplifies the integration of Vault into Spring Boot applications. This integration enables applications to securely retrieve and manage secrets stored in Vault, bolstering security practices and eliminating the need to hardcode sensitive information in application configurations.

- *Spring Cloud OpenFeign* for request/response communication between microservices

- *Spring Cloud Gateway* with some of its features such as filters with *Load Balancing* and *Circuit Breakers*

I chose these Spring Cloud technologies because they are the most commonly used in the IT industry for cloud development.

For purposes of this discussion, assume that we are using either Amazon Web Services, Google Cloud, or Microsoft Azure cloud infrastructure and using only compute instances (virtual machines) to deploy our apps (Users App and My Retro Apps). We need to have our apps in a distributed way and with multiple instances because we require high availability, access to external configuration, and security for sensitive information (for example, by requiring credentials to connect to a database). Figure 13-1 depicts the overall cloud architecture that we'll implement in this chapter.

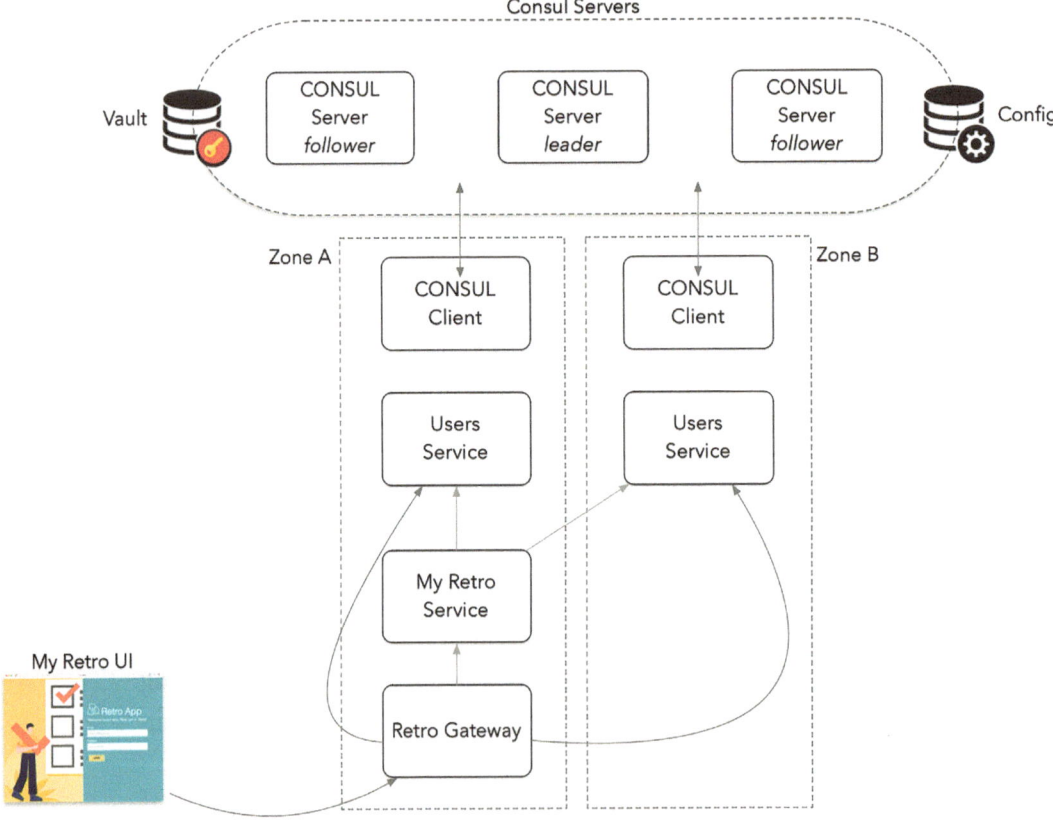

Figure 13-1. *Users App and My Retro App cloud architecture*

Figure 13-1 shows that we need two instances of the Users App Service, and a gateway that can help us to reach out to these services, so the UI or Client only know one address to do some requests, and not every single service address and port.

HashiCorp Consul and Spring Cloud Consul

Spring Cloud Consul technology is integrated with HashiCorp Consul technology, so let's start by understanding what Consul is and exploring some of its use cases, features, and benefits.

Note   In this discussion, "Consul" without "HashiCorp" or "Spring Cloud" preceding it always refers to HashiCorp Consul.

HashiCorp Consul

HashiCorp Consul (`https://www.consul.io/`) is a service networking platform designed to manage secure connectivity between services across diverse environments, including on-premises, multi-cloud, and various runtimes. It serves as a central control plane for service discovery, secure communication, and network automation.

The following are some use cases for Consul:

- *Microservices architectures*: Easily connect and manage microservices with dynamic service discovery and health checks.

- *Multi-cloud and hybrid deployments*: Ensure consistent service connectivity across different cloud providers and on-premises infrastructure.

- *API gateways and service meshes*: Implement secure service-to-service communication with features like automatic TLS encryption and identity-based authorization.

- *Network automation*: Automate network infrastructure configuration and updates based on service changes.

Consul includes the following features (among others):

- *Service discovery*: Register and locate services using DNS, HTTP, or gRPC interfaces.

- *Health checks*: Monitor service health and automatically deregister unhealthy instances.

- *Secure communication*: Enable service-to-service encryption with mutual TLS (mTLS) and identity-based authorization.

- *Key-value store*: Store configuration and secrets securely within Consul.

- *Multi-datacenter support*: Scale Consul across multiple data centers or regions seamlessly.

- *Service mesh integration*: Integrate with existing service meshes like Linkerd or Istio.

- *API gateway*: Manage traffic and access control for services within Consul Service Mesh.

- *Network automation*: Automate network infrastructure configuration based on service changes.

Benefits of Consul include

- *Simplified service management*: Centrally manage service discovery, health checks, and security.

- *Increased agility*: Accelerate service deployment and scaling with multi-platform support.

- *Improved operational efficiency*: Automate tasks and gain centralized visibility into services.

- *Enhanced security*: Enforce least privilege access and enable secure communication.

Using HashiCorp Consul

In this section, we are going to use Docker to start the Consul server and a Consul client. Consul requires a server-client architecture that enables it to provide features like distributed consensus, high availability, security, and efficient management, which wouldn't be possible with just clients alone. This architecture ensures reliability, scalability, and security for service discovery and networking in various scenarios.

To start the Consul server, execute the following command:

```
docker run \
    -d \
    -p 8500:8500 \
    -p 8600:8600/udp \
    --rm \
    --name=consul-server \
    consul:1.15.4 agent -server -ui -node=server-1 -bootstrap-expect=1
    -client=0.0.0.0
```

Note   At the time of writing, the Consul version is 1.15.4. You can check Docker Hub for the latest version: `https://hub.docker.com/_/consul`.

Next, let's find out the address the client requires to connect to the server. Execute the following command:

```
docker exec consul-server consul members
```

The output of this command is the default docker IP address (normally `172.17.0.2`). Then, you can start the client and use the IP address:

```
docker run \
    --name=consul-client --rm -d \
    consul:1.15.4 agent -node=client-1 -retry-join=172.17.0.2
```

Because we are going to use Consul as an external configuration mechanism, we need to add some keys. To add some keys, you can either use Consul's REST API or install the Consul CLI tool to interact with it (`https://developer.hashicorp.com/consul/docs/install`).

The following command adds the necessary key/value pairs:

```
curl -X PUT -d 'admin' http://localhost:8500/v1/kv/config/users-service/db/
username
curl -X PUT -d 'mysecretpassword' http://localhost:8500/v1/kv/config/users-
service/db/password
```

It's important to note that we are using a very specific way to add these key/value pairs. We are following the Spring Cloud Consul convention, which requires this syntax:

```
config/<application-name>/<your-properties>
config/<application-name>,<profiles>/<your-properties>
```

So, in this case, we are using users-service as the application name (spring. application.name) and db.username and db.password as properties. These properties are for our database.

Also, add the following properties (we are using the consul CLI in this case):

```
consul kv put config/users-service/user/reportFormat PDF
consul kv put config/users-service/user/emailSubject 'Welcome to the Users
Service!'
consul kv put config/users-service/user/emailFrom 'users@email.com'
consul kv put config/users-service/user/emailTemplate '
Thanks for choosing Users Service
We have a REST API that you can use to integrate with your Apps.
Thanks from the Users App team.'
```

Also, you can use the Web UI, shown in Figure 13-2, by pointing your browser to http://localhost:8500.

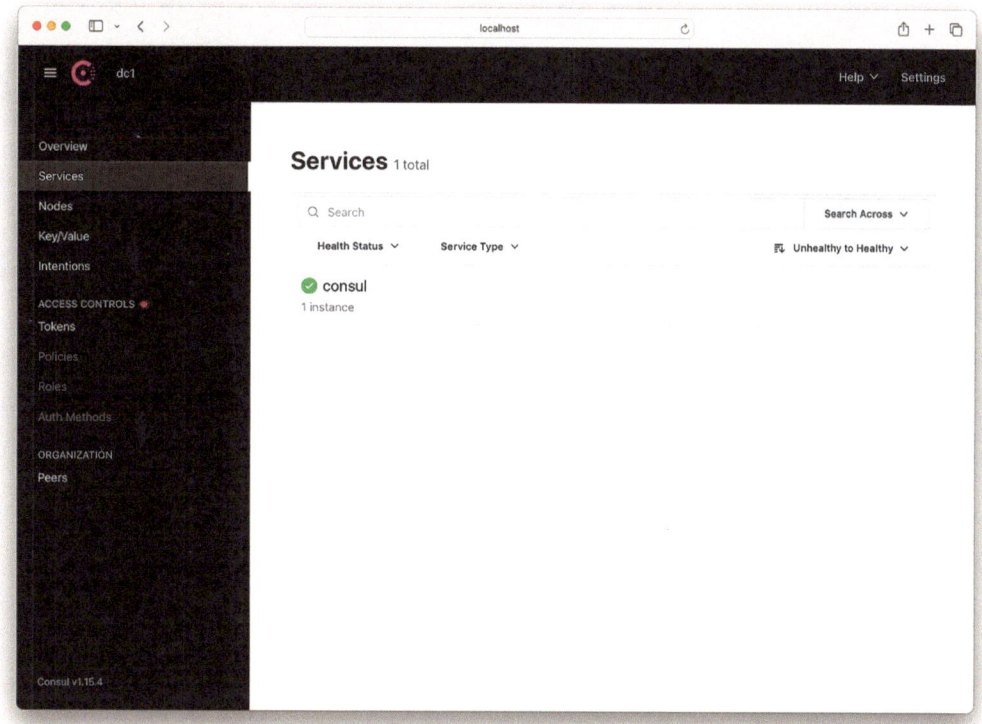

Figure 13-2. *HashiCorp Consul Web UI (http://localhost:8500)*

If you look at the Key/Value section, you should see the db.* and user.* properties defined, as shown in Figure 13-3.

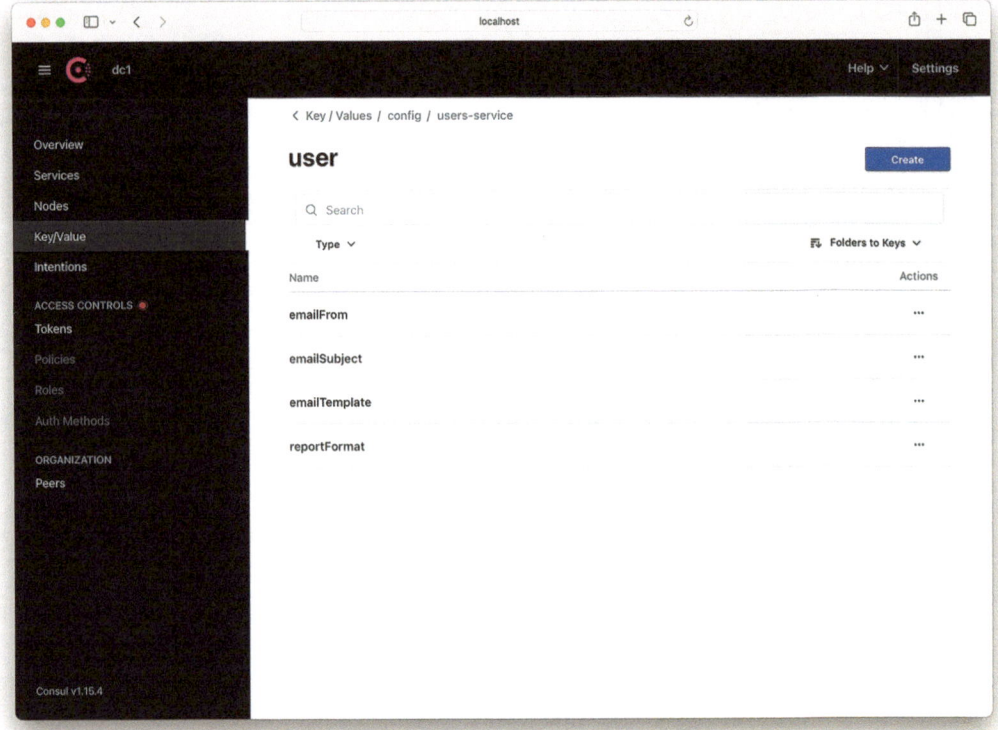

Figure 13-3. *Consul Key/Value section (config/users-service/user properties)*

Now, let's review what Spring Cloud Consul technology is and how it can help us to use HashiCorp Consul.

Spring Cloud Consul

Spring Cloud Consul is a library that seamlessly integrates Spring Boot applications with HashiCorp Consul. It leverages Consul's capabilities to simplify service discovery, communication, and configuration management within microservices architectures built with Spring.

Some of the key features of Spring Cloud Consul are

- *Simplified service discovery*:

    - Automatically register and discover Spring Boot applications with Consul.

- Utilize annotations to integrate service discovery features without manual configuration.

- Supports Spring Cloud LoadBalancer for intelligent routing and Ribbon for client-side load balancing.

- Integrates with Spring Cloud Gateway for dynamic API gateway routing.

- *Robust configuration management*:

  - Leverage Consul's key-value store for centralized configuration management.

  - Use Spring Environment to access configuration values stored in Consul.

  - Dynamically update configurations across all services without individual redeploys.

- *Secure service communication*:

  - Enable strong mutual TLS (mTLS) encryption by leveraging Consul's built-in security features.

  - Enforce identity-based authorization for secure access control.

- *Distributed control bus*:

  - Utilize Consul's events to trigger actions across the microservice environment.

  - Send and receive events for coordination and notification purposes.

- *Integration with other Spring Cloud projects*:

  - Works seamlessly with other Spring Cloud projects, such as Spring Cloud Netflix for advanced features like Hystrix resilience.

The benefits of using Spring Cloud Consul include the following:

- *Reduced development time*: Simplifies service discovery and configuration management through annotations and autoconfiguration.

- *Improved developer experience*: Provides a familiar Spring approach to working with Consul, minimizing the learning curve.

- *Enhanced reliability and scalability*: Leverages Consul's high availability and multi-datacenter support for robust microservices.

- *Increased security*: Enforces secure communication and authorization within your microservices architecture.

In essence, Spring Cloud Consul bridges the gap between Spring Boot applications and Consul, empowering developers to build resilient, scalable, and secure microservices with ease.

Using Spring Cloud Consul in the Users App

In this section, we are going to add Spring Cloud Consul to our Users App and review the service discovery feature and the external configuration. External configuration using tools such as Consul and Spring Cloud Consul provides significant advantages for microservices architectures. It fosters increased agility, enhanced security, improved maintainability, and simplified scaling, contributing to the overall success of your microservice ecosystem.

Let's start with the Users App. The source code for this chapter is in the `13-cloud/users` folder. If you want to start from scratch with the Spring Initializr (`https://start.spring.io`), set the Group field to `com.apress` and the Artifact and Name fields to `users` and add as dependencies JPA, PostgreSQL, Web, Validation, Actuator, Consul Configuration, Consul Discovery, and Lombok. Leave all other settings to their default. Then, generate and download the project, unzip it, and import it into your favorite IDE.

Let's start by opening the `build.gradle` file See Listing 13-1.

Listing 13-1. build.gradle

```
plugins {
    id 'java'
    id 'org.springframework.boot' version '3.2.2'
    id 'io.spring.dependency-management' version '1.1.4'
    id 'org.hibernate.orm' version '6.1.7.Final'
    id 'org.graalvm.buildtools.native' version '0.9.20'
}
```

```
group = 'com.apress'
version = '0.0.1-SNAPSHOT'

java {
    sourceCompatibility = '17'
}

configurations {
    compileOnly {
        extendsFrom annotationProcessor
    }
}

ext {
    set('springCloudVersion', "2023.0.0")
}

repositories {
    mavenCentral()
}

dependencyManagement {
    imports {
        mavenBom "org.springframework.cloud:spring-cloud-dependencies:${spr
        ingCloudVersion}"
    }
}

dependencies {
    implementation 'org.springframework.boot:spring-boot-starter-web'
    implementation 'org.springframework.boot:spring-boot-starter-
    validation'
    implementation 'org.springframework.boot:spring-boot-starter-data-jpa'
    implementation 'org.springframework.boot:spring-boot-starter-actuator'

    // Consult
  implementation 'org.springframework.cloud:spring-cloud-starter-
  consul-config'
      implementation 'org.springframework.cloud:spring-cloud-starter-consul-
discovery'
```

```
  runtimeOnly 'org.postgresql:postgresql'

    compileOnly 'org.projectlombok:lombok'
    annotationProcessor 'org.projectlombok:lombok'
    annotationProcessor 'org.springframework.boot:spring-boot-
    configuration-processor'

    // Web
    implementation 'org.webjars:bootstrap:5.2.3'

    // Test
    testImplementation 'org.springframework.boot:spring-boot-starter-test'
}

tasks.named('test') {
    useJUnitPlatform()
}

hibernate {
    enhancement {
        lazyInitialization true
        dirtyTracking true
        associationManagement true
    }
}
```

Listing 13-1 shows that we are using the Maven Bom (bill of materials) that allows us to get the spring-cloud-dependencies that we need, which in this case are the spring-cloud-starter-consul-config (for external configuration) and the spring-cloud-starter-consul-discovery (for service discovery).

Next, create/open the UserProperties class. See Listing 13-2.

Listing 13-2. src/main/java/com/apress/users/config/UsersProperties.java

```
package com.apress.users.config;

import lombok.Data;
import org.springframework.boot.context.properties.ConfigurationProperties;
```

```
@Data
@ConfigurationProperties(prefix = "user")
public class UserProperties {

    private String reportFormat;
    private String emailSubject;
    private String emailFrom;
    private String emailTemplate;

}
```

Listing 13-2 shows that we are marking the UserProperties class as @ConfigurationProperties, meaning that the value will be bound at startup. These values are kept in an external storage, in this case, the Consul server. That's why we needed to create, for example, the config/users-service/user/reportFormat key and its value PDF.

Next, open/create the application.yaml file. See Listing 13-3.

Listing 13-3. src/main/resources/application.yaml.

```
spring:
  application:
    name: users-service
  config:
    import: consul://

  datasource:
    url: jdbc:postgresql://localhost:5432/users_db?sslmode=disable
    username: ${db.username}
    password: ${db.password}
  jpa:
    generate-ddl: true
    show-sql: true
    hibernate:
      ddl-auto: update

info:
  developer:
    name: Felipe
    email: felipe@email.com
```

```
  api:
    version: 1.0
management:
  endpoints:
    web:
     exposure:
        include: health,info,event-config,shutdown,configprops,beans
  endpoint:
    configprops:
      show-values: always
    health:
      show-details: always
      status:
        order: events-down, fatal, down, out-of-service, unknown, up
    shutdown:
      enabled: true

  info:
    env:
      enabled: true
server:
  port: ${PORT:8080}
```

The application.yaml file includes the following:

- spring.application.name: This property is a *must* because the
  configuration is based on a naming convention, so the name must be
  set here. In this case, we set the name to users-service.

- spring.config.import: This property informs Spring Boot that part
  of the configuration is located in the Consul server, so the value is set
  to consul://.

- spring.datasource.username and spring.datasource.password:
  These properties refer to ${db.username} and ${db.password},
  respectively. The values will be taken from the key/value config/
  users-service/db/username and config/users-service/db/
  password, respectively.

- `management.endpoint.configprops.show-values`: By default, the
  `/configpros` actuator endpoint masks all the properties' values, but
  we are setting this key to `always` to show the properties because we
  are going to see a very cool feature for the configuration properties.

We've previously covered the other properties in Listing 13-3, so we won't revisit
them here.

Running the Users App with Spring Consul

To run the Users App, we need to run the PostgreSQL database first. So, execute the
following command to start the PostgreSQL database:

```
docker run --name postgres --rm \
 -p 5432:5432 \
 --platform linux/amd64 \
 -e POSTGRES_PASSWORD=mysecretpassword \
 -e POSTGRES_USER=admin \
 -e POSTGRES_DB=users_db \
 -d postgres
```

After the DB is up and running, you can run your application. Set the PORT to 8091,
either by using your IDE (every IDE has a way to set the environment variables) or by
executing the following command:

```
PORT=8091 ./gradlew bootRun
```

Everything should work! If you look at the Services section of the Consul UI (http://
localhost:8500/ui/dc1/services), you should see users-service listed (with "1
instance" below it). See Figure 13-4.

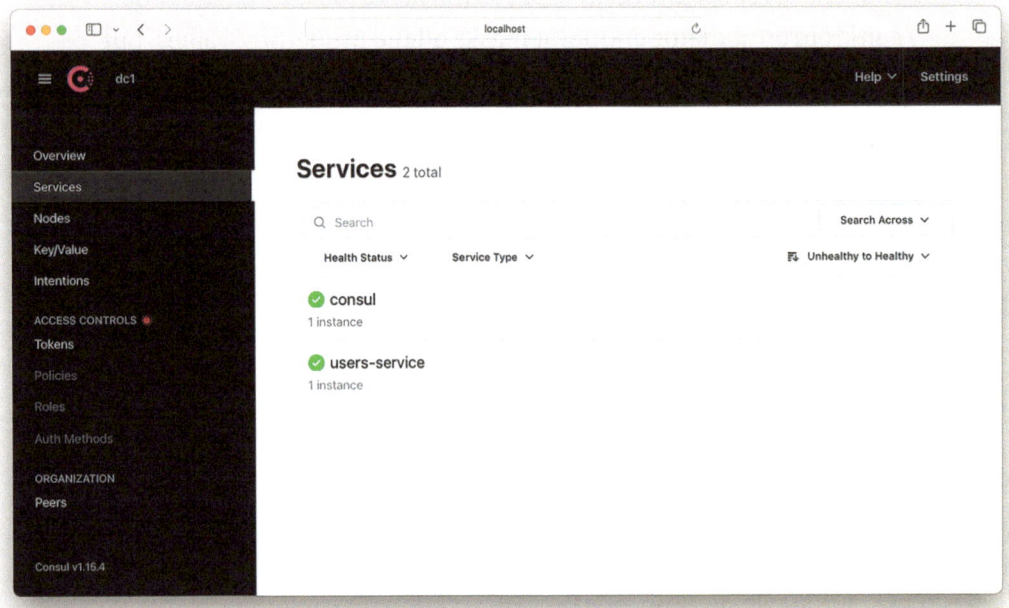

Figure 13-4. *Consul UI Services section (`http://localhost:8500/ui/dc1/`*
`services`)

So, what happened? Because we added the `spring-cloud-starter-consul-`
`discovery` dependency, Spring Boot and the Spring Cloud Consul auto-configuration
automatically registered this service in the HashiCorp Consul, making this available to
be discovered by other services. And because we added the `spring-cloud-starter-`
`consul-config` dependency and we set the `spring.config.import` key to `consul://`,
it will get all the properties from the Consul Config storage based on the naming
convention `config/users-service/*` properties. So, when it sees the `db.username` the
value will be picked up from `consul://config/users-service/db/username` or when
it sees the `user.reportFormat` (from the `UserProperties` class) will get its value from
`consul://config/users-service/user/reportFormat`. Very nice, right?

Now, going back to the browser, if you click `users-service`, you will see something
like Figure 13-5.

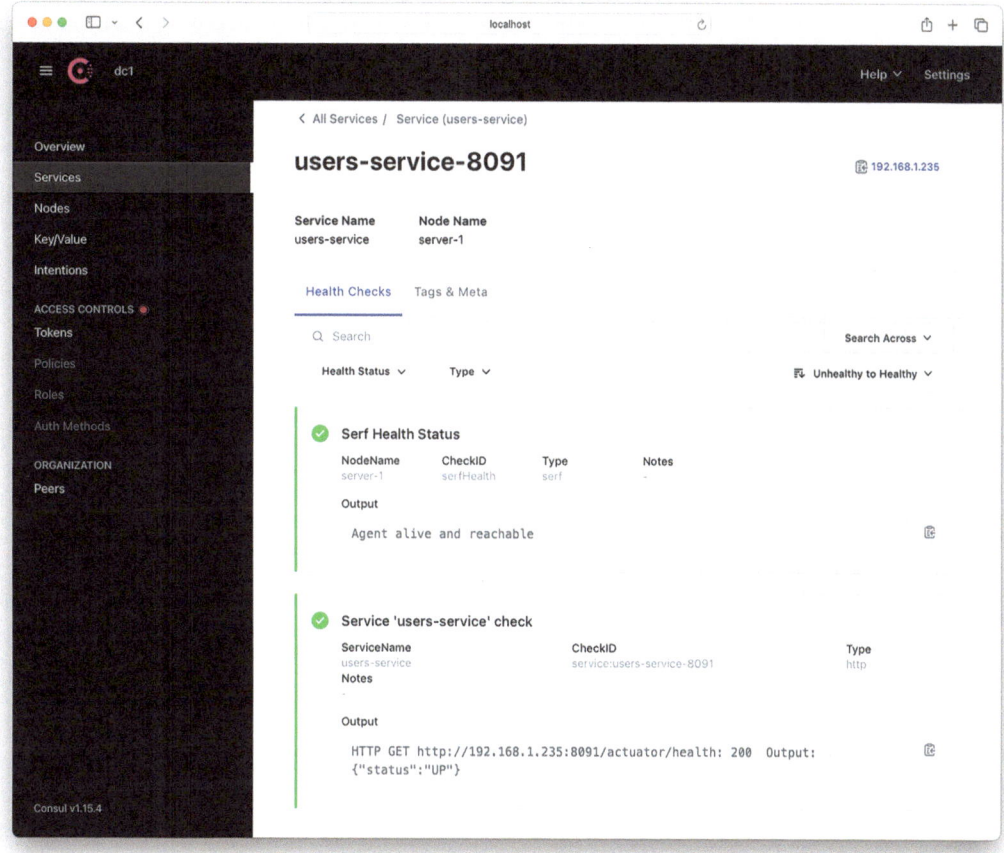

Figure 13-5. *Consul UI displaying users-service-8091*

Consul uses the `spring-boot-actuator` endpoint (`/actuator/health`) to know if the service is alive or not.

So, what happens if you have multiple instances of the Users App? Well, you can open another terminal and run another instance using the following command:

```
PORT=8092 ./gradlew bootRun
```

If you take a look at the Services section of the Consul UI again, you will see it updated to appear as shown in Figure 13-6, with "2 instances" under `users-service`. Click `users-service` and you will see the two services listed, as shown in Figure 13-7.

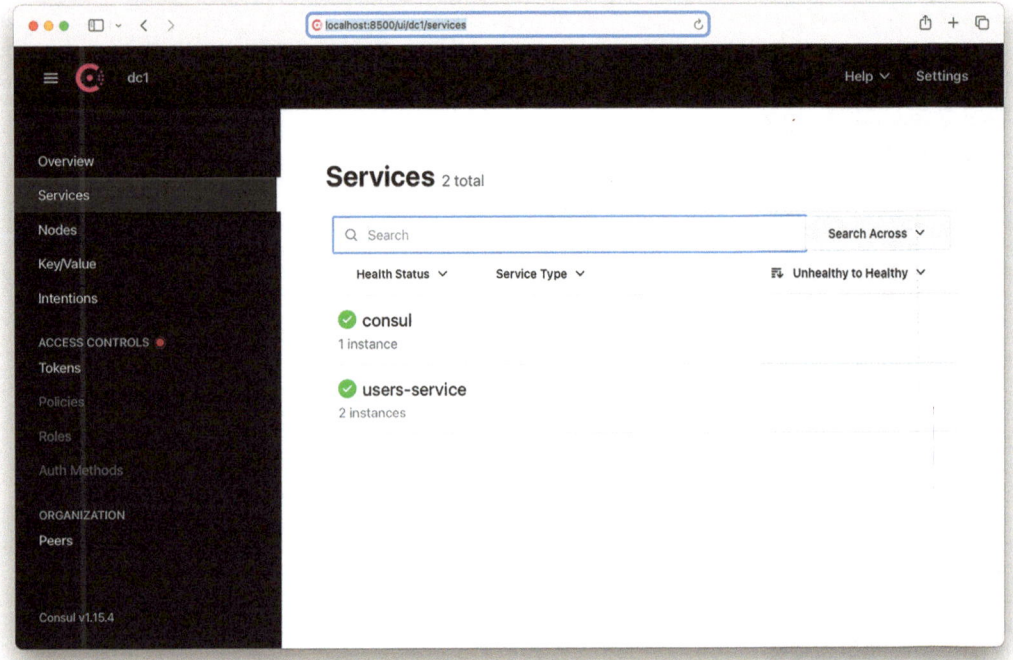

Figure 13-6. *Updated Consul UI Services section showing users-service (2 instances)*

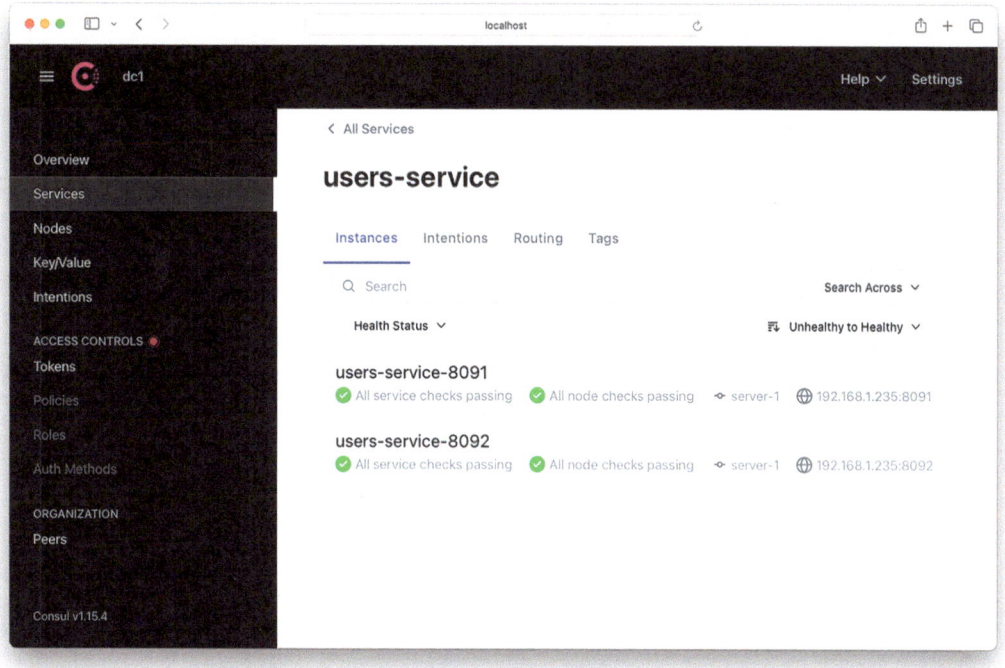

Figure 13-7. *Consul UI displaying users-services-8091 and users-service-8092*

Reviewing the UserProperties?

Next, let's review our UserProperties class using the actuator. Point your browser to the http://localhost:8091/actuator/configprops endpoint. As shown in Figure 13-8, you should see the values we entered before (In Consul).

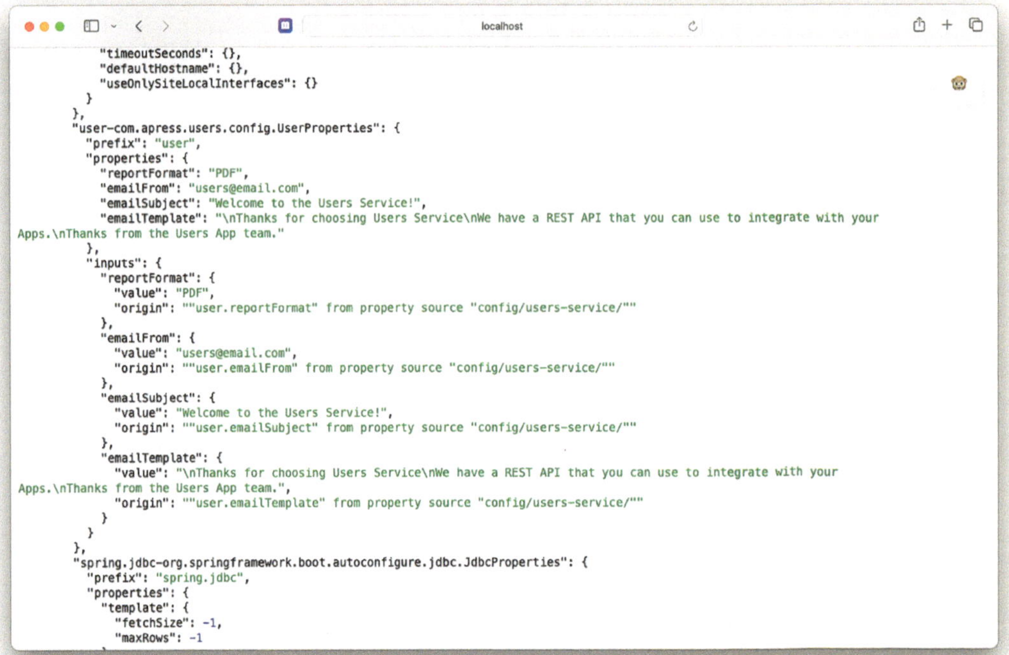

Figure 13-8. `http://localhost:8091/actuator/configprops`

If you need to change any property when the app is running, you will need to restart the app. But one of the nice features of using the combination of Spring Boot and Spring Cloud Consul is that Spring Boot provides the @RefreshScope annotation that allows you to change the value of a property and Spring Boot will take care of re-creating the bean that has this marker and reflect the new value. So, for example, suppose you have something like the following:

```
@Configuration
public class UserConfigurationExample {

    @Value("${message}")
    private String message;

    ...
}
```

This configuration will look for config/users-service/message, and it won't change even though you changed it in Consul. This is because it was set and used at the beginning of the app. So, if you need to change this behavior and use a new value set, then you need to use the @RefreshScope annotation:

```
@RefreshScope
@Configuration
public class UserConfigurationExample {

    @Value("${message}")
    private String message;

    ...
}
```

Using a YAML format Instead of a Key/Value Pair As a Configuration

If you review the application.yaml file again (see Listing 13-3), you might wonder whether you can format everything as YAML in Consul. The answer is *yes*, you can add a whole YAML blob instead of key/value pairs. To activate this feature, you need to set the property format to YAML in the application.yaml file:

```
spring:
  cloud:
    consul:
      config:
        format: YAML
```

Then you need to add a data key and set all the YAML. So, you need to add the config/users-service/data key and add the following as a value:

```
spring:
  datasource:
    url: jdbc:postgresql://localhost:5432/users_db?sslmode=disable
    username: admin
    password: mysecretpassword
```

```yaml
  jpa:
    generate-ddl: true
    show-sql: true
    hibernate:
      ddl-auto: update

info:
  developer:
    name: Felipe
    email: felipe@email.com
  api:
    version: 1.0

management:
  endpoints:
    web:
     exposure:
       include: health,info,event-config,shutdown,configprops,beans
    endpoint:
      configprops:
        show-values: always
      health:
        show-details: always
        status:
          order: events-down, fatal, down, out-of-service, unknown, up
      shutdown:
        enabled: true

    info:
      env:
        enabled: true

user:
  reportFormat: PDF
  emailSubject: 'Welcome to the User Services'
  emailFrom: 'user@email.com'
  emailTemplate: 'Thanks for choosing Users Service'
```

Now we are adding the database `username` and `password` in the yaml (`config/users-service/data value`) and the `user.*` properties. See Figure 13-9.

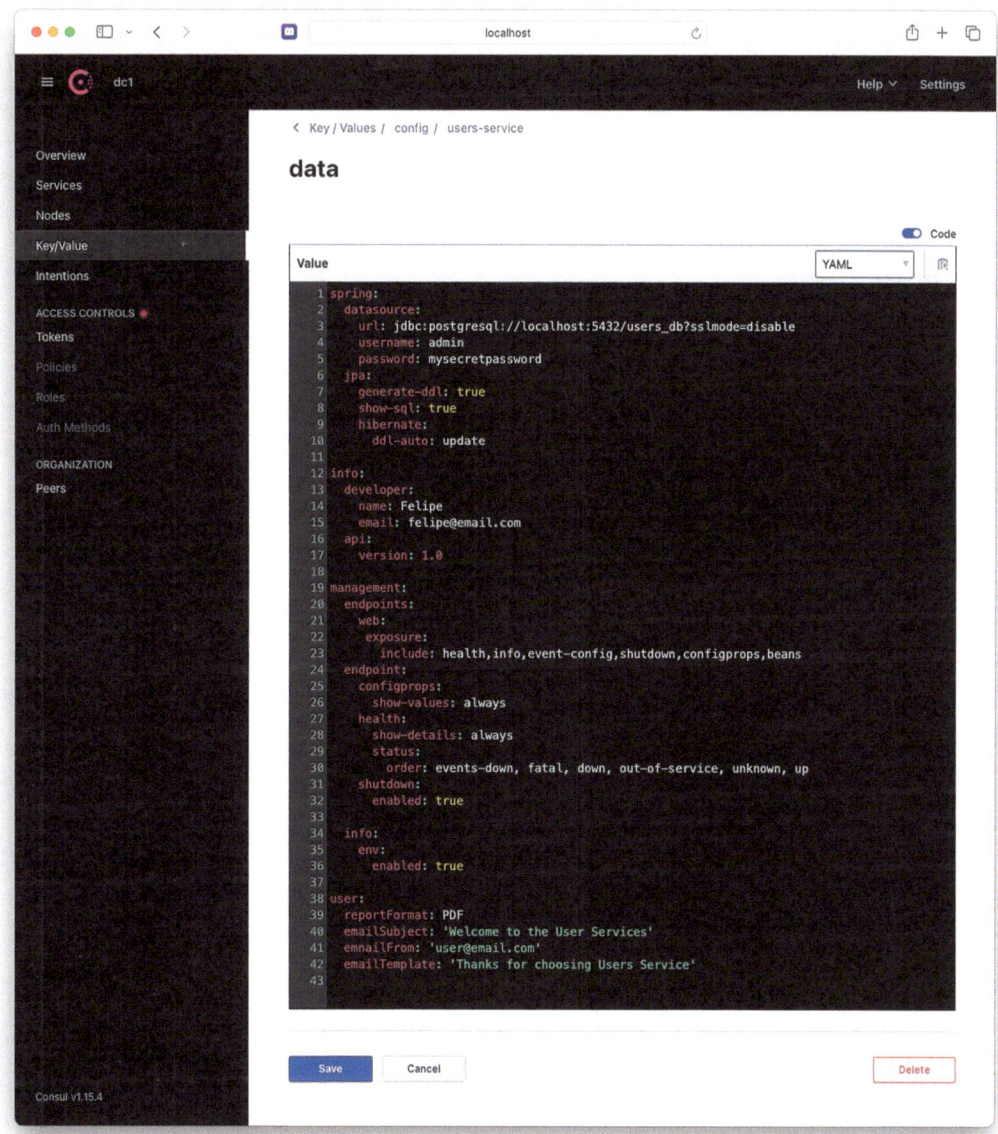

Figure 13-9. *Consul UI Key/Value section displaying config/users-service/ data key*

Your `application.yaml` file will have the content shown in Listing 13-4.

Listing 13-4. src/main/resources/application.yaml

```
spring:
  application:
    name: users-service
  config:
    import: consul://
  cloud:

    config:
      format: YAML

server:
  port: ${PORT:8080}
```

As you can see, which a minimal configuration, everything is externalized now. And, of course, you can play with the Spring Profiles (e.g. test, dev, prod, etc.) and make this more configurable.

Using Spring Cloud Consul in the My Retro App

Now we are going to add Spring Cloud Consul capabilities to the My Retro App. The source code is in the 13-cloud/myretro folder. We are going to reuse the code from the Spring Boot Actuator chapter (Chapter 11) and make a few modifications. If you want to start from scratch with the Spring Initializr (https://start.spring.io), set the Group field to com.apress and the Artifact and Name fields to myretro, and add as dependencies JPA, H2, Web, Validation, Actuator, Consul Configuration, Consul Discovery, OpenFeign, and Lombok. Leave all other settings at their default. Then, generate and download the project, unzip it, and import it into your favorite IDE.

Start by opening the build.gradle file. See Listing 13-5.

Listing 13-5. build.gradle

```
plugins {
    id 'java'
    id 'org.springframework.boot' version '3.2.2'
    id 'io.spring.dependency-management' version '1.1.4'
```

```
    id 'org.hibernate.orm' version '6.1.7.Final'
    id 'org.graalvm.buildtools.native' version '0.9.20'
}

group = 'com.apress'
version = '0.0.1-SNAPSHOT'
sourceCompatibility = '17'

configurations {
    compileOnly {
        extendsFrom annotationProcessor
    }
}

repositories {
    mavenCentral()
}

ext {
    set('springCloudVersion', "2023.0.0")
}

dependencies {
    implementation 'org.springframework.boot:spring-boot-starter-web'
    implementation 'org.springframework.boot:spring-boot-starter-
    validation'
    implementation 'org.springframework.boot:spring-boot-starter-data-jpa'
    implementation 'org.springframework.boot:spring-boot-starter-actuator'

    // Consul
    implementation 'org.springframework.cloud:spring-cloud-starter-
    consul-config'
    implementation 'org.springframework.cloud:spring-cloud-starter-consul-
    discovery'

    // OpenFeign
    implementation 'org.springframework.cloud:spring-cloud-starter-
    openfeign'
```

```
    // Kubernetes
    implementation 'org.springframework.cloud:spring-cloud-starter-
    kubernetes-client-all'

    runtimeOnly 'com.h2database:h2'
    runtimeOnly 'io.micrometer:micrometer-registry-prometheus'

    compileOnly 'org.projectlombok:lombok'
    annotationProcessor 'org.projectlombok:lombok'
    annotationProcessor 'org.springframework.boot:spring-boot-
    configuration-processor'

    testImplementation 'org.springframework.boot:spring-boot-starter-test'
}

dependencyManagement {
    imports {
        mavenBom "org.springframework.cloud:spring-cloud-dependencies:${spr
        ingCloudVersion}"
    }
}

tasks.named('test') {
    useJUnitPlatform()
}
```

Listing 13-5 shows that we are adding to build.gradle not only the spring-cloud-start-consul dependencies but also the spring-cloud-starter-openfeign and spring-cloud-starter-kubernetes-client-all dependencies. We are going to discuss the Kubernetes client in the last section of the chapter, which will clarify why we needed to include it here.

If you remember, the My Retro App has a way to communicate with the Users App, and in Chapter 12, we used the @HttpExchange and @GetExchange annotations that simplify the way to connect to other services' APIs. This time we are going to use a different approach.

Using OpenFeign in My Retro App

Spring Cloud OpenFeign is a library that simplifies consuming RESTful web services by declaratively defining client interfaces annotated with Spring MVC annotations. It automatically generates clients based on these interfaces, handling tasks like these:

- HTTP request mapping based on annotations like @GetMapping and @PostMapping

- Parameter decoding and encoding using Spring's HttpMessageConverters

- Feign interceptor integration for custom request/response processing

- Hystrix fault tolerance for resilient service calls

The following are some of the key features of Spring Cloud OpenFeign:

- *Declarative interface-based clients*: Write interfaces, not low-level HTTP code, for a cleaner and more concise approach.

- *Spring MVC annotations*: Leverage familiar Spring MVC annotations for intuitive syntax and integration.

- *Default encoder and decoder support*: Uses Spring's default message converters for seamless data handling.

- *Support for Feign interceptors*: Integrate custom interceptors for additional processing and logging.

- *Hystrix integration*: Automatically enables Hystrix fault tolerance for resilient service calls.

- *Load balancing support*: Integrates with Spring Cloud LoadBalancer for intelligent service discovery and routing.

- *Service discovery integration*: Works with various service discovery tools like Eureka and Consul.

- *OAuth2 Support*: Simplify secure communication with OAuth2-protected services.

- *Micrometer support*: Monitor and collect metrics from Feign clients using Micrometer.

- *Spring Data support*: Use Spring Data repositories directly with Feign clients for convenient data access.

- *Spring @RefreshScope support*: Dynamically update Feign client configurations while the application is running.

Some of the benefits of Spring Cloud OpenFeign are

- *Reduced development time*: Declarative approach and annotations simplify client development.

- *Improved maintainability*: Cleaner code with a clearer separation of concerns.

- *Increased reusability*: Interfaces promote code reuse and easier sharing.

- *Automatic Hystrix resilience*: Built-in fault tolerance for robust service communication.

- *Seamless Spring integration*: Leverages familiar Spring concepts and tools.

Spring Cloud OpenFeign helps developers build resilient, maintainable, and scalable RESTful clients for microservices architectures. If you want to know more about it, go to `https://spring.io/projects/spring-cloud-openfeign`.

We'll use Consul for the My Retro App as well, and it needs a way to connect to the Users App. For that purpose, we are going to take advantage of the service discovery and load balancing features that OpenFeign and Consul provide. Keep in mind that we have two instances of the Users Service App up and running; the My Retro App will connect to both of them.

Next, open/create the `UserClient` interface. See Listing 13-6.

Listing 13-6. src/main/java/com/apress/myretro/client/UserClient.java

```
package com.apress.myretro.client;

import com.apress.myretro.client.model.User;
import org.springframework.cloud.openfeign.FeignClient;
import org.springframework.http.ResponseEntity;
import org.springframework.web.bind.annotation.GetMapping;
import org.springframework.web.bind.annotation.PathVariable;
```

```
@FeignClient(name = "users-service")
public interface UserClient {

    @GetMapping("/users")
        ResponseEntity<Iterable<User>> getAllUsers();

        @GetMapping("/users/{email}")
        ResponseEntity<User> getById(@PathVariable String email);
}
```

Let's review the annotations:

- @FeignClient: This annotation passes the name of the service, in this case users-service, the service registered in Consul.

- @GetMapping: You already know this annotation, and as you can see, we are defining the same as previously in the UsersController, in this case just two methods, one to get all users and the other to find users by email.

Next, open/create the RetroBoardConfig class. See Listing 13-7.

Listing 13-7. src/main/java/com/apress/myretro/config/ RetroBoardConfing.java

```
package com.apress.myretro.config;

import com.apress.myretro.board.Card;
import com.apress.myretro.board.CardType;
import com.apress.myretro.board.RetroBoard;
import com.apress.myretro.service.RetroBoardAndCardService;
import org.springframework.boot.CommandLineRunner;
import org.springframework.cloud.openfeign.EnableFeignClients;
import org.springframework.context.annotation.Bean;
import org.springframework.context.annotation.Configuration;

import java.util.Arrays;

@EnableFeignClients(basePackages = "com.apress.myretro.client")
@Configuration
```

```
public class RetroBoardConfig {

    @Bean
    CommandLineRunner init(RetroBoardAndCardService service){
        return args -> {
            //... some code here
        };
    }

}
```

Listing 13-7 shows that the RetroBoardConfig class has a new annotation:
@EnableFeignClients(basePackages = "com.apress.myretro.client"). This a way
to tell the auto-configuration where all the clients are, and even though there is only one
client, this is how you set it, or you can use the clients parameter and add the actual
class UserClient.class as the value.

Next, we need to configure the UserClient interface to enable the My Retro App to
communicate with the Users App. Open/create the RetroBoardAndCardService class.
See Listing 13-8.

Listing 13-8. src/main/java/com/apress/myretro/service/
RetroBoardAndCardService.java

```
package com.apress.myretro.service;

import com.apress.myretro.board.RetroBoard;
import com.apress.myretro.client.UserClient;
import com.apress.myretro.events.RetroBoardEvent;
import com.apress.myretro.events.RetroBoardEventAction;
import com.apress.myretro.exceptions.RetroBoardNotFoundException;
import com.apress.myretro.persistence.RetroBoardRepository;
import io.micrometer.core.instrument.Counter;
import lombok.AllArgsConstructor;
import org.springframework.context.ApplicationEventPublisher;
import org.springframework.stereotype.Service;

import java.time.LocalDateTime;
import java.util.UUID;
```

```java
@AllArgsConstructor
@Service
public class RetroBoardAndCardService {

    private RetroBoardRepository retroBoardRepository;
    private ApplicationEventPublisher eventPublisher;
    private Counter retroBoardCounter;

    private UserClient userClient;

    // more code here...

    // External Call
    public void getAllUsers(){
                userClient.getAllUsers().getBody().forEach(System.
                out::println);
        }
}
```

The important piece in Listing 13-8 is that we are injecting the UserClient interface
and we just using it in the getAllUsers method (and we are just printing the values).

Next, open/create the RetroBoardController class. This is the method that exposes
the /retros/users endpoint with the following code (no need to show everything):

```java
@GetMapping("/users")
    public ResponseEntity getAllUsers(){
        this.retroBoardAndCardService.getAllUsers();
        return ResponseEntity.ok(Map.of("status",200,"message","Users
        retrieved. Should be in the console","time",LocalDateTime.now()));
    }
```

This code will call the service and print out the users in the console.

Next, open the application.properties file. See Listing 13-9.

Listing 13-9. src/main/resources/application.properties

```properties
## DataSource
spring.h2.console.enabled=true
spring.datasource.generate-unique-name=false
spring.datasource.name=test-db
```

```
spring.jpa.show-sql=true

## Server
server.port=${PORT:8080}
```

Consul
spring.config.import=consul://

```
## Application
```
spring.application.name=my-retro-app
```
logging.pattern.correlation=[${spring.application.name:},%X{traceId:-
},%X{spanId:-}]

## Actuator Info
info.developer.name=Felipe
info.developer.email=felipe@email.com
info.api.version=1.0
management.endpoint.env.enabled=true

## Actuator
management.endpoints.web.exposure.include=health,info,metrics,prometheus

## Actuator Observations
management.observations.key-values.application=${spring.application.name}

## Actuator Metrics
management.metrics.distribution.percentiles-histogram.http.server.
requests=true

## Actuator Tracing
management.tracing.sampling.probability=1.0

## Actuator Prometheus
management.prometheus.metrics.export.enabled=true
management.metrics.use-global-registry=true
```

The important part here in the application.properties is to specify the name of the app, spring.application.name=my-retro-app, and the spring.config. import=consul:// that will use Consul as external configuration (even though we haven't specified anything).

Running the My Retro App

It's time to run the My Retro App. Make sure the Consul server and client are still up and running as well as the two instances of the Users Service App.

Set the PORT to 8081 (in case we need more than one instance). You can run it using your IDE (don't forget to set the PORT environment variable) or you can use the command line:

```
PORT=8081 ./gradlew bootRun
```

Once it is up and running, you can look at the Consul UI at http://localhost:8500/ui/dc1/services. See Figure 13-10.

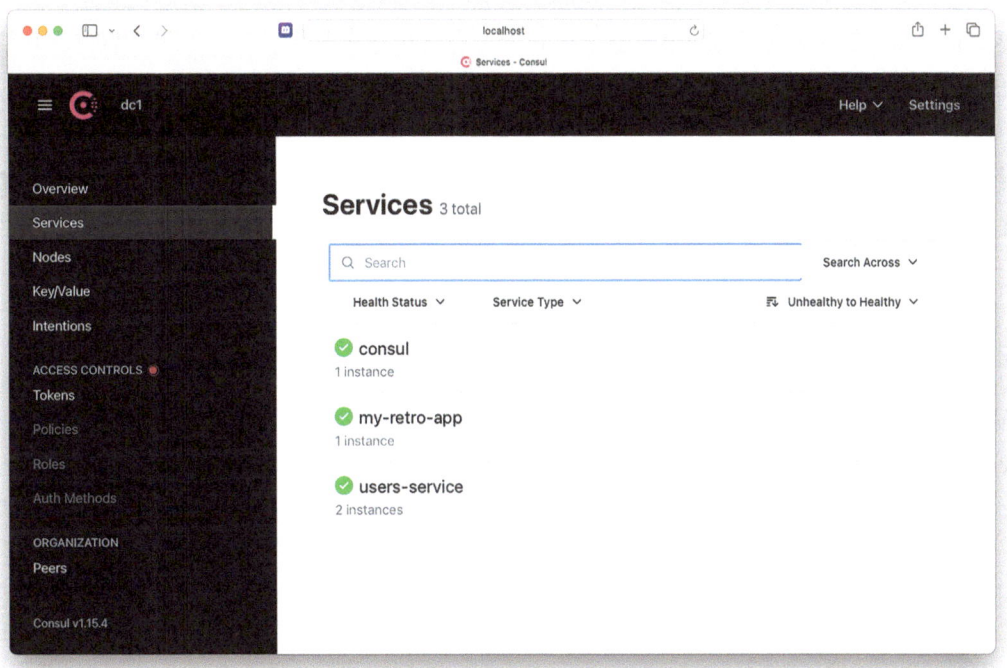

Figure 13-10. *Consul UI Services section showing both my-retro-app and users-service*

Now my-retro-app is listed. And if you call the /retros/users endpoint, you will see the users have been printed out.

So, how does OpenFeign know where to locate `users-service`? If there are two instances of the Users App, which one will be selected? This is the beauty of Spring Boot, Spring Cloud Consul, and Spring Cloud OpenFeign; the only thing you need to declare is the name of the service (`users-service`) that you want to consume, then Consul will do the service discovery and tell OpenFeign where to go, and in this case, because there are two instances, behind the scenes OpenFeign will use a load balancer using round robin to reach these instances. This is so amazing! It requires minimal code—just let Spring Boot do the job!

HashiCorp Vault and Spring Cloud Vault

Next, let's take a look at HashiCorp Vault (`https://www.vaultproject.io/`) and Spring Cloud Vault and how they can help us with our solution. First we'll review what HashiCorp Vault is, and then use it in our two projects.

Note   In this discussion, "Vault" without "HashiCorp" or "Spring Cloud" preceding it always refers to HashiCorp Vault.

HashiCorp Vault

HashiCorp Vault is an identity-based secrets and encryption management system. It serves as a central repository for securely storing, accessing, and controlling access to sensitive data such as:

- API keys and passwords

- Certificates and encryption keys

- Database credentials

- Infrastructure credentials

Vault addresses various security challenges across diverse environments, including:

- *Microservices*: Securely store and manage service credentials, API keys, and database connections.

- *Cloud deployments*: Rotate and manage cloud platform credentials centrally across different providers.

- *Data encryption*: Encrypt data at rest and in transit using managed encryption keys.

- *Automated infrastructure*: Securely automate infrastructure provisioning and configuration with dynamic secrets.

- *Compliance and auditing*: Maintain auditable access logs and enforce granular access control for secrets.

Vault offers a rich set of features for secure secrets management:

- *Secure storage*: Utilizes encryption at rest and in transit to protect sensitive data.

- *Identity-based access control*: Enforces granular access control based on user identities and roles.

- *Dynamic secrets*: Generates and rotates secrets automatically based on predefined policies.

- *Secret leasing*: Issues temporary access to secrets with defined expiration times.

- *Audit logging*: Logs all access attempts and data revisions for auditing and compliance.

- *Integration with diverse tools*: Connects with popular DevOps and infrastructure platforms.

- *Multi-datacenter deployments*: Scales and replicates across multiple data centers for high availability.

Some of Vault's benefits are

- *Enhanced security*: Centralized control and strong access control mitigate security risks.

- *Operational efficiency*: Streamlines secrets management and automate workflows.

- *Improved compliance*: Simplified audit logging and access control enforcement.

- *Increased scalability*: Scales seamlessly to accommodate growing needs.

> **Caution**   Before continuing, *remove* the db.username and db.password
> properties from Consul.

Using HashiCorp Vault with Credentials Creation for PostgreSQL

Let's start now with HashiCorp Vault. The idea is to have a special user role that can be created and rotate their credentials often (depending on our configuration), so we can make sure this service is secure and the credentials are not shared. Practically, Vault will rotate the credentials and Spring Cloud Vault will get the new credentials and use them accordingly. We are still using Consul, and we can configure Vault to use its storage and take advantage of the high availability that Consul provides.

Start the Vault server with the following command:

```
docker run --cap-add=IPC_LOCK -d --rm --name vault -p 8200:8200 \
-e 'VAULT_LOCAL_CONFIG={"backend": {"consul": {"address": "host.docker.inte
rnal:8500","path":"vault/"}}}' \
-e 'VAULT_DEV_ROOT_TOKEN_ID=my-root-id' \
vault:1.13.3
```

This command starts Vault with Consul as storage. Also, we are defining a TOKEN_ID with the value my-root-id. This is just development, but normally you need to generate this key (TOKEN_ID) and its value, and use them as an authentication mechanism. If you take a look at the Consul UI, you should see Vault listed as a service. See Figure 13-11.

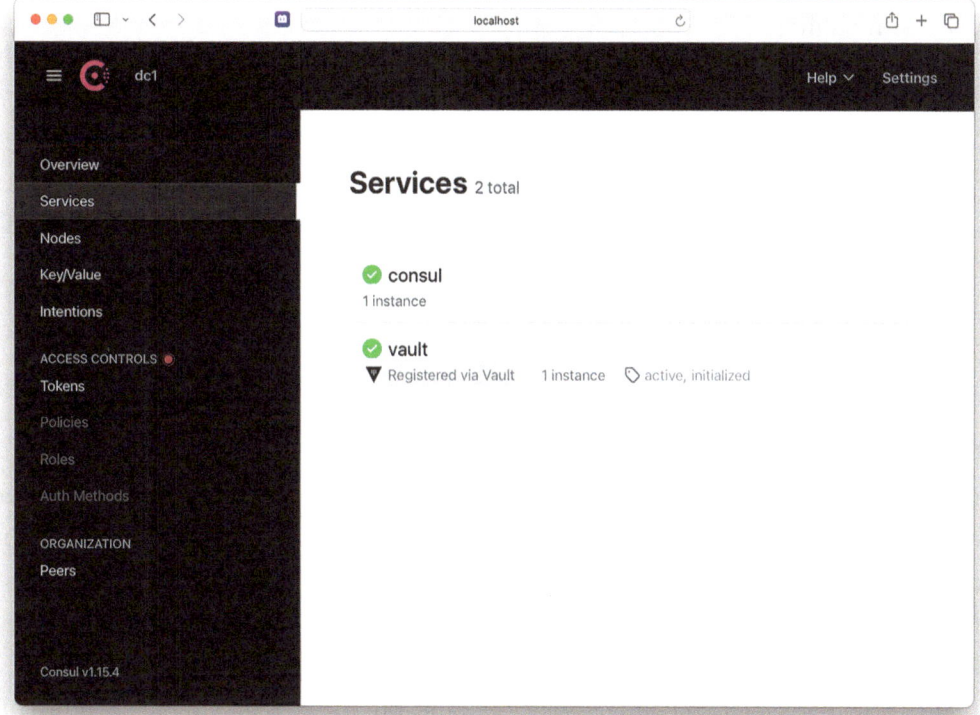

Figure 13-11. *Consul UI Service section showing Vault listed (`http://
localhost:8500/ui/dc1/services`)*

And if you take a look at the Key/Value section in Consul, you should see the *Vault*
key configurations.

Next, we need to configure Vault to use a database secret and rotate our credentials
based on a default policy. To do this, you can either use cURL to point to its rest API
or use the `vault` CLI tool. You can install the CLI tool from `https://developer.
hashicorp.com/vault/docs/install`. Then you can use the following command to log
in to Vault:

```
vault login -address="http://127.0.0.1:8200"
```

The TOKEN will be `my-root-id`. This login is required to execute the next
commands.

Next, enable the Database Secrets engine with

```
vault secrets enable -address="http://127.0.0.1:8200" database
```

Next, create the connection to the database and use the PostgreSQL plugin that comes within Vault:

```
vault write -address="http://127.0.0.1:8200" \
    database/config/users_db \
    plugin_name=postgresql-database-plugin \      connection_
    url="postgresql://{{username}}:{{password}}@host.docker.internal:5432/
    users_db?sslmode=disable" \
    allowed_roles="*" \
    username="admin" \
    password="mysecretpassword"
```

One of the main things to notice is that connection_url is pointing to the host.docker.internal, and this is because we are reaching out outside the container, so we need to know the docker host IP. It can't be localhost, because PostgreSQL has its IP but is exporting the access through port 5432.

Next, we need to create a role that will do all the operations in our database, and its credentials will be rotated by Vault based on its default policy (in this case, just a few minutes). Of course, you can change these defaults, but for now just to have a proof of concept here, we are going to leave the default. Execute the following command:

```
vault write -address="http://127.0.0.1:8200" \
    database/roles/users-role db_name=users_db \
    creation_statements="CREATE ROLE \"{{name}}\" WITH LOGIN PASSWORD
    '{{password}}' VALID UNTIL '{{expiration}}'; GRANT ALL PRIVILEGES ON
    DATABASE users_db TO \"{{name}}\"; GRANT USAGE ON SCHEMA public TO
    \"{{name}}\"; GRANT ALL PRIVILEGES ON ALL TABLES IN SCHEMA public TO
    \"{{name}}\"; ALTER DATABASE users_db OWNER TO \"{{name}}\";" \
    default_ttl="30s" \
    max_ttl="1m"
```

The preceding command basically says create a role (users-role) and enable it to operate on the users_db. We are using SQL statements to grant some privileges to this user role that will be created when needed.

Vault will rotate its credentials, meaning that the user will no longer be useful if you use its previous credentials. If you want to know about the users and their credentials, you can execute the following command:

```
vault read -address="http://127.0.0.1:8200" \
database/creds/users-role

Key                 Value
---                 -----
lease_id            database/creds/users-role/t8vRNfXvONqy15u41EQqqX3Q
lease_duration      30s
lease_renewable     true
password            ci8HZxkgJJn-KIvMRWoe
username            v-token-users-ro-2wtaGdop4qbR63NMsI1L-1707512396
```

In the output, a username and a password were created. Now, you are all set. You can access the Vault UI by going to `http://localhost:8200`. See Figure 13-12.

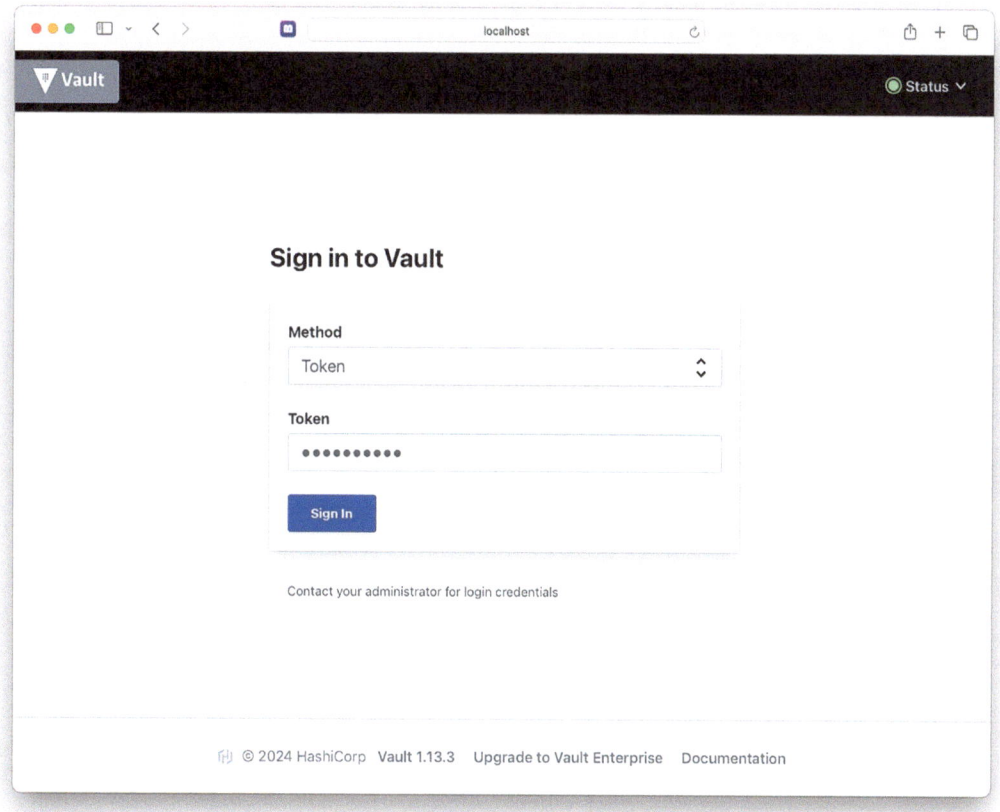

Figure 13-12. *Vault UI login page (`http://localhost:8200` - Token/my-root-id)*

The Token is `my-root-id`. Then, you can browse around and see the Secrets Engines (see Figure 13-13) and the Database Engine (see Figure 13-14).

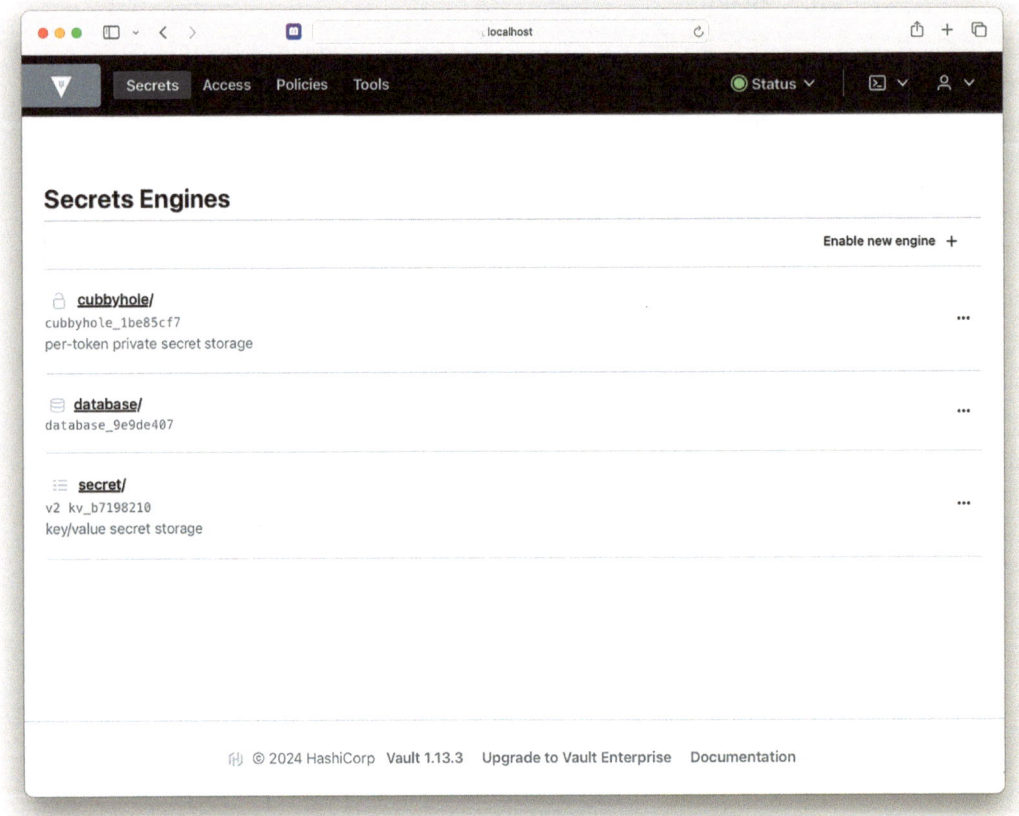

Figure 13-13. *Vault UI Secret Engines page*

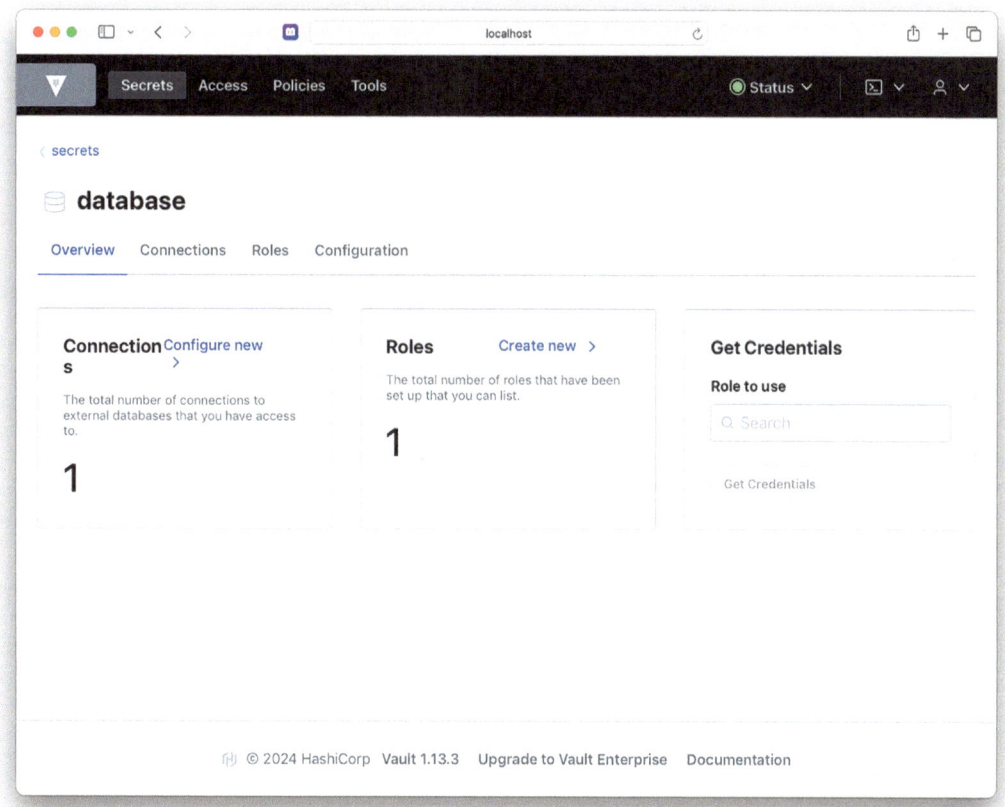

Figure 13-14. *Vault UI Database Engine*

On the Secrets Engines page, click the Enable New Engine link to see what other secrets engines are available (see Figure 13-15). In our case, we are just going to use the standard database.

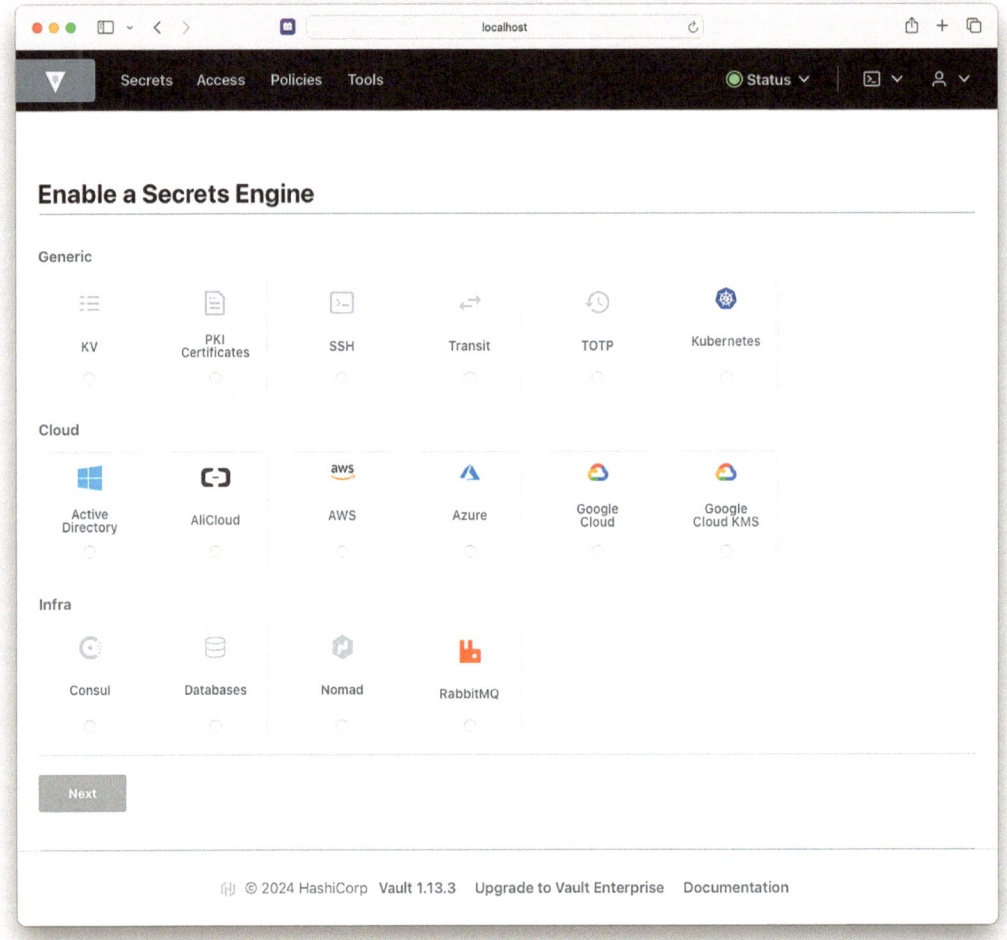

Figure 13-15. *Vault UI showing list of available secrets engines (*`http://`
`localhost:8200/ui/vault/settings/mount-secret-backend`*)*

Spring Cloud Vault

Now that we have Vault up and running, let's review what Spring Cloud Vault is and how it can help us with our applications.

Spring Cloud Vault is a project that helps you integrate your Spring applications with HashiCorp Vault to manage secrets securely. Essentially, Spring Cloud Vault provides client-side support for externalized configuration using Vault as the central source. Let's dive into its benefits and features:

Spring Cloud Vault offers the following benefits (among others):

- *Improved security*: By storing sensitive information like passwords, API keys, and other credentials in Vault instead of in your application code, you significantly reduce the risk of exposure and potential breaches.

- *Centralized management*: Vault offers a single platform to manage secrets across all your applications and environments, simplifying operations and ensuring consistency.

- *Dynamic provisioning*: Spring Cloud Vault can dynamically generate credentials for various services like databases, cloud platforms, and more, ensuring they are always fresh and secure.

- *Reduced code complexity*: You don't need to embed secrets directly in your code, eliminating the need for hardcoding and improving code maintainability.

- *Enhanced configurability*: Easily manage configuration changes and rollbacks without impacting your application deployments.

And these are some of its features:

- *Secret retrieval*: Retrieves secrets from Vault and automatically refreshes them based on configured policies.

- *Environment initialization*: Initializes the Spring environment with remote property sources populated from Vault.

- *Secure communication*: Supports secured communication with Vault using SSL and various authentication mechanisms.

- *Credential generation*: Generates dynamic credentials for different services like MySQL, PostgreSQL, AWS, and others.

- *Multiple authentication methods*: Supports tokens, AppId, AppRole, Client Certificate, Cubbyhole, AWS EC2 & IAM, and Kubernetes authentication for secure access.

- *Cloud Foundry integration*: Enables integration with HashiCorp's Vault service broker through Spring Cloud Vault Connector.

Implementing Credential Generation on the Users Service App Using HashiCorp Vault

We are going to use the Users App with some modifications. Remember that you have access to the source code in the 13-cloud/users folder so you can follow along.

So, the first thing we need to do is add the necessary dependencies to the build. gradle file. In this case, add the spring-cloud-vault-config-databases and spring-cloud-starter-vault-config dependencies, as shown in Listing 13-10.

Listing 13-10. build.gradle

```
plugins {
    id 'java'
    id 'org.springframework.boot' version '3.2.2'
    id 'io.spring.dependency-management' version '1.1.4'
    id 'org.hibernate.orm' version '6.1.7.Final'
    id 'org.graalvm.buildtools.native' version '0.9.20'
}

group = 'com.apress'
version = '0.0.1-SNAPSHOT'

java {
    sourceCompatibility = '17'
}

configurations {
    compileOnly {
        extendsFrom annotationProcessor
    }
}

ext {
    set('springCloudVersion', "2023.0.0")
}

repositories {
    mavenCentral()
}
```

```
dependencyManagement {
    imports {
        mavenBom "org.springframework.cloud:spring-cloud-dependencies:${spr
        ingCloudVersion}"
    }
}

dependencies {
    implementation 'org.springframework.boot:spring-boot-starter-web'
    implementation 'org.springframework.boot:spring-boot-starter-
    validation'
    implementation 'org.springframework.boot:spring-boot-starter-data-jpa'
    implementation 'org.springframework.boot:spring-boot-starter-actuator'

    // Consult
    implementation 'org.springframework.cloud:spring-cloud-starter-
    consul-config'
    implementation 'org.springframework.cloud:spring-cloud-starter-consul-
    discovery'

    // Vault
    implementation 'org.springframework.cloud:spring-cloud-vault-config-
    databases'
            implementation 'org.springframework.cloud:spring-cloud-starter-
            vault-config'

    runtimeOnly 'org.postgresql:postgresql'

    compileOnly 'org.projectlombok:lombok'
    annotationProcessor 'org.projectlombok:lombok'
    annotationProcessor 'org.springframework.boot:spring-boot-
    configuration-processor'

    // Web
    implementation 'org.webjars:bootstrap:5.2.3'

    // Test
    testImplementation 'org.springframework.boot:spring-boot-starter-test'
}
```

```
tasks.named('test') {
    useJUnitPlatform()
}

tasks.named("bootBuildImage") {
    builder = "dashaun/builder:tiny"
    environment = ["BP_NATIVE_IMAGE" : "true"]
}

hibernate {
    enhancement {
        lazyInitialization true
        dirtyTracking true
        associationManagement true
    }
}
```

Next, let's modify the application.yaml. Add/replace the content as shown in Listing 13-11.

Listing 13-11. src/main/resources/application.yaml

```
spring:
  application:
    name: users-service
  config:
    import: consul://,vault://

  cloud:
    vault:
      authentication: TOKEN
      token: ${VAULT_TOKEN}  # my-root-id
      scheme: http
      database:
        enabled: true
        role: users-role
      config:
        lifecycle:
          enabled: true
```

```
        min-renewal: 10s
        expiry-threshold: 1m
      fail-fast: true

  datasource:
    url: jdbc:postgresql://localhost:5432/users_db?sslmode=disable

  jpa:
    generate-ddl: true
    show-sql: true
    hibernate:
      ddl-auto: update

info:
  developer:
    name: Felipe
    email: felipe@email.com
  api:
    version: 1.0

management:
  endpoints:
    web:
     exposure:
       include: health,info,event-config,shutdown,configprops,beans
    endpoint:
      configprops:
        show-values: always
      health:
        show-details: always
        status:
          order: events-down, fatal, down, out-of-service, unknown, up
      shutdown:
        enabled: true

    info:
      env:
        enabled: true
```

```
logging:
  level:
    org.springframework.vault: ERROR

server:
  port: ${PORT:8080}
```

Let's review what is new and what changed in the `application.yaml` file:

- `spring.config.import`: Because we are using Vault, we need to get the credentials (set in the `spring.datasource.username` and `spring.datasource.password` properties) from `vault://` as well, and we are still using Consul for the other properties, so we can declare both such as `consul://`,`vault://`.

- `spring.cloud.vault.*`: All of this section is for Vault. First, we need to tell Vault how we are going to authenticate, which in this case is using `TOKEN` and the token value that points to a `${VAULT_TOKEN}` environment variable that has the value `my-root-id`. Then, we have the `database.enabled` and `database.role` properties that uses the `users-role` we previously created using the command line. And some extra properties are considered as default.

- `spring.datasource.url`: This URL only points to the database `users_db`, and if you look closely, there is no *username* or *password*, because Vault will generate the credentials and grant permissions to the user created, and Spring Cloud Vault will gather these credentials and add them to the `spring.datasource.username` and `spring.datasource.password` properties automatically.

The other properties remain the same. And that's it. It was more configuration than any code, right?

Running the Users App with Vault

Before you run the Users App, make sure to set the `VAULT_TOKEN=my-root-id` environment variable either in your IDE (if you are running it from there) or by executing the following command:

```
PORT=8091 VAULT_TOKEN=my-root-id ./gradlew bootRun
```

Everything should work now! The credentials for the database were created by Vault and Spring Cloud Vault set them in our applications. But wait! What happens if you execute a `curl` command or point your browser to `http://localhost:8091/users`?

After a few minutes, it will fail with an error like this:

```
org.postgresql.util.PSQLException: ERROR: permission denied for
table people
```

Do you know why? That's right. Vault has the credential rotation feature, so it has already changed it after a minute or two; this is based on a policy that you can change in Vault. So, how can we fix this issue?

Fixing Credential Rotation by Adding a Listener

One of the cool features of Vault and Spring Cloud Vault is that there are listeners that can send events about any change, and in this case, we can add one when a credential rotation happens. Spring Cloud Vault provides the `SecretLeaseContainer` class that will help to detect when Vault renews its lease/rotation.

Open/create the `UserConfiguration` class. See Listing 13-12.

Listing 13-12. src/main/java/com/apress/users/config/UserConfiguration.java

```java
package com.apress.users.config;

import com.apress.users.model.User;
import com.apress.users.model.UserRole;
import com.apress.users.service.UserService;
import com.zaxxer.hikari.HikariConfigMXBean;
import com.zaxxer.hikari.HikariDataSource;
import com.zaxxer.hikari.HikariPoolMXBean;
import jakarta.annotation.PostConstruct;
import lombok.RequiredArgsConstructor;
import lombok.extern.slf4j.Slf4j;
import org.springframework.beans.factory.annotation.Value;
import org.springframework.boot.CommandLineRunner;
import org.springframework.boot.context.properties.
EnableConfigurationProperties;
import org.springframework.context.annotation.Bean;
```

```java
import org.springframework.context.annotation.Configuration;
import org.springframework.vault.core.lease.SecretLeaseContainer;
import org.springframework.vault.core.lease.domain.RequestedSecret;
import org.springframework.vault.core.lease.event.SecretLeaseCreatedEvent;
import org.springframework.vault.core.lease.event.SecretLeaseExpiredEvent;

import java.util.Arrays;

@Slf4j
@RequiredArgsConstructor
@Configuration
@EnableConfigurationProperties({UserProperties.class})
public class UserConfiguration {

    @Bean
    CommandLineRunner init(UserService userService){
        return args -> {
            userService.saveUpdateUser(new User("ximena@email.
            com","Ximena","https://www.gravatar.com/avatar/23bb62a7d
            0ca63c9a804908e57bf6bd4?d=wavatar","aw2s0meR!", Arrays.
            asList(UserRole.USER),true));
            userService.saveUpdateUser(new User("norma@email.com","Norma" ,
            "https://www.gravatar.com/avatar/f07f7e553264c9710105edebe
            6c465e7?d=wavatar", "aw2s0meR!", Arrays.asList(UserRole.USER,
            UserRole.ADMIN),false));

        };
    }

    @Value("${spring.cloud.vault.database.role}")
    private String databaseRoleName;
    private final SecretLeaseContainer secretLeaseContainer;
    private final HikariDataSource hikariDataSource;

    @PostConstruct
    private void postConstruct() {
        final String vaultCredentialsPath = String.format("database/
        creds/%s", databaseRoleName);
```

```
secretLeaseContainer.addLeaseListener(event -> {
    log.info("[SecretLeaseContainer]> Received event: {}", event);

    if (vaultCredentialsPath.equals(event.getSource().getPath())) {
        if (event instanceof SecretLeaseExpiredEvent &&
                event.getSource().getMode() == RequestedSecret.
                Mode.RENEW) {
            log.info("[SecretLeaseContainer]> Let's replace the
            RENEWED lease by a ROTATE one.");
            secretLeaseContainer.requestRotatingSecret(vaultCredent
            ialsPath);
        } else if (event instanceof SecretLeaseCreatedEvent
        secretLeaseCreatedEvent &&
                event.getSource().getMode() == RequestedSecret.
                Mode.ROTATE) {
            String username = secretLeaseCreatedEvent.getSecrets().
            get("username").toString();
            String password = secretLeaseCreatedEvent.getSecrets().
            get("password").toString();

            updateHikariDataSource(username, password);
        }
    }
});
}

private void updateHikariDataSource(String username, String password) {

    log.info("[SecretLeaseContainer]> Soft evict the current database
    connections");
    HikariPoolMXBean hikariPoolMXBean = hikariDataSource.
    getHikariPoolMXBean();
    if (hikariPoolMXBean != null) {
        hikariPoolMXBean.softEvictConnections();
    }
```

```
        log.info("[SecretLeaseContainer]> Update database credentials with
        the new ones.");
        HikariConfigMXBean hikariConfigMXBean = hikariDataSource.
        getHikariConfigMXBean();
        hikariConfigMXBean.setUsername(username);
        hikariConfigMXBean.setPassword(password);
    }
}
```

The UserConfiguration class will help to identify the renewed credentials and rotate them. Let's review it:

- @RequiredArgsConstructor: This annotation (introduced in Chapter 3) from the Lombok library allows us to define final fields that will be used as parameters for the constructor, and they will be injected by Spring. The fields are databaseRoleName, SecretLeaseContainer, and HikariDataSource (this is the Spring Boot default when a database is used).

- @Value("${spring.cloud.vault.database.role}"): This is used when we are going to get the vault://database/creds/<role>; in this case, the value is the users-role defined in the application. yaml file and its content that we created with the vault CLI.

- SecretLeaseContainer: This class is an event-based container that requests secrets from Vault and renews the lease that is associated with the secret. We are adding a lease listener so that we get a notification from Vault every time there is a new/renewed lease. We need to ask for the SecretLeaseExpiredEvent and the secret Mode (RENEW or ROTATE); if it's RENEW, we need to request the rotating secret, and if it's ROTATE, we need to get the new username and password and update the DataSource.

- HikariDataSource/HikariPoolMXBean: By default, Spring Boot uses the HikariDataSource as DataSource, so we need to use HikariPoolMXBean to update the username and password.

Then, if you run the Users App again and wait for a minute or two, you will see the logs about receiving the `SecretLeaseContainer` event, and then the logic to renew the rotated credentials by updating it using the `DataSource` (`HikariDataSource`). This is so cool! Now, you know how to rotate credentials.

If you want to know more about the policies you can apply to Vault, take a look here: `https://developer.hashicorp.com/vault/docs/concepts/policies`.

Next, it's time to finish the cloud architecture (shown previously in Figure 13-1) by adding an API gateway.

Spring Cloud Gateway

Spring Cloud Gateway is an API gateway built on top of the Spring ecosystem, and its purpose is to simplify API routing and add valuable features to your microservices architecture. We can define it as

- A Java-based API gateway built on Spring Framework, Spring Boot, and Project Reactor

- Acts as a single entry point for your microservices, routing requests to the appropriate service based on various criteria

- Provides "cross-cutting concerns" like security, monitoring, and resiliency at the gateway level

Some of Spring Cloud Gateway's key features are

- *Routing*: Defines routes based on URL path, headers, methods, and more.

- *Predicates*: Control which requests match specific routes using built-in or custom predicates.

- *Filters*: Add preprocessing and post-processing logic to requests and responses.

- *Security*: Integrate with Spring Security for authentication and authorization.

- *Monitoring*: Tracks metrics and provides health checks for the gateway and downstream services.

- *Resiliency*: Integrates with *Spring Cloud Circuit Breaker* for fault tolerance and fallback mechanisms.

- *Discovery*: Integrates with Spring Cloud `DiscoveryClient` for automatic service discovery of your microservices.

- *Other features*: Includes rate limiting, path rewriting, Hystrix integration, and much more.

Some of the Spring Cloud Gateway use cases are

- *Single entry point*: Simplifies access to your microservices by providing a unified API endpoint.

- *Security*: Enforces centralized security for all API access.

- *Monitoring*: Provides insights into API traffic and health of your microservices.

- *Resiliency*: Protects against service failures and ensures high availability.

- *Load balancing*: Distributes traffic among multiple instances of your microservices.

- *API management*: Enables you to control access, rate limit requests, and collect usage data.

To summarize, Spring Cloud Gateway is a powerful tool for managing and enhancing your microservices architecture. It offers flexible routing, robust security, and valuable monitoring capabilities to streamline your API interactions.

Creating a My Retro Gateway

You can find this section's code in the `13-cloud/myretro-gateway` folder. If you want to start from scratch with Spring Initializr (`https://start.spring.io`), set the Group field to `com.apress` and the Artifact and Name fields to `myretro-gateway` and add as dependencies Actuator, Consul Configuration, Consul Discovery, Reactive Gateway, and Resilience4J Circuit Breaker. Leave all other settings at their default. Then, generate and download the project, unzip it, and import it into your favorite IDE.

Let's review the `build.gradle` file. See Listing 13-13.

Listing 13-13. build.gradle

```
plugins {
    id 'java'
    id 'org.springframework.boot' version '3.2.2'
    id 'io.spring.dependency-management' version '1.1.4'
}

group = 'com.apress'
version = '0.0.1-SNAPSHOT'

java {
    sourceCompatibility = '17'
}

repositories {
    mavenCentral()
}

ext {
    set('springCloudVersion', "2023.0.0")
}

dependencies {
    // Actuator
    implementation 'org.springframework.boot:spring-boot-starter-actuator'

    // Gateway
    implementation 'org.springframework.cloud:spring-cloud-starter-gateway'
    implementation 'org.springframework.cloud:spring-cloud-starter-
    circuitbreaker-reactor-resilience4j'

    // Consul
    implementation 'org.springframework.cloud:spring-cloud-starter-
    consul-config'
    implementation 'org.springframework.cloud:spring-cloud-starter-consul-
    discovery'
```

```
    // Kubernetes
    implementation 'org.springframework.cloud:spring-cloud-starter-
kubernetes-fabric8-all'

    testImplementation 'org.springframework.boot:spring-boot-starter-test'
    testImplementation 'io.projectreactor:reactor-test'
}

dependencyManagement {
    imports {
        mavenBom "org.springframework.cloud:spring-cloud-dependencies:${spr
        ingCloudVersion}"
    }
}

tasks.named('test') {
    useJUnitPlatform()
}
```

Listing 13-13 shows that we need to include the dependencies `spring-cloud-starter-gateway`, `spring-cloud-starter-circuitbreaker-reactor-resilience4j` (discussed a bit later in the chapter), and `spring-cloud-starter-kubernetes-fabric8-all` (which becomes relevant in the final section, "Using a Cloud Platform: Kubernetes"). Also, we added the Consul dependencies so that it gets registered into the Consul server automatically. One of the cool features of Spring Cloud Gateway is that if you already are using Consul—it can take advantage of local services without any programming.

Spring Cloud Gateway was created with WebFlux in mind, so basically you have the advantage of using this technology either in a regular MVC or in a Reactive web application.

Next, open/create the `application.yaml` file. See Listing 13-14.

Listing 13-14. src/main/resources/application.yaml

```
spring:
  application:
    name: myretro-gateway
```

```
config:
  import: optional:consul://
cloud:
  gateway:
    routes:
      - id: users
        uri: lb://users-service
        predicates:
          - Path=/users/**
      - id: myretro
        uri: lb://my-retro-app
        predicates:
          - Path=/retros/**

server:
  port: ${PORT:8080}
```

Let's review `application.yaml`:

- `spring.application.name`: Remember that you always need this property when you want to use service discovery, and because we are using Consul, this is a must. The value set is `myretro-gateway` that automatically will be registered in Consul with that name.

- `spring.config.import`: When adding the `spring-cloud-starter-consul-config` dependency, we need to specify this property even if we don't use it at all. In our example, we are not going to use Consul as an external configuration, so we can declare it as `optional:`, and that's why we added that at the beginning of its value.

- `spring.cloud.gateway.*`: This includes all the Gateway declaration configurations. In this case, we are defining two routes, one for the `/users` endpoint and the other for the `/retros` endpoint. The `uri` property for the user's route is using the name of the service (in this case `users-service`) but with a prefix `lb://`, meaning that this is how we need to declare the usage of a load balancer; and remember that we have two Users Services apps instances running (on ports 8091 and 8092). We are declaring the `predicates.Path` with the value /users/** and /retros/** for each service. This means that when we access the Gateway with /users, it will redirect to http:// users-service/users (http://localhost:8091/users or http:// localhost:8092/users, depending on the load balancer, but normally is a round-robin scenario).

Of course, there are more configuration options, but we've covered the minimal ones that involve service discovery and load balancing.

And that's it! No need of any programming or anything else.

Running the My Retro Gateway

Before running the My Retro Gateway, make sure you have all the services (consul, vault, users-service, and my-retro-app) up and running and visible in Consul.

To run the My Retro Gateway App, you can use either your IDE or the following command:

```
./gradlew bootRun
```

By default, the app runs on port 8080. If you open the Consul UI (http:// localhost:8500), you can see all the services. See Figure 13-16.

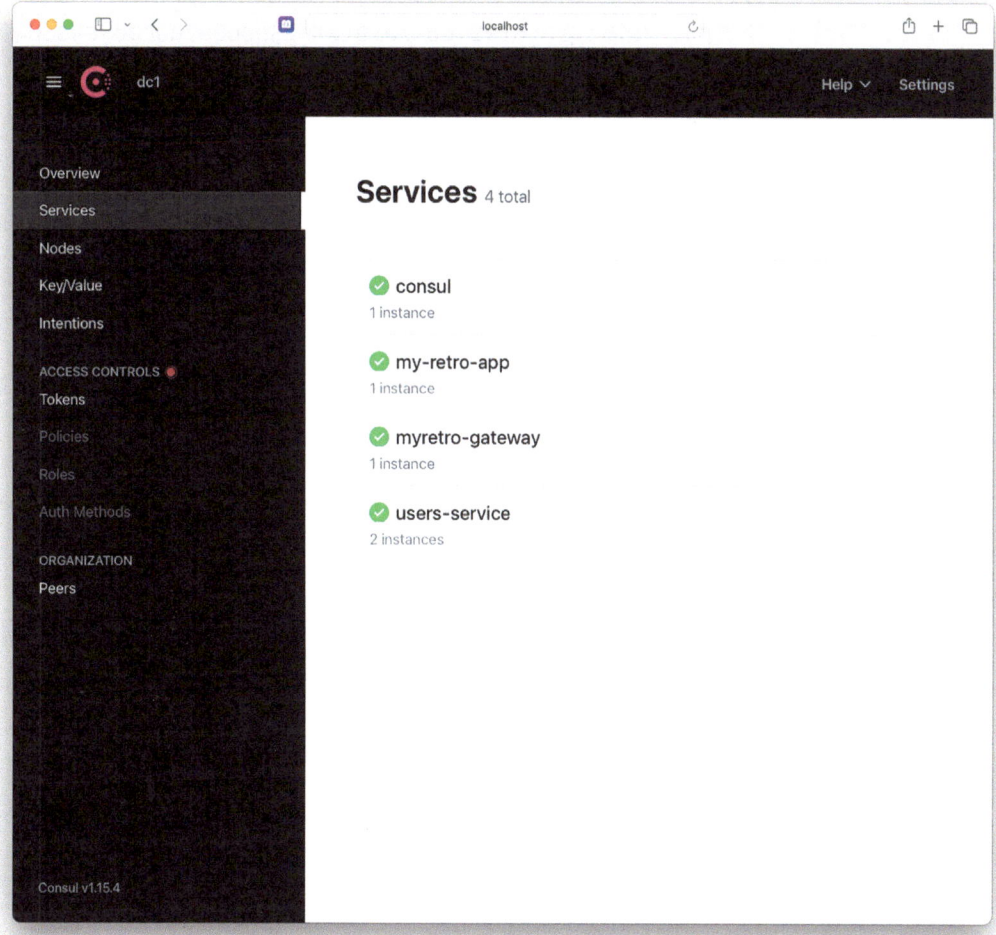

Figure 13-16. *Consul UI Services section with list of all services (`http://localhost:8500/ui/dc1/services`)*

Now that you have the My Retro Gateway up and running, it's time to test it. You can use the browser or the terminal. If you are using the terminal, you can execute the following command to see the users:

```
curl -s http://localhost:8080/users | jq

[
  {
    "email": "ximena@email.com",
```

```
    "name": "Ximena",
    "gravatarUrl": "https://www.gravatar.com/avatar/23bb62a7d0ca63c9a80490
    8e57bf6bd4?d=wavatar",
    "password": "aw2s0meR!",
    "userRole": [
      "USER"
    ],
    "active": true
  },
  {
    "email": "norma@email.com",
    "name": "Norma",
    "gravatarUrl": "https://www.gravatar.com/avatar/f07f7e553264c9710105ede
    be6c465e7?d=wavatar",
    "password": "aw2s0meR!",
    "userRole": [
      "USER",
      "ADMIN"
    ],
    "active": false
  }
]
```

If you execute the following command, you will get the retros:

```
curl -s http://localhost:8080/retros | jq
```

```
[
  {
    "retroBoardId": "4efaf18e-2141-40c5-abff-0c0951d62938",
    "name": "Spring Boot 3 Retro",
    "cards": [
      {
        "cardId": "d9df505f-3564-4104-a339-415db9f4a29a",
        "comment": "Nice to meet everybody",
```

```
        "cardType": "HAPPY",
        "created": "2024-02-12 13:17:30",
        "modified": "2024-02-12 13:17:30"
      },
      {
        "cardId": "3906a3cb-69cc-412f-8965-42979ac91052",
        "comment": "When are we going to travel?",
        "cardType": "MEH",
        "created": "2024-02-12 13:17:30",
        "modified": "2024-02-12 13:17:30"
      },
      {
        "cardId": "fdbce1c8-f9ec-45fd-873b-b4ab43c2aff5",
        "comment": "When are we going to travel?",
        "cardType": "SAD",
        "created": "2024-02-12 13:17:30",
        "modified": "2024-02-12 13:17:30"
      }
    ],
    "created": "2024-02-12 13:17:30",
    "modified": "2024-02-12 13:17:30"
  }
]
```

But what happens if you turn off both users-service apps? To find out, do so, and then execute

```
curl -s http://localhost:8080/users | jq
```

```
{
  "timestamp": "2024-02-12T20:14:12.361+00:00",
  "path": "/users",
  "status": 503,
  "error": "Service Unavailable",
  "requestId": "2a02d2b8-2"
}
```

So, how can we add a default message or a dummy user in case of service disruption? With the Circuit Breaker pattern, discussed in the following section.

More Gateway Features

Spring Cloud Gateway has many more features, such as predicates and filters. Examples of predicates include `After`, `Before`, `Between`, `Cookie`, `Header`, `Host`, `Method`, `Path`, `Query`, `ReadBody`, `RemoteAddr`, `XForwardedRemoteAddr`, `Weight`, `CloudFoundryRouteService`, and custom. The list of filters is long. The following is just a sampling, presented in alphabetical order through the letter *J*: `AddRequestHeader`, `AddRequestHeaderIfNotPresent`, `AddRequestParameter`, `AddResponseHeader`, `CircuitBreaker`, `CacheResponseBody`, `DedupeResponseHeader`, `FallbackHeaders`, `JsonToGrpc`. You also have the option to create a custom filter if none of the filters provide the business logic that you need. Let's review some of these predicates and filters.

CircuitBreaker Filter

To understand the purpose of the `CircuitBreaker` filter, it's helpful to first understand the Circuit Breaker pattern. The Circuit Breaker pattern is a resilience mechanism to protect systems from cascading failures when a dependent service or resource becomes unavailable or overwhelmed. It acts like an automatic switch and can be in one of three states:

- *Closed*: Normal operation; requests are forwarded to the service.

- *Open*: Service failure threshold reached, requests are immediately short-circuited and directed to a fallback mechanism (e.g., cached response, alternative service).

- *Half-Open*: After a timeout, a single request probes the service. If successful, the circuit closes; if not, it stays open longer.

This pattern prevents further burdening the failing service and allows the system to continue functioning gracefully while the issue is resolved. The `CircuitBreaker` filter is based on the Circuit Breaker pattern.

Adding the CircuitBreaker Filter to My Retro Gateway App

Let's add the `CircuitBreaker` filter to the `application.yaml` file in the My Retro Gateway app. You can add the following snippet in the `spring.cloud.gateway.routes users` section:

```
filters:
  - name: CircuitBreaker
    args:
      name: users
      fallbackUri: forward:/fallback/users
```

We are defining the name of the filter to use `CircuitBreaker`, then we are declaring some of the arguments it requires, a name and `fallbackUri`, which indicates that if the `/users` endpoint is not reachable, it will default to the `/fallback/users` endpoint. The complete configuration is presented in Listing 13-15.

Listing 13-15. src/main/resources/application.yaml with CircuitBreaker

```
spring:
  application:
    name: myretro-gateway
  config:
    import: optional:consul://
  cloud:
    gateway:
      routes:
        - id: users
          uri: lb://users-service
          predicates:
            - Path=/users/**
          filters:
            - name: CircuitBreaker
              args:
                name: users
                fallbackUri: forward:/fallback/users
        - id: myretro
          uri: lb://my-retro-app
          predicates:
            - Path=/retros/**
          filters:
```

```
          - name: CircuitBreaker
            args:
              name: retros
              fallbackUri: forward:/fallback/retros

server:
  port: ${PORT:8080}
```

Listing 13-15 shows the final version of the application.yaml file. Talking about the fallbackUri, see that we are declaring a forward to a new route /fallback/users and /fallback/retros. This means that we need to implement these fallbacks. So, open/ create the MyRetroFallbackController class. See Listing 13-16.

Listing 13-16. src/main/java/com/apress/myretrogateway/ MyRetroFallbackController.java

```
package com.apress.myretrogateway;

import org.springframework.http.ResponseEntity;
import org.springframework.web.bind.annotation.GetMapping;
import org.springframework.web.bind.annotation.RequestMapping;
import org.springframework.web.bind.annotation.RestController;

import java.time.LocalDateTime;
import java.util.Map;

@RestController
@RequestMapping("/fallback")
public class MyRetroFallbackController {

    @GetMapping("/users")
    public ResponseEntity userFallback() {
        return ResponseEntity.ok(Map.of(
                "status","Service Down",
                "message","/users endpoint is not available at this moment",
                "time",LocalDateTime.now(),
                "data", Map.of(
                        "email","dummy@email.com",
                        "name","Dummy",
```

```
                    "password","dummy",
                    "active",false)
            ));
    }

    @GetMapping("/retros")
    public ResponseEntity retroFallback() {
        return ResponseEntity.ok(Map.of("status","Service
        Down","message","/retros endpoint is not available at this
        moment","time",LocalDateTime.now()));
    }
}
```

Listing 13-16 shows the MyRetroFallbackController class. You already know this, so no need to explain. Practically you are using a default response, that not necessarily can be with a Map like we have it here, also you can reach out to a different service that you know is never down.

So, if you want to test this, you can shut down the two instances of the users-service app and rerun the My Retro Gateway with the new configuration. Execute the following command over the /users endpoint:

```
curl -s http://localhost:8080/users | jq


{
  "message": "/users endpoint is not available a this moment",
  "status": "Service Down",
  "data": {
    "active": false,
    "name": "Dummy",
    "email": "dummy@email.com",
    "password": "dummy"
  },
  "time": "2024-02-12T16:00:34.087459"
}
```

Now, you can try to shut down the my-retro-app service and call it with /retros and see the message from the controller.

Integraging the Cloud Enbvironment in Docker Compose

So far, we have been dealing with every component in a separate way, so now let's work on everything in this single file. Let's use Docker Compose to set up the whole environment.

Next, let's define the `compose.yaml` file. You can find this file in the `13-cloud/docker-compose` folder. See Listing 13-17.

Listing 13-17. compose.yaml

```
services:

  ## HashiCorp Consul
  consul-server:
    hostname: consul-server
    container_name: consul-server
    image: consul:1.15.4
    restart: always
    networks:
      - cloud
    healthcheck:
      test: ["CMD", "curl", "-X", "GET", "localhost:8500/v1/status/leader"]
      interval: 1s
      timeout: 3s
      retries: 60
    command: "agent -server -ui -node=server-1 -bootstrap-expect=1
    -client=0.0.0.0"

  consul-client:
    hostname: consul-client
    container_name: consul-client
    image: consul:1.15.4
    restart: always
    networks:
      - cloud
    command: "agent -node=client-1 -join=consul-server -retry-
    join=172.17.0.2"
```

```
consul-init:
  hostname: consul-init
  container_name: consul-init
  image: consul:1.15.4
  depends_on:
    consul-server:
      condition: service_healthy
  volumes:
    - ./consul-init.sh:/tmp/consul-init.sh
  networks:
    - cloud
  command: |
    sh -c "/tmp/consul-init.sh"

## PostgreSQL
postgres:
  hostname: postgres
  container_name: postgres
  image: postgres
  platform: linux/amd64
  restart: always
  networks:
    - cloud
  environment:
    POSTGRES_PASSWORD: mysecretpassword
    POSTGRES_USER: admin
    POSTGRES_DB: users_db
  ports:
    - "5432:5432"
  healthcheck:
    test: pg_isready
    interval: 10s
    timeout: 5s
    retries: 5
```

```
  ## Users Service
users-service:
  image: users
    build:
      context: ../users
      dockerfile: Dockerfile
    restart: always
    environment:
      - SPRING_DATASOURCE_URL=jdbc:postgresql://postgres:5432/users_db
      - SPRING_CLOUD_CONSUL_HOST=consul-server
    networks:
      - cloud
    depends_on:
      consul-server:
        condition: service_healthy
      postgres:
        condition: service_healthy

  ## My Retro App
my-retro-app:
  image: myretro
    build:
      context: ../myretro
      dockerfile: Dockerfile
    environment:
      - SPRING_CLOUD_CONSUL_HOST=consul-server
    networks:
      - cloud
    depends_on:
      consul-server:
        condition: service_healthy

  ## Gateway
myretro-gateway:
  image: myretro-gateway
    build:
```

```
    context: ../myretro-gateway
    dockerfile: Dockerfile
  networks:
    - cloud
  environment:
    - SPRING_CLOUD_CONSUL_HOST=consul-server
  depends_on:
    consul-server:
      condition: service_healthy
  ports:
    - "8080:8080"
## Networks
networks:
  cloud:
    name: cloud
    #external: true
```

Listing 13-17 shows that we don't need to define any export port, just the gateway. Before continuing, review Listing 13-17 carefully and note that we need to add the SPRING_CLOUD_CONSUL_HOST environment variable to each Spring Boot service and the SPRING_DATASOURCE_URL for the users-service. Also, we have a consul-init service that will initialize the properties we need.

Now you can run the environment with

```
docker compose up
```

The first time you run this command, it will take some time to complete because we are building every service. Subsequent runs should run very fast. After this is up and running, you can test it by trying to make some requests to the /users and /retros endpoints:

```
curl -s http://localhost:8080/users | jq
```

```
curl -s http://localhost:8080/retros | jq
```

To stop the services, you can open a new terminal and change to the directory where the compose.yaml is and execute

```
docker compose down
```

In every project (users, myretro, myretro-gateway), I created the following Dockerfile:

```
FROM eclipse-temurin:17-jdk-jammy AS build
WORKDIR /workspace/app

COPY . /workspace/app
RUN --mount=type=cache,target=/root/.gradle ./gradlew clean build -x test
RUN mkdir -p build/dependency && (cd build/dependency; jar -xf ../libs/*-
SNAPSHOT.jar)

FROM eclipse-temurin:17-jdk-jammy
VOLUME /tmp
ARG DEPENDENCY=/workspace/app/build/dependency
COPY --from=build ${DEPENDENCY}/BOOT-INF/lib /app/lib
COPY --from=build ${DEPENDENCY}/META-INF /app/META-INF
COPY --from=build ${DEPENDENCY}/BOOT-INF/classes /app
ENTRYPOINT ["java","-cp","app:app/lib/*","com.apress.users.
UsersApplication"]
```

This file is used by the docker-compose to build the images. So, it will create each image and they should be in your local Docker registry. You can inspect with

```
docker images | grep -E "retro|users"
```

If you don't get anything, you can build the images using the following commands (on each project folder):

```
# cd users
docker build -t users .

# cd myretro
docker build -t myretro .

# cd myretro-gateway
docker build -t myretro-gateway .
```

Creating Multi-Architecture Images

Docker with multi-architecture allows you to create a single Docker image that works seamlessly on different computer systems, like those with ARM or x86 processors. It's like having a universal remote that can control different TV brands – one image can run on various devices without needing separate versions. This is achieved by packaging multiple versions of your application (each compiled for a specific architecture) into a single image. Docker then automatically selects the right version for the device it's running on, making deployment simpler and more flexible. I'm using an M3 Mac, so by default, some of the architecture layers are based on an ARM64, but how can we do some multi-architecture, for example, a Linux AMD64?

The newest versions of Docker come with the `--platform` parameter that allows us to create a multi-architecture image. To create ARM64 and AMD64/Intel-based images, you can execute the following command:

```
docker build \
--push \
--platform linux/arm64,linux/amd64 \
--tag <your-username>/<your-image-name>:<your-tag> .
```

This command will not only create your image, but also add the necessary OS architecture layers (ARM64/AMD64) and push the image to the Docker Hub registry (`--push`). This means that you need to log in into a Docker registry (I recommend creating an account at `https://hub.docker.com/`, which is free). Of course, you can remove the `--push`, but we are going to need it in the next section.

Note   If you get an error, "ERROR: Multi-platform build is not supported for the docker driver," when executing the previous command, you need to enable the `containerd` feature. Check out this link: `https://docs.docker.com/desktop/containerd/#build-multi-platform-images`.

So, for the users project, we can execute

```
docker build \
--push \
--platform linux/arm64,linux/amd64 \
--tag felipeg48/users:latest .
```

For the myretro project:

```
docker build \
--push \
--platform linux/arm64,linux/amd64 \
--tag felipeg48/myretro:latest .
```

For the myretro-gateway project:

```
docker build \
--push \
--platform linux/arm64,linux/amd64 \
--tag felipeg48/myretro-gateway:latest .
```

You can look at Docker Hub and review your images. For example, see Figure 13-17.

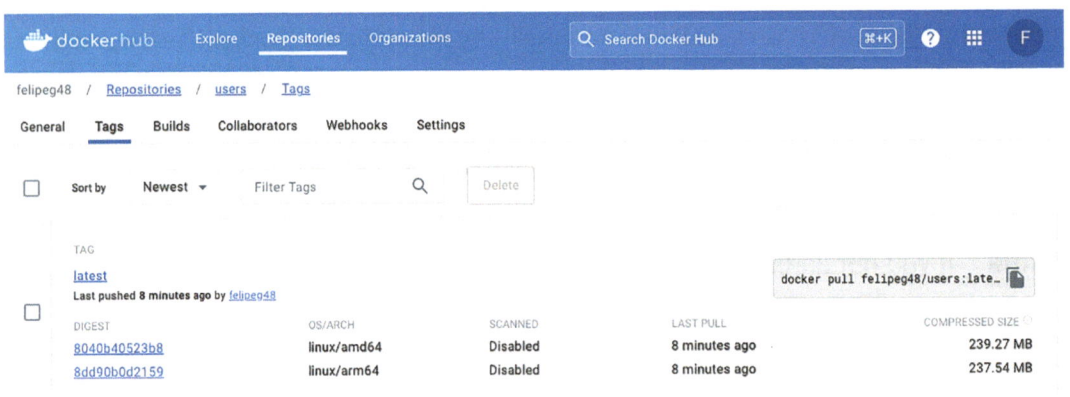

Figure 13-17. *Viewing images in Docker Hub (https://hub.docker.com/ repository/docker/felipeg48/users/tags)*

Now we are ready to talk about a cloud Platform!

Using a Cloud Platform: Kubernetes

Kubernetes (`https://kubernetes.io/`), often abbreviated as K8s, is an open source platform for managing containerized applications. Think of it as a conductor for an orchestra, but instead of managing instruments, it manages containers across a cluster of machines. Now, let's see how Kubernetes helps with microservices:

Benefits of using Kubernetes for microservices include the following:

- *Scalability*: Kubernetes enables you to easily scale individual services up or down based on demand, eliminating the need to provision entire servers. This lets you adapt quickly to changing traffic patterns and avoid wasting resources.

- *High availability*: Kubernetes keeps your services running even if a container or machine fails. It automatically restarts crashed containers and reschedules them on healthy nodes, ensuring your application remains available.

- *Decoupling and autonomy*: Kubernetes helps manage each microservice independently, allowing development teams to work autonomously and deploy updates faster without impacting other services.

- *Resource management*: Kubernetes enables you to define resource limits and quotas for each service, ensuring fair sharing and preventing resource starvation. This optimizes resource utilization and helps maintain system stability.

- *Deployment and rollbacks*: Streamline deployments and rollbacks for individual services with ease. This reduces risk and makes updates less disruptive.

- *Observability and monitoring*: You can monitor the health and performance of each service within Kubernetes, helping you identify and troubleshoot issues quickly.

The following are some key features of Kubernetes that aid microservices:

- *Pods*: These are group-related containers that share resources and storage. Think of them as mini environments for your services.

- *Deployments*: You can define the desired state for your services and Kubernetes ensures they reach that state, handling scaling and updates automatically.

- *Services*: You can expose your microservices to other services or external clients and Kubernetes will handle load balancing and service discovery.

- *Namespaces*: Kubernetes enables you to isolate resources and configurations for different teams or environments, promoting security and organization.

Kubernetes provides a powerful platform for deploying and managing microservices, resulting in the following benefits:

- Increased agility and faster development cycles

- Improved scalability and high availability

- Efficient resource utilization and cost savings

- Simplified operations and management

Kubernetes isn't the only microservices platform, of course. Other popular microservices platforms include Docker Swarm (`https://docs.docker.com/engine/swarm/`), HashiCorp Nomad (`https://www.nomadproject.io/`), Marathon (`https://mesosphere.github.io/marathon/`), Rancher (`https://www.rancher.com/`), Cattle (`https://github.com/rancher/cattle`), and Cloud Foundry (`https://www.cloudfoundry.org/`).

Kubernetes is a strong contender for orchestrating your containerized landscape. Kubernetes has a learning curve, but the benefits of using it to manage distributed systems, especially at scale, are significant.

Using Kubernetes

As previously described, Kubernetes offers service discovery, configuration, secrets, reliability, observability, high availability, and much more. So, let's use Kubernetes as our cloud platform and see how we can deploy our apps.

In this section we will use Minikube as our Kubernetes cluster that we can test locally on your computer. You can use the same steps if you already have a Kubernetes cluster on Google Cloud, AWS, or Microsoft Azure.

If you are following along, you will find the files for this section in the 13-cloud/ k8s folder.

Prerequisites: Installation

The following are the prerequisites to run Minikube:

1. Install Minikube from https://minikube.sigs.k8s.io/ docs/start/.

2. Install the kubectl CLI from https://kubernetes.io/docs/ tasks/tools/. If you already have Docker Desktop installed, kubectl typically comes as part of the installation.

3. Optionally, install K9s from https://k9scli.io/. This is a text-based UI that can help visualize all the Kubernetes components without typing too much.

Starting Minikube

To start Minikube, make sure you have a minimum of 4GB of RAM. You can start the default configuration with

```
minikube start
```

If you have more memory, you can use the --memory parameter. For example, if you have 8GB total in your computer, you can start minikube with

```
minikube start --memory 4096
```

There are other parameters, such as --cpu, which allows you to specify how much CPU core to use. To learn more about the minikube CLI, check out the documentation here: https://minikube.sigs.k8s.io/docs/.

Creating the Postgres Deployment

Our Users App needs to use PostgreSQL as a database, so we need to start it using the declarations described next.

ConfigMap

We need to create the username and password. Kubernetes has a way to save them and use them when we need them; in this case, there is a component named ConfigMap, declared as follows:

```
apiVersion: v1
kind: ConfigMap
metadata:
  name: postgres-config
  labels:
    app: postgres
data:
  POSTGRES_DB: users_db
  POSTGRES_USER: admin
  POSTGRES_PASSWORD: mysecretpassword
```

PersistentVolume and PersistentVolumeClaim

We need to store the data correctly, and in Kubernetes, *PersistentVolume* (PV) is the actual storage space, like a hard drive, while *PersistentVolumeClaim* (PVC) is a user's request for a specific amount of that storage. Think of it as a parking lot (PV) where users request parking spaces (PVC). The *PersistentVolume* declaration is as follows:

```
kind: PersistentVolume
apiVersion: v1
metadata:
  name: postgres-pv-volume
  labels:
    type: local
    app: postgres
```

```
spec:
  storageClassName: manual
  capacity:
    storage: 5Gi
  accessModes:
    - ReadWriteMany
  hostPath:
    path: "/mnt/data"
```

The *PersistentVolumeClaim* declaration is as follows:

```
kind: PersistentVolumeClaim
apiVersion: v1
metadata:
  name: postgres-pv-claim
  labels:
    app: postgres
spec:
  storageClassName: manual
  accessModes:
    - ReadWriteMany
  resources:
    requests:
      storage: 1Gi
```

Deployment

Deployment in Kubernetes is your autopilot for pods, ensuring desired number of instances running, updating smoothly, and rolling back if needed. The PostgreSQL *Deployment* is as follows:

```
apiVersion: apps/v1
kind: Deployment
metadata:
  name: postgres  # Sets Deployment name
spec:
  replicas: 1
```

```yaml
selector:
  matchLabels:
    app: postgres
template:
  metadata:
    labels:
      app: postgres
  spec:
    containers:
      - name: postgres
        image: postgres
        imagePullPolicy: "IfNotPresent"
        ports:
          - containerPort: 5432  # Exposes container port
        envFrom:
          - configMapRef:
              name: postgres-config
        volumeMounts:
          - mountPath: /var/lib/postgresql/data
            name: postgredb
        resources:
          limits:
            cpu: "1"
            memory: "1Gi"
          requests:
            cpu: "0.5"
            memory: "512Mi"
    volumes:
      - name: postgredb
        persistentVolumeClaim:
          claimName: postgres-pv-claim
```

Before continuing, make sure to review this declaration carefully.

Service

A *Service* in Kubernetes is a single entry point for reaching a group of pods, hiding their individual details, and ensuring high availability. The PostgreSQL *Service* is as follows:

```
apiVersion: v1
kind: Service
metadata:
  name: postgres
  labels:
    app: postgres
spec:
  type: NodePort
  ports:
    - port: 5432
  selector:
    app: postgres
```

Installing the Declarations

All of these declarations are in a file named postgresql.yaml. To install it in your Kubernetes, you can execute the following command:

```
kubectl apply -f postgresql.yaml
```

Deployin Microservices in Kubernetes: Users App, My Retro App, My Retro Gateway App Deployments

In this section, we are going to deploy all our microservices. Recall that we used in the My Retro APP and My Retro Gateway projects, respectively, the spring-cloud-starter-kubernetes-client-all and spring-cloud-starter-kubernetes-fabric8-all dependencies. With these dependencies, we can integrate our apps seamlessly within Kubernetes and take advantage of the Kubernetes features in our Spring Boot apps. We are going to talk about some of these features next.

ClusterRole and ClusterRoleBinding

ClusterRole defines cluster-wide permissions, while *ClusterRoleBinding* assigns them to users or groups across all namespaces. Think of it as setting access rules (*ClusterRole*) and then attaching them to users (*ClusterRoleBinding*).

We need to declare both because the Spring Cloud Kubernetes dependencies require access to *Services, Endpoints, ConfigMap, Pods*, and more, to make this an easy integration without any code in the Spring Boot side. The *ClusterRole* declaration is as follows:

```
apiVersion: rbac.authorization.k8s.io/v1
kind: ClusterRole
metadata:
  name: cluster-read-role
rules:
- apiGroups:
  - ""
  resources:
  - endpoints
  - pods
  - services
  - configmaps
  verbs:
  - get
  - list
  - watch
```

The *ClusterRoleBinding* declaration is as follows:

```
apiVersion: rbac.authorization.k8s.io/v1
kind: ClusterRoleBinding
metadata:
  name: cluster-read-rolebinding
subjects:
- kind: ServiceAccount
  name: default
  namespace: default
```

```
roleRef:
  kind: ClusterRole
  name: cluster-read-role
  apiGroup: rbac.authorization.k8s.io
```

We these declarations, the Spring Boot apps that are using the Spring Cloud Kubernetes dependencies (My Retro App and My Retro Gateway) can make use of the integration features that we will discuss next.

Users Deployment and Service

Let's start with the Users App Deployment declaration:

```
apiVersion: apps/v1
kind: Deployment
metadata:
  labels:
    app: users
  name: users
spec:
  replicas: 1
  selector:
    matchLabels:
      app: users
  template:
    metadata:
      labels:
        app: users
    spec:
      containers:
        - name: users
          image: felipeg48/users:latest
          ports:
          - containerPort: 8080
          env:
          - name: spring.cloud.consul.enabled
            value: "false"
```

```
    - name: spring.datasource.url
      value: "jdbc:postgresql://postgres.default.svc.cluster.
      local:5432/users_db"
    - name: spring.datasource.username
      valueFrom:
        configMapKeyRef:
          name: postgres-config
          key: POSTGRES_USER
    - name: spring.datasource.password
      valueFrom:
        configMapKeyRef:
          name: postgres-config
          key: POSTGRES_PASSWORD
  resources:
    limits:
      cpu: "0.5"
      memory: "768Mi"
  livenessProbe:
    httpGet:
      path: /actuator/health
      port: 8080
    initialDelaySeconds: 30
    periodSeconds: 10
    timeoutSeconds: 5
    failureThreshold: 6
```

Let's review this declaration:

- `felipeg48/users:latest`: We are using this image that we generated previously. You can switch to yours, but this image is publicly available.

- `spring.cloud.consul.enabled`: We are declaring environment variables in the env section. We are disabling Consul, because we are not going to use Consul for service discovery and configuration but instead use the Kubernetes defaults. So, we set this value to `false`.

- `spring.datasource.url`: This property holds the URL (yes, we are overwriting its value) using the standard way to locate services. The syntax is `<service-name>.<namespace>.svc.cluster.local`. We can just use `postgres`, but I want it to show you the Kubernetes DNS usage.

- `spring.datasource.username`/`spring.datasource.password`: We are also overwriting these properties. We are using the reference (`configMapKeyRef`) to the *ConfigMap* we defined earlier with `postgres-config`, by using the keys.

- `livenessProbe`: We are using this declaration to use the `/actuator/health` endpoint to make sure our app is healthy. Also, we are using some resource limits that help us minimize the overall usage of our Kubernetes installation.

Next, let's check the Users App *Service* declaration:

```
apiVersion: v1
kind: Service
metadata:
  labels:
    app: users
  name: users-service
spec:
  type: NodePort
  selector:
    app: users
  ports:
  - name: http
    protocol: TCP
    port: 80
    targetPort: 8080
```

Here we are defining a *NodePort,* and we are going to use port 80 that will redirect to the 8080 in the container.

My Retro Deployment and Service

Next, let's check out the My Retro Deployment:

```
apiVersion: apps/v1
kind: Deployment
metadata:
  labels:
    app: myretro
  name: myretro
spec:
  replicas: 1
  selector:
    matchLabels:
      app: myretro
  template:
    metadata:
      labels:
        app: myretro
    spec:
      containers:
      - name: myretro
        image: felipeg48/myretro:latest
        ports:
        - containerPort: 8080
        env:
          - name: spring.cloud.consul.enabled
            value: "false"
        resources:
          limits:
            cpu: "0.5"
            memory: "768Mi"
        livenessProbe:
          httpGet:
            path: /actuator/health
            port: 8080
```

```
initialDelaySeconds: 30
periodSeconds: 10
timeoutSeconds: 5
failureThreshold: 6
```

We are using my image, but you are welcome to change to yours. Also, we are disabling the Consul configuration with the `spring.cloud.consul.enabled` property by setting the value to `false`.

Remember that this app uses OpenFeign to connect to the `users-service`, and because we used the Spring Cloud Kubernetes dependency, behind the scenes Kubernetes will handle the *Service Discovery* and *Load Balancing* using/reaching to the service using something like `users-service.default.svc.cluster.local` as host.

The Users *Service* declaration is as follows:

```
apiVersion: v1
kind: Service
metadata:
  labels:
    app: myretro
  name: my-retro-app
spec:
  type: NodePort
  selector:
      app: myretro
  ports:
  - name: http
    protocol: TCP
    port: 80
    targetPort: 8080
```

My Retro Gateway ConfigMap, Deployment, and Service

Next, let's look at the My Retro Gateway declaration. First, let's look at the *ConfigMap*:

```
apiVersion: v1
kind: ConfigMap
metadata:
  name: myretro-gateway
```

```yaml
data:
  application.yaml: |-
    management:
      endpoint:
        gateway:
          enabled: true
      endpoints:
        web:
          exposure:
            include: "health,info,gateway,configprops,conditions,env,beans"
    spring:
      cloud:
        gateway:
          routes:
            - id: users
              uri: http://users-service
              predicates:
                - Path=/users/**
              filters:
                - name: CircuitBreaker
                  args:
                    name: users
                    fallbackUri: forward:/fallback/users
            - id: myretro
              uri: http://my-retro-app
              predicates:
                - Path=/retros/**
              filters:
                - name: CircuitBreaker
                  args:
                    name: retros
                    fallbackUri: forward:/fallback/retros
```

Let's review it:

- `application.yaml`: In the `data` section of the *ConfigMap*, we are declaring a new `application.yaml` file. Yes, we are going to overwrite all the `application.yaml` that comes in the image. Here we are defining new properties, but the most important are next.

- `spring.cloud.gateway.routes.*`: We are redefining the `routes`, but the only change is the `uri` key, where instead of the `lb://` (load balancer) we are using just the `http://`. In this case, we are using the `http://` because the load balancing comes from the Kubernetes services, so there's no need to add this prefix (lb://). Also, we are not using Consul anymore. We are using the Kubernetes defaults (service discovery and load balancing).

Next, the *Deployment* declaration:

```
apiVersion: apps/v1
kind: Deployment
metadata:
  labels:
    app: myretro-gateway
  name: myretro-gateway
spec:
  replicas: 1
  selector:
    matchLabels:
      app: myretro-gateway
  template:
    metadata:
      labels:
        app: myretro-gateway
    spec:
      containers:
      - name: myretro-gateway
        image: felipeg48/myretro-gateway:latest
        ports:
        - containerPort: 8080
```

```
    env:
    - name: spring.config.import
      value: "kubernetes:"
    - name: spring.cloud.consul.enabled
      value: "false"
    resources:
      limits:
        cpu: "0.5"
        memory: "768Mi"
    livenessProbe:
      httpGet:
        path: /actuator/health
        port: 8080
      initialDelaySeconds: 30
      periodSeconds: 10
      timeoutSeconds: 5
      failureThreshold: 6
```

Here we are disabling the Consul (with `spring.cloud.consul.enabled=false`). And we are using a known property, the `spring.config.import` with value `kubernetes://`; and just by adding this, the Spring Cloud Kubernetes auto-configuration will look for a *ConfigMap* that matches the name of the application, in this case the `myretro-gateway` *ConfigMap* that we previously declared, and it will apply/overwrite the `application.yaml` file to the app with the new values. This is so cool! Now, if you happen to have a different name for the `ConfigMap`, you can still use it, but you must follow some naming conventions. If you want to know more, go here: `https://docs.spring.io/spring-cloud-kubernetes/reference/property-source-config/configmap-propertysource.html`.

Next, let's review the *Service*:

```
apiVersion: v1
kind: Service
metadata:
  labels:
    app: myretro-gateway
  name: myretro-gateway-service
```

```
spec:
  type: NodePort
  selector:
    app: myretro-gateway
  ports:
  - name: http
    protocol: TCP
    port: 8080
    targetPort: 8080
```

Here we are using port 8080.

In the source code `13-cloud/k8s` folder, you can find a single `myretro.yaml` file that contains all the declarations we have just reviewed. To apply it, you can use the following command:

```
kubectl apply -f myretro.yaml
```

That's it. Now we can use the My Retro Gateway service to access our applications. You can check if the pods are up and running and the services with the following commands:

```
kubectl get pods
kubectl get services
```

Keep in mind that I'm showing you one alternative; you can use Consul as well, or some other component such as Istio or even a special *Spring Cloud Gateway for Kubernetes* that can be installed as Operator/Helm (with commercial support), and it's even better than using it alone. If you want to know more, take a look at `https://docs.vmware.com/en/VMware-Spring-Cloud-Gateway-for-Kubernetes/index.html`.

Access to the My Retro Gateway: Using Port-Forward

Next, let's access our apps by using the Kubernetes `port-forward` and the `myretro-gateway-service`. In a terminal, execute the following command:

```
kubectl port-forward svc/myretro-gateway-service 8080:8080
```

With this, you can either go to your browser or you can use the following commands to reach the apps:

```
curl -s http://localhost:8080/users | jq
```

```
curl -s http://localhost:8080/retros | jq
```

```
curl -s http://localhost:8080/retros/users | j
```

If you want to take a look at some of the logs, first get the names of the pods with

```
kubectl get pods
```

Then, you can look at the logs with

```
kubectl logs <pod-name>
```

That's it! Now you know how to deploy microservices in Kubernetes.

If you are interested in using Consul instead of the default Kubernetes service discovery, configuration, and load balancing, you can use Consul for Kubernetes, which makes it even easier to deploy microservices. Check out its documentation here: https://developer.hashicorp.com/consul/docs/k8s.

Cleaning Up

The easiest way to shut down everything is just by executing the following command, which removes the VM created for Kubernetes:

```
minikube delete --all --purge
```

If you want just to stop it, you can use

```
minikube stop
```

I encourage you to continue experimenting with all the features that Spring Cloud offers, not only for Spring Cloud Consul and Spring Cloud Vault, but also for other projects.

Summary

In this chapter, we reviewed Spring Cloud Consul for service discovery and configuration; Spring Cloud Vault for secrets and configuration; Spring Cloud Gateway for routing, filtering, and fallback by using the Circuit Breaker pattern; and some of the HashiCorp products such as Consul and Vault.

We also looked at Docker Compose and Kubernetes and saw how we can use this cloud platform to have multiple instances, service discovery, high availability, and much more out of the box; and thanks to Spring Cloud Kubernetes, the integration is easier.

There are so many more Spring Cloud technologies that I would need an entire second book to describe them. In my book *Spring Boot Messaging* (Apress, 2017), I cover Spring Cloud Stream, and in *Spring Cloud Data Flow* (Apress, 2020), I cover other technologies that would complement your cloud journey.

CHAPTER 14

Extending Spring Boot

In this chapter, we will review how we can extend Spring Boot by creating a custom
Spring Boot starter. We'll discuss the benefits of a custom starter, review the annotations
required to create a custom starter, identify the requirements of the custom starter that
we'll create, and survey the general rules for creating a custom starter. Then, we'll create
a custom starter for the My Retro App project. We'll wrap up by using our custom starter.

Benefits of a Custom Starter

The following are the potential benefits of creating a custom starter:

- *Increased development speed*:

  - *Reduced boilerplate*: By encapsulating common dependencies
    and configurations into a starter, you eliminate the need to
    manually include them in each project, saving time and effort.

  - *Consistent setup*: A starter ensures a consistent environment
    across all your projects, simplifying onboarding and maintenance
    for your team.

  - *Auto-configuration*: Spring Boot will automatically configure
    beans based on the dependencies included in your starter,
    further reducing manual work.

- *Improved code reusability*:

  - *Shareable code*: If your custom functionality is valuable to others,
    you can publish your starter as a library for broader use.

  - *Modular design*: Starters promote modularity, making it easier to
    integrate specific features into different projects as needed.

© Felipe Gutierrez 2024
F. Gutierrez, *Pro Spring Boot 3*, https://doi.org/10.1007/978-1-4842-9294-5_14

- *Dependency management*: You control the exact versions of dependencies included in your starter, avoiding conflicts and ensuring consistency.

- *Enhanced maintainability*:

  - *Centralized changes*: Updating one starter impacts all projects utilizing it, simplifying maintenance and upgrades.

  - *Reduced complexity*: Smaller projects with focused functionality generally lead to improved maintainability.

  - *Clear documentation*: Having well-documented starters clarifies expected behavior and simplifies troubleshooting.

Creating a custom Spring Boot starter can be beneficial for managing common configurations, promoting code reusability, and improving maintainability for projects with shared needs. However, it's crucial to weigh the benefits against the added complexity and ensure that the custom starter aligns with your specific project requirements.

So, let's begin with a review of what we discussed in Chapter 1 regarding the Spring Boot internals and features, specifically how Spring Boot is doing the auto-configuration and how it uses certain annotations that can help to determine what Spring beans need to be created.

Revisiting @Conditional and @Enable

As you've seen throughout the previous chapters, the @Conditional annotation in Spring is a powerful tool for managing your application's configuration based on specific conditions. It allows you to control whether a bean, configuration class, or even an entire profile is included in the Spring application context, depending on various factors.

As a refresher, this is how the @Conditional annotation works:

- You annotate a bean, configuration class, or method with @Conditional(MyCondition.class).

- MyCondition must implement the Condition interface and define a matches method.

- This `matches` method receives information about the bean and the application context and returns `true` if the condition is met and `false` otherwise.

- Spring checks the condition before creating the bean or loading the configuration.

Common usages of the `@Conditional` annotation include the following:

- *Environment-based configuration*: Load beans based on specific environment variables, like `@ConditionalOnProperty("spring.profiles.active=dev")`.

- *Dependency checks*: Create beans only if certain libraries or classes are present in the classpath.

- *Testing*: Exclude beans that depend on external resources during unit tests.

- *Custom conditions*: Implement your own logic for specific needs (e.g., checking OS version).

The important part to remember when using this annotation is that `@Conditional` offers immense flexibility for conditional configuration in Spring Boot

In Spring and Spring Boot, the `@Enable` annotation serves as a convenient way to activate specific features or configurations within your application. It simplifies the process by taking care of various tasks under the hood, making your code concise and readable.

The `@Enable` annotation serves the following purposes:

- *Automatic configuration*: The primary function of `@Enable` annotations is to trigger auto-configuration based on provided classes or conditions. This saves you from manually writing boilerplate code and leverages Spring's built-in capabilities.

- *Declarative approach*: Instead of explicitly configuring every aspect of a feature, you use the `@Enable` annotation to declare your intent, and Spring handles the specifics. This leads to cleaner and more maintainable code.

Here are some common use cases of @Enable annotations:

- *Enabling features*: Some examples are

  - @EnableWebMvc: Enables Spring MVC web application support.

  - @EnableAsync: Allows asynchronous method execution using @Async.

  - @EnableJpaRepositories: Enables automatic scanning for JPA repositories.

  - @EnableCaching: Activates Spring's caching framework.

Note   All the preceding annotations (except for @EnableAsync) are enabled by default in a Spring Boot app when the auto-configuration starts and the dependency is in the classpath. But, you already knew that!

- *Customizing features*:

  - Many @Enable annotations accept configuration attributes to tailor their behavior.

  - @EnableScheduling(fixedRate = 1000): Schedules tasks with a fixed rate of 1 second.

  - @EnableJpaRepositories(basePackages = "com.apress. myretro.repository"): Scans for repositories only in the specified package.

- *Creating custom starters*:

  - You can develop your own @Enable annotations within custom Spring Boot starters.

  - This allows you to package reusable configurations and simplify feature activation across projects.

By effectively using @Enable annotations, you can streamline configuration, promote code reusability, and leverage Spring's powerful features efficiently.

Having reviewed the annotations, it's time to identify the requirements of our own custom Spring Boot starter.

Requirements of Our Custom Spring Boot Starter

As you already know from Chapter 10, both projects (Users App and My Retro App) have events that use the Spring Events mechanism to log what is happening in the services or in any other part of the system. Right now, it is an adequate solution, but imagine that you need this kind of event or audit in more modules that will be built in the next release, so you will be duplicating and doing the same (copy/paste) all over.

The solution is to modularize, and the best way to implement that solution is to create a custom Spring Boot starter that can do the following:

- Create a generic event that contains a timestamp of when the event occurred, the method that was called, the arguments of the method with values, a result if that method returns something, and a message that allows us to see what happened.

- Add a configurable way to log the events Before, After, or Both.

- Add a configurable way to log the event in plain text or a JSON format.

- Add a configurable way to do a pretty print when printing to Console, such as the JSON format.

- Add a configurable way to persist into a Database the Events, or just the Console.

- Add a configurable property to add a prefix to the event.

- Add a configurable way to generate logs or just standard output.

Tip   As a suggestion, you can use the @Enable* annotation, a custom annotation, and the ability to add properties and override them if necessary.

With these requirements, we are going to be able to create a custom starter that allows us to share the same practices and modules with other developers. In other words, after we create our custom starter, any new developer who wants to use its functionality in their own project will simply have to add the dependency and use the required annotations in the program to create an event or audit. So, in their `build.gradle` file, they would have something like:

```
implementation 'com.apress:myretro-spring-boot-starter:0.0.1'
```

Rules and Guidelines for a Custom Spring Boot Starter

When creating a custom Spring Boot starter, we need to follow certain rules:

- *Structure*: Divide your project into two modules:

  - *Autoconfigure module*: Contains auto-configuration classes and properties classes.

  - *Starter module*: Includes the autoconfigure module as a dependency, along with the library your starter supports and any additional dependencies.

  In our case, we are going to have only one project, to make things simpler, so we are going to create only one project that holds the auto-configuration and the starter module.

- *Auto-configuration class*: Use the `@EnableAutoConfiguration` annotation on your main auto-configuration class:

  - Extend the `SpringBootApplication` class (optional, especially for complex configurations).

  - Use conditional checks and bean factory methods to create beans conditionally based on classpath presence and properties.

  - Consider `@ConfigurationProperties` for custom configuration options.

- *Properties class*:

  - Define a class annotated with `@ConfigurationProperties` to hold configurable properties for your library.

  - Use descriptive names and consider sensible defaults.

- *Starter pom*:

  - In the starter pom, define your project's `groupId`, `artifactId`, and `version`.

  - Add dependencies for the autoconfigure module, the supported library, and any additional dependencies.

- *Enable auto-configuration*:

  - Define your auto-configuration classpath in the `META-INF/spring/org.springframework.boot.autoconfigure.AutoConfiguration.imports` file in your starter module.

  - Use the following format: `[package].YourAutoConfigurationClass`

- *Additional considerations*:

  - Use Spring Boot best practices for code style and documentation.

  - Test your starter thoroughly under various configurations.

  - Consider providing examples and documentation for users.

- *Gradle-specific considerations*:

  - Use appropriate Gradle plugins for build management and testing.

  - Leverage Gradle features like multi-project builds and tasks customization.

- *Follow the naming convention*:

  - Your module should never start with `spring-boot`; this is a specific rule from the Spring Team that is intended to avoid any conflicts with the Spring and Spring Boot packages.

  - Our project is `myretro`, so we will name out module `myretro-spring-boot-starter`.

Creating My Retro Spring Boot Starter

You can locate the source code for the projects we are going to use in the `14-extending/` folder. You can import the project `myretro-spring-boot-starter` into your favorite IDE. If you want to start from scratch with the Spring Initializr (`https://start.spring.io`), set the Group field to `com.apress`, the Artifact and Name fields to `myretro-spring-boot-starter`, and the Package field to `com.apress.myretro`. Add as dependencies Web, JPA, Processor, Lombok, and H2. Generate and download the project, unzip it, and import it into your favorite IDE.

Let's start with our build.gradle file. See Listing 14-1.

Listing 14-1. build.gradle

```
plugins {
    id 'java'
    id 'org.springframework.boot' version '3.2.2' apply false
    id 'io.spring.dependency-management' version '1.1.4'
}

group = 'com.apress'
version = '0.0.1'

java {
    sourceCompatibility = '17'
}

configurations {
    compileOnly {
        extendsFrom annotationProcessor
    }
}

tasks.named('compileJava') {
    inputs.files(tasks.named('processResources'))
}

repositories {
    mavenCentral()
}

dependencyManagement {
    imports {
        mavenBom org.springframework.boot.gradle.plugin.SpringBootPlugin.
        BOM_COORDINATES
    }
}
```

```
dependencies {
    implementation 'org.springframework.boot:spring-boot-starter-web'
    implementation 'org.springframework.boot:spring-boot-starter-aop'
    implementation 'org.springframework.boot:spring-boot-starter-data-jpa'
    implementation 'com.fasterxml.jackson.datatype:jackson-datatype-jsr310'

    runtimeOnly 'com.h2database:h2'

    compileOnly 'org.projectlombok:lombok'

    annotationProcessor 'org.springframework.boot:spring-boot-
    configuration-processor'
    annotationProcessor 'org.projectlombok:lombok'

    testImplementation 'org.springframework.boot:spring-boot-starter-test'
}
tasks.named('test') {
    useJUnitPlatform()
}
```

Let's review the build.gradle file:

- plugins: Note that the id org.springframework.boot is using the apply false statement. This means that this specific plugin will not be applied to the current build. While it's declared and available, its functionalities won't be accessible during this build. These are the common reasons for using the apply false statement:

  - *Conditional application*: You might have a build property that determines whether to apply the plugin based on specific conditions.

  - *Conflict avoidance*: If another plugin provides similar functionality and might cause conflicts, using apply false can prevent issues.

  - *Build simplification*: For builds that don't need specific functionalities, excluding unnecessary plugins can keep the configuration cleaner.

But why is this necessary? Well, remember that our custom starter will be a library and not an application, so we need just to build it, not run it.

- `tasks.named('compileJava')`: This refers to the `compileJava` task, responsible for compiling Java source code in our project. `inputs.files(tasks.named('processResources'))` instructs the `compileJava` task to consider the files produced by the `processResources` task as part of its inputs. We need this task because we are going to create custom properties to enable to configure the an output prefix, and the usage of a Logger for the console output; and we are going to provide some metadata that allows other IDEs to have hints of what our properties mean.

- `dependencyManagement`: This section serves as a central location for declaring dependencies and their versions without including them in our project. Think of it as a template for dependencies.

  - `imports { }`: We are importing dependencies defined elsewhere.

  - `mavenBom org.springframework.boot.gradle.plugin.SpringBootPlugin.BOM_COORDINATES`: This specific line imports a bill of materials (BOM) from the Spring Boot Gradle plugin. A BOM is a compressed list of dependencies with specific versions, ensuring consistency and compatibility across projects.

Next, let's check/define the model that we are going to use as an event; in the end, this event will be persistent in a database. Listing 14-2 shows the `MyRetroAuditEvent` class.

Listing 14-2. src/main/java/com/apress/myretro/model/MyRetroAuditEvent.java

```
package com.apress.myretro.model;

import com.fasterxml.jackson.annotation.JsonFormat;
import jakarta.persistence.Entity;
import jakarta.persistence.GeneratedValue;
import jakarta.persistence.GenerationType;
```

```
import jakarta.persistence.Id;
import lombok.AllArgsConstructor;
import lombok.Data;
import lombok.NoArgsConstructor;

import java.time.LocalDateTime;

@AllArgsConstructor
@NoArgsConstructor
@Data
@Entity
public class MyRetroAuditEvent {

    @Id
    @GeneratedValue(strategy = GenerationType.AUTO)
    private Long id;
    @JsonFormat(pattern="yyyy-MM-dd HH:mm:ss")
    private LocalDateTime timestamp = LocalDateTime.now();
    private String interceptor;
    private String method;
    private String args;
    private String result;
    private String message;
}
```

The MyRetroAuditEvent class will be our main event with the fields we need, the timestamp (when the event occurred), a method (the method being executed), args (any arguments passed to the method), a result (if the method returns something, we need to know the result), a message (a text that can be used to identify the event), and an interceptor (the name of what action happens, a BEFORE, AFTER, AROUND). And, of course, this class will be persistent in a database, so we need to use the @Entity and @Id annotations. The @Id is an auto-increment feature.

Next, check/define the repository pattern we are going to use. Listing 14-3 shows the MyRetroAuditEventRepository interface.

Listing 14-3. src/main/java/com/apress/myretro/model/
MyRetroAuditEvent.java

```
package com.apress.myretro.model;

import org.springframework.data.repository.CrudRepository;

public interface MyRetroAuditEventRepository extends CrudRepository<MyRetro
AuditEvent, Long> {
}
```

The `MyRetroAuditEventRepository` interface will be the JPA repository model that
will be persistent in the database.

Now that we set our model and repository, open/create the `MyRetroAudit`
annotation. See Listing 14-4.

Listing 14-4. src/main/java/com/apress/myretro/annotations/
MyRetroAudit.java

```
package com.apress.myretro.annotations;

import com.apress.myretro.formats.MyRetroAuditOutputFormat;

import java.lang.annotation.ElementType;
import java.lang.annotation.Retention;
import java.lang.annotation.RetentionPolicy;
import java.lang.annotation.Target;

@Retention(RetentionPolicy.RUNTIME)
@Target({ElementType.METHOD, ElementType.TYPE})
public @interface MyRetroAudit {
    boolean showArgs() default false;
    MyRetroAuditOutputFormat format() default MyRetroAuditOutputFormat.TXT;

    MyRetroAuditIntercept intercept() default MyRetroAuditIntercept.BEFORE;

    String message() default "";

    boolean prettyPrint() default false;
}
```

Listing 14-4 shows the MyRetroAudit custom annotation that will be used in any method we want to happen the audit/event. Let's review the parameters it's using:

- showArgs: This parameter will be a Boolean, meaning that if it's set to true, the event will show the arguments passed to the method being audited as part of the event; false will be otherwise. The default value is set to false.

- format: This is an Enum value (from MyRetroAuditOutputFormat enum) that has only JSON or TXT values. The default value is set to MyRetroAuditOutputFormat.TXT.

- intercept: This is an Enum value (from MyRetroAuditIntercept enum); this enum has the BEFORE, AFTER, and AROUND values. This parameter will set the event behavior if the audited method is called and will generate the event before, after, or during its execution. The default value is set to MyRetroAuditIntercept.BEFORE.

- message: This is an arbitrary message that can be set as part of the event. We can use it just to enable some convention for future reference, perhaps in a logging system (such as Splunk or PaperTrail) to find events and group them quickly. The default value is set to an empty string.

- prettyPrint: This is a Boolean value that will set the way it prints the event in the console. Normally, if the output is a JSON string, you will want it to see in a pretty form and not in a single line. The default value is set to false.

Listings 14-5 and 14-6 show the Enums MyRetroAuditOutputFormat and MyRetroAuditIntercept.

Listing 14-5. src/main/java/com/apress/myretro/annotations/MyRetroAuditOutputFormat.java

```
package com.apress.myretro.formats;

public enum MyRetroAuditOutputFormat {
    JSON, TXT
}
```

Listing 14-6. src/main/java/com/apress/myretro/annotations/
MyRetroAuditIntercept.java

```
package com.apress.myretro.annotations;

public enum MyRetroAuditIntercept {
    BEFORE, AFTER, AROUND
}
```

Next, open/create the EnableMyRetroAudit annotation. See Listing 14-7.

Listing 14-7. src/main/java/com/apress/myretro/annotations/
EnableMyRetroAudit.java

```
package com.apress.myretro.annotations;

import com.apress.myretro.configuration.MyRetroAuditConfiguration;
import org.springframework.context.annotation.Import;

import java.lang.annotation.ElementType;
import java.lang.annotation.Retention;
import java.lang.annotation.RetentionPolicy;
import java.lang.annotation.Target;

@Retention(RetentionPolicy.RUNTIME)
@Target(ElementType.TYPE)
@Import(MyRetroAuditConfiguration.class)
public @interface EnableMyRetroAudit {
    MyRetroAuditStorage storage() default MyRetroAuditStorage.DATABASE;
}
```

The EnableMyRetroAudit annotation will be required to configure the
@MyRetroAudit annotation. It is our @Enable* feature! The parameter we are using is
the storage, which is a MyRetroAuditStorage enum with values CONSOLE, DATABASE,
and FILE. This will allow to persist the event (DATABASE), or just to console (CONSOLE,
not persistent) or in a file (FILE). The default value is set to MyRetroAuditStorage.
DATABASE. Listing 14-8 shows the MyRetroAuditStorage enum.

Listing 14-8. src/main/java/com/apress/myretro/annotations/
MyRetroAuditStorage.java

```
package com.apress.myretro.annotations;

public enum MyRetroAuditStorage {
    CONSOLE, DATABASE, FILE
}
```

Next, to be able to know at runtime what storage value was set in the
@EnableMyRetroAudit annotation, we need to find out the value at runtime; there are
many ways to do so in Spring, so I'll demonstrate only one of them.

So, open/create the EnableMyRetroAuditValueProvider class. See Listing 14-9.

Listing 14-9. src/main/java/com/apress/myretro/annotations/
EnableMyRetroAuditValueProvider.java

```
package com.apress.myretro.annotations;

import org.springframework.beans.BeansException;
import org.springframework.beans.factory.config.BeanFactoryPostProcessor;
import org.springframework.beans.factory.config.
ConfigurableListableBeanFactory;
import org.springframework.stereotype.Component;

import java.util.Arrays;

@Component
public class EnableMyRetroAuditValueProvider implements
BeanFactoryPostProcessor {

    private static MyRetroAuditStorage storage = MyRetroAuditStorage.
    DATABASE;

    @Override
    public void postProcessBeanFactory(ConfigurableListableBeanFactory
    beanFactory) throws BeansException {

        String beanName = Arrays.stream(beanFactory.getBeanNamesFor
        Annotation(EnableMyRetroAudit.class)).findFirst().orElse(null);
```

```
    if (beanName != null) {
        storage = beanFactory.findAnnotationOnBean(beanName,
        EnableMyRetroAudit.class).storage();
    }
}

public static MyRetroAuditStorage getStorage() {
    return storage;
}
}
```

Let's review the EnableMyRetroAuditValueProvider class:

- BeanFactoryPostProcessor: The BeanFactoryPostProcessor
  shines in Spring for customizing Spring bean definitions before
  they're instantiated. Think of it as a handy tool for fine-tuning your
  application's configuration after Spring reads initial settings. It's great
  for the following:

  - *Dynamic configuration*: Adapt bean definitions based on
    environment variables, external files, or runtime conditions.

  - *Centralized property handling*: Inject shared properties into
    multiple beans without repetitive configuration.

  - *Conditional bean registration*: Create beans only if specific
    conditions are met, making your application flexible.

  - *Custom bean modification*: Add custom logic or modify bean
    definitions before bean creation, tailoring behavior to your needs.

- postProcessBeanFactory (ConfigurableListableBeanFactory
  beanFactory): This is the method to implement. Note that with the
  ConfigurableListableBeanFactory; we can find the annotations
  declared in Spring beans. We are using the EnableMyRetroAudit
  to be found in all beans, then if the auto-configuration find this
  annotation (@EnableMyRetroAudit), we can get the value used in the
  storage parameter; if not found, then it will use the default value,
  the MyRetroAuditStorage.DATABASE.

Also notice that this class is marked using the @Component annotation, so it will be picked up during the Spring bean life cycle and initialization. So, at any time, we can use the following code to retrieve the value set:

```
EnableMyRetroAuditValueProvider.getStorage()
```

Next, let's talk about the formats we are going to be using. Recall that the @MyRetroAudit can use the format parameter. So, let's create a strategy that can help us with the format (pretty print or not) of the event received, that will be easy to use and extend if necessary.

Open/create the MyRetroAuditFormatStrategy interface. See Listing 14-10.

Listing 14-10. src/main/java/com/apress/myretro/formats/ MyRetroAuditFormatStrategy.java

```
package com.apress.myretro.formats;

import com.apress.myretro.model.MyRetroAuditEvent;

public interface MyRetroAuditFormatStrategy {
    String format(MyRetroAuditEvent event);
    String prettyFormat(MyRetroAuditEvent event);
}
```

We are defining two methods in the MyRetroAuditFormatStrategy interface, format and prettyFormat, that accept the MyRetroAuditEvent and return a String type.

Next, open/create the MyRetroAuditFormatStrategyFactory class, which, as the name implies, will help us to create the strategy we need based on the MyRetroAuditOutputFormat enum. See Listing 14-11.

Listing 14-11. src/main/java/com/apress/myretro/formats/ MyRetroAuditFormatStrategyFactory.java

```
package com.apress.myretro.formats;

import com.apress.myretro.annotations.MyRetroAuditOutputFormat;

public class MyRetroAuditFormatStrategyFactory {
```

```
    public static MyRetroAuditFormatStrategy getStrategy(MyRetroAuditOutput
    Format outputFormat) {
        switch (outputFormat) {
            case JSON:
                return new JsonOutputFormatStrategy();
            case TXT:
            default:
                return new TextOutputFormatStrategy();
        }
    }
}
```

As you can see, this is a very simple Factory implementation. Next, let's check the two strategy implementations. Open/create the JsonOutputFormatStrategy and TextOutputFormatStrategy classes that will implement our strategy MyRetroAuditFormatStrategy interface. See Listings 14-12 and 14-13.

Listing 14-12. src/main/java/com/apress/myretro/formats/
JsonOutputFormatStrategy.java

```
package com.apress.myretro.formats;

import com.apress.myretro.model.MyRetroAuditEvent;
import com.fasterxml.jackson.databind.ObjectMapper;
import com.fasterxml.jackson.datatype.jsr310.JavaTimeModule;
import lombok.SneakyThrows;

public class JsonOutputFormatStrategy implements
MyRetroAuditFormatStrategy {

    final ObjectMapper objectMapper = new ObjectMapper();

    public JsonOutputFormatStrategy(){
        objectMapper.registerModule(new JavaTimeModule());
    }

    @SneakyThrows
    @Override
    public String format(MyRetroAuditEvent event) {
```

```
        return objectMapper
                .writeValueAsString(event);
    }

    @SneakyThrows
    @Override
    public String prettyFormat(MyRetroAuditEvent event) {
        return "\n\n" + objectMapper
                .writerWithDefaultPrettyPrinter()
                .writeValueAsString(event) + "\n";
    }
}
```

Listing 14-13. src/main/java/com/apress/myretro/formats/
TextOutputFormatStrategy.java

```
package com.apress.myretro.formats;

import com.apress.myretro.model.MyRetroAuditEvent;

public class TextOutputFormatStrategy implements
MyRetroAuditFormatStrategy {
    @Override
    public String format(MyRetroAuditEvent event) {
        return event.toString();
    }

    @Override
    public String prettyFormat(MyRetroAuditEvent event) {
        return "\n\n" + event.toString() + "\n";
    }
}
```

As you can see, these classes are very straightforward, but you should still review
them before continuing.

Next, open/create the MyRetroAuditAspect class, shown in Listing 14-14. This class,
as the name says, will implement an AOP Aspect that allows us to intercept calls and do
much more, and in this case, it will do all the logic based on the requirements.

Listing 14-14. src/main/java/com/apress/myretro/aop/
MyRetroAuditAspect.java

```
package com.apress.myretro.aop;

import com.apress.myretro.annotations.MyRetroAudit;
import com.apress.myretro.annotations.MyRetroAuditIntercept;
import com.apress.myretro.annotations.MyRetroAuditStorage;
import com.apress.myretro.configuration.MyRetroAuditProperties;
import com.apress.myretro.formats.MyRetroAuditFormatStrategy;
import com.apress.myretro.formats.MyRetroAuditFormatStrategyFactory;
import com.apress.myretro.model.MyRetroAuditEvent;
import com.apress.myretro.model.MyRetroAuditEventRepository;
import lombok.AllArgsConstructor;
import lombok.extern.slf4j.Slf4j;
import org.aspectj.lang.ProceedingJoinPoint;
import org.aspectj.lang.annotation.Around;
import org.aspectj.lang.annotation.Aspect;

import java.util.Arrays;

@Slf4j(topic = "MyRetroAudit")
@AllArgsConstructor
@Aspect
public class MyRetroAuditAspect {

    private MyRetroAuditEventRepository eventRepository;
    private MyRetroAuditProperties properties;
    private MyRetroAuditStorage storage;

    @Around("@annotation(audit)")
    public Object auditAround(ProceedingJoinPoint joinPoint,
    MyRetroAudit audit) throws Throwable {
        MyRetroAuditEvent myRetroEvent = new MyRetroAuditEvent();
        myRetroEvent.setMethod(joinPoint.getSignature().getName());
        myRetroEvent.setArgs(audit.showArgs() ? Arrays.toString(joinPoint.
        getArgs()) : null);
        myRetroEvent.setMessage(audit.message());
```

```
    if (audit.intercept() == MyRetroAuditIntercept.BEFORE) {
        myRetroEvent.setInterceptor(MyRetroAuditIntercept.BEFORE.
        name());
    } else if (audit.intercept() == MyRetroAuditIntercept.AROUND) {
        myRetroEvent.setInterceptor(MyRetroAuditIntercept.AROUND.
        name());
    }

    Object result = joinPoint.proceed(joinPoint.getArgs());
    myRetroEvent.setResult(result.toString());

    if (audit.intercept() == MyRetroAuditIntercept.AFTER) {
        myRetroEvent.setInterceptor(MyRetroAuditIntercept.AFTER.
        name());
    }

    // Database, Console or File
    if (storage == MyRetroAuditStorage.DATABASE) {
        myRetroEvent = eventRepository.save(myRetroEvent);
    }

    // Logger or Console
    String formattedEvent = formatEvent(audit, myRetroEvent);
    if (properties.getUseLogger()) {
        log.info("{}{}", properties.getPrefix(), formattedEvent);
    } else {
        System.out.println(properties.getPrefix() + formattedEvent);
    }

    return result;
}

private String formatEvent(MyRetroAudit audit, MyRetroAuditEvent
myRetroEvent) {
    MyRetroAuditFormatStrategy strategy =
    MyRetroAuditFormatStrategyFactory.getStrategy(audit.format());
    return audit.prettyPrint() ? strategy.prettyFormat(myRetroEvent) :
    strategy.format(myRetroEvent);
}
}
```

In the `MyRetroAuditAspect` class, we are using the `MyRetroAuditEventRepository`, `MyRetroAuditProperties`, and `MyRetroAuditStorage` as fields that are required when this class is constructed. The `MyRetroAuditEventRepository` will help to persist the event in the database, the `MyRetroAuditProperties` will be used to get the properties' values from the `prefix`, `file`, and `useLogger` fields from this class that will talk about it later, and the `MyRetroAuditStorage` that will know the value from the `EnableMyRetroAuditValueProvider` class; but let's review it more in detail:

- `@Aspect`: This annotation indicates that this class is an AOP Aspect, and there are some configurations that need to be set.

- `@Around("@annotation(audit)")`: This annotation is an around advice that will be executed when the `@MyRetroAudit` annotation is found. With the around advice, we have control over what to call, what to send to the actual method call, and how to respond (use this carefully; with this, you can become a hacker!).

- `auditAround(ProceedingJoinPoint joinPoint, MyRetroAudit audit)`: This method is the *pointcut* that will be executed when the method is about to be called. For an around advice, will always require the `ProceedingJoinPoint`, and in this case, the `@MyRetroAudit` annotation, and this is because we need its parameter values, if any; if not, we are taking the defaults.

Before continuing, review this class and see the sections of the Database and the Logger.

Next, open/create the `MyRetroAuditProperties` class. See Listing 14-15.

Listing 14-15. src/main/java/com/apress/myretro/configuration/MyRetroAuditProperties.java

```
package com.apress.myretro.configuration;

import lombok.Data;
import org.springframework.boot.context.properties.ConfigurationProperties;

@Data
@ConfigurationProperties(prefix = "myretro.audit")
public class MyRetroAuditProperties {
```

```
/**
 * The prefix to use for all audit messages.
 */
private String prefix = "[AUDIT] ";
/**
 * The file to use for audit messages.
 */
private String file = "myretro.events";
/*
 * User logger instead of standard print out.
 */
private Boolean useLogger = false;
}
```

Listing 14-15 shows that in the @ConfigurationProperties annotation, we are defining the configurable properties that will be exposed as myretro.audit.* (prefix, file, useLogger). Look that we are creating some comments/documentation on each field, and this is because we want to be used as a hint/description when other developers use this custom starter. We are going to talk more about these hints in the following sections because we need to ensure that these hints are processed so they can be used by editors.

Next, open/create the MyRetroAuditConfiguration class. This will be the magic behind our custom starter; this is when the auto-configuration will happen. See Listing 14-16.

Listing 14-16. src/main/java/com/apress/myretro/configuration/ MyRetroAuditConfiguration.java

```
package com.apress.myretro.configuration;

import com.apress.myretro.annotations.EnableMyRetroAuditValueProvider;
import com.apress.myretro.aop.MyRetroAuditAspect;
import com.apress.myretro.model.MyRetroAuditEventRepository;
import org.springframework.boot.autoconfigure.AutoConfiguration;
import org.springframework.boot.autoconfigure.domain.EntityScan;
import org.springframework.boot.context.properties.
EnableConfigurationProperties;
```

```
import org.springframework.context.annotation.Bean;
import org.springframework.context.annotation.ComponentScan;
import org.springframework.context.annotation.Conditional;
import org.springframework.data.jpa.repository.config.
EnableJpaRepositories;

@EnableConfigurationProperties(MyRetroAuditProperties.class)
@Conditional(MyRetroAuditCondition.class)
@EnableJpaRepositories(basePackages = "com.apress")
@ComponentScan(basePackages = "com.apress")
@EntityScan(basePackages = "com.apress")
@AutoConfiguration
public class MyRetroAuditConfiguration {

    @Bean
    public MyRetroAuditAspect myRetroAuditAspect(MyRetroAuditEvent
    Repository myRetroAuditEventRepository, MyRetroAuditProperties
    properties) {
        return new MyRetroAuditAspect(myRetroAuditEventRepository,
        properties, EnableMyRetroAuditValueProvider.getStorage());
    }

}
```

Again, the `MyRetroAuditConfiguration` class is the magic that we are providing to developers who will be using this custom starter. The key piece here is the `@Conditional` annotation, which checks out if a condition is met, and in this case, that condition is from the `MyRetroAuditCondition` class (discussed shortly). If this condition is not met, then any declaration from this class is omitted/skipped.

The other annotations are standard to our custom starter. The `@AutoConfiguration` annotation indicates that a class provides configuration that Spring Boot can automatically apply. This annotation is a subclass (or composed annotation) of `@Configuration`, and it has the parameter `proxyBeanMethods` set to `false`.

Spring creates proxies for `@Bean` methods in configuration classes. These proxies ensure consistent bean lifecycle management and enforce singleton behavior even if you call the `@Bean` method directly. However, using `proxyBeanMethods = false` disables this behavior. With the `proxyBeanMethods = false`:

- No proxies are created for @Bean methods.

- Calling a @Bean method directly creates a new instance of the bean. This can improve performance, as proxy creation and invocation have some overhead. However, singleton behavior is not enforced if you call the method directly.

And this is what we want, because we are creating the @Bean MyRetroAuditAspect! In Listing 14-16, we are creating a new instance and passing the necessary fields, MyRetroAuditEventRepository, MyRetroAuditProperties, and the value of the EnableMyRetroAuditValueProvider.getStorage().

Again, we are providing the magic behind using the @EnableMyRetroAudit annotation when it's declared. So, open/create the MyRetroAuditCondition class. See Listing 14-17.

Listing 14-17. src/main/java/com/apress/myretro/configuration/ MyRetroAuditCondition.java

```
package com.apress.myretro.configuration;

import com.apress.myretro.annotations.EnableMyRetroAudit;
import org.springframework.context.annotation.Condition;
import org.springframework.context.annotation.ConditionContext;
import org.springframework.core.type.AnnotatedTypeMetadata;

public class MyRetroAuditCondition implements Condition {
    @Override
    public boolean matches(ConditionContext context, AnnotatedTypeMetadata
    metadata) {
        return context.getBeanFactory().getBeansWithAnnotation(EnableMy
        RetroAudit.class).size() > 0;
    }
}
```

In the MyRetroAuditCondition class, if the call of this condition (which checks for the @EnableMyRetroAudit being declared) is false, then the auto-configuration skips it and doesn't continue with the rest of the configuration, and it goes to the next auto-configuration. But if the call returns true, the logic or statements of the auto-configuration continues.

Note that the MyRetroAuditCondition#matches the method using the context.
getBeanFactory().getBeansWithAnnotation to see if it's in the Spring registry map.

Next, we already defined our auto-configuration, and per the rules of creating a
custom starter, we need to tell Spring Boot where to get this auto-configuration. So, we
need to create the org.springframework.boot.autoconfigure.AutoConfiguration.
imports file in the META-INF/spring folder with the name of our auto-configuration
class. See Listing 14-18.

Listing 14-18. src/main/resources/META-INF/spring/org.springframework.
boot.autoconfigure.AutoConfiguration.imports

```
com.apress.myretro.configuration.MyRetroAuditConfiguration
```

We are defining the complete name (package.name) of the auto-configuration class.
Automatically, Spring Boot will see that we defined this org.springframework.boot.
autoconfigure.AutoConfiguration.imports file, and it will execute any logic behind
this auto-configuration class. You can add multiple classes; each class has to be on its
own line (with a carriage return). If you need to, you can take a peek at the spring-boot-
autoconfigure JAR and see this file.

As a final step, we need to give our developers hints about using the custom starter
properties. So, open/create the META-INF/additional-spring-configuration-
metadata.json file. See Listing 14-19.

Listing 14-19. src/main/resources/META-INF/additional-spring-configuration-
metadata.json

```
{
  "groups": [
    {
      "name": "myretro.audit",
      "type": "com.apress.myretro.configuration.MyRetroAuditProperties",
      "sourceType": "com.apress.myretro.configuration.
      MyRetroAuditProperties"
    }
  ],
  "properties": [
```

```
  {
    "name": "myretro.audit.prefix",
    "type": "java.lang.String",
    "sourceType": "com.apress.myretro.configuration.
    MyRetroAuditProperties",
    "defaultValue": "[AUDIT] ",
    "description": "Prefix for audit messages"
  },
  {
    "name": "myretro.audit.file",
    "type": "java.lang.String",
    "sourceType": "com.apress.myretro.configuration.
    MyRetroAuditProperties",
    "defaultValue": "myretro.events",
    "description": "File to write audit messages to"
  },
  {
    "name": "myretro.audit.useLogger",
    "type": "java.lang.Boolean",
    "sourceType": "com.apress.myretro.configuration.
    MyRetroAuditProperties",
    "defaultValue": "false",
    "description": "Enable audit logging"
  }
],
"hints": [
  {
    "name": "myretro.audit.prefix",
    "values": [
      {
        "value": "[AUDIT] ",
        "description": "Prefix for audit messages"
      },
      {
        "value": ">>> ",
```

```json
            "description": "Prefix for audit messages"
        }
      ]
    },
    {
      "name": "myretro.audit.file",
      "values": [
        {
          "value": "myretro.events",
          "description": "File to write audit messages to"
        },
        {
          "value": "myretro.log",
          "description": "File to write audit messages to"
        }
      ]
    },
    {
      "name": "myretro.audit.useLogger",
      "values": [
        {
          "value": "true",
          "description": "Enable audit logging"
        },
        {
          "value": "false",
          "description": "Disable audit logging, just print out console"
        }
      ]
    }
  ]
}
```

This JSON file is very straightforward; we just define some values and descriptions for our properties (`prefix`, `file`, `useLogger`). When we compile our project automatically, the `configuration-processor` will generate the necessary metadata for editors. This is very helpful for developers when they are required to know what those properties mean and what possible values they can be set to.

Building the myretro-spring-boot-starter Custom Starter

Next, let's compile our custom starter. If everything went smoothly in the previous sections, then you can build the custom starter with the following command:

```
./gradlew build
```

This command generates the `myretro-spring-boot-starter-0.0.1.jar` file in the `build/libs` folder, ready to be used! If you want to skip the following section of publishing the JAR to a Maven repository, you can do so and use this JAR *only* in the following form in your Users App project or My Retro App project (in your Users App or My Retro App `build.gradle` file):

```
//...
dependencies {
    //...
    implementation files('../myretro-spring-boot-starter/build/libs/
    myretro-spring-boot-starter-0.0.1.jar')
    //...
}
```

This declaration assumes that you have the `myretro-spring-boot-starter` project one level below.

But the idea behind a custom starter is that every developer has access to it, right? So, we need to publish it to a Maven repository.

Publishing the Custom Starter in GitHub As a Maven Artifact

To publish our starter, we can use GitHub as a Maven artifact repository. The good part is that GitHub gives you those publishing features, not only for Maven but also for Docker images.

To publish a Maven artifact, we need to do three things:

1. Create a GitHub token that allows us to write/publish to our repositories. Consult this documentation to create your personal token: https://docs.github.com/en/authentication/keeping-your-account-and-data-secure/managing-your-personal-access-tokens.

2. Create a new empty repository in GitHub as a regular project. In my case, I created a new repo here: https://github.com/felipeg48/myretro-spring-boot-starter.

3. Add the necessary declarations to the build.gradle file to publish the artifact.

Next, let's modify our build.gradle file. See Listing 14-20.

Listing 14-20. build.gradle

```
plugins {
    id 'java'
    id 'org.springframework.boot' version '3.2.2' apply false
    id 'io.spring.dependency-management' version '1.1.4'
    id 'maven-publish'
}

group = 'com.apress'
version = '0.0.1'

java {
    sourceCompatibility = '17'
}

configurations {
    compileOnly {
        extendsFrom annotationProcessor
    }
}
```

```
tasks.named('compileJava') {
    inputs.files(tasks.named('processResources'))
}

repositories {
    mavenCentral()
}

dependencyManagement {
    imports {
        mavenBom org.springframework.boot.gradle.plugin.SpringBootPlugin.
        BOM_COORDINATES
    }
}

dependencies {
    implementation 'org.springframework.boot:spring-boot-starter-web'
    implementation 'org.springframework.boot:spring-boot-starter-aop'
    implementation 'org.springframework.boot:spring-boot-starter-data-jpa'
    implementation 'com.fasterxml.jackson.datatype:jackson-datatype-jsr310'

    runtimeOnly 'com.h2database:h2'

    compileOnly 'org.projectlombok:lombok'

    annotationProcessor 'org.springframework.boot:spring-boot-
    configuration-processor'
    annotationProcessor 'org.projectlombok:lombok'

    testImplementation 'org.springframework.boot:spring-boot-starter-test'
}

tasks.named('test') {
    useJUnitPlatform()
}

publishing {
    publications {
        mavenJava(MavenPublication) {
            from components.java
```

```
    artifactId = 'myretro-spring-boot-starter'

    versionMapping {
        usage('java-api') {
            fromResolutionOf('runtimeClasspath')
        }
        usage('java-runtime') {
            fromResolutionResult()
        }
    }

    pom {
        name = 'My Retro Starter'
        description = 'A spring-boot-starter library example'

        licenses {
            license {
                name = 'The Apache License, Version 2.0'
                url = 'http://www.apache.org/licenses/
                LICENSE-2.0.txt'
            }
        }
        developers {
            developer {
                id = 'felipeg48'
                name = 'Felipe'
                email = ''
            }
        }
        scm {
            connection = 'scm:git:git://github.com/felipeg48/
            myretro-spring-boot-starter.git'
            developerConnection = 'scm:git:ssh://github.com/
            felipeg48/myretro-spring-boot-starter.git'
            url = 'https://github.com/felipeg48/myretro-spring-
            boot-starter'
        }
```

```
        }

    }

}
repositories {
    maven {
        name = "GitHubPackages"
        url = uri("https://maven.pkg.github.com/felipeg48/myretro-
        spring-boot-starter")
        credentials {
            username = project.findProperty("GITHUB_USERNAME") ?:
            System.getenv("GITHUB_USERNAME")
            password = project.findProperty("GITHUB_TOKEN") ?:
            System.getenv("GITHUB_TOKEN")
        }
    }
}

}
```

Let's review the modified build.gradle file per section:

- plugins: Here, we added the maven-publish plugin, which has the
  logic to publish the artifact to a Maven repository.

- publishing: This section helps to define the metadata necessary for
  the Maven repository; it also identifies how to sign in if the repository
  is private or if it requires credentials or some authentication
  mechanism.

- publishing.publications. Here, we are defining the metadata that
  the artifact needs to be registered in the Maven repository. Reviewing
  it is very straightforward.

- publishing.repositories: This section defines where the repository
  is located. We have a special URL, https://maven.pkg.github.
  com/felipeg48/myretro-spring-boot-starter. This URL is just
  a declaration and it is used to find the packages. It also defines
  any credentials needed, and in this case, we are looking for the

username and password using the `GITHUB_USERNAME` and `GITHUB_` `TOKEN` variables (which can be set in the `$HOME/.gradle/gradle.` `properties` file or with environment variables).

Now that we have everything in place, let's publish our artifact. Remember that you need to create an empty repository; in my case, I created the `myretro-spring-boot-starter` in GitHub (as a project). Execute the following command to publish the artifact in your repo (see Figure 14-1):

```
./gradlew publish
```

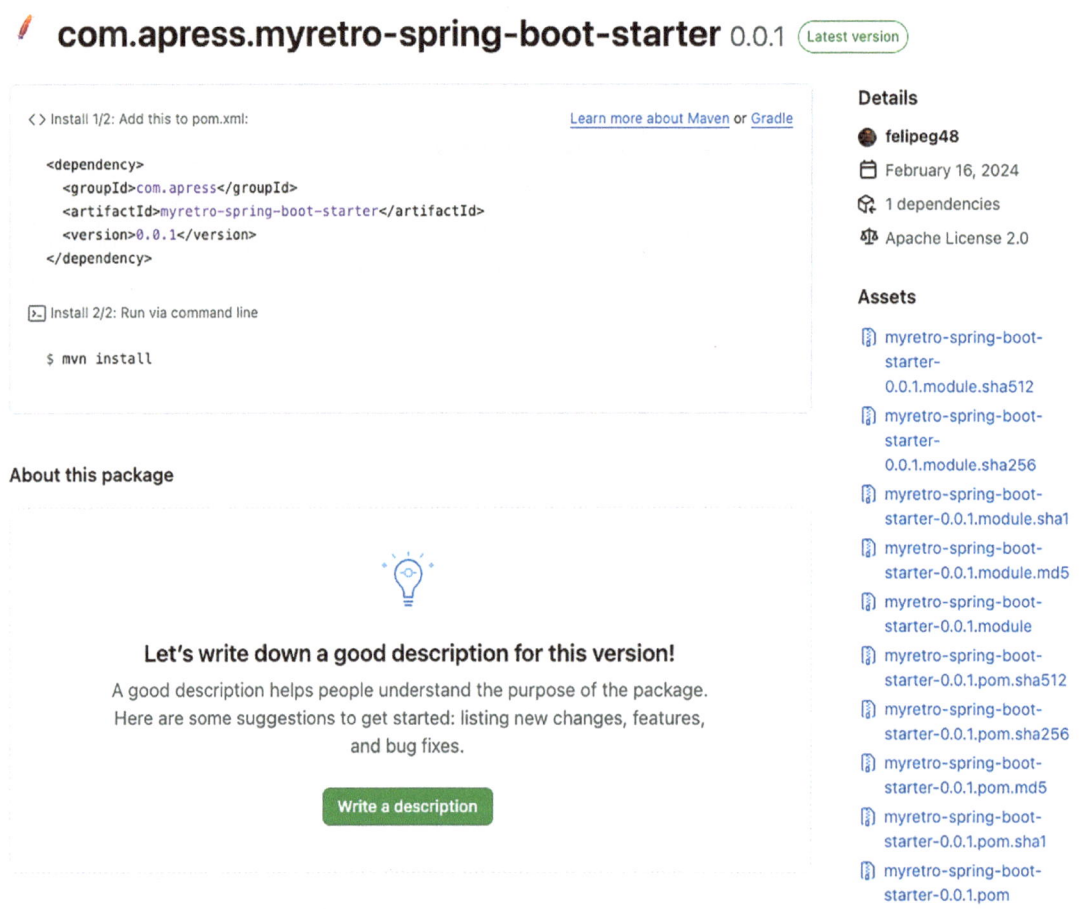

Figure 14-1. *Package in GitHub (`https://github.com/felipeg48/myretro-spring-boot-starter/packages`)*

As you can see, our package was deployed successfully. Now, it's time to use it.

Using the myretro-spring-boot-starter Custom Starter

Let's open the Users App project; you can import it into your favorite IDE from the 14-extending/users directory. Or if you want to start from scratch with the Spring Initializr (https://start.spring.io), set the Group field to com.apress and the Artifact and Name fields to users. Add as dependencies Web, JPA, Processor, Validation, Actuator, Lombok, and H2. Generate and download the project, unzip it, and import it into your favorite IDE.

Next, open the build.gradle file. See Listing 14-21.

Listing 14-21. build.gradle

```
plugins {
    id 'java'
    id 'org.springframework.boot' version '3.2.2'
    id 'io.spring.dependency-management' version '1.1.4'
    id 'org.hibernate.orm' version '6.4.1.Final'
}

group = 'com.apress'
version = '0.0.1-SNAPSHOT'

java {
    sourceCompatibility = '17'
}

configurations {
    compileOnly {
        extendsFrom annotationProcessor
    }
}

repositories {
    mavenCentral()
    maven {
```

```
        url 'https://maven.pkg.github.com/felipeg48/myretro-spring-
        boot-starter'
        credentials {
            username = project.findProperty("GITHUB_USERNAME") ?:
            System.getenv("GITHUB_USERNAME")
            password = project.findProperty("GITHUB_TOKEN") ?:
            System.getenv("GITHUB_TOKEN")
        }
    }
}

dependencies {
    implementation 'org.springframework.boot:spring-boot-starter-web'
    implementation 'org.springframework.boot:spring-boot-starter-
    validation'
    implementation 'org.springframework.boot:spring-boot-starter-data-jpa'
    implementation 'org.springframework.boot:spring-boot-starter-actuator'

    // H2 runtime only
    runtimeOnly 'com.h2database:h2'

    // My Retro Starter
        implementation 'com.apress:myretro-spring-boot-starter:0.0.1'

    compileOnly 'org.projectlombok:lombok'
    annotationProcessor 'org.projectlombok:lombok'
    annotationProcessor 'org.springframework.boot:spring-boot-
    configuration-processor'

    // Web
    implementation 'org.webjars:bootstrap:5.2.3'

    // Test
    testImplementation 'org.springframework.boot:spring-boot-starter-test'
}

tasks.named('test') {
    useJUnitPlatform()
}
```

Let's review the build.gradle file:

- repositories.maven: In this statement, we are using the url that we set to https://maven.pkg.github.com/felipeg48/myretro-spring-boot-starter, and then in the credentials section we are looking at the GITHUB_USERNAME and GITHUB_TOKEN variables (which can be set in the $HOME/.gradle/gradle.properties file or with environment variables). If your repository is public, then you can omit this credentials section.

- implementation 'com.apress:myretro-spring-boot-starter:0.0.1': Here we are using the artifact that we just published! Remember that if you need to test the JAR alone, you can use the following (assuming you have the projects in the same folder):

  implementation files('../myretro-spring-boot-starter/build/libs/myretro-spring-boot-starter-0.0.1.jar')

If you are using the source code from 14-extending/users, then you can continue following along, but if you are creating this from scratch, you can use the code from the JPA chapter by removing the events package and fixing the UserService where we are using/publishing the events. This will change now.

Next, open/create the UserService class. See Listing 14-22.

Listing 14-22. src/main/java/com/apress/users/UserService.java

```java
package com.apress.users.service;

import com.apress.myretro.annotations.MyRetroAudit;
import com.apress.myretro.annotations.MyRetroAuditOutputFormat;
import com.apress.users.actuator.LogEventEndpoint;
import com.apress.users.model.User;
import com.apress.users.repository.UserRepository;
import lombok.AllArgsConstructor;
import org.springframework.context.ApplicationEventPublisher;
import org.springframework.stereotype.Service;

import java.util.Optional;
```

```
@AllArgsConstructor
@Service
public class UserService {

    private UserRepository userRepository;
    private ApplicationEventPublisher publisher;
    private LogEventEndpoint logEventsEndpoint;

    public Iterable<User> getAllUsers() {
        return this.userRepository.findAll();
    }

    public Optional<User> findUserByEmail(String email) {
        return this.userRepository.findById(email);
    }

    @MyRetroAudit(showArgs = true, message = "Saving or updating user",
    format = MyRetroAuditOutputFormat.JSON, prettyPrint = false)
    public User saveUpdateUser(User user) {
        User userResult = this.user repository.save(user);
        return userResult;
    }

    public void removeUserByEmail(String email) {
        this.userRepository.deleteById(email);
    }
}
```

Listing 14-22 shows that we are using our @MyRetroAudit annotation, and we are using some parameters there.

Now, if you try to run the application (with ./gradlew bootRun), nothing will happen. There are no logs; there are no records of the events in the database. We are missing the @EnableMyRetroAudit annotation. So, let's add it. Open/create the UserConfiguration class. See Listing 14-23.

Listing 14-23. src/main/java/com/apress/users/config/UserConfiguration.java

```java
package com.apress.users.config;

import com.apress.myretro.annotations.EnableMyRetroAudit;
import com.apress.users.model.User;
import com.apress.users.model.UserRole;
import com.apress.users.service.UserService;
import org.springframework.boot.CommandLineRunner;
import org.springframework.boot.context.properties.
EnableConfigurationProperties;
import org.springframework.context.annotation.Bean;
import org.springframework.context.annotation.Configuration;

import java.util.List;

@EnableMyRetroAudit
@Configuration
@EnableConfigurationProperties({UserProperties.class})
public class UserConfiguration {

    @Bean
    CommandLineRunner init(UserService userService) {
        return args -> {
            userService.saveUpdateUser(new User("ximena@email.com",
                "Ximena", "https://www.gravatar.com/avatar/23bb62a7d0ca63c
                9a804908e57bf6bd4?d=wavatar", "aw2sOmeR!", List.of(UserRole.
                USER), true));
            userService.saveUpdateUser(new User("norma@email.com", "Norma",
                "https://www.gravatar.com/avatar/f07f7e553264c9710105edebe6c46
                5e7?d=wavatar", "aw2sOmeR!", List.of(UserRole.USER, UserRole.
                ADMIN), false));
        };
    }
}
```

Listing 14-23 shows that now we are using our @EnableMyRetroAudit annotation
(with no parameters, meaning that it will take the default value).

Before we run it, let's add some properties to our `application.yaml` file. See Listing 14-24.

Listing 14-24. src/main/resources/application.yaml

```
spring:
  application:
    name: users-service
  h2:
    console:
      enabled: true

  jpa:
    generate-ddl: true
    show-sql: true
    hibernate:
      ddl-auto: update

  datasource:
    url: jdbc:h2:mem:users_db

info:
  developer:
    name: Felipe
    email: felipe@email.com
  api:
    version: 1.0

management:
  endpoints:
    web:
     exposure:
        include: health,info,event-config,shutdown,configprops,beans
  endpoint:
    configprops:
      show-values: always
    health:
      show-details: always
```

```
    status:
      order: events-down, fatal, down, out-of-service, unknown, up
    shutdown:
      enabled: true

  info:
    env:
      enabled: true

server:
  port: ${PORT:8091}

myretro:
  audit:
    useLogger: true
    prefix: '>>> '
```

Listing 14-24 shows that we are using the myretro.audit.* properties to set the useLogger and prefix. If you play around with the properties in an IDE, you should be able to see the help, description, and hints that we added in the additional-spring-configuration-metadata.json file.

Now we are ready to run the app.

Running the Users App with myretro-spring-boot-starter

To run the Users App, you can use your IDE or use the following command:

./gradle bootRun

When you run it, in the UserConfiguration#init method, we are using the UserService to save two users, so you should see the following output in your console:

...
...
INFO 19475 --- [users-service] [main] **MyRetroAudit: >>>**
{"id":1,"timestamp":"2024-02-20 18:34:38","interceptor":"BEFORE","metho
d":"saveUpdateUser","args":"[User(email=ximena@email.com, name=Ximena,
gravatarUrl=https://www.gravatar.com/avatar/23bb62a7d0ca63c9a804908e57bf6bd
4?d=wavatar, password=aw2s0meR!, userRole=[USER], active=true)]","result":"

User(email=ximena@email.com, name=Ximena, gravatarUrl=https://www.gravatar.
com/avatar/23bb62a7d0ca63c9a804908e57bf6bd4?d=wavatar, password=aw2sOmeR!,
userRole=[USER], active=true)","message":"Saving or updating user"}

. . . .

. . . .

INFO 19475 --- [users-service] [main] **MyRetroAudit**: **>>>**
{"id":2,"timestamp":"2024-02-20 18:34:38","interceptor":"BEFORE","met
hod":"saveUpdateUser","args":"[User(email=norma@email.com, name=Norma,
gravatarUrl=https://www.gravatar.com/avatar/f07f7e553264c9710105edebe6c46
5e7?d=wavatar, password=aw2sOmeR!, userRole=[USER, ADMIN], active=false)]
","result":"User(email=norma@email.com, name=Norma, gravatarUrl=https://
www.gravatar.com/avatar/f07f7e553264c9710105edebe6c465e7?d=wavatar,
password=aw2sOmeR!, userRole=[USER, ADMIN], active=false)","message":"Savin
g or updating user"}

. . . .

. . . .

Yeah! We have our custom starter using the logger and the JSON format.

Now, let's check if the events were persisted in the database. Because in the
application.yaml file, we enabled the H2 console (with spring.h2.console.
enabled=true). Open your browser and go to http://localhost:8091/h2-console.
In the url field, enter jdbc:h2:mem:users_db, and then click Connect. You will see two
tables, PEOPLE and MY_RETRO_AUDIT_EVENT. Select the MY_RETRO_AUDIT_EVENT table and
run the following SQL statement:

SELECT * FROM MY_RETRO_AUDIT_EVENT

You should see two rows listed, as shown in Figure 14-2.

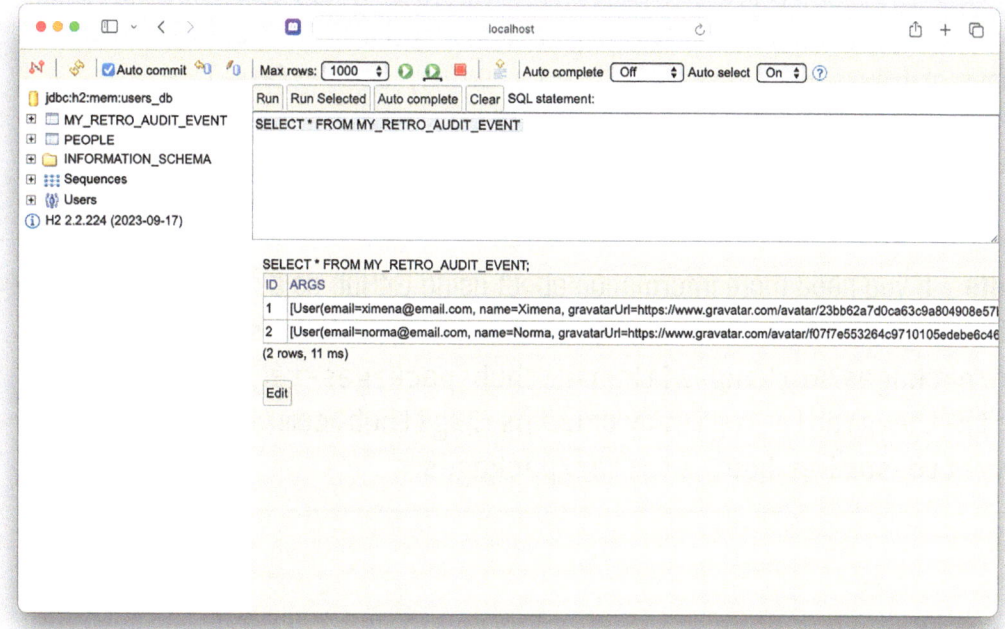

Figure 14-2. *http://localhost:8091/h2-console*

Congrats! You have just created your Spring Boot starter!

You can experiment with all the setting we added; for example, if you change the `prettyPrint = true` and rerun the app, you should see something like the following output (in a pretty JSON format!):

```
2024-02-20T18:45:26.405-05:00  INFO 20644 --- [users-service] [      main]
MyRetroAudit                              : >>>

{
  "id" : 2,
  "timestamp" : "2024-02-20 18:45:26",
  "interceptor" : "BEFORE",
  "method" : "saveUpdateUser",
  "args" : "[User(email=norma@email.com, name=Norma, gravatarUrl=https://
  www.gravatar.com/avatar/f07f7e553264c9710105edebe6c465e7?d=wavatar,
  password=aw2s0meR!, userRole=[USER, ADMIN], active=false)]",
```

```
  "result" : "User(email=norma@email.com, name=Norma, gravatarUrl=https://
  www.gravatar.com/avatar/f07f7e553264c9710105edebe6c465e7?d=wavatar,
  password=aw2sOmeR!, userRole=[USER, ADMIN], active=false)",
  "message" : "Saving or updating user"
}
```

Note   If you need more information about using GitHub as a registry (not only for Maven but also for Docker images), check out `https://docs.github.com/en/packages/working-with-a-github-packages-registry/working-with-the-gradle-registry` or `https://github.com/felipeg48/myretro-spring-boot-starter/packages`.

Summary

In this chapter, you learned how to create your own Spring Boot starter. You also learned more about the `@Conditional*`, `@Enable*`, and `@AutoConfiguration` annotations that perform the magic behind the scenes to create what you need.

You learned about the `Condition` interface and how you can modify it to go through even more configuration and logic to create your own beans or skip to the next auto-configuration.

You learned that you need to declare your auto-configuration classes in the `META-INF/spring/org.springframework.boot.autoconfigure.AutoConfiguration.imports`. You also learned more about the `BeanFactoryPostProcessor` and how to use it to initialize your beans or find information like you did in your `@EnableMyRetroAudit` annotation.

Now that you have a clearer understanding of how Spring Boot works and what you can do with it, in Chapter 15 we are going to review two new Spring projects: Spring Modulith and Spring AI.

Spring Boot New Projects

In this chapter, we are going to review two new Spring projects that are based on Spring Boot and deliver new technologies: *Spring Modulith* and *Spring AI*. Spring Modulith can help you to build modular applications with Spring Boot, and Spring AI can help you to create an out-of-the-box interface with OpenAI/ChatGPT.

Spring Modulith

Imagine building a complex city, not brick by brick, but by assembling predesigned districts, each with its own function and character. This is the essence of Spring Modulith (`https://spring.io/projects/spring-modulith`) for building software. Just as Spring Boot is an opinionated runtime (as discussed in Chapter 1), Spring Modulith is an opinionated a set of tools with which to organize your application, not just technically but also functionally.

Like Spring Boot provides a blueprint for the technical foundation, Spring Modulith guides you in structuring your app's core functionalities as distinct, interacting modules. This approach makes your application more modular and adaptable, allowing you to easily swap or update individual modules as your business needs evolve.

In short, Spring Modulith helps you to build software that's easier to change and grow with your business, just like a city adapts to its residents' needs.

Comparing Spring Modulith with Microservices

Both Spring Modulith and microservices aim to build complex applications efficiently, but they take different approaches. To understand the differences, consider the analogy of constructing a metropolis.

© Felipe Gutierrez 2024
F. Gutierrez, *Pro Spring Boot 3*, https://doi.org/10.1007/978-1-4842-9294-5_15

The Spring Modulith approach resembles building well-defined districts within a single city. Each district has its own function (e.g., shopping, residential, industrial) and interacts with other districts through defined channels (roads, bridges). In Spring Modulith, each module has its own function and interacts with other modules through defined channels. The benefits of this approach include

- *Easier development and deployment*: You build and deploy the application, simplifying initial setup and maintenance.

- *Reduced complexity*: Communication between modules happens internally, avoiding the overhead of network calls in microservices.

- *Faster iteration*: Updates can be made within modules without affecting the entire system, allowing quicker changes.

By comparison, the microservices approach resembles building entirely independent cities. Each city (microservice) is self-sufficient and interacts with other cities (microservices) through, the equivalent of APIs for interaction between cities (APIs). Benefits of this approach include the following:

- *High scalability*: Each service can scale independently based on its needs, making the overall system more flexible.

- *Technology independence*: Different services can use different technologies, fostering innovation and flexibility.

- *Resilience*: Failure in one service doesn't bring down the entire system, improving fault tolerance.

However, microservices also come with the following drawbacks:

- *Increased complexity*: Development, deployment, and communication between services are more involved.

- *Performance overhead*: Network calls between services can add latency and complexity.

- *Distributed complexity*: Debugging and monitoring become more challenging across multiple services.

Choosing the Right Approach

The best approach depends on your specific needs. Spring Modulith is ideal for:

- *Smaller applications or initial stages of development*: Its simplicity and faster iteration make it great for starting projects.

- *Applications with tightly coupled functionalities*: When modules rely heavily on each other, Spring Modulith's internal communication can be more efficient.

- *Limited technical resources*: Spring Modulith's centralized deployment and configuration are easier to manage with smaller teams.

Microservices are better suited for:

- *Large, complex applications with independent functionalities*: When each service can operate independently, microservices offer better scalability and resilience.

- *Teams with diverse technical expertise*: Microservices allow different technologies for different services, leveraging team strengths.

- *Need for high availability and fault tolerance*: The distributed architecture of microservices minimizes the impact of failures in individual services.

Ultimately, the best approach depends on your project's needs and constraints. Consider the trade-offs between simplicity and flexibility before building your software metropolis!

Fundamentals

To use Spring Modulith in your projects, you must add the following to your build. gradle file and some of the Spring Modulith libraries you will be using:

```
dependencyManagement {
    imports {
        mavenBom 'org.springframework.modulith:spring-modulith-bom:1.1.2'
    }
}
```

```
dependencies {
//...

implementation 'org.springframework.modulith:spring-modulith-starter-core'
implementation 'org.springframework.modulith:spring-modulith-starter-jpa'
testImplementation 'org.springframework.modulith:spring-modulith-
starter-test'

//...
}
```

Spring Modulith helps developers organize Spring Boot applications into logical building blocks called *modules*. It provides tools to

- *Validate the structure*: Ensure the modules are well-organized and adhere to best practices.

- *Document the arrangement*: Create clear documentation of how the modules interact.

- *Test modules independently*: Conduct integration tests on individual modules without relying on the entire application.

- *Monitor module interactions*: Observe how modules communicate and behave during runtime.

- *Promote loose coupling*: Encourage interactions between modules that avoid tight dependencies.

Spring Boot applications can be organized into modules, each focused on a specific function. These modules have three key parts:

- *Public interface*: This is like a service menu, offering functionalities (implemented as Spring beans) and events that other modules can access.

- *Internal workings*: This is the "kitchen" where the module's magic happens, hidden from other modules.

- *Dependencies*: Like ingredients needed for a recipe, modules rely on functionalities (beans), events, and configuration settings provided by other modules.

Spring Modulith offers various ways to build these modules with different levels of complexity. This lets developers start simple and gradually add more advanced features as needed.

Understanding Module Packages in Spring Boot

Let's review how Spring Modulith takes in interprets your code packages:

- *Main package*: This is where your main application class lives, usually annotated with @SpringBootApplication and containing the main method that starts the application.

- *Sub-packages*: Any package directly under the main package is considered an *application module package*. If there are no sub-packages, the main package itself becomes the module.

- *Code visibility*: Using Java's package scope, code within a module package is hidden from other packages. This means classes cannot be directly injected into other modules, which promotes loose coupling.

- *Module API*: By default, the public classes in a module package form its API, accessible to other modules. This API defines how the module can be interacted with.

As an example, consider the Users App structure:

```
Users
└── src/main/java
    ├── com.apress.users
    │   └── UsersApplication.java
    └── com.apress.users.model
        ├── User.java
        └── UserRole.java
```

Spring Modulith will take the `com.apress.users.*` as an application module `users`. Let's continue with the complete solution to see Spring Modulith in action.

Using Spring Modulith in My Retro Solution

Throughout the book, we have been working on two packages, Users App and My Retro App. Now it's time to create a Modulith with them and see how Spring Modulith can help us to create a robust modular single application.

Our My Retro App solution will consist of merging both projects into one. The complete source code is available in the `15-new-techs/myretro-modulith` folder, from which you can import it into your favorite IDE.

If you want to start from scratch, you must create a folder named `myretro-modulith`, copy the `myretro` and `users` packages into the same `src/` folder, and move/create the `MyretroApplication` class in the `com.apress` package level. Remove any other class that contains the `@SpringBootApplication` annotation (such as `UsersApplication`). You should end up with a structure like that shown in Figure 15-1.

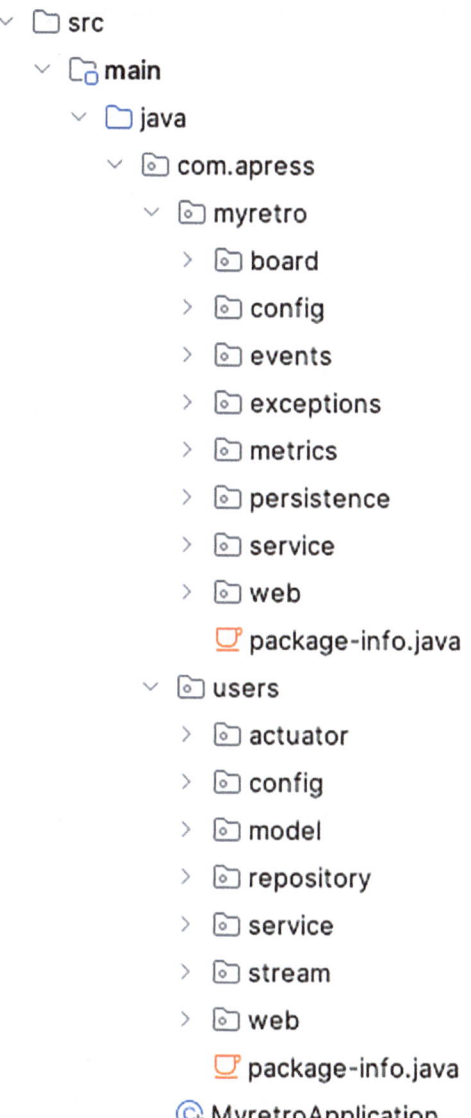

Figure 15-1. *Structure of myretro-modulith solution*

Figure 15-1 shows the final structure we are going to use. Open/create the `build.gradle` file. See Listing 15-1.

Listing 15-1. build.gradle

```
plugins {
    id 'java'
    id 'org.springframework.boot' version '3.2.2'
    id 'io.spring.dependency-management' version '1.1.4'
    id 'org.hibernate.orm' version '6.4.1.Final'
}

group = 'com.apress'
version = '0.0.1-SNAPSHOT'

java {
    sourceCompatibility = '17'
}

configurations {
    compileOnly {
        extendsFrom annotationProcessor
    }
}

repositories {
    mavenCentral()
    maven {
        url 'https://maven.pkg.github.com/felipeg48/myretro-spring-
        boot-starter'
        credentials {
            username = project.findProperty("GITHUB_USERNAME") ?: System.
            getenv("GITHUB_USERNAME")
            password = project.findProperty("GITHUB_TOKEN") ?: System.
            getenv("GITHUB_TOKEN")
        }
    }
}
```

```
dependencies {
    implementation 'org.springframework.boot:spring-boot-starter-web'
    implementation 'org.springframework.boot:spring-boot-starter-
    validation'
    implementation 'org.springframework.boot:spring-boot-starter-data-jpa'
    implementation 'org.springframework.boot:spring-boot-starter-actuator'

    // H2 runtime only
    runtimeOnly 'com.h2database:h2'

    // Modulith
    implementation 'org.springframework.modulith:spring-modulith-
    starter-core'
    implementation 'org.springframework.modulith:spring-modulith-
    starter-jpa'

    compileOnly 'org.projectlombok:lombok'
    annotationProcessor 'org.projectlombok:lombok'
    annotationProcessor 'org.springframework.boot:spring-boot-
    configuration-processor'

    // Web
    implementation 'org.webjars:bootstrap:5.2.3'

    // Test
    testImplementation 'org.springframework.boot:spring-boot-starter-test'
    testImplementation 'org.springframework.modulith:spring-modulith-
    starter-test'
}

dependencyManagement {
    imports {
        mavenBom 'org.springframework.modulith:spring-modulith-bom:1.1.2'
    }
}

tasks.named('test') {
    useJUnitPlatform()
}
```

```
test {
    testLogging {
        events "passed", "skipped", "failed"

        showExceptions true
        exceptionFormat "full"
        showCauses true
        showStackTraces true

        // Change to `true` for more verbose test output
        showStandardStreams = true
    }
}
```

Listing 15-1 shows that we are using the dependencyManagement section, and we are declaring the spring-modulith-starter-core, spring-modulith-starter-jpa, and spring-modulith-starter-test dependencies that will help to make sure our application follows the modular structure we need.

Reviewing Figure 15-1 again, note that the package-info java at the top of the com. apress.myretro and com.apress.users packages. This is useful information that can help based on Java 9 modularity, to use this information as part of your modular application. See Listings 15-2 and 15-3.

Listing 15-2. src/main/java/com/apress/myretro/package-info.java

```
@org.springframework.lang.NonNullApi
package com.apress.myretro;
```

Listing 15-3. src/main/java/com/apress/users/package-info.java

```
@org.springframework.lang.NonNullApi
package com.apress.users;
```

Normally, a package-info.java can help to specify package-level visibility modifiers or declare custom annotations for your own framework or library, as well as setting default annotations for tools like FindBugs, Lombok, Spring Security, and Spring Modulith.

Next, one of the benefits of using Spring Modulith in your Spring Boot app is that it provides module validation tests. Let's look. Create/open the ModularityTests class. See Listing 15-4.

Listing 15-4. src/main/test/com/apress/ModularityTests.java

```
package com.apress;

import org.junit.jupiter.api.Test;
import org.springframework.modulith.core.ApplicationModules;
import org.springframework.modulith.docs.Documenter;

public class ModularityTests {

    ApplicationModules modules = ApplicationModules.
    of(MyretroApplication.class);

    @Test
    void verifiesModularStructure() {
        modules.verify();
    }

    @Test
    void createApplicationModuleModel() {
        ApplicationModules modules = ApplicationModules.of
        (MyretroApplication.class);
        modules.forEach(System.out::println);
    }

    @Test
    void createModuleDocumentation() {
        new Documenter(modules).writeDocumentation();
    }
}
```

The first thing to review in Listing 15-4 is the ApplicationModules class, which will set up everything that we need to know about our apps and how modular our app is. We need to pass the name of our main app, in this case, the MyretroApplication class (where the @SpringBootApplication is declared).

Let's run the tests, one by one, and see the results. Let's start with the
verifiesModularStructure test:

```
./gradlew test --tests ModularityTests.verifiesModularStructure

> Task :test

ModularityTests > verifiesModularStructure() PASSED

BUILD SUCCESSFUL in 2s
5 actionable tasks: 5 executed
```

Recall that, normally, we can communicate from the My Retro App to the Users App
by requesting the getAllUsers by reaching out to the /users endpoint. Well, we can
certainly still communicate from the My Retro App to the Users App, but now we can do
it directly, without the help of the /users endpoint. The question is, is this okay?

Let's imagine that for every new User (saved into the database), we emit an event
that a user was saved with the actual use, right?

So, in the com.apress.users.service.UserService class, in the saveUpdateUser
method, use the following code:

```
private ApplicationEventPublisher events;

@Transactional
public User saveUpdateUser(User user) {
    User userResult = this.userRepository.save(user);

    // Only when the user is saved do we publish the event
    events.publishEvent(user);

    return userResult;
}
```

You already know about the ApplicationEventPublisher class and how to publish
an event, and in this case, we are just publishing the User.

Then, in the `com.apress.myretro.service.RetroBoardAndCardService` class, add the following code:

```
@Async
@EventListener
public void newSavedUser (User user){
    log.info("New user saved: {} {}",user.getEmail(), LocalDateTime.now());
}
```

You already know how to use the `EventListener`. Now, if we run the tests again, we will have the following output:

```
./gradlew tests --tests ModularityTests.verifiesModularStructure

> Task :compileJava
> Task :test FAILED

ModularityTests > verifiesModularStructure() FAILED
    org.springframework.modulith.core.Violations: - Module 'myretro'
    depends on non-exposed type com.apress.users.model.User within module
    'users'!
    User declares parameter User.newSavedRetroBoard(User) in
    (RetroBoardAndCardService.java:0)
    - Module 'myretro' depends on non-exposed type com.apress.users.model.
      User within module 'users'!
    Method <com.apress.myretro.service.RetroBoardAndCardService.
    newSavedRetroBoard(com.apress.users.model.User)> calls method <com.
    apress.users.model.User.getEmail()> in (RetroBoardAndCardService.
    java:64)
    - Module 'myretro' depends on non-exposed type com.apress.users.model.
      User within module 'users'!
    Method <com.apress.myretro.service.RetroBoardAndCardService.
    newSavedRetroBoard(com.apress.users.model.User)> has parameter of type
    <com.apress.users.model.User> in (RetroBoardAndCardService.java:0)
```

We have this legend, a failure due to the following violation: `Module 'myretro'` depends on non-exposed type `com.apress.users.model.User` within module `'users'`!

We are in a modular error because the User lives in an internal and private package that shouldn't be a dependency of any other class. So, how can we fix this? There are different ways to avoid these conflicts, so let's use Spring Modulith to help with all of this

First, let's create a UserEvent class that will hold some important information; at the end, probably, we don't want to pass all User's info (password, gravatar, active, etc), right? See Listing 15-5.

Listing 15-5. src/main/java/com/apress/users/UsersEvent.java

```
package com.apress.users;

import lombok.AllArgsConstructor;
import lombok.Data;
import lombok.NoArgsConstructor;

@AllArgsConstructor
@NoArgsConstructor
@Data
public class UserEvent {
    private String email;
    private String action;
}
```

Listing 15-5 shows that we are creating the UserEvent class in the com.apress.users package level. We are making sure that all other sublevels are kept private.

Next, in the UsersService#saveUpdateUser, change the event that instead of a User, you send and UserEvent:

```
events.publishEvent(new UserEvent(user.getEmail(), "save"));
```

And in the RetroBoardAndCardService#newSavedUser, change it using the following code:

```
@Async
@TransactionalEventListener
public void newSavedUser(UserEvent userEvent){
  log.info("New user saved: {} {} {}",userEvent.getEmail(), userEvent.
  getAction(), LocalDateTime.now());
}
```

We are now using a Spring Modulith annotation, @TransactionalEventListener. This is an annotation that inherits from @EventListner, and it has more logic within Spring Modulith that allows us to persist the event into the database if the dependency is added.

Now that the changes have been made, let's rerun our test:

```
./gradlew clean test --tests ModularityTests.verifiesModularStructure

> Task :compileJava
> Task :test

ModularityTests > verifiesModularStructure() PASSED

BUILD SUCCESSFUL in 2s
5 actionable tasks: 5 executed
```

Yes, our solution is modular! Now, what happens if we have some heavy dependencies that we cannot avoid? We can do the following in the package-info.java:

```
@org.springframework.lang.NonNullApi
@org.springframework.modulith.ApplicationModule(
  allowedDependencies = "users"
)
package com.apress.myretro;
```

In this case, code in the myretro module was only allowed to refer to code in the users module (and code not assigned to any module in the first place).

Spring Modulith can print out your modules and dependencies, and not only that, you can generate documentation as well. So, next, run the ModularityTests#createApp licationModuleModel test:

```
./gradlew clean test --tests ModularityTests.createApplicationModuleModel

> Task :compileJava
> Task :test

ModularityTests STANDARD_OUT
    15:07:18.494 [Test worker] INFO com.tngtech.archunit.core.PluginLoader
    -- Detected Java version 17.0.9
```

ModularityTests > createApplicationModuleModel() STANDARD_OUT

    # Myretro
    > Logical name: myretro
    > Base package: com.apress.myretro
    > Spring beans:
      o ....config.RetroBoardConfig
      o ....events.RetroBoardLog
      o ....metrics.RetroBoardMetrics
      o ....persistence.RetroBoardRepository
      o ....service.RetroBoardAndCardService
      o ....web.RetroBoardController
      o io.micrometer.core.instrument.Counter
      o io.micrometer.observation.aop.ObservedAspect
      o org.springframework.boot.CommandLineRunner
      o org.springframework.web.servlet.handler.MappedInterceptor

    # Users
    > Logical name: users
    > Base package: com.apress.users
    > Spring beans:
      o ....actuator.EventsHealthIndicator
      o ....actuator.LogEventEndpoint
      o ....config.UserConfiguration
      o ....config.UserProperties
      o ....repository.UserRepository
      o ....service.UserService
      o ....stream.UserProcessor
      o ....stream.UserSource
      o ....web.UsersController
      o java.util.function.Function
      o java.util.function.Supplier
      o org.springframework.boot.CommandLineRunner

```
ModularityTests > createApplicationModuleModel() PASSED

BUILD SUCCESSFUL in 1s
5 actionable tasks: 5 executed
```

The output shows you the modules, and its submodules. With this output, you can better understand what to do when a dependency is needed.

Next, let's run the `ModularityTests#createModuleDocumentation` test:

```
./gradlew clean test --tests ModularityTests.createModuleDocumentation

> Task :compileJava
> Task :test

ModularityTests STANDARD_OUT
    15:09:46.442 [Test worker] INFO com.tngtech.archunit.core.PluginLoader
    -- Detected Java version 17.0.9

ModularityTests > createModuleDocumentation() PASSED

BUILD SUCCESSFUL in 1s
5 actionable tasks: 5 executed
```

This test generates the documentation in AsciiDoc format (https://asciidoctor. org/), similar to a markdown language. And it will generate a PlantUML (https:// plantuml.com/) code so that you can generate the graph of your modules. The generated docs and code are in the build/spring-modulith-docs folder. See Figure 15-2.

MyretroApplication

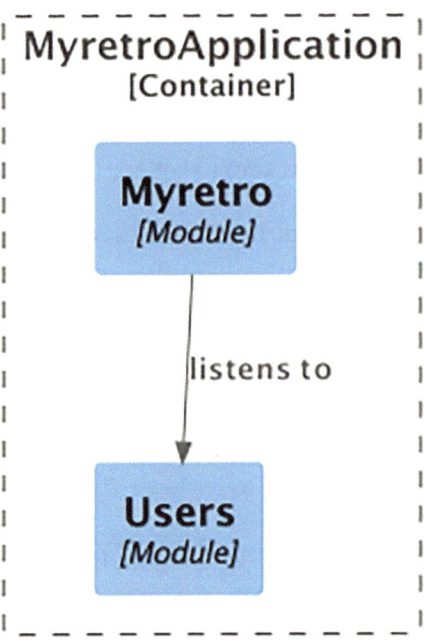

Legend
component
□ container boundary (dashed)

Figure 15-2. Spring Modulith docs

Note   In JetBrains IntelliJ IDEA, there is a plugin for AsciiDoc and PlantUML. You also need to install GraphViz (`https://plantuml.com/graphviz-dot`) so that you can visualize the graphs.

Running the My Retro Solution

Run the application either using your IDE or using the following command:

```
./gradlew bootRun
```

Now check the `/h2-console` endpoint. Use the URL `jdbc:h2:mem:myretro_db` and click Connect, and you will see not only the `PEOPLE`, `CARD`, and `RETRO_BOARD` tables but also the `EVENT_PUBLICATION` table, which includes everything that Spring Modulith does when there is an event. See Figure 15-3.

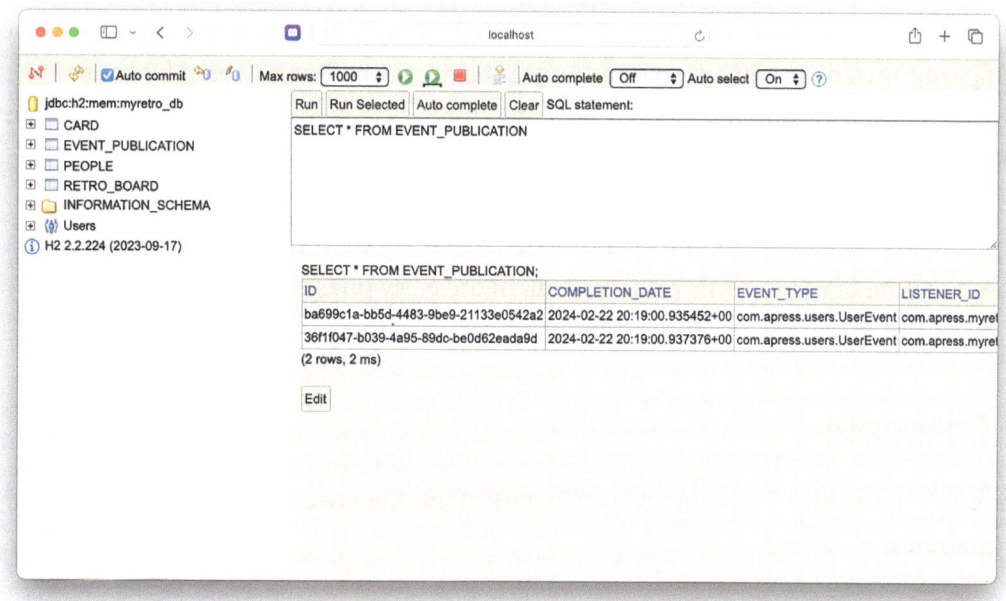

Figure 15-3. *`http://localhost:8080/h2-console` - URL: jdbc:h2:mem:myretro_db*

This section has provided a sneak peek of Spring Modulith and how it can help you create a modular application with Spring Boot. Spring Modulith is in General Availability (GA), and at the time of this writing the version is 1.1.2. If you want to learn more about Spring Modulith, visit `https://docs.spring.io/spring-modulith/reference/index.html`.

Spring AI

Spring AI (`https://spring.io/projects/spring-ai`) helps developers build AI-powered applications without getting bogged down in complexity. Inspired by Python projects like LangChain and LlamaIndex, Spring AI is specifically designed for developers using various programming languages, not just Python.

At its core, Spring AI provides building blocks (abstractions) for different AI tasks. These blocks can be easily swapped out, allowing you to switch between different tools (like OpenAI, Azure OpenAI, Hugging Face, etc.) with minimal code changes. This makes your applications flexible and adaptable.

Beyond basic blocks, Spring AI offers prebuilt solutions for common tasks. Imagine asking your own documentation questions or having a chatbot powered by your documentation! As your needs grow, Spring AI integrates with other Spring tools (such as Spring Integration and Spring Data) to handle complex workflows.

Spring AI focuses on AI models that process language input and provide language output. Some of the models Spring AI uses are GPT-3.5, GPT-4, and much more.

At the time of writing, this project is still in the early phases and is in version 0.9.0-SNAPSHOT, but it's mature enough to try.

AI Concepts

This section provides a brief overview of some of the AI concepts that Spring AI incorporates.

Models

Think of AI models as powerful tools that learn from tons of data. They analyze information, identify patterns, and even mimic how we think. This enables them to create things like text, images, and predictions, helping humans in many ways across different fields.

There are many types of AI models, each with its own specialty. Think of ChatGPT, which generates text based on what you type. But there are also models that turn text into images, like Midjourney and Stable Diffusion!

Spring AI currently focuses on models that understand and respond in language. Think of it like talking to a friend who's super good at understanding and replying. We're starting with OpenAI and Azure OpenAI, but there's more to come!

What makes models like ChatGPT special? They're already pretrained, like having a head start in learning. This makes them easier to use, even if you're not an AI expert!

Prompts

Imagine talking to a super smart friend, but they need specific instructions to understand you. That's where *prompts* come in! They tell AI models like ChatGPT what to do with your words.

- *More than just text*: Unlike asking, "Hey Google, what's the capital of France?", prompts in ChatGPT can have different parts like "Tell me a story" (user role) and "Once upon a time…" (system role) to set the scene.

- *Crafting prompts is an art*: It's not just about typing. Like talking to a friend, you must be clear and guide the AI model in the right direction.

- *Learning to speak "AI"*: Since interacting with AI models is different from asking questions, like in SQL, experts are learning how to best talk to AI models. This is called *prompt engineering*, and it's helping us get better results!

- *Sharing tips and tricks*: People are even sharing their best prompts, and researchers are studying how to make them even more effective.

- *It's not always easy*: Just like learning a new language, it takes practice to master prompts. Even the best models like ChatGPT 3.5 might not understand us perfectly, but we're improving daily!

- *Remember*: Prompts are the key to unlocking the full potential of AI models. By understanding them better, we can build even more amazing things together!

Prompt Templates

Crafting good prompts for AI models is like setting the stage for a play. You need to

1. *Set the scene*: Explain what you want the AI model to do by providing context.

2. *Fill in the blanks*: Replace parts of your request with specific details from the user's input.

The following...

- *Think of templates like scripts*: We use a tool called `StringTemplate` to create templates with placeholders for those details.

  Example: Imagine a template that says, `"Tell me a {funny} joke about {cats}."` When someone asks for a joke, we replace the placeholders with their input (e.g., "Tell me a silly joke about puppies").

- *Templates are like views in apps*: They provide a structure, and we fill it with data (like a map) to create the final prompt for the AI model.

- *Prompts are getting more complex*: They used to be simple text, but now they can have multiple parts with different roles for the AI model.

Embeddings

Imagine you're a Java developer and you want to add AI features to your app. You might come across the term "embeddings," which sounds complex. But worry not! You don't need to be a math genius to use embeddings.

Here's the gist: Embeddings take text (like sentences) and turn it into numbers (like arrays). This helps AI models "understand" the meaning of the text. Think of it like translating words into a language the AI model speaks.

This transformation is particularly useful for tasks like:

- *Finding similar things*: Imagine you have a product recommendation system. Embeddings can help the AI model group similar products together based on their "meaning" in the text descriptions.

- *Classifying text*: Want to automatically categorize emails as either spam or important? Embeddings can help the AI model understand the meaning of the email content and classify it correctly.

Think of it like a map: Instead of words, the AI model uses these numbers (embeddings) to navigate a map of meaning. Similar words and sentences are closer on this map, making it easier for the AI model to identify connections and relationships.

So, while the "how" behind embeddings is complex, understanding their "what" and "why" is crucial for using them effectively in your Java applications. They can be powerful tools for adding intelligence and functionality to your projects!

Tokens

AI models are somewhat like word processors:

- *Input*: The AI model breaks down sentences into smaller units called *tokens*, similar to words. One token is roughly 3/4 of a word.

- *Processing*: Internally, the AI model uses these tokens to understand the meaning of your text.

- *Output*: Finally, it converts the tokens back into words to give you the answer.

Think of tokens as the "currency" for AI models:

- You pay for using AI models based on the number of tokens you use, both for input and output.

- Each model has a limit on how many tokens it can handle at once, called the *context window*. Imagine it as being like a short message box!

- Different models have different limits: ChatGPT3 has a 4,000-token limit, while others offer options like 8,000 or even 100,000!

- This token limit means that if you want to analyze massive amounts of text, like Shakespeare's works, you need to break it down into smaller chunks to fit within the model's limit.

- Spring AI helps you with this! It provides tools to chop up your data and present it to the model in the right way, maximizing efficiency and avoiding hitting those limits.

Teaching AI New Tricks: Beyond the Training Dataset

Imagine you have an AI model trained on a massive dataset, like GPT-3.5/4.0. The model is great, but its knowledge stops in September 2021! What if you need answers about newer events? Here are your options to "equip" the model with beyond-the-training data:

- *Fine-tuning (expert mode)*: Think of this as rewiring the AI model's brain. You feed it your specific data and adjust its internal workings. This is powerful but complex, requiring machine learning expertise and many resources, especially for large models like GPT. Plus, not all models offer this option.

- *Prompt stuffing (more practical)*: This is like whispering hints to the AI model. Instead of rewiring it, you embed your data within the question or prompt you give it. But there's a catch: the AI model has limited attention (like a short message box). Techniques are needed to fit your data within the model's context window. Think of it as "stuffing the prompt" with relevant info.

- *Spring AI to the rescue*: Don't worry about the stuffing techniques! Spring AI provides tools to help you present your data effectively within the context window, maximizing what the AI model can learn from your prompts. It's like having a helper to whisper the right information at the right time.

And there are more AI concepts that Spring AI uses, but for now, I think this will be enough to continue with the following example.

Create a Chat GPT Client

Spring AI is still in progress, not yet GA, but mature enough to do a lot of AI applications. One of the easy and out-of-the-box solutions is that Spring AI has a Chat GPT client ready to use. So, let's start with that.

There is no Spring Initializr yet to create or include a Spring AI. The Spring AI team relies on the (also pre-release) *Spring CLI* (`https://spring.io/projects/spring-cli`) project that supports creating different Spring Boot apps with ease. So, let's start with these simple steps:

1. Install the Spring CLI:

    If you are on a Mac, you can install the Spring CLI with

    ```
    brew tap spring-cli-projects/spring-cli
    brew install spring-cli
    ```

    If you are on Windows, visit this link: `https://github.com/ spring-projects/spring-cli/releases/tag/early-access`. And make sure to do so something like this to execute the Spring CLI: `java -jar <YOUR-INSTALLATION>/spring-cli-0.8.1.jar'` On Windows, you can create a `spring.cmd` that executes the JAR. The idea is to have the `spring` command.

2. Create the Spring AI app with the following command:

    ```
    spring boot new --from ai --name myretro-ai
    ```

    This will generate the `myretro-ai` folder structure with Maven as a build tool by default. Now, you can import your project in your favorite IDE.

Note   At the time of writing, the Spring CLI supports only single-module Maven projects. Support for single-module Gradle projects is planned for the 1.0 release. No timeline is defined for supporting multi-module projects.

By default, the above command (spring boot new) will generate multiple files; first look at the README.md file, which includes all the instructions on how to run it. It will also create a package: `org.springframework.ai.openai.samples.helloworld`. It will generate the `Application` class and the `SimpleAiController`. You can open the `SimpleAiController` class as shown in Listing 15-6.

Listing 15-6. src/main/java/org/springframework/ai/openai/samples/ helloworld/simple/SimpleAiController.java

```
package org.springframework.ai.openai.samples.helloworld.simple;

import org.springframework.ai.chat.ChatClient;
import org.springframework.beans.factory.annotation.Autowired;
import org.springframework.web.bind.annotation.GetMapping;
```

```
import org.springframework.web.bind.annotation.RequestParam;
import org.springframework.web.bind.annotation.RestController;

import java.util.Map;

@RestController
public class SimpleAiController {

    private final ChatClient chatClient;

    @Autowired
    public SimpleAiController(ChatClient chatClient) {
        this.chatClient = chatClient;
    }

    @GetMapping("/ai/simple")
    public Map<String, String> completion(@RequestParam(value = "message",
    defaultValue = "Tell me a joke") String message) {
        return Map.of("generation", chatClient.call(message));
    }
}
```

The only important part to note in the SimpleAiController class is that we are using a ChatClient class. This class has all the logic to call to ChatGPT.

Running the App

To run the app, it is important to set the SPRING_AI_OPENAI_API_KEY with your key. You can get it from https://platform.openai.com/api-keys (after you sign up, of course; it's free). Then you can run it with

```
SPRING_AI_OPENAI_API_KEY=<yourkey> ./gradlew bootRun
```

You can execute a cURL command to get the result of using ChatGPT in a Spring Boot app with

```
curl localhost:8080/ai/simple
```

```
Why did the cow go to space?
Because it wanted to see the mooooon!
```

Congrats! You created a simple ChatGPT app!

Summary

In this chapter, we reviewed two of the most anticipated technologies from the Spring team: Spring Modulith, which enables you to create modular applications with Spring Boot and ensure your architecture is good and in the right place, and Spring AI, which helps you to create an out-of-the-box interface with OpenAI/ChatGPT.

APPENDIX A

Spring Boot Migration

If you need to migrate Spring Boot 2.x apps into the most recent version of Spring Boot 3.x, this appendix shows you how to do so. We will work on code snippets and not complete solutions here, but at least it will give you an idea of what you need to prepare for a migration. I'll present the code snippets in a From/To format for easy comparison and understanding.

Let's start with Spring Security and see what change and what you need to do.

Migrating Spring Security

It's important to know that since Spring Security 5.7, the WebSecurityConfigurerAdapter usage has been deprecated. So, you need to do the following when replacing the extends of this class.

Using HttpSecurity

In Spring Security 5.4, the Spring Security team introduced a way to configure the HttpSecurity class by creating a SecurityFilterChain bean. Example:

From:

```
@Configuration
public class UserSecurityConfig extends WebSecurityConfigurerAdapter{

    @Override
    protected void configure(HttpSecurity http) throws Exception {
        http
            .authorizeHttpRequests((auth) -> auth
                .anyRequest().authenticated()
            )
```

© Felipe Gutierrez 2024
F. Gutierrez, *Pro Spring Boot 3*, https://doi.org/10.1007/978-1-4842-9294-5

```
            .httpBasic(withDefaults());
    }

}
```

   To:

```
@Configuration
public class UserSecurityConfig {

    @Bean
    SecurityFilterChain filterChain(HttpSecurity http) throws Exception {
        http
                .authorizeHttpRequests( auth -> auth.anyRequest().
                 authenticated())
                .httpBasic(Customizer.withDefaults());

        return http.build();
    }
}
```

WebSecurity Configuration

Spring Security 5.4 also introduced the WebSecurityCustomizer. This is a callback interface for customizing WebSecurity. Beans of this type will automatically be used by WebSecurityConfiguration to customize WebSecurity.

   From:

```
@Configuration
public class UserSecurityConfig extends WebSecurityConfigurerAdapter{

    @Override
    public void configure(WebSecurity web) {
        web.ignoring().antMatchers("/about", "/docs");
    }

}
```

To:

```
@Configuration
public class UserSecurityConfig {

    @Bean
    public WebSecurityCustomizer webSecurityCustomizer() {
        return (web) -> web.ignoring().antMatchers("/about", "/docs");
    }

}
```

LDAP Authentication

Spring Security 5.7 introduced the EmbeddedLdapServerContextSourceFactoryBean, LdapBindAuthenticationManagerFactory, and LdapPasswordComparison AuthenticationManagerFactory classes. These classes can help to create an embedded LDAP server instead of the hassle of using one and configuring it (which presents a lot of issues, in my experience) and use an AuthenticationManager that performs the LDAP authentication.

From:

```
@Configuration
public class UserSecurityConfig extends WebSecurityConfigurerAdapter{

    @Override
    protected void configure(AuthenticationManagerBuilder auth) throws
    Exception {
        auth
            .ldapAuthentication()
            .userDetailsContextMapper(new PersonContextMapper())
            .userDnPatterns("uid={0},ou=people")
            .contextSource()
            .port(0);
    }
}
```

To:

```
@Configuration
public class UserSecurityConfig {
    @Bean
    public EmbeddedLdapServerContextSourceFactoryBean
    contextSourceFactoryBean() {
        EmbeddedLdapServerContextSourceFactoryBean
        contextSourceFactoryBean =
            EmbeddedLdapServerContextSourceFactoryBean.
            fromEmbeddedLdapServer();
        contextSourceFactoryBean.setPort(0);
        return contextSourceFactoryBean;
    }

    @Bean
    AuthenticationManager ldapAuthenticationManager(
            BaseLdapPathContextSource contextSource) {
        LdapBindAuthenticationManagerFactory factory =
            new LdapBindAuthenticationManagerFactory(contextSource);
        factory.setUserDnPatterns("uid={0},ou=people");
        factory.setUserDetailsContextMapper(new PersonContextMapper());
        return factory.createAuthenticationManager();
    }
}
```

In-Memory Authentication

As of Spring Security 5.X, configuring in-memory authentication is easier than
ever before.

From:

```
@Configuration
public class UserSecurityConfig extends WebSecurityConfigurerAdapter{
    @Override
```

```
    protected void configure(AuthenticationManagerBuilder auth) throws
    Exception {
        UserDetails user = User.withDefaultPasswordEncoder()
            .username("admin")
            .password("admin")
            .roles("ADMIN")
            .build();
        auth.inMemoryAuthentication()
            .withUser(user);
    }
}
```

To:

```
@Configuration
public class UserSecurityConfig {
    @Bean
    public InMemoryUserDetailsManager userDetailsService() {
        UserDetails admin = User
                .builder()
                .username("admin")
                .password(passwordEncoder.encode("admin"))
                .roles("ADMIN", "USER")
                .build();          return new InMemoryUserDetailsManager
                (admin);
    }

    @Bean
    PasswordEncoder passwordEncoder(){
        return new BCryptPasswordEncoder();
    }
}
```

JDBC Authentication

From:

```
@Configuration
public class UserSecurityConfig extends WebSecurityConfigurerAdapter{
    @Bean
    public DataSource dataSource() {
        return new EmbeddedDatabaseBuilder()
            .setType(EmbeddedDatabaseType.H2)
            .build();
    }

    @Override
    protected void configure(AuthenticationManagerBuilder auth) throws
    Exception {
        UserDetails user = User.withDefaultPasswordEncoder()
            .username("admin")
            .password("admin")
            .roles("ADMIN")
            .build();
        auth.jdbcAuthentication()
            .withDefaultSchema()
            .dataSource(dataSource())
            .withUser(user);
    }
}
```

To:

```
@Configuration
public class SecurityConfiguration {
    @Bean
    public DataSource dataSource() {
        return new EmbeddedDatabaseBuilder()
            .setType(EmbeddedDatabaseType.H2)
```

```
                .addScript(JdbcDaoImpl.DEFAULT_USER_SCHEMA_DDL_LOCATION)
                .build();
    }

    @Bean
    public UserDetailsManager users(DataSource dataSource) {
        UserDetails admin = User
                .builder()
                .username("admin")
                .password(passwordEncoder.encode("admin"))
                .roles("ADMIN", "USER")
                .build();

JdbcUserDetailsManager users = new JdbcUserDetailsManager(dataSource);
        users.createUser(admin);
        return users;
    }

    @Bean
    PasswordEncoder passwordEncoder(){
        return new BCryptPasswordEncoder();
    }

}
```

Upgrade to Spring Boot 3

In this section, I will enumerate what is new in Spring Boot 3 and what you need to do to migrate from the previous version of Spring Boot to the latest. Remember, these will be just snippets and a bit of explanation if necessary.

- Java 17 baseline and Java 19 support: Spring Boot 3.0 requires Java 17 as a minimum version.

- Spring Boot requires GraalVM 22.3 (https://www.graalvm.org/) or later and Native Build Tools Plugin 0.9.17 or later.

- Spring Framework 6:

  - Entire framework codebase based on Java 17 source code level now.

  - Migration from `javax` to `jakarta` namespace for Servlet, JPA, etc.

  - Runtime compatibility with Jakarta EE 9 as well as Jakarta EE 10 APIs.

  - Compatible with latest web servers: Tomcat 10.1, Jetty 11, Undertow 2.3.

  - Early compatibility with virtual threads (in preview as of JDK 19).

  - Upgrade to ASM 9.4 and Kotlin 1.7.

  - Complete CGLIB fork with support for capturing CGLIB-generated classes.

  - Comprehensive foundation for Ahead Of Time (AOT) transformations.

  - First-class support for GraalVM native images (see related Spring Boot 3 blog post).

  - `RSocket` interface client based on `@RSocketExchange` service interfaces.

  - Early support for Reactor Netty 2 based on Netty 5 alpha.

  - Support for Jakarta WebSocket 2.1 and its standard `WebSocket` protocol upgrade mechanism.

  - HTTP interface client based on `@HttpExchange` service interfaces.

  - Support for RFC 7807 problem details.

  - Unified HTTP status code handling.

  - Support for Jackson 2.14.

  - Alignment with Servlet 6.0 (while retaining runtime compatibility with Servlet 5.0).

- Improved `@ConstructorBinding` detection.

- Micrometer updates:

  - New `ObservationRegistry` interface can be used to create observations, which provides a single API for both metrics and traces.

  - Spring Boot now auto-configures *Micrometer Tracing* for you.

  - When there is a Micrometer Tracing `Tracer` bean, and Prometheus is on the classpath, a `SpanContextSupplier` is now auto-configured.

- More flexible auto-configuration for Spring Data JDBC.

- Enabling async acks with Apache Kafka: A new property, `spring.kafka.listener.async-acks`, for enabling async acks with Kafka has been added.

- Elasticsearch Java Client. Auto-configuration for the new Elasticsearch Java Client has been introduced. It can be configured using the existing `spring.elasticsearch.*` configuration properties.

- Auto-configuration of `JdkClientHttpConnector`.

- `@SpringBootTest` with main methods. The `@SpringBootTest` annotation can now use the main method of any discovered `@SpringBootConfiguration` class if it's available. This means that tests can now pick up any custom `SpringApplication` configuration performed by your main method.

- Testcontainers: Support for using Testcontainers to manage external services at development time has been introduced.

- Docker Compose:

  - A new module, `spring-boot-docker-compose`, provides integration with Docker Compose. When your app is starting up, the Docker Compose integration will look for a configuration file in the current working directory: `compose.yaml`, `compose.yml`, `docker-compose.yaml`, and `docker-compose.yml` are supported.

  - To use a non-standard file, set the `spring.docker.compose.file` property.

- Auto-configuration for Spring Authorization Server: Support has been added for the Spring Authorization Server project along with a new `spring-boot-starter-oauth2-authorization-server` starter.

- Docker image building: The `spring-boot:build-image` Maven goal and `bootBuildImage` Gradle task are available.

- `RestClient` support: Spring Boot 3.2 supports the new `RestClient` interface introduced in Spring Framework 6.1. This interface provides a functional-style blocking HTTP API with a like design to `WebClient`.

- Support for `JdbcClient`: Auto-configuration for `JdbcClient` has been added based on the presence of a `NamedParameterJdbcTemplate`. If the latter is auto-configured, properties of `spring.jdbc.template.*` are considered.

- Support for virtual threads:

  - Spring Boot 3.2 ships support for virtual threads. To use virtual threads, you must run on Java 21 and set the property `spring.threads.virtual.enabled` to `true`.

  - When virtual threads are enabled, Tomcat and Jetty will use virtual threads for request processing.

  - Spring WebFlux's support for block execution is auto-configured to use the `applicationTaskExecutor` bean when it is an `AsyncTaskExecutor`.

  - When virtual threads are enabled, the applicationTaskExecutor bean will be a `SimpleAsyncTaskExecutor` configured to use virtual threads.

  - When virtual threads are enabled, the `taskScheduler` bean will be a `SimpleAsyncTaskScheduler` configured to use virtual threads. The `spring.task.scheduling.thread-name-prefix` property and `spring.task.scheduling.simple.*` properties are applied. Other `spring.task.scheduling.*` properties are ignored as they are specific to a pool-based scheduler.

As you can see, there are a lot of new improvements to take into consideration when migration is a must.

Using Configuration Properties Migration

With all these changes introduced in Spring Boot 3.0, it should include a tool or library that can help us migrate, right? Because, we need to make a few configuration properties changes due they have been renamed or removed, so your trusty `application.properties` or `application.yml` needs an update.

But don't worry; Spring Boot has your back! Introducing the `spring-boot-properties-migrator` module. Add it to your project, and it will

- *Scan your environment*: Like a detective, it sniffs out all your properties.

- *Print helpful messages*: It tells you exactly which properties need attention.

- *Temporarily fix things*: No need for immediate code changes! It automatically adjusts your properties at runtime for a smooth transition.

If you are using Maven, this is what you need to include in your `pom.xml`:

```
<dependency>
    <groupId>org.springframework.boot</groupId>
    <artifactId>spring-boot-properties-migrator</artifactId>
    <scope>runtime</scope>
</dependency>
```

If you are using Gradle:

```
runtime("org.springframework.boot:spring-boot-properties-migrator")
```

That's it your app,. Run the app, and these properties will analyze and they will tell you what to change and what to do.

So, you can migrate easily and keep your Spring Boot application running smoothly with the `spring-boot-properties-migrator` module!

Spring Boot Migrator

Even though there is the Configuration Properties Migration library, this is only for your properties. Is there anything better? Yes, there is! It's called Spring Boot Migrator, which you can find here: `https://github.com/spring-projects-experimental/spring-boot-migrator`.

So, if you have a very *old* Java app (not Spring) and Spring Boot version 2.x or earlier, the Spring Boot Migrator can help you with the migration.

Summary

In this appendix, you learned the basics of migrating to the latest version of Spring Boot 3. You saw some of the features and changes and discovered that there are tools available that can help you with your migration, such as Configuration Properties Migrator and the Spring Boot Migrator project.

APPENDIX B

Spring Boot GraphQL

This appendix introduces GraphQL and describes how you can use it with Spring Boot. Let's start with some basic concepts of what GraphQL is and how it can help in some use cases and then see how easy it is to implement logic around GraphQL using Spring Boot.

What Is GraphQL?

GraphQL (`https://graphql.org/`) is an open source query language and server-side runtime for APIs. It allows clients to request exactly the data they need, unlike REST APIs, which often return entire datasets. Using GraphQL leads to a more efficient and flexible experience for both developers and users.

The following are some use cases for GraphQL:

- *Mobile apps*: With limited bandwidth and resources, GraphQL's ability to fetch only necessary data shines.

- *Single-page applications (SPAs)*: SPAs often need dynamic data fetching, and GraphQL's flexibility makes it a good fit.

- *Complex data structures*: When dealing with interconnected data, GraphQL's ability to traverse relationships easily is helpful.

- *Content management systems (CMSs)*: GraphQL empowers content editors to retrieve specific content sections efficiently.

GraphQL offers the following performance benefits:

- *Reduced data transfer*: GraphQL sends only requested data, minimizing network traffic and improving load times.

- *Client-side caching*: Clients can cache specific queries, reducing server calls and improving responsiveness.

- *Batching*: Multiple queries can be combined into a single request, further boosting performance.

© Felipe Gutierrez 2024
F. Gutierrez, *Pro Spring Boot 3*, https://doi.org/10.1007/978-1-4842-9294-5

Table B-1 provides a comparison of REST and GraphQL.

Table B-1. *REST vs. GraphQL*

Feature	REST	GraphQL
Data fetching	Predefine endpoints	Client-specific queries
Data granularity	Entire datasets returned	Only requested data returned
Flexibility	Limited	Highly flexible
Performance	Can be inefficient	Can be more performant

Consider the following recommendations for using GraphQL:

- *Start with a good use case*: GraphQL isn't a silver bullet. Choose it when its benefits align with your project's needs.

- *Design a clear schema*: Define your data structure clearly for efficient client queries.

- *Use a GraphQL client library*: Simplify data fetching and error handling on the client side.

- *Consider security*: Implement proper authentication and authorization mechanisms.

- *Monitor performance*: Track query response times and optimize as needed.

Remember that GraphQL is a powerful tool, but it's not always the right choice. Evaluate your project's needs and weigh the pros and cons before diving in.

Spring for GraphQL

Spring for GraphQL (`https://spring.io/projects/spring-graphql`) is a library that brings a bunch of advantages to developers building GraphQL APIs on the Spring platform. Here's what it offers:

- *Simplified development*:

  - *Annotation-based approach*: Instead of manual configuration, you define data fetching methods using annotations like @ QueryMapping and @MutationMapping, making the code cleaner and more readable.

  - *Leverages Spring ecosystem*: Integrates seamlessly with other Spring libraries you're already familiar with, like Spring Security and Spring Data, reducing development time.

  - *Built-in features*: Offers automatic schema generation, data validation, and error handling, freeing you from boilerplate code.

- *Enhanced performance*:

  - *Efficient data fetching*: Optimizes data retrieval by leveraging Spring's caching and data access capabilities, improving API performance.

  - *Batching*: Combines multiple queries into one request, reducing roundtrips and boosting speed.

  - *Data loaders*: Allows you to prefetch related data, further minimizing database calls and improving response times.

- *Improved maintainability*:

  - *Modular design*: Separates schema definition from data fetching logic, making code easier to understand and maintain.

  - *Testing tools*: Provides built-in tools for testing GraphQL resolvers and mutations, ensuring code quality and stability.

  - *Reactive support*: Offers reactive programming capabilities for building scalable and responsive APIs.

- *Additional benefits*:

  - *Community support*: Backed by a large and active community, providing resources, tutorials, and help when needed.

  - *Regular updates*: Continuously updated with new features and improvements, keeping your API modern and secure.

Spring for GraphQL and Spring Boot: A Match Made in Developer Heaven

Both Spring for GraphQL and Spring Boot are built on the Spring platform, making them perfectly compatible and complementary. Here's how they play together:

- *Seamless integration*:

  - *Spring Boot Starter*: Spring for GraphQL offers a Spring Boot Starter (`spring-boot-starter-graphql`) that simplifies setup and configuration. Just add the starter dependency to your Spring Boot project, and you're ready to go.

  - *Auto-configuration*: Spring Boot's auto-configuration magic applies to Spring for GraphQL as well, automatically detecting and configuring beans based on your project setup.

  - *Existing Spring components*: You can reuse existing Spring components like Spring Data and Spring Security with Spring for GraphQL, leveraging their features and expertise.

- *Enhanced development experience*:

  - *Developer-friendly*: Both Spring Boot and Spring for GraphQL are known for their developer-friendly approach, simplifying complex tasks and offering clear documentation.

  - *Rapid prototyping*: Spring Boot's fast startup time and opinionated conventions make it ideal for rapid prototyping of GraphQL APIs.

  - *Production-ready*: Once you're ready for production, Spring Boot and Spring for GraphQL provide robust features and security for reliable deployments.

Using GraphQL in the Users App

You have access to this code in the `appendix-b-graphql/users` folder. But if you want to start from scratch with the Spring Initializr (`https://start.spring.io`), set the Group field to `com.apress` and the Artifact and Name fields to `users` with and add the following dependencies: Web, GraphQL, JPA, Validation, H2, PostgreSQL, and Lombok. Click Generate and download the project, unzip it, and import it into your favorite IDE.

Let's start by reviewing the build.gradle file. See Listing B-1.

Listing B-1. build.gradle

```
plugins {
    id 'java'
    id 'org.springframework.boot' version '3.2.2'
    id 'io.spring.dependency-management' version '1.1.4'
}

group = 'com.apress'
version = '0.0.1-SNAPSHOT'
sourceCompatibility = '17'

repositories {
    mavenCentral()
}

dependencies {
    implementation 'org.springframework.boot:spring-boot-starter-web'
    implementation 'org.springframework.boot:spring-boot-starter-
    validation'
    implementation 'org.springframework.boot:spring-boot-starter-graphql'

    implementation 'org.springframework.boot:spring-boot-starter-data-jpa'
    runtimeOnly 'com.h2database:h2'
    runtimeOnly 'org.postgresql:postgresql'

    compileOnly 'org.projectlombok:lombok'
    annotationProcessor 'org.projectlombok:lombok'

    // Web
    implementation 'org.webjars:bootstrap:5.2.3'

    testImplementation 'org.springframework.boot:spring-boot-starter-test'
    testImplementation 'org.springframework.graphql:spring-graphql-test'
}
```

```
tasks.named('test') {
    useJUnitPlatform()
}
```

Listing B-1 show that we are adding the `spring-boot-starter-graphql` dependency to `build.gradle`. Spring Boot will auto-configure all the necessary beans for GraphQL and set up the GraphiQL app (`https://github.com/graphql/graphiql`). You will have access using the `/graphics` endpoint.

Next, open/create and review the User class. See Listing B-2.

Listing B-2. src/main/java/com/apress/users/User.java

```
package com.apress.users;

import jakarta.persistence.Entity;
import jakarta.persistence.Id;
import jakarta.persistence.PrePersist;
import jakarta.validation.constraints.NotBlank;
import jakarta.validation.constraints.Pattern;
import lombok.*;

import java.util.Collections;
import java.util.List;

@Builder
@AllArgsConstructor
@NoArgsConstructor
@Data
@Entity(name="USERS")
public class User {

    @Id
    @NotBlank(message = "Email can not be empty")
    private String email;

    @NotBlank(message = "Name can not be empty")
    private String name;

    private String gravatarUrl;
```

```
@Pattern(message = "Password must be at least 8 characters long and
contain at least one number, one uppercase, one lowercase and one
special character",
        regexp = "^(?=.*[0-9])(?=.*[a-z])(?=.*[A-Z])(?=.*[@#$%^&+=!])
        (?=\\S+$).{8,}$")
private String password;

@Singular("role")
private List<UserRole> userRole;

private boolean active;

@PrePersist
private void prePersist(){
    if (this.gravatarUrl == null)
        this.gravatarUrl = UserGravatar.
         getGravatarUrlFromEmail(this.email);

    if(this.userRole == null){
        this.userRole = Collections.singletonList(UserRole.INFO);
    }
}
}
```

As you already know, we need to add the @Entity and use the @Id annotations to make the User class persistent in the database. Also, we are adding some validation annotations. Next, open/create and check the UserRole enum. See Listing B-3.

Listing B-3. src/main/java/com/apress/users/UserRole.java

```
package com.apress.users;

public enum UserRole {
    USER, ADMIN, INFO
}
```

Next, open/create and review the UserRepository interface. See Listing B-4.

Listing B-4. src/main/java/com/apress/users/UserRepository.java

```java
package com.apress.users;

import org.springframework.data.jpa.repository.JpaRepository;
import org.springframework.data.repository.CrudRepository;

public interface UserRepository extends JpaRepository<User,String> {
}
```

Next, let's add some users. Open/create and review the UserConfiguration class.
See Listing B-5.

Listing B-5. src/main/java/com/apress/users/UsersController.java

```java
package com.apress.users;

import org.springframework.boot.context.event.ApplicationReadyEvent;
import org.springframework.context.ApplicationListener;
import org.springframework.context.annotation.Bean;
import org.springframework.context.annotation.Configuration;

@Configuration
public class UserConfiguration {

    @Bean
    ApplicationListener<ApplicationReadyEvent> init(UserRepository
    userRepository) {
        return applicationReadyEvent -> {
            userRepository.save(User.builder()
                    .email("ximena@email.com")
                    .name("Ximena")
                    .gravatarUrl("https://www.gravatar.com/avatar/23bb62a7d
                    0ca63c9a804908e57bf6bd4?d=wavatar")
                    .password("aw2sOmeR!")
                    .role(UserRole.USER)
                    .active(true)
                    .build());
```

```
        userRepository.save(User.builder()
                .email("norma@email.com")
                .name("Norma")
                .gravatarUrl("https://www.gravatar.com/avatar/f07f
                7e553264c9710105edebe6c465e7?d=wavatar")
                .password("aw2sOmeR!")
                .role(UserRole.USER)
                .role(UserRole.ADMIN)
                .active(true)
                .build());
    };
}

}
```

Next, open/create and review the UserGravatar class that helps to collect the Gravatar for the user based on their email address. See Listing B-6.

Listing B-6. src/main/java/com/apress/users/UserGravatar.java

```
package com.apress.users;

import java.io.UnsupportedEncodingException;
import java.security.MessageDigest;
import java.security.NoSuchAlgorithmException;

public class UserGravatar {

    public static String getGravatarUrlFromEmail(String email){
        return String.format("https://www.gravatar.com/
        avatar/%s?d=wavatar", md5Hex(email));
    }

    private static String hex(byte[] array) {
        StringBuffer sb = new StringBuffer();
        for (int i = 0; i < array.length; ++i) {
            sb.append(Integer.toHexString((array[i]
                    & 0xFF) | 0x100).substring(1, 3));
        }
```

```
        return sb.toString();
    }

    private static String md5Hex(String message) {
        try {
            MessageDigest md =
                    MessageDigest.getInstance("MD5");
            return hex(md.digest(message.getBytes("CP1252")));
        } catch (NoSuchAlgorithmException e) {
        } catch (UnsupportedEncodingException e) {
        }
        return "23bb62a7d0ca63c9a804908e57bf6bd4";
    }
}
```

All these classes are part of the JPA persistence package introduced in Chapter 5. So, the next section is where GraphQL will shine!

Users GraphQL Controller

Next, open/create and review the UsersController class. See Listing B-7.

Listing B-7. src/main/java/com/apress/users/UsersController.java

```
package com.apress.users;

import jakarta.validation.Valid;
import lombok.AllArgsConstructor;
import org.springframework.graphql.data.method.annotation.Argument;
import org.springframework.graphql.data.method.annotation.MutationMapping;
import org.springframework.graphql.data.method.annotation.QueryMapping;
import org.springframework.http.HttpStatus;
import org.springframework.stereotype.Controller;
import org.springframework.validation.FieldError;
import org.springframework.web.bind.MethodArgumentNotValidException;
import org.springframework.web.bind.annotation.ExceptionHandler;
import org.springframework.web.bind.annotation.ResponseStatus;
```

```java
import java.time.LocalDateTime;
import java.time.format.DateTimeFormatter;
import java.util.HashMap;
import java.util.Map;

@AllArgsConstructor
@Controller
public class UsersController {

    private UserRepository userRepository;

    @QueryMapping
    public Iterable<User> users() {
        return this.userRepository.findAll();
    }

    @QueryMapping
    public User user(@Argument String email) throws Throwable {
        return this.userRepository.findById(email).orElseThrow(() -> new
        RuntimeException("User not found"));
    }

    @MutationMapping
    public User createUser(@Argument @Valid User user) {
        user.setGravatarUrl(UserGravatar.getGravatarUrlFromEmail(user.
        getEmail()));
        return userRepository.save(user);
    }

    @MutationMapping
    public User updateUser(@Argument @Valid User user) {
        User userToUpdate = userRepository.findById(user.getEmail()).
          orElseThrow(() -> new RuntimeException("User not found"));
        userToUpdate.setName(user.getName());
        userToUpdate.setPassword(user.getPassword());
        userToUpdate.setUserRole(user.getUserRole());
        userToUpdate.setActive(user.isActive());
        return userRepository.save(userToUpdate);
    }
```

```
@MutationMapping
public boolean deleteUser(@Argument String email) {
    userRepository.deleteById(email);
    return true;
}

@ExceptionHandler({MethodArgumentNotValidException.class})
@ResponseStatus(HttpStatus.BAD_REQUEST)
public Map<String, Object> handleValidationExceptions(MethodArgumentNot
ValidException ex) {
    Map<String, Object> response = new HashMap<>();

    response.put("msg","There is an error");
    response.put("code",HttpStatus.BAD_REQUEST.value());
    response.put("time", LocalDateTime.now().format(DateTimeFormatter.
    ofPattern("yyyy-MM-dd HH:mm:ss")));

    Map<String, String> errors = new HashMap<>();
    ex.getBindingResult().getAllErrors().forEach((error) -> {
        String fieldName = ((FieldError) error).getField();
        String errorMessage = error.getDefaultMessage();
        errors.put(fieldName, errorMessage);
    });
    response.put("errors",errors);

    return response;
}
}
```

The UsersController class includes the following annotations:

- @Controller: Like in Spring MVC, we are using the @Controller
  annotation (we are no longer in REST city!) to mark the class as a
  bean for Spring management, but instead of handling HTTP requests,
  it identifies methods that fetch data for GraphQL fields. These
  methods are annotated with @QueryMapping or @MutationMapping.
  This means there is no view rendering; Spring for GraphQL is not

designed for view rendering. The data returned from the methods annotated with @QueryMapping and @MutationMapping is part of the GraphQL response directly.

- @MutationMapping: This annotation defines methods that handle mutations in a GraphQL schema. Mutations represent actions that modify data on the server, like creating, updating, or deleting data. Methods annotated with @MutationMapping typically return the newly created object or the updated object after the data modification.

- @QueryMapping: This annotation marks methods that handle queries in the GraphQL schema. Queries represent requests for data from the server. Methods annotated with @QueryMapping typically return the requested data, either a single object or a collection of objects. The method name often corresponds to the field name in the schema, but the name can be explicitly specified in the annotation.

- @Argument: This annotation is used on method parameters to bind them to arguments in a GraphQL field definition. It specifies which parameter receives the value corresponding to a specific argument in the GraphQL query. By default, the argument name and the parameter name are matched, but you can specify the argument name explicitly in the annotation.

Other annotations used with Spring for GraphQL include the following:

- @SchemaDirective: Used to define custom directives that modify the behavior of the schema.

- @DataFetcher: An alternative to @QueryMapping and @MutationMapping that allows defining a DataFetcher directly on a field in the schema.

- @SubscriptionMapping: Used to define methods that handle subscriptions, a real-time data stream feature in GraphQL.

- @PathVariable: Like @Argument but used with path variables in the URL that map to method parameters.

Users GraphQL Schema

Next, we need to declare a GraphQL schema. This is the key for Spring for GraphQL to work, because the GraphQL schema does the following:

1. Defines a contract between the client and server:

   - The schema acts as a contract outlining the available data types, fields, queries, and mutations within your API.

   - It explicitly specifies what data the client can access and how they can request it, ensuring consistency and clarity.

2. Enables introspection and validation:

   - Clients can use the schema to introspect your API, dynamically discovering available fields and their types.

   - This allows the client to build self-documenting interfaces and validate their queries against the schema before sending them to the server.

3. Facilitates efficient data fetching:

   - Spring for GraphQL leverages the schema to optimize data fetching.

   - It knows what data each query requests based on the schema, allowing it to efficiently fetch only the necessary information.

4. Improves developer experience:

   - A well-defined schema improves code readability and maintainability.

   - Developers can easily understand the available data and how to access it, making code navigation and updates smoother.

5. Encourages consistent API design:

   - By having a centralized schema, you can ensure consistency and predictability in your API design.

   - This helps maintain a consistent experience for client applications interacting with your API.

While Spring for GraphQL allows automatic schema generation from code, it's generally recommended to explicitly define your schema for clarity and maintainability. You can use schema definition tools like GraphQL Schema Language (SDL) or dedicated libraries to define your schema in a structured way.

To create a GraphQL schema, you need to do the following:

1. Define data types:

   - Use types like Int, String, and Boolean for basic data.

   - Define custom types (objects) to represent your domain entities (e.g., User, UserRole).

2. Specify fields in types:

   - Define fields for each custom type, representing the data they contain (e.g., name and email for User).

   - Specify the field's type.

   - Optionally, use arguments to provide additional filtering or sorting options for queries.

3. Define queries and mutations:

   - Define @QueryMapping and @MutationMapping annotated methods to handle data fetching and manipulation requests.

   - These methods return the requested data or a response for mutations.

Next, open/create and review the schema.graphqls file, which is mandatory to have. This file must live in the resources/graphql folder. See Listing B-8.

Listing B-8. src/main/java/resources/graphql/schema.graphqls

```
type Query {
    users: [User]
    user(email: String): User
}
```

```
type Mutation {
    createUser(user: UserInput!): User
    updateUser(user: UserInput!): User
    deleteUser(email: String!): Boolean
}

type User {
    email: String!
    name: String!
    gravatarUrl: String
    password: String!
    userRole: [UserRole]!
    active: Boolean!
}

input UserInput {
    email: String!
    name: String!
    password: String!
    userRole: [UserRole]!
    active: Boolean!
}

enum UserRole {
    USER
    ADMIN
    INFO
}
```

Listing B-8 shows that the schema.graphqls file includes the following data types, queries, and mutations:

- *Data types*:

  - User: Represents individual users with fields for email, name, gravatarUrl, password, userRole (an array representing roles), and active status.

- UserInput: Used for inputting user data during creation and updates.

- UserRole (Enum): Defines possible user roles: USER, ADMIN, and INFO.

- *Queries*:

  - users: Retrieves a list of all users.

  - user(email: String): Fetches a specific user based on their email address.

- *Mutations*:

  - createUser(user: UserInput!): Creates a new user with the provided input.

  - updateUser(user: UserInput!): Updates an existing user's information based on the input.

  - deleteUser(email: String!): Deletes a user with the specified email address.

The ! in the mutations means that the field is non-nullable. In other words, the field must always have a value and cannot be null.

Next, let's add a particular property that will allow us to have the GraphiQL app. Listing B-9 shows the application.properties.

Listing B-9. src/main/java/resources/application.properties

```
spring.h2.console.enabled=true
spring.datasource.generate-unique-name=false
spring.datasource.name=test-db
```

spring.graphql.graphiql.enabled=true

Setting the spring.graphql.graphiql.enabled=true property activates the /graphiql endpoint.

Running the Users App

You can run the Users App in your IDE or use the following command:

```
./gradlew bootRun
```

Once the Users App is up and running, direct your browser to `http://localhost:8080/graphiql` and you will see the GraphiQL app. Enter the following query and click the Run button. See Figure B-1.

```
query {
    users {
        email
        name
        userRole
        active
    }
}
```

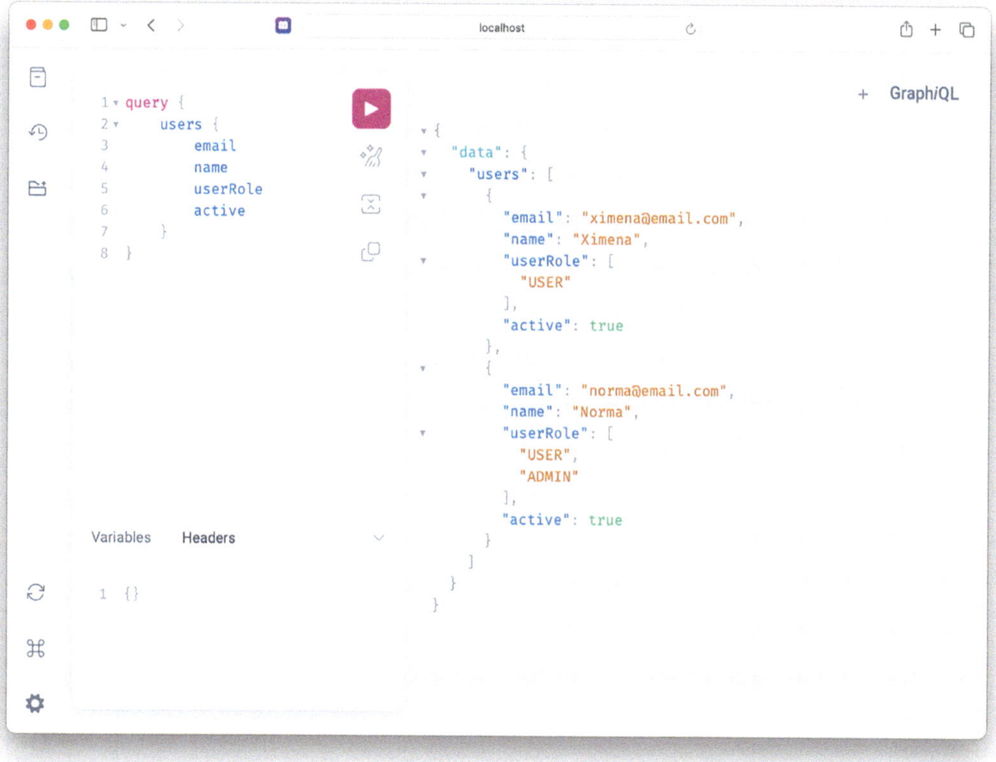

Figure B-1. *Running the Users App in the GraphiQL app (`http://localhost:8080/graphiql`)*

Figure B-1 shows the /graphiql endpoint where you can add your queries and mutations. We are using the query { } keyword and adding the model with the fields we need back, so you can use

```
query {
    users {
        email
        name
    }
}
```

and you should get back the following:

```
{
  "data": {
    "users": [
      {
        "email": "ximena@email.com",
        "name": "Ximena"
      },
      {
        "email": "norma@email.com",
        "name": "Norma"
      }
    ]
  }
}
```

Next, try to do the queries and mutations. Query the method user(@Argument String email) with

```
query {
    user(email: "ximena@email.com") {
        email
        name
        userRole
        active
    }
}
```

and you will get back:

```
{
  "data": {
    "user": {
      "email": "ximena@email.com",
      "name": "Ximena",
```

```
      "userRole": [
        "USER"
      ],
      "active": true
    }
  }
}
```

Use the mutation of User createUser(@Argument @Valid User user) with

```
mutation {
    createUser(user: {
        email: "dummy@email.com"
        name: "Dummy"
        password: "awesome!R2D2"
        userRole: [USER]
        active: true
    }) {
        email
        name
        gravatarUrl
        userRole
        active
    }
}
```

Remember that a mutation regularly returns something, so you need to specify what field you want back from the created user. So, the previous mutation will give you:

```
{
  "data": {
    "createUser": {
      "email": "dummy@email.com",
      "name": "Dummy",
      "gravatarUrl": "https://www.gravatar.com/avatar/fb651279f4712e20999
      1e05610dfb03a?d=wavatar",
```

```
      "userRole": [
        "USER"
      ],
      "active": true
    }
  }
}
```

You can test the mutation User updateUser(@Argument @Valid User user) with

```
mutation {
    updateUser(user: {
        email: "dummy@email.com"
        name: "Dummy"
        password: "awesome!C3PO"
        userRole: [USER, ADMIN]
        active: true
    }) {
        email
        name
        gravatarUrl
        userRole
        active
    }
}
```

and you will get back:

```
{
  "data": {
    "updateUser": {
      "email": "dummy@email.com",
      "name": "Dummy",
      "gravatarUrl": "https://www.gravatar.com/avatar/fb651279f4712e20999
      1e05610dfb03a?d=wavatar",
```

```
      "userRole": [
        "USER",
        "ADMIN"
      ],
      "active": true
    }
  }
}
```

Finally, you can test the mutation deleteUser(@Argument String email) with

```
mutation {
    deleteUser(email: "dummy@email.com")
}
```

and you will get back:

```
{
  "data": {
    "deleteUser": true
  }
}
```

Congrats! Now you know how GraphQL works and how Spring for GraphQL and Spring Boot!

Using GraphQL in the My Retro App

Using GraphQL in the My Reto App App is basically the same idea as using it in the Users App. You can get any of the other chapter projects and use the spring-boot-starter-graphql dependency to activate GraphQL. But if you have already downloaded the code, you can import it from the appendix-b-graphql/myretro folder.

So, in this section I'm just going to show for the My Retro App the controller (Listing B-10), the GraphQL schema (Listing B-11), and the queries and mutations that you are able to do (Listing B-12).

Listing B-10. src/main/java/com/apress/myretro/web/
RetroBoardController.java

```java
package com.apress.myretro.web;

import com.apress.myretro.board.Card;
import com.apress.myretro.board.RetroBoard;
import com.apress.myretro.service.RetroBoardService;
import jakarta.validation.Valid;
import lombok.AllArgsConstructor;
import org.springframework.graphql.data.method.annotation.Argument;
import org.springframework.graphql.data.method.annotation.MutationMapping;
import org.springframework.graphql.data.method.annotation.QueryMapping;
import org.springframework.http.HttpStatus;
import org.springframework.http.ResponseEntity;
import org.springframework.stereotype.Controller;
import org.springframework.validation.FieldError;
import org.springframework.web.bind.MethodArgumentNotValidException;
import org.springframework.web.bind.annotation.ExceptionHandler;
import org.springframework.web.bind.annotation.PathVariable;
import org.springframework.web.bind.annotation.RequestBody;
import org.springframework.web.bind.annotation.ResponseStatus;
import org.springframework.web.servlet.support.ServletUriComponentsBuilder;

import java.net.URI;
import java.time.LocalDateTime;
import java.time.format.DateTimeFormatter;
import java.util.HashMap;
import java.util.Map;
import java.util.UUID;

@AllArgsConstructor
@Controller
public class RetroBoardController {

    private RetroBoardService retroBoardService;
```

```java
@QueryMapping
public Iterable<RetroBoard> retros(){
    return retroBoardService.findAll();
}

@MutationMapping
public RetroBoard createRetro(@Argument String name){
    return retroBoardService.save(RetroBoard.builder().id(UUID.
    randomUUID()).name(name).build());
}

@QueryMapping
public RetroBoard retro(@Argument UUID retroId){
    return retroBoardService.findById(retroId);
}

@QueryMapping
public Iterable<Card> cards(@Argument UUID retroId){
    return retroBoardService.findAllCardsFromRetroBoard(retroId);
}

@MutationMapping
public Card createCard(@Argument UUID retroId,@Argument @Valid
Card card){
    return retroBoardService.addCardToRetroBoard(retroId,card);
}

@QueryMapping
public Card card(@Argument UUID cardId){
    return retroBoardService.findCardByUUID(cardId);
}

@MutationMapping
public Card updateCard(@Argument UUID cardId, @Argument @Valid
Card card){
    Card result = retroBoardService.findCardByUUID(cardId);
    result.setComment(card.getComment());
    return retroBoardService.saveCard(result);
}
```

```java
@MutationMapping
public Boolean deleteCard(@Argument UUID cardId){
    retroBoardService.removeCardByUUID(cardId);
    return true;
}

@ExceptionHandler(MethodArgumentNotValidException.class)
@ResponseStatus(HttpStatus.BAD_REQUEST)
public Map<String, Object> handleValidationExceptions(MethodArgumentNot
ValidException ex) {
    Map<String, Object> response = new HashMap<>();

    response.put("msg","There is an error");
    response.put("code",HttpStatus.BAD_REQUEST.value());
    response.put("time", LocalDateTime.now().format(DateTime
    Formatter.ofPattern("yyyy-MM-dd HH:mm:ss")));

    Map<String, String> errors = new HashMap<>();
    ex.getBindingResult().getAllErrors().forEach((error) -> {
        String fieldName = ((FieldError) error).getField();
        String errorMessage = error.getDefaultMessage();
        errors.put(fieldName, errorMessage);
    });
    response.put("errors",errors);

    return response;
}
}
```

Listing B-11. src/main/java/resources/graphql/schema.graphqls

```graphql
type Query {
    retros: [RetroBoard]
    retro(retroId: ID!): RetroBoard
    cards(retroId: ID!): [Card]
    card(cardId: ID!): Card
}
```

```
type Mutation {
    createRetro(name: String!): RetroBoard
    createCard(retroId: ID!, card: CardInput!): Card
    updateCard(cardId: ID!, card: CardInput!): Card
    deleteCard(cardId: ID!): Boolean
}

type RetroBoard {
    id: ID!
    name: String!
    cards: [Card]
}

type Card {
    id: ID!
    comment: String!
    cardType: CardType!
}

enum CardType {
    HAPPY
    MEH
    SAD
}

input CardInput {
    comment: String!
    cardType: CardType!
}
```

Queries

Listing B-12. Queries and Mutations

```
query {
  retros {
    id
    name
```

```
    cards {
      id
      comment
      cardType
    }
  }
}

query {
  retro(retroId: "1") {
    id
    name
    cards {
      id
      comment
      cardType
    }
  }
}

query {
  cards(retroId: "1") {
    id
    comment
    cardType
  }
}

query {
  card(cardId: "1") {
    id
    comment
    cardType
  }
}
```

Mutations

```
mutation {
  createRetro(name: "Retro 1") {
    id
    name
  }
}

mutation {
  createCard(retroId: "1", card: {comment: "Great job team!", cardType:
  HAPPY}) {
    id
    comment
    cardType
  }
}

mutation {
  updateCard(cardId: "1", card: {comment: "Great job team!", cardType:
  HAPPY}) {
    id
    comment
    cardType
  }
}

mutation {
  deleteCard(cardId: "1")
}
```

Happy GraphQL!

Summary

In this appendix, you learned about GraphQL and how to use it with Spring for GraphQL and Spring Boot. You identified what you needed to do to create the queries and mutations. You learned about the `@QueryMapoping` and `@Muttation` mapping annotations and learned how to create your GraphQL schema.

Index

A

© Felipe Gutierrez 2024
F. Gutierrez, *Pro Spring Boot 3*, https://doi.org/10.1007/978-1-4842-9294-5

T

U

V

W, X, Y, Z